The B-29 Superfortress

ALSO BY ROBERT A. MANN
AND FROM MCFARLAND

The B-29 Superfortress Chronology, 1934–1960 (2009)

Aircraft Record Cards of the United States Air Force: How to Read the Codes (2008)

The B-29 Superfortress

A Comprehensive Registry of the Planes and Their Missions

ROBERT A. MANN

McFarland & Company, Inc., Publishers
Jefferson, North Carolina, and London

The present work is a reprint of the illustrated case bound edition of The B-29 Superfortress: A Comprehensive Registry of the Planes and Their Missions, *first published in 2004 by McFarland.*

LIBRARY OF CONGRESS CATALOGUING-IN-PUBLICATION DATA

Mann, Robert A., 1930–
The B-29 Superfortress : a comprehensive registry of the planes and their missions / Robert A. Mann.
    p.   cm.
Includes bibliographical references.

ISBN 978-0-7864-4458-8
softcover : 50# alkaline paper ∞

1. B-29 bomber — Registers. 2. B-29 bomber — History. I. Title.
UG1242.B6M284 2009   358.4'283'0973 — dc22   2004016414

British Library cataloguing data are available

©2004 Robert A. Mann. All rights reserved

No part of this book may be reproduced or transmitted in any form or by any means, electronic or mechanical, including photocopying or recording, or by any information storage and retrieval system, without permission in writing from the publisher.

Cover photograph ©1997 Bill Crump

Manufactured in the United States of America

McFarland & Company, Inc., Publishers
Box 611, Jefferson, North Carolina 28640
www.mcfarlandpub.com

To the crews of 44-61640, 44-87756 and 44-69770,
Who died in the service of their country,
"Flying the Weather"

44-61640	44-87756	44-69770
2/26/52	4/5/52	10/26/52
Krueger, Walter 1Lt	Acebidio, Bruce Maj	Harrell, Sterling Maj
Gendusa, Vincent 2Lt	Hopkins, G. Capt	Knickmeyer, C. 1Lt
Shaw, Robert 2Lt	Kizer, Robert Capt	Burchell, W. P. 1Lt
Leach, Frank M/Sgt	Winstead, L.B. Capt	Polack, Frank Capt
Parker, Donald Sgt	Lam, August 2Lt	Baird, Donald Capt
Toland, Francis Cpl	Fultz, Edwin M/Sgt	Fontaine, Edward M/Sgt
	Shook, George T/Sgt	Brewton, Alton A/1c
	Schulz, Hayden S/Sgt	Fasullo, Tony A/1c
	Fose, Carlton S/Sgt	Colgan, William A/1c
	King, Elbert S/SGT	Verrill, Rodney A/3c

Acknowledgments

So many people, so few names written down. Over the years I have talked on the phone, e-mailed and snail-mailed dozens of veterans who contributed a name, a serial number, a unit, to my database — and, sadly, I no longer have their names.

The people listed below are those I did make a note of, people who contributed heavily, lightly, indirectly — and for that I thank them.

Francis Gulling	Henry Sakaida	Terry Lindall
Tom Robison	Lee Florence	Otha Spencer
Bud Farrell	Sparky Corradina	Hurth Tompkins
P. Schifferli	Bill Copeland	Jean Allen
Ron Ellison	Tom Britton	W. Lancaster
Chris Howlett	Ray Brashear	Mark Styling
Charles Frey	Sallyann Wagoner	S. Smisek
Patrick Stinson	Chester Marshall	F. Crowell
Pete Weiler	Harry Changdon	Bill Howe

Also I want to thank the USAFHRC archivists and all those bomb wing and group historical officers and statistical control officers for their help.

And, again, to all those others whose names have fallen through the cracks: I sincerely apologize.

A special acknowledgment has to go to Joe Baugher, whose magnificent web site on military aircraft serial numbers I visited over and over again seeking confirmation of data I had received on individual planes. Anyone doing military aircraft research must visit http://home.att.net/ ~jbaugher. It will be a visit well spent.

CONTENTS

Acknowledgments vii
Preface xi
Introduction 1
Abbreviations, Column Headings and Acronyms 3

I. Master List 9

II. Names List 89
Names with Known Group Assignments but Unknown Serial Numbers *96*
Unnamed Nose Art with Known Serial Number or Group Assignment *98*
Names without Known Serial Numbers or Group Assignments *99*

III. Block Numbers and Construction Numbers 100

IV. Design Specifications and Performance Parameters 103

V. B-29 Variants 104
F-13 Photo Reconnaissance Configuration *105*
KB-29M Hose and Reel Tanker Configuration *106*
KB-29P Boom Tanker Configuration *107*
SB-29 Rescue Configuration *107*
WB-29 Weather Reconnaissance Configuration *108*

VI. The XX Bomber Command 112
Mission List *113*
Mission Details *114*
Losses *138*

VII. The XXI Bomber Command   139

    Bomber Command Mission List *140*
    Bomb Group Mission Lists *172*
    Bomb Group Plane Rosters *195*
    Targets Listed by Name *199*
    Targets Listed by Number *215*
    The Fire Raids on Japan *229*
    Missions Against Kyushu Airfields *236*
    Operation STARVATION: The Aerial Mining of the Japanese Home Waters *241*
    "Pumpkin" Missions of the 509th Composite Group *245*

VIII. Strategic Air Command   246

    B-29 Strength, by Year *246*
    RB-29 Strength, by Year *247*
    Bombardment and Reconnaissance Units Assigned B-29s *248*

IX. The B-29 in Korea   250

    Bomb Group Plane Rosters *251*
    Korean Tactical Drop Sample: 19th Bomb Group, May 2–14, 1951 *252*
    Known B-29 Losses in Korea, 1950–1953 *253*
    South and North Korean Air Base Codes, 1950–1953 *254*

X. B-29s in the Royal Air Force   255

Appendix: Serial Key Matrix   257
Bibliography   275

PREFACE

Way back in 1982, I asked myself a simple question: "I wonder what ever happened to old 094?" Old 094 was WB-29 44-62094, a plane I loved and hated, sweet-talked and threatened, gently stroked and hit with hammers, praised and cursed to the outer reaches of hell. In short, my plane.

That simple question was the seed from which this book grew. The examination of uncounted frames of microfilm, pages of books, phone conversations, and e-mails, not to mention web surfing, had resulted in a B-29 database listing every one of the 3960 B-29s manufactured.[1] Those planes with a combat record, I had records for each group they served with in World War II or Korea (or both). The database contained some 5,700 records, each with up to 23 fields of information.

Ancillary databases spun off from the mother database, covering the combat records of specific airplanes, rosters of planes by group, planes participating on specific missions, names of individual B-29s, and more.

Over the years I gained some small notoriety as a person who could answer B-29 questions and received a small but steady stream of inquiries. Not being real fast out of the starting gate, it finally occurred to me that since I was looking up the same data for the third or fourth time, I should just print out a hard copy and put it in a reference binder. This promptly grew like Topsy.

I finally realized that there was no system to the way things were put into the binder, so I decided to really organize it. When I finished I paged through the final result and said, paraphrasing Mel Brooks' 2000 Year Old Man, "I think there is a book here." And away we went.

---

[1] There is some controversy as to how many B-29s were built, various sources citing 3960, 3962, 3965 and 3970 delivered. In an attempt to settle the issue, the author performed an analysis using Joe Baugher's definitive listing of all USAAF serial numbers.

All individual serial numbers for each type — XB-29, YB-29, B-29, B-29A and B-29B — were input into a spreadsheet and the line record number on the left side of the spreadsheet noted for the last serial number of each type. These were: XB-29-BO, 3; YB-29-BW, 14; B-29-BW, 1620; B-29-BA, 357; B-29-MO, 536; B-29A-BN, 1119; B-29B-BA, 311; Total Built, 3960.

Some of the confusion in the build totals may be caused by five serial numbers in the midst of B-29 sequences that were cancelled. These serial numbers were: 42-65314; 44-84140; 44-84150; 44-84153 (assigned to Christopher XAG-1CH); 44-84154 (assigned to Christopher XAG-1CH). (*Source:* http://home.att.net/~jbaugher/usafserials.html. Click USAAS/USAAC/USAAF/USAF Aircraft Serials.)

INTRODUCTION

This is a book about the Boeing B-29 Superfortress. It covers not the manufacturing process, not the performance parameters and characteristics, but the operational histories of the 3960 individual planes: when and where assigned, combat missions flown, when and why removed from USAAF/USAF inventory.

The main text begins with the Master List, which provides the core information for each plane — serial number, name, identification (or tail code), date delivered or assigned, assignment, date off inventory and circumstances of removal from inventory. Detailed explanation of each aspect of the core information will be found in the introduction to that list, which is arranged by serial number.

Most crews named their planes, and names are easier to remember than serial numbers, so this book also includes an alphabetized Names List. Beside each name is listed that plane's serial number, which the reader can use to look up the plane in the Master List. Following the Names List is a list connecting serial numbers with block and construction numbers that link the planes to their manufacturers.

The next section is a brief description of the physical characteristics and performance parameters of the B-29.

Next comes a section listing the B-29 variants, with special attention to the five produced in significant numbers: the F-13A Reconnaissance version, the KB-29M and KB-29P Tankers, the SB-29 Rescue planes and the WB-29 Weather Reconnaissance models. (Note: The B-29s known as Silverplates were production aircraft modified for delivery of atomic weapons and were not true variants like the F-13A, the KB-29M, and so on. For this reason they are not included in this book.) A detailed description of each variant is followed by a list of serial numbers for all planes of that type.

The next section sets forth details of the XX Bomber Command's activities in the China-Burma-India Theater of Operations. Included are a summary list of the missions flown and more detailed lists for each of the 49 CBI missions, including serial numbers of planes on each mission and a list of the B-29s lost by the XX Bomber Command.

Next follows a lengthy section with much detailed information on the XXI Bomber Command, who flew 331 separate missions. This section includes a composite list of all XXI Bomber Command Missions; individual bomb group mission lists; rosters showing the planes assigned to each group; target lists organized by name and number; and lists of four specialized missions: the fire raids on Japan, the raids on the Kyushu airfields, the "Operation Starvation" mining missions in Japan's home waters, and the 509th Composite Group's "Pumpkin" missions, practice atomic strikes on various potential atomic targets.

Representing the postwar period is a section that examines the strength of the Strategic Air Command. The next section covers the use of the B-29 in Korea. The B-29 units participating in the conflict are identified. Plane rosters for each of the five groups are presented, followed by a information on losses and a list of the Korean Air Base codes.

The United States loaned the Royal Air Force ninety-one B-29s in 1950 and 1951. The final section of the main text identifies these planes by USAF serial number. By checking the serial numbers against the Master List, one may determine when the plane was returned to the USAF.

Following the main text is an appendix offering a matrix reference system for determining full serial numbers from the last three digits.

The book closes with a bibliography.

Abbreviations, Column Headings and Acronyms

2BLURUDSTRP — CBI Tail Code of 793BS, 468BG; 2 oblique blue rudder stripes
2REDRUDSTRP — CBI Tail Code of 794BS, 468BG; 2 oblique red rudder stripes
2WHTRUDSTRP — CBI Tail Code of 792BS, 468BG; 2 oblique white rudder stripes
2YLWRUDSTRP — CBI Tail Code of 795BS, 468BG; 2 oblique yellow rudder stripes
A SQ — XXIBC Tail Code for 497BG; "A" over Hollow Square over ID number
A/C — Aircraft Commander
A/F — Air Field
A-1 — Air Field at Hsinching, China; forward base of the 40th Bomb Group
A-3 — Air Field at Kwanghan, China; forward base of the 444th Bomb Group
A-5 — Air Field at Kiunglai, China; forward base of the 462nd Bomb Group
A-7 — Air Field at Pengshan, China; forward base of the 468th Bomb Group
AAC — Army Air Corps
AAF — Army Air Forces
AAF — Army Air Field
AB — Air Base
ABIE — Radio call sign for 16th Bomb Group, August 1945
ABND — Abandoned; crew bail out
AC — Aircraft
AC — AFHRC microfilm containing Aircraft Record Cards of planes still active in 1951
AC UP — Number of aircraft airborne for a specific mission
ACA — AFHRC microfilm containing Aircraft Record Cards of planes active 1955 to 1964
ACR — AFHRC microfilm containing Aircraft Record Cards of planes inactive by 1951
ACTOR — Radio call sign for 40th Bomb Group, August 1945
AEROSCOPE — XXIBC Target Name for Kushira Air Field
AFB — Air Force Base
AFHRC — Air Force Historical Research Center, Maxwell AFB, Alabama
AGHAST — XXIBC Target Name for Tachiari Air Field
AIR ABORT — Number of aircraft unable to continue mission to target for mechanical or personnel problems
AIRCRAFT RECORD CARDS — Logistics Command system for tracking movements/assignments of individual aircraft
AIREDALE — 313th Bomb Wing B-29 Super Dumbo (Rescue Plane)
AKA — Also Known As
ALBATROSS — Radio call sign for 504th Bomb Group, May 1945
ANAPHASE — Designation for 58th Bomb Wing Field Order #2
APU — Auxiliary Power Unit in rear unpressurized section of B-29; see "Putt-Putt"
ARC — See "AIRCRAFT RECORD CARDS"
ARC — Aircraft Resupply & Communications (Squadron)
ARRANGE — XXIBC Target Name for Tokyo Urban Area
ASSIGN — Assignment (group to which plane was assigned)
ASSY — Assembly; a collection of parts into a functioning unit
ATC — Air Transport Command
B-1 — Air Field at Kharagpur, India. Rear base of the 468th Bomb Group

B-2 — Air Field at Piordoba, India. Rear base of the 462nd Bomb Group
B-3 — Air Field at Charra, later Dudhkundi, India. Rear bases of the 444th Bomb Group
B-4 — Air Field at Chakulia, India. Rear base of the 40th Bomb Group
BALD EAGLE — Radio call sign for 330th Bomb Group
BALIFF — Radio call sign for 501st Bomb Group, August 1945
BARRANCA — XXIBC Target Name for Kokubu Air Field
BAYWOOD — Radio call sign for 331st Bomb Group, May 1945
BC — Bomber Command
BG — Bomb Group
BIRDDOG — Life Guard surface vessel stationed off Japan 1945
BLACKJACK — Radio call sign for 39th Bomb Group, May 1945
BLOCK NUMBER — Designates a group of planes in the manufacturing sequence built to same engineering configuration
BLK SQ K — XXIBC Tail code for 330th Bomb Group. Yellow "K" in solid black square on vertical stabilizer
BLK SQ M — XXIBC Tail code for 19th Bomb Group. Unpainted "M" in solid black square on vertical stabilizer
BLK SQ O — XXIBC Tail code for 29th Bomb Group. Unpainted "O" in solid black square on vertical stabilizer
BLK SQ P — XXIBC Tail code for 39th Bomb Group. Unpainted "P" in solid black square on vertical stabilizer
BLOWZY — XXIBC Target Name for Usa Air Field
BLU — Blue
BLUE PLATE — Radio call sign for 16th Bomb Group, May 1945
BOMCOM — Bomber Command
BOMWG — Bomb Wing
BOMB GROUP — Organizational unit consisting of three or four Bomb Squadrons. Redesignated 'Wing' in June, 1953
BOMB OTHER — Number of aircraft, of those airborne for a specific mission, bombing a target other than the primary
BOMB PRIMARY — Number of aircraft, of those airborne for a specific mission, bombing the primary target
BOMB SQUADRON — Organizational unit making up part of Bomb Group; consisted of 12 -15 aircraft
BOMB WING — Organizational unit whose combat components were three or four Bomb Groups
BOX KITE — Designator for Super Dumbo missions, 1945
BRISKET — XXIBC Target Name for Kawasaki
BROOKLYN — XXIBC Target Name for Tokyo Industrial Area
BS — Bomb Squadron
BU — (Air) Base Unit; Organization "owning" planes assigned to Base
BULLISH — XXIBC Target Name for Izumi Air Field
BUSHING — XXIBC Target Name for Nittigahara Air Field
BUTTERBALL — XXIBC Target Name for Hodagaya Chemical Works, Koriyama
BUZZARD — Designator for weather tracks flown by 56th WRS, Yokota AFB
BW — Bomb Wing
C/O — Checkout
CALIB — Calibration
CAMLET — XXIBC Target Name for Oita Air Field
CAMPBELL — 314th BW Code to execute landings any friendly field due to weather
CARNATION — A Guam Air Field; not North Field
CATCALL — XXIBC Target Name for Hitachi Aircraft Factory
CAVU — Ceiling And Visibility Unlimited
CBI — China-Burma-India Theater of Operations
CFC — Central Fire Control. Gunner in upper dorsal sighting blister on B-29 controlled assignments of turrets
CG — Commanding General
CHARACTER — Surface Life Guard vessel on B-29 return track
CHARRA — First India base of 444th Bomb Group. Unsuitable
CHECKBOOK — XXIBC Target Name for Kanoya Air Field
CHICKEN — Radio call sign for fighters escorted by B-29
CIRC E — XXIBC Tail code for 504th Bomb Group from April 1945. "E in a circle"
CIRC R — XXIBC Tail code for 6th Bomb Group from April 1945. "R in a circle"
CIRC W — XXIBC Tail code for 505th Bomb Group from April 1945. "W in a circle"
CIRC X — XXIBC Tail code for 9th Bomb Group from April 1945. "X in a circle"
CIRCUMSTANCES — Circumstances surrounding plane's removal from inventory
CLASS O-1A — USAF Property Class for complete aircraft
CLASS O-1Z — Entire airframe taken off inventory by transfer to a school as a ground instructional airframe
CLASS 26 — Entire airframe taken off inventory by transfer to a school as a ground instructional airframe
CLASS 28 — Expendable training aid
CLASS 32 — Transfer to USAF Museum
CO — Commanding Officer
COCKCROW — XXIBC Target Name for Saeki Air Field
COLL — Collision
COMP — Complete
CONDEMNED — Aircraft sent to salvage
CONUS — Continental United States
CONV — Converted
CORNLET — XXIBC Target Name for Oita Air Field
COUNTERFEIT — Aerial Sea mining missions, Japanese Home Waters, 1945
CROSSTOWN — Radio call sign for 9th Bomb Group, August 1945
CUCKOO — Radio call sign for 6th Bomb Group, August 1945

CURIOUS — Radio call sign for 19th Bomb Group, August 1945
CYCLONE — XXIBC Super Dumbo call sign
DAREDEVIL — Radio call sign for 6th Bomb Group, May 1945
DARKEYES — Submarine life guards on B-29 return track
DAVY JONES — Survivors in water without life vests
DELV — Delivery date
DIAM B — B in a diamond on vertical stabilizer; XXIBC Tail code for 16th Bomb Group, 315th Bomb Wing
DIAM H — H in a diamond on vertical stabilizer; XXIBC Tail code for 331st Bomb Group, 315th Bomb Wing
DIAM L — L in a diamond on vertical stabilizer; XXIBC Tail code for 501st Bomb Group, 315th Bomb Wing
DIAM Y — Y in a diamond on vertical stabilizer; XXIBC Tail code for 502nd Bomb Group, 315th Bomb Wing
DID NOT BOMB — Number of aircraft, airborne for a specific mission, returning to base with bombs or jettisoning
DIMPLES — Radio call sign for 509th Composite Group
DITCHED — Landed in sea
DMG — Damage
DMG'D — Damaged
DOMINO — Radio call sign for 9th Bomb Group, May 1945
DRACULA — Radio call sign for 29th Bomb Group, 314th Bomb Wing
DRAGON — Life guard submarine
DREAMBOAT — Radio call sign for navigational B-29 escorting fighters
DRIPPER — XXIBC Target Name for Miyakonojo Air Field
DROPSONDE — AN/AMT Radiosondes. Small droppable unit that recorded and transmitted weather data
DSTRY — Destroyed
"DUDKHINDI, INDIA" — Permanent India base of the 444th Bomb Group, 58th Bomb Wing, XX Bomber Command
E TRI — "E" over hollow triangle over ID number, tail code for 504th Bomb Group, 313th Bomb Wing
E/A — Enemy aircraft
EIELSON AFB — Located 26 miles southeast of Fairbanks, Alaska
ELMENDORF AFB — Located adjacent to Anchorage, Alaska
ENG — Engine
ENKINDLE — XXIBC Target Name for Nakajima Aircraft Plant, Tokyo
ERADICATE — XXIBC Target Name for Mitsubishi Aircraft Factory, Nagoya
EVERGREEN — Dye marker used by survivors in the water
EXPLD — Exploded
FAINTER — XXIBC Target Name for Otaki Oil Refinery
FAMISH — XXIBC Target Name for Kanoya East Air Field
FE — Flight Engineer
FEAF — Far East Air Force

FEARLESS — XXIBC Target Name for Omura Aircraft Factory
FLAK — Anti-aircraft fire
FLT — Flight
FRACTION — XXIBC Target Name for Nakajima Aircraft Plant at Ota
FRUITCAKE — XXIBC Target Name for Kawasaki Aircraft Plant, Akashi, Nagoya
FTR — Fighter
FURIOUS — XXIBC Target Name for Nakajima Aircraft Factory at Koizuma
GAD — Guam Air Depot
GMT — Greenwich Mean Time - time at Zero Longitude; Zulu Time
GOLDBUG — Radio call sign for 313th Bomb Wing
GOODYEAR — Survivors in life rafts
GROUND ABORT — Number of aircraft scheduled for mission unable to take off
GULFBIRD — Radio call sign for 504th Bomb Group, August 1945
HAPPY — Radio call sign for 497th Bomb Group, May 1945
HELLHOG — Radio call sign for 3rd Photo Recon Squadron, May 1945
HESITATION — XXIBC Target Name for Mitsubishi Aircraft Factory, Nagoya
HICKAM AFB — Currently shares runways with Honolulu International Airport
HUMP — Himalaya Mountains between India and China
IDENTIFICATION — Identifying marks on tail (tail code)
IFF — Identification, Friend or Foe. An airborn radio system that emitted a preset signal when queried
INFIRMARY — XXIBC Target Name for Chiran Air Field
INOP — Inoperative
INV — Inventory
INVINCIBLE — Surface lifeguard vessel on B-29 return track
IP — Initial Point; start of bomb run
IS — Island, Islands
IWO — The island of Iwo Jima
K TIME — Marianas time. Add 10 hours to GMT (Zulu) time
KB-29 — Designation for Tanker configuuration
KHARAGPUR — Permanent India base of 468th BG, XX Bomber Command
KIA — Killed In Action
KINDLEY AFB — Located in Bermuda
KINGBIRD — Radio call sign for 19th Bomb Group, May 1945
LADD AFB — Located adjacent to Fairbanks, Alaska
LARK — Designator for weather tracks flown by 374th Recon Sq (VLR) Weather, Fairfield-Suisun AFB
LD — Landing, landed
LEAFSTALK — XXIBC Target Name for Kawanishi
LOON — Designator for weather tracks flown by 375th Recon Sq (VLR) Weather, Eielson AFB

LORAN — Long Range Navigational Aid., a navigation system
LOYAL — Radio call sign for 58th Bomb Wing, August 1945
LUNCHROOM — XXIBC Target Name for Koriyama
MACR — Missing Air Crew Report. Contains details regarding the loss of a plane/air crew
MALF — Malfunction
MASCOT — Radio call sign for 498th Bomb Group, May 1945
MASHNOTE — Radio call sign for 444th Bomb Group, May 1945
MATTERHORN — Plan for B-29's to bomb Japan from bases around Chengtu, China
MEETINGHOUSE — XXIBC Target Name for Tokyo
MEMPHIS — XXIBC Target Name for Mitsubishi Engine Plant, Nagoya
M/F — Microfilm
MI — Miles
MIA — Missing In Action
MICROSCOPE — XXIBC Target Name for Urban Nagoya
MIDDLEMAN — XXIBC Target Name for Kobe Urban Area
MINE FIELD ABLE — Kobe/Osaka Approaches and harbors
MINE FIELD BAKER — Aki Nada; main Inland Sea shipping lane between Hzuma Hana and Kajitoro Hana
MINE FIELD CHARLIE — Fukuoka Bay and Harbor
MINE FIELD DOG — North of Shodo Shima
MINE FIELD EASY — Bisan Seto; south of Shodo Shima
MINE FIELD FOX — Bingo Nada between Imabari and Mi Saki
MINE FIELD GEORGE — Hamada
MINE FIELD HOW — Main Harbor facilities at Kure Naval base and anchorage between Kereko and Nishinomi Shima
MINE FIELD ITEM — Main shipping channel to Kure through Hashirashima Suido
MINE FIELD JIG — Entrance to Hiroshima Harbor between Itsuku Shima and Nishinomi Strait; northern Approach to Hiroshima and Kure
MINE FIELD KING — Channel leading to the Port of Tokuyama
MINE FIELD LOVE — Outer Eastern Approaches to Shimonoseki Strait and Moji Area
MINE FIELD MIKE — Inner Shimonoseki Approaches and Strait; Moji Harbor; Western Approaches to Shimonoseki Kaikyo
MINE FIELD NAN — Fushiki, Nanao Bay
MINE FIELD OBOE — Tokyo
MINE FIELD ROGER — Western Approach to Sasebo Naval Base and Mura Wan
MINE FIELD TARE — Nagoya, across mouth of Ise Wan
MINE FIELD UNCLE — Niigata;Sakata
MINE FIELD WILLIAM — Fungkoshi, Funakawa
MINE FIELD X-RAY — Senzaki, Hagi, oura
MINE FIELD YOKE — Sakai
MINE FIELD ZEBRA — Maizuru Bay, Tsuruga Bay and Harbor, Miyazu

MINE FIELD RASHIN — Korea
MINE FIELD SEISHIN — Korea
MINE FIELD MASAN — Korea
MINE FIELD FUSAN — Korea
MINE FIELD PUSAN — Korea
MINE FIELD GEIJITSU — Korea
MINGTOY — Radio call sign for 468th Bomb Group, May 1945
MISER — Radio call sign for 39th Bomb Group, August 1945
MIZPAH — Radio call sign for 330th Bomb Group, August 1945
MODELLER — XXIBC Target Name for Tachikawa Aircraft Company, Tachikawa
MODEX — Radio call sign for 444th Bomb Group, CBI Theater
MOPISH — XXIBC Target Name for Matsugama
MSN — Mission
NAV — Navigator or navigational
NE — Northeast
NECKCLOTH — XXIBC Target Name for Miyazaki Air Field
NR — Near
NW — Northwest
OD — Olive Drab, (color of many army clothes)
OFF INV — "OFF INVENTORY"; the date plane was removed from inventory for any reason
OGP — Original Group Plane
OGRE — Radio call sign for 444th Bomb Group, August 1945
OOF — Out Of Fuel
O'SEAS — Overseas
OUTFIELD — Radio call sign for 58th Bomb Wing, May 1945
PANHANDLE — XXIBC Target Name for Moen Island Air Fields, Truk
PATHWAY — Radio call sign for 501st Bomb Group, May 1945
PEACHBLOW — XXIBC Target Name for Urban Osaka
PERDITION — XXIBC Target Name for Tokyo Arsenal
PETREL — Designator for weather tracks flown by 57th WRS, Hickam AFB
PIARDOBA — Permanent India base of the 462nd Bomb Group
PILOT — Known as 'Copilot' on other planes. Sits in right cockpit seat
PLAYMATE — 313th Bomb Wing Super Dumbo, 1945
PLUTO — Radio call sign for 500th Bomb Group, May 1945
POSS — Possible
POW — Prisoner of War
PRS — Photo Reconnaissance Squadron
PUMPKIN — High explosive bomb of weight and shape of Fat Man atomic weapon
PUTT-PUTT — B-29 on-board auxiliary power unit; provides electrical power for starting engines
RAF — Royal Air Force

Abbreviations, Column Headings and Acronyms

RAMPAGE — Radio call sign for 314th Bomb Wing, May 1945
RANGER TOWER — North Field, Guam, tower call sign
RB-29 — Designation for reconnaissance configured B-29
RCM — Radar Countermeasures
RCN — Reconnaissance
RDC — Radar Calibration Squadron
RECD — Received
RECLAIMED — Removed from AF inventory for scrapping; date turned over to scrapper
RECON — Reconnaissance
RETD — Returned
ROBUST — Radio call sign for 40th Bomb Group, May 1945
RG — Right Gunner on B-29 crew
RO — Radio Operator on B-29 crew
ROCKCRUSHER — XXIBC Target Name for Iwo jima Air Fields
RPTD — Reported
RUD — Rudder
RUDSTRP — Stripe on Rudder
SALV — Salvaged
SALVAGED — Pulled usable parts before scrapping plane
SAN ANTONIO — XXIBC Target Name for Tokyo
SANDY — Radio call sign for 499th Bomb Group, May 1945
SB-29 — Designation for Rescue configured B-29
SCRAPPED — Plane beyond economical repair, sent to scrapyard
SE — Southeast
SERVICE TEST AIRCRAFT — Followed prototype in build sequence; used for field testing
SETTING SUN — Plan to construct B-29-capable air bases on a 400 mile axis N/S Changsha, China
SHEEPSKIN — Surface lifeguard vessel May 1945
SHYSTER — Radio call sign for 497th Bomb Group, August 1945
SKEEZIX — Radio call sign for 505th Bomb Group, May 1945
SKEWER — XXIBC Target Name for Tomitaki Air field
SKOOKUM — Radio call sign for 468th Bomb Group, August 1945
SKYBLUE — Radio call sign for 505th Bomb Group, August 1945
SLEDGEHAMMER — XXIBC Target Name for Iwo Jima Air Fields
SLICKER — Radio call sign for 331st Bomb Group, August 1945
SOL — Solid
SOLBLURUD — Solid blue rudder; tail code of 771st BS, 462nd Bomb Group, in the CBI
SOLGRNRUD — Solid green rudder; tail code of 770th BS, 462nd Bomb Group, in the CBI
SOLREDRUD — Solid red rudder; tail code of 768th BS, 462nd Bomb Group, in the CBI
SOLRUD — Solid Colored rudder

SOLYLWRUD — Solid yellow rudder; tail code of 769th BS, 462nd Bomb Group, in the CBI
SQ — Squadron
STARLIT — XXIBC Target Name for Iwo Jima Air Fields
STARVATION — Operation name for aerial mining of Japanese Home Waters
STOPWATCH — Radio call sign for 502nd Bomb Group
STRP — Stripe
SURVEYED — Surveyed
SURVEYED — Report of Survey written on plane declaring it damaged beyond economical repair
SWPAC — SouthWest Pacific
SYSTEM — Functional unit made up of assemblies
T — After April, 1945, a 12' "T" on the vertical stabilizer denoted the 498th Bomb Group
T SQ — "T" over hollow square over ID# denoted tail code for 498th Bomb Group prior to April 1945
T/O — Takeoff
TDY — Temporary Duty away from home base
TG — Tail Gunner
TIMBER — Radio call sign for 502nd Bomb Group, August 1945
TK — Symbol replaces Class O-1Z
TO — Target of Opportunity
TOBY — Radio call sign for 29th Bomb Group, August 1945
TREADLE — XXIBC Target Name for Chiran Air Field
TRI — Triangle
TRI I — Triangle containing "I"; XXIBC tail code of 468th Bomb Group on Tinian
TRI N — Triangle containing "N"; XXIBC tail code of 444th Bomb Group on Tinian
TRI S — Triangle containing "S"; XXIBC tail code of 40th Bomb Group on Tinian
TRI U — Triangle containing "U"; XXIBC tail code of 462nd Bomb Group on Tinian
TRNG — Training
TW — Test Wing
TWILIGHT — Gen. Stilwell alternative to SETTING SUN — Changsha advanced bases
UPCAST — XXIBC Target Name for Shizuoka Aircraft Engine Plant
V — From April 1945 12 ft "V" on vertical stabilizer designated 499th Bomb Group
V SQ — V over hollow square over ID#, tail code 499th Bomb Group prior to April, 1945
VAMOOSE — XXIBC Target Name for Omura Air Field
VLR — Very Long Range
VULTURE — Designator for weather tracks flown by 54th WRS, Andersen AFB
WALNUT — Central Field, Iwo Jima
WARNING — Radio call sign for 498th Bomb Group, August 1945
WB-29 — Designation for weather reconnaissance B-29 configuration

WIA — Wounded in Action
WICKED — Radio call sign for 462nd Bomb Group, May 1945
WISDOM — Radio call sign for 500th Bomb Group, August 1945
WOLFE PLAN — Modified TWILIGHT — 100 B-29's in China, 150 India transports supporting
XFER — Transfer

YELLOW JACKETS — Survivors in water in life jackets
YLW — Yellow
Z — From April 1945 12 ft "Z" on vertical stabilizer designated 500th Bomb Group
Z SQ — Z over hollow square over ID#, tail code 500th Bomb Group prior to April, 1945
ZI — Zone of the Interior; continental United States
ZK — Symbol replacing Class O-1Z

I

MASTER LIST

The Master List contains at least one entry for each of the 3,960 B-29s built. Multiple entries are included for those aircraft serving combat tours in both World War II and Korea, as well as aircraft assigned to more than one combat organization in either conflict.

The Master List contains the core data necessary to identify any B-29: serial number, name, tail code (identifying markings on tail), delivery date, assignment(s), off inventory date, and circumstances of removal from inventory.

Serial Number

The list is arranged in sequential order by serial number. The serial number consists of a two-digit number representing the fiscal year in which the aircraft was ordered (e.g., 41=1941), followed by a four- or five-digit number designating the place of the airplane in the sequential builds of the B-29 contracts then in effect.

The serial number almost always appeared on the exterior of the aircraft in two places. The first was on the vertical stabilizer. In this location the serial number was limited to fiscal year and the four- or five-digit sequential number, e.g., 6343, 24888 or 62094. In rare instances, only the last four digits were shown.

The serial number was also stenciled, in one-inch letters and numbers, below the aircraft commander's window on the left side of the nose in what is known as the technical data block, or TDB. The TDB was the only place on the aircraft that presented both the model designation and serial number.

A third, less common, application of part of the serial number was in use just before and during the Korean War. In this case, the last three or four digits, preceded by a "BF," signifying B-29, were painted in large numbers on the nose, the rear fuselage, or both. This was known as a "buzz number"; purportedly it would facilitate identification of low-flying, or "buzzing," airplanes.

A fourth application of the last four digits of the serial number was on the nose wheel doors. This application was most often found on aircraft of the 98th Bomb Group, stationed at Yokota Air Base during the Korean Conflict.

Name

The name of the aircraft was usually selected by the flight crew or, less often, by the ground crew. The large expanse of virgin aluminum on the B-29's nose gave rise, in 1944 and early 1945, to some truly magnificent examples of nose art. Some of the art was risqué, and, inevitably, some censorship took place.

The two most blatant acts of censorship involved the 314th and 73rd Bomb Wings. With the 73rd, the rumor persists that Eleanor Roosevelt, who was on Saipan on some sort of "inspection" tour, took offense at some of the nose art. The 73rd hierarchy ordered the art removed and replaced by standardized artwork consisting of a ball pierced by a spear with the name of the plane written along the shaft of the spear. This standardized artwork was derisively referred to as "the chicken on a spit."

The 314th decreed that standard names and artwork would be applied to all 314th planes. Each plane would be named for a United States city, and the artwork would consist of a circle containing an outline of North America with a flagpole rising from the appropriate map coordinates. The flag on the pole would proclaim the plane to be "City of _____."

The 314th command ran into determined opposition from the flight crews. The most prevalent form of resistance was to seek out real cities with names of a sexual or scatological nature, e.g., "The City of Intercourse, Pennsylvania."

Finally, command and crews reached a compromise: The "city" name would be painted on the right side of the nose and the crew's private name for the plane on the left. All official photos of the crews would be taken on the right side of the plane.

Censorship of nose art raised it head again during the Korean conflict when, the story goes, the wife of the 19th Bomb Group's commander was reportedly the instigator of a nose art censorship crusade. What she might have been doing on the flight line in order to see the nose art has never been explained.

Some of the names are followed by boldface numbers in brackets. These indicate notes at the end of the Master List.

Identification

With large numbers of planes filling the air in the chaos of bombing attacks, it became apparent very early in the air war in Europe that highly visible identification was a necessity. Drawing on this institutional experience, when the 58th Bomb Wing deployed to India in the spring of 1944, a rudimentary but effective system of tail markings was put into effect.

The 40th Bomb Group used four horizontal stripes high on the vertical stabilizer. Red stripes and a matching stabilizer tip identified the 25th Bomb Squadron; blue, the 44th Bomb Squadron; yellow, the 45th Squadron; and black, the 395th Squadron (deactivated 10/44). A single letter was painted below the stripes to identify individual aircraft within the squadron.

In the master list, these tail codes are identified with abbreviations. For example, for plane "C" from the 45th Squadron, with four yellow stripes, the tail code abbreviation would by FOURYLWSTRP C.

The 444th Bomb Group used a diamond on the upper half of the vertical stabilizer containing a two-digit number for plane identification. A colored fuselage band aft of the gun blisters identified the Squadron: 676BS — green; 677BS — yellow; 678BS — red. A typical 444th tail code abbreviation might be DIAM# 36.

The 462nd Bomb Group used solid-colored rudders — red for the 768th, yellow for the 769th, and green for the 770th. A large letter on the stabilizer identified the individual plane. A typical tail code abbreviation for the 462nd might be SOLBLURUD B.

The 468th Bomb Group was identified by two oblique stripes on the rudder — white for the 792nd BS, blue for the 793rd, and yellow for the 794th. Plane identification was accomplished by painting the last three digits of the serial number on the stabilizer in large numbers. A tail code abbreviation for a 792nd plane would be 2WHTRUDSTRP.

The XXI Bomber Command showed a little more consistency in the use of tail codes: Each group was assigned a different letter. Thus in the 73rd bomb wing, the 497th bomb group was identified by an "A" over a hollow square over an individual plane number; the 498th by a "T"; the 499th by a "V"; and the 500th by a "Z." In April of 1945 these codes were replaced by twelve-foot A's, T's, V's and Z's, with the plane identification number on the fuselage in the general location of the rear unpressurized compartment.

The 313th Bomb Wing was assigned an "L" over a hollow triangle over the plane number for the 6th Bomb Group; an "X" for the 9th Bomb Group; "E" for the 504th Bomb Group; and "K" for the 505th. In April 1945, the tail codes were change to an "R" in a circle for the 6th, "X" in a circle for the 9th, "E" in a circle for the 504th, and "W" in a circle for the 505th.

The 314th Bomb Wing used an unpainted "M" in a black square for the 19th Bomb Group; an unpainted "O" for the 29th Bomb Group; an unpainted "P" for the 39th; and an unpainted or yellow "K" for the 330th.

The 315th Bomb Wing used a "B" in a diamond for the 16th bomb group; an "L" for the 331st; a "Y" for the 501st; and an "H" for the 502nd.

Sometimes a number in parentheses follows the tail code, e.g. TRI S36 (1). The number in parentheses indicates how many times that particular tail code was assigned to a new aircraft following the loss of a plane that previously bore that code.

Delivery

For all aircraft with a single entry in the Master List, this column cites the date the airplane was delivered by the manufacturer to the USAF or the USAAF. For those aircraft with more than one entry, the date in the "Delv" column reflects the date that the organization cited received the airplane; information in the "Off Inv." and "Circumstances" columns continues the story. Here is a typical example:

Serial #	Delv	Assign	Off Inv	Circumstances
44–62135	8/14/45	44BG	6/15/51	Transferred to RAF
44–62135	6/15/51	RAF	3/16/54	Returned to USAF
44–62135	3/16/54	USAF	7/14/54	Salvaged at Davis-Monthan

In the first line, 8/14/45 is the date the plane was delivered to the USAF. When assigned to the 44BG it was transferred to the RAF (Royal Air Force) on 6/15/51.

The second line shows the RAF received the plane on 6/15/45 and returned it to the USAF on 3/16/54. The third line shows the USAF gaining 44–62135 on 3/16/54 and turning it over to salvage at Davis-Monthan AFB on 7/14/54.

Assignment

The *"Assign"* (Assignments) column notes the organizational unit to which the aircraft was assigned during its combat period. Some planes were transferred from one unit to another; in those cases, the Assignments column includes an entry for each assignment.

Off Inventory

There are numerous ways an aircraft can be removed from inventory: combat losses, crashes and scrapping are the primary means. In the vast majority of these cases the dates of removal can be definitively established by USAF/USAAF records, such as examination of Aircraft Record Cards, Consolidated Mission Summary reports and other historical documents found in the Bomb Wing Histories, and Missing Air Crew Summary Reports (MACR).

Aircraft Record Cards are forms used by the Army Air Forces to track the movement and assignment of aircraft. Working with their elaborate codes can be a difficult and daunting task.

Bomb Wing Histories are available on microfilm from the Air Force Historical Research Center at Maxwell Air Force Base, Alabama. They contain a plethora of details on day-to-day operations. There is a drawback to this type of information: It is only as good as the Historical Officer wanted it to be. Sometimes a wing history contains frame upon frame upon frame of personnel data and a minimal amount of information on operations. Two months later the situation could be the reverse as a new Historical Officer takes over.

On the other hand, the reports dealing with operations mandated by wing, such as the Consolidated Mission Reports, present hard facts, uncolored by personality or interpretation.

The MACR is also straightforward, recording date, serial number, group, an alpha code giving the cause of loss, aircraft commander, number of survivors if any, and a MACR file number.

Circumstances

This column contains a brief description as to why the aircraft was removed from inventory. This information comes from the documents described above or a variety of other sources, such as Japanese records (very reliable), squadron and personal histories, and other publications.

Serial #	Name	Identification	Delv	Assign	Off Inv	Circumstances
41-002					5/11/1948	Scrapped
41-003					2/14/1943	Crashed
41-18335	THE FLYING GUINEA PIG					Unknown
41-36954	SPIRIT OF LINCOLN					Unknown
41-36955				40BG		Unknown
41-36956				40BG	1/19/1944	Wright Field, Ohio
41-36957						Unknown
41-36958				444BG		Unknown
41-36959	AMARILLO'S FLYING SOLENOID		9/20/1943	468BG	11/6/1943	Smoky Hill Kansas
41-36960						Unknown
41-36961			9/20/1943	468BG	10/24/1943	Crashed on takeoff
41-36962					2/24/1944	Great Bend Kansas
41-36963	HOBO QUEEN	SOLYLWRUD	4/6/1944	462BG		Unknown
41-36964						
41-36965					1/2/1945	
41-36966			9/6/1943	40BG	8/21/1944	Crashed, structural failure
41-36967			9/11/1943	462BG	1/29/1944	Out of fuel, crashed, surveyed
42-6205			9/11/1943	40BG	12/31/1949	Reclaimed Pyote
42-6206			9/23/1943	444BG	9/24/1949	Reclaimed Hill
42-6207			9/23/1943	468BG	8/9/1944	Class 26 Alamogordo
42-6208	PIONEER	2BLURUDSTRP	9/23/1943	468BG	12/21/1949	Reclaimed Pyote
42-6209		SOLGRNRUD	9/30/1943	462BG	5/22/1945	Class 26 Tinian
42-6210			10/27/1943	462BG	7/4/1944	Surveyed Clovis, structural failure
42-6211		SOLBLURUD	10/26/1943	462BG	5/23/1944	Surveyed, crash, fire on takeoff
42-6212		DIAM #	11/17/1943	444BG	9/8/1944	Crashed 3 miles west of Hsian, out of fuel
42-6213		SOLYLWRUD	12/16/1943	462BG	12/29/1949	Reclaimed Pyote
42-6214			12/18/1943	40BG	10/17/1946	Reclaimed Clovis
42-6215	DEACON'S DISCIPLES	DIAM #	10/27/1943	444BG	11/9/1944	Abandoned Dudhkindi, engine failure
42-6216			11/16/1943	468BG	10/8/1944	Class 26 at Clovis NM
42-6217	PEPPER	2WHTRUDSTRP	11/28/1943	468BG	12/21/1949	Reclaimed Pyote
42-6218			12/6/1943	40BG	5/31/1946	Reclaimed Amarillo

Serial #	Name	Identification	Delv	Assign	Off Inv	Circumstances
42-6219	PETROL PACKIN' MAMA	SOLGRNRUD	12/7/1943	462BG	5/31/1946	Class 01Z Roswell
42-6220		DIAM #	12/13/1943	444BG	6/15/1944	Crashed Kiangyu, China on photo mission
42-6221		2RUDSTRP	12/13/1943	468BG	11/21/1944	Surveyed Clovis, engine failure
42-6222	DEUCES WILD	FOURYLWSTRP	12/30/1943	40BG	9/11/1944	Crashed at B-4, no fuel. Surveyed
42-6223	LADY BOOMERANG	SOLBLURUD B 3	12/16/1943	462BG	8/25/1944	Surveyed, crash landed B-6, no fuel
42-6224	ANNIE	224	11/15/1943	468BG	4/30/1946	Reclaimed Davis-Monthan
42-6225	DING HOW	DIAM #	12/16/1943	444BG	11/1/1948	Reclaimed
42-6226		SOLRUD	11/5/1943	462BG	6/20/1944	Caught fire and was destroyed
42-6227			11/1/1943	462BG	11/17/1943	Class 26 Victoria Kansas
42-6228		DIAM #	11/4/1943	444BG	8/2/1944	Crashed, burned near Syhlet, China
42-6229		2BLURUDSTRP	2/24/1944	468BG	6/15/1944	Crashed on takeoff to Yawata
42-6230	LIMBER DUGAN	2WHTRUDSTRP	2/19/1944	468BG	6/15/1944	Shot down over Yawata
42-6231		2REDRUDSTRP	1/26/1944	468BG	6/16/1944	Hit mountain returning from Yawata
42-6232	KICKAPOO II	2WHTRUDSTRP	1/26/1944	468BG	12/21/1949	Reclaimed Pyote
42-6233			2/20/1944	4141BU	12/21/1949	Reclaimed Pyote
42-6234		DIAM #	12/19/1943	444BG	9/8/1944	Crashed China from Anshan
42-6235		2YLWRUDSTRP	1/2/1944	468BG	6/26/1944	Surveyed CBI-Gear came up taxiing
42-6236			4/6/1944	4136BU	5/24/1949	Reclaimed Tinker
42-6237	SIR TROFREPUS	FOURREDSTRP	12/23/1943	40BG	11/12/1944	Flak damage at Nanking
42-6238		2WHTRUDSTRP	11/8/1943	468BG	10/1/1944	Crashed, burned Omei, China on Hump msn
42-6239			11/12/1943	468BG	5/26/1944	Surveyed Smoky Hill
42-6240	HAP'S HOPE	FOURBLKSTRP	11/6/1943	40BG	7/30/1944	Crashed in weather on Hump mission
42-6241	HUMP HAPPY MAMMY	FOURBLUSTRP	11/4/1943	40BG	12/21/1949	Reclaimed Pyote
42-6241	MALFUNCTION JUNCTION	FOURBLUSTRP	11/4/1943	40BG	12/21/1949	Reclaimed Pyote
42-6241	YELLOW ROSE OF TEXAS	FOURBLUSTRP	11/4/1943	40BG	12/21/1949	Reclaimed Pyote
42-6242	ESSO EXPRESS	2YLWRUDSTRP	11/3/1943	468BG	7/7/1946	Reclaimed Lowry
42-6243	RODGER THE LODGER	2WHTRUDSTRP	1/31/1944	468BG	8/11/1944	Abandoned, crashed Hwaning, China
42-6244			11/19/1943	462BG	2/3/1944	Surveyed Amarillo
42-6245			3/1/1944	498BG	6/7/1944	Surveyed Wichita, crashed on takeoff
42-6246		DIAM #	11/15/1943	444BG	6/28/1944	#1 & #3 engine failures over Hump
42-6247		FOURREDSTRP	11/11/1943	40BG	4/19/1944	Crashed Cairo en route to India
42-6248	GREEN DRAGON	SOLREDRUD	11/12/1943	462BG	12/21/1949	Reclaimed Pyote
42-6249		FOURREDSTRP	11/11/1943	40BG	4/19/1944	Crashed on T/O from Cairo en route CBI
42-6250	BOMBIN' BUGGY	FOURBLUSTRP	11/18/1943	40BG	12/21/1949	Reclaimed Pyote
42-6251	OLE BATTLER	DIAM# 72	11/17/1943	444BG	12/21/1949	Reclaimed Pyote
42-6252	RAGGED BUT RIGHT	SOLBLURUD	11/19/1943	462BG	12/21/1949	Reclaimed Pyote
42-6253	WINDY CITY	2YLWRUDSTRP	11/28/1943	468BG	10/11/1944	Surveyed, bellied in from Yawata
42-6254	HUMP HAPPY PAPPY	FOURYLWSTRP	12/6/1943	40BG	12/21/1949	Reclaimed Pyote
42-6255		2WHTRUDSTRP	12/13/1943	468BG	4/19/1944	Abandoned enroute A-7 from India
42-6256	RAMP TRAMP	SOLBLURUD	11/28/1943	462BG	7/29/1944	Interned Vladivostok
42-6257	DUMBO	DIAM #	11/28/1943	444BG	12/21/1949	Reclaimed Pyote
42-6258		DIAM #	11/30/1943	444BG	8/21/1945	Reclaimed Roswell
42-6259		SOLRUD	11/30/1943	462BG	5/11/1948	Reclaimed Fort Worth
42-6260			12/11/1943	2132BU	12/21/1949	Reclaimed Pyote
42-6261	STOCKETT'S ROCKET	FOURYLWSTRP	12/13/1943	40BG	6/14/1944	Crashed Hump 1st leg 2nd combat mission
42-6262	ROUND TRIP TICKET	DIAM # EE	12/7/1943	444BG	12/7/1944	Missing on Hump
42-6263	SIR TROFREPUS	SOLRUD	12/7/1943	462BG	7/8/1944	Ditched from Sasebo
42-6264	O'REILLY'S DAUGHTER	2WHTRUDSTRP	12/16/1943	468BG	8/21/1944	Crashed Occupied China from Yawata
42-6265	RAIDEN MAIDEN	2BLURUDSTRP	12/10/1943	468BG	6/27/1949	Reclaimed Davis-Monthan
42-6266	TARFU	SOLREDRUD	12/21/1943	462BG	7/2/1946	Reclaimed Roswell
42-6267	WINNIE	DIAM# 80	12/19/1943	444BG	12/21/1949	Reclaimed Pyote
42-6268	GOOD HUMPIN'	FOURBLUSTRP D	12/19/1943	40BG	9/18/1944	Surveyed B-4; local flt, 2 eng on fire
42-6269	SMOKEY STOVER	FOURREDSTRP C-E	12/21/1943	40BG	12/21/1949	Reclaimed Pyote
42-6270	CASE ACE	SOLGRNRUD K7	12/18/1943	462BG	12/21/1949	Reclaimed Pyote
42-6271		2WHTRUDSTRP	12/19/1943	468BG	6/7/1944	Abandoned enroute to A-7
42-6272	MISS MINETTE	2YLWRUDSTRP	12/18/1943	468BG	12/21/1949	Reclaimed
42-6272	THE OLD CAMPAIGNER	2YLWRUDSTRP	12/18/1943	468BG	12/21/1949	Reclaimed
42-6273	OLD-BITCH-U-AIRY BESS	SOLYLWRUD	12/23/1943	462BG	12/21/1949	Reclaimed Pyote
42-6274	LADY HAMILTON	2YLWRUDSTRP	12/20/1943	468BG	7/29/1944	Crashed Occupied China from Anshan
42-6275	SNAFUPER BOMBER	FOURYLWSTRP	12/23/1943	40BG	11/23/1944	Crew abandoned 20 miles north of A-1
42-6276	BLACK MAGIC	FOURBLKSTRP A	12/31/1943	40BG	12/19/1944	Crew abandoned over Hump — out of fuel
42-6276	OLD CRACKER KEG	FOURBLKSTRP A	12/31/1943	40BG	12/19/1944	Crew abandoned over Hump — out of fuel
42-6277	SHANGHAI LIL	DIAM# 24	12/30/1943	444BG	9/7/1944	Crash landed Dudhkundi
42-6278		SOLBLURUD	12/31/1943	462BG	11/21/1944	Crashed in sea from Omura
42-6279	POSTVILLE EXPRESS	2WHTRUDSTRP	12/29/1943	468BG	12/21/1949	Reclaimed

I. Master List

13

Serial #	Name	Identification	Delv	Assign	Off Inv	Circumstances
42-6280		DIAM #	12/30/1943	444BG	10/14/1944	Exploded on Formosa mission
42-6281	20TH CENTURY UNLIMITED	FOURYLWSTRP	12/24/1943	40BG	10/25/1944	Crew abandoned Laohokow from Omura
42-6281	HEAVENLY BODY	FOURYLWSTRP	12/24/1943	40BG	10/25/1944	Crew abandoned Laohokow from Omura
42-6282		FOURBLKSTRP	12/23/1943	40BG	6/5/1944	Ditched Bay of Bengal — out of fuel
42-6283			12/24/1943	40BG	12/21/1949	Reclaimed Pyote
42-6284	THE CHALLENGER	2YLWRUDSTRP	12/31/1943	468BG		Unknown
42-6285	OURS	SOLYLWRUD	12/24/1943	462BG	12/21/1949	Reclaimed Pyote
42-6286	PRAYING MANTIS	DIAM #	12/31/1943	444BG	8/21/1944	MIA from Yawata, crashed China
42-6287		SOLGRNRUD	12/31/1943	462BG	6/26/1945	Taxi accident in ZI
42-6288		FOURREDSTRP	12/31/1943	40BG	10/22/1944	Crashed on takeoff from A-1, photo msn
42-6289	NIPPON NIPPER	FOURBLUSTRP A	1/6/1944	40BG	7/8/1944	Burned on ground Chakulia
42-6290	WEMPY'S BLITZBURGER	FOURBLUSTRP	1/10/1944	40BG	11/23/1944	Hit by 444BG plane at Lianshan
42-6291		FOURBLKSTRP	1/9/1944	40BG	7/26/1944	Crash Midnapare after takeoff from B-4
42-6292	BLACK JACK	DIAM #	1/10/1944	444BG	5/19/1944	Missing
42-6293		DIAM #	1/6/1944	444BG	6/16/1944	Destroyed on ground, forced landing
42-6294	MONSOON	FOURREDSTRP K	1/11/1944	40BG	12/21/1949	Reclaimed Pyote
42-6295	MONSOON MINNIE	FOURYLWSTRP B	1/11/1944	40BG	12/21/1949	Reclaimed
42-6296		DIAM #	1/14/1944	444BG	5/19/1944	Surveyed — training mission
42-6297	OL' 297	FOURBLUSTRP W	1/16/1944	40BG	12/21/1949	Reclaimed Pyote
42-6298	KATIE	FOURREDSTRP S	1/18/1944	40BG	11/3/1944	Crash landing — blew tire, ground looped
42-6299	HUMPIN HONEY	SOLGRNRUD E 3	1/11/1944	462BG	12/7/1944	Rammed over Mukden
42-6300		DIAM #	1/14/1944	444BG	11/11/1944	MIA Omura
42-6301		FOURREDSTRP	1/10/1944	40BG	8/20/1944	Crashed west of Hsinching
42-6302		DIAM #	1/18/1944	444BG	7/15/1944	Crashed Chabua, abandoned on Hump msn
42-6303	TYPHOON MCGOON	FOURBLKSTRP L	1/20/1944	40BG	6/27/1949	Reclaimed Davis-Monthan
42-6304		FOURBLKSTRP	1/18/1944	40BG	6/5/1944	Ditched from Bangkok — flak damage
42-6305	STARDUSTER	SOLYLWRUD	1/16/1944	462BG	10/1/1944	Surveyed — crashed takeoff A-5 8/20/44
42-6306	PAMPERED LADY	FOURBLKSTRP D	1/15/1944	40BG	12/21/1949	Reclaimed Pyote
42-6307	BLUE BONNET BELLE	DIAM #	1/16/1944	444BG	11/12/1944	Crash landed on return
42-6308	FEATHER MERCHANT	FOURYLWSTRP	1/19/1944	40BG	8/20/1944	Crew abandoned Laohokow from Yawata
42-6309		2BLURUDSTRP	1/16/1944	468BG	9/2/1944	Missing en route China from India
42-6310	HUMP HAPPY JR	FOURREDSTRP	1/20/1944	40BG	12/21/1949	Reclaimed Pyote
42-6311	NIGHT MARE	SOLBLURUD	1/20/1944	462BG	6/27/1949	Reclaimed Davis-Monthan
42-6312	MYSTERIOUS MISTRESS	SOLREDRUD H1	1/19/1944	462BG	5/2/1949	Reclaimed Pyote
42-6313	DING HOW	FOURYLWSTRP R	1/20/1944	40BG	12/21/1949	Reclaimed Pyote
42-6314		2BLURUDSTRP	1/20/1944	468BG	6/9/1944	Lost enroute A-7, crashed Choating
42-6315		DIAM #	1/20/1944	444BG	12/21/1949	Reclaimed Pyote
42-6316	EXCALIBUR	SOLREDRUD	1/21/1944	462BG	6/27/1949	Reclaimed Davis-Monthan
42-6317	HOMBICRISMUS	DIAM #	1/20/1944	444BG	12/21/1949	Reclaimed Pyote
42-6318		FOURBLKSTRP	1/22/1944	40BG	6/5/1944	Crashed on Takeoff from B-4 to Bangkok
42-6319	HIMALAYA HUSSY	FOURBLUSTRP I	1/21/1944	40BG	12/21/1949	Reclaimed Pyote
42-6320	FLYING STUD	DIAM #	1/24/1944	444BG	8/21/1944	Abandoned, engine fire
42-6321	THIS IS IT	DIAM# 53	1/21/1944	444BG	11/21/1944	Crashed Ankang from Omura
42-6322	BONNIE LEE	FOURBLUSTRP B	1/24/1944	40BG	12/23/1944	Cr 50 miles from Ankang, lost two engines
42-6323	PRINCESS EILEEN	DIAM #	1/24/1944	444BG	6/25/1944	Crashed, burned, China on Hump mission
42-6324	THREE FEATHERS	DIAM #	1/24/1944	444BG	12/21/1949	Reclaimed Pyote
42-6325	B-SWEET	DIAM #	1/25/1944	444BG	12/21/1949	Reclaimed Pyote
42-6325	NO PAPA	DIAM #	1/25/1944	444BG	12/21/1949	Reclaimed Pyote
42-6326		FOURYLWSTRP	1/25/1944	40BG	9/11/1944	Surveyed -emer landing Hsinching 6/18/44
42-6327		DIAM #	1/25/1944	444BG	5/1/1944	Surveyed, engine fire, crash landing
42-6328		SOLREDRUD	1/31/1944	462BG	6/27/1944	Crashed, burned Kiunglai on Hump mission
42-6329	DING HAO	SOLREDRUD N1	1/27/1944	462BG	12/21/1949	Reclaimed Pyote
42-6330		DIAM #	1/28/1944	444BG	8/21/1944	Crashed 10 miles east of A-3 from Yawata
42-6331	GONE WITH THE WIND	FOURREDSTRP	1/28/1944	40BG	12/20/1944	Shot down by British Beaufighter
42-6332	BELLE OF BALTIMORE	SOLREDRUD	1/28/1944	462BG	8/20/1944	Flak at Yawata — crashed China
42-6333	CAMEL CARAVAN	2BLURUDSTRP	1/31/1944	468BG	3/2/1949	Reclaimed Pyote
42-6334	GERTRUDE C	2REDRUDSTRP	2/4/1944	468BG	8/20/1944	Rammed at Yawata
42-6335		SOLRUD	2/2/1944	462BG	10/18/1944	Surveyed — exploded on ground
42-6336		SOLBLURUD	2/3/1944	462BG	6/5/1944	Surveyed; crashed Dum-Dum first msn
42-6337		SOLGRNRUD	1/31/1944	462BG	12/21/1949	Reclaimed Pyote
42-6338	PATCHES	SOLYLWRUD	2/7/1944	462BG	12/21/1949	Reclaimed Pyote
42-6339			2/26/1944	248BU	6/27/1949	Reclaimed Davis-Monthan
42-6340		DIAM #	2/4/1944	444BG	12/21/1949	Reclaimed Pyote
42-6341		DIAM #	2/4/1944	444BG	12/21/1949	Reclaimed Pyote
42-6342	SISTER SUE	FOURBLKSTRP T	2/8/1944	40BG	10/17/1944	Engine fire, abnd SW of A-1

Serial #	Name	Identification	Delv	Assign	Off Inv	Circumstances
42-6342	TABOOMA	FOURREDSTRP	2/8/1944	40BG	10/17/1944	Engine fire, abnd SW of A-1
42-6343	MARY'S LIL LAMBS	DIAM #	2/4/1944	444BG	12/27/1944	Hit mountain Kwandhan, China
42-6344	OLD FIRING BUTT	FOURREDSTRP	2/5/1944	40BG	12/21/1949	Reclaimed Pyote
42-6344	PRINCESS PATSY	FOURREDSTRP	2/5/1944	40BG	12/21/1949	Reclaimed Pyote
42-6345		SOLRUD	2/9/1944	462BG	6/26/1944	Crashed Karachi 4/21/44, salvaged
42-6346	MAN-O-WAR	SOLREDRUD A1	2/4/1944	462BG	12/21/1949	Reclaimed Pyote
42-6347	KING SIZE	SOLYLWRUD A2	2/8/1944	462BG	12/21/1949	Reclaimed Pyote
42-6348	BENGAL LANCER	FOURBLUSTRP E	2/4/1944	40BG	12/21/1949	Reclaimed Pyote
42-6349			2/24/1944	4121BU	12/21/1949	Reclaimed Pyote
42-6350		SOLREDRUD	2/11/1944	462BG	4/13/1944	Destroyed by fire at Marrakech
42-6351		FOURREDSTRP	2/12/1944	40BG	7/29/1944	Crashed, exploded landing at A-1
42-6352	FU-KEMAL	DIAM# 21	2/11/1944	444BG	12/21/1949	Reclaimed Pyote
42-6353	BIG POISON	DIAM# 30	12/12/1944	444BG	12/21/1949	Reclaimed Pyote
42-6354	MISS LACE	SOLYLWRUD	2/19/1944	462BG	4/1/1947	Reclaimed
42-6355			2/24/1944	40BG	6/15/1944	Surveyed at Clovis
42-6356	THE GUSHER	2WHTRUDSTRP	2/23/1944	468BG	8/26/1944	Crashed Fulin from A-7 to Kharagpur
42-6357			2/23/1944	611BEG	12/21/1949	Reclaimed Pyote
42-6358	DING HAO	2YLWRUDSTRP	2/23/1944	468BG	11/21/1944	Interned Vladivostok
42-6359	MISSOURI QUEEN	SOLBLURUD	2/18/1944	462BG	12/7/1944	Missing
42-6360		SOLGRNRUD	2/18/1944	462BG	9/11/1944	Surveyed belly landing Lachokow 9/8/44
42-6361		DIAM #	2/19/1944	444BG	6/5/1944	Abandoned Kunming, China from Bangkok
42-6362		2WHTRUDSTRP	2/22/1944	468BG	11/21/1944	Crashed, exploded on takeoff from A-7
42-6363					3/31/1945	Class 01Z
42-6364			2/26/1944	500BG	6/27/1949	Reclaimed Davis-Monthan
42-6365	GENERAL HH ARNOLD SPCL	2YLWRUDSTRP	2/24/1944	468BG	11/11/1944	Interned Vladivostok from Omura
42-6366			2/24/1944	234BU	11/20/1948	Reclaimed Robins
42-6367			2/24/1944	234BU	4/25/1949	Reclaimed Robins
42-6368	CALAMITY SUE	2REDRUDSTRP	2/24/1944	468BG	8/20/1944	Hit by pieces of 42-6334
42-6369			2/21/1944	468BG	4/21/1944	Crashed Karachi, India, enroute to join
42-6370	LETHAL LADY	2BLURUDSTRP	2/19/1944	468BG	11/5/1944	MIA Singapore
42-6371			2/26/1944	234BU	11/20/1946	Class 01Z Clovis
42-6372			2/23/1944	500BG	6/29/1944	Class 26 Walker
42-6373			2/26/1944	500BG	7/13/1944	Surveyed at Walker
42-6374			2/23/1944	499BG	9/14/1948	Reclaimed Hill
42-6375			2/23/1944	234BU	6/4/1944	Surveyed at Clovis
42-6376			2/26/1944	611EXGP	6/27/1949	Reclaimed Davis-Monthan
42-6377			3/1/1944	498BG	12/21/1949	Reclaimed Pyote
42-6378			3/10/1944	499BG	6/27/1949	Reclaimed Davis-Monthan
42-6379			2/26/1944	500BG	9/18/1944	Surveyed at Walker
42-6380	QUEEN OF THE AIR		2/26/1944	234BU	12/21/1949	Reclaimed Pyote
42-6381			3/1/1944	234BU	9/14/1948	Reclaimed Victorville
42-6382	BIG CHIEF	SOLGRNRUD	3/8/1944	462BG	6/27/1949	Reclaimed Davis-Monthan
42-6382		SOLBLURUD	3/8/1944	462BG	6/27/1949	Reclaimed Davis-Monthan
42-6383			3/14/1944	40BG	6/10/1944	Ditched — fire enroute to join
42-6384			3/6/1944	497BG	8/27/1944	Class 26 at Pratt
42-6385			3/4/1944	499BG	4/25/1949	Reclaimed Chanute
42-6386			3/7/1944	2532BU	12/21/1949	Reclaimed Pyote
42-6387			3/9/1944	497BG	6/14/1944	Class 26 Pratt — crashed
42-6388			3/4/1944	343BU	9/14/1948	Reclaimed Victorville
42-6389	PARTY GIRL	2WHTRUDSTRP	3/2/1944	468BG	12/7/1944	Hit Mountain from Mukden
42-6390	GALLOPIN' GOOSE	2YLWRUDSTRP	3/4/1944	468BG	12/7/1944	Rammed at Mukden
42-6391			4/2/1944	497BG	12/21/1949	Reclaimed Pyote
42-6392			3/19/1944	231BU	9/14/1948	Reclaimed Victorville
42-6393			3/14/1944	497BG	12/21/1949	Reclaimed Pyote
42-6394			3/7/1944		12/21/1949	Reclaimed Pyote
42-6395			3/20/1944	29BG	10/8/1944	Surveyed at Pratt
42-6396			3/9/1944	6BG	5/14/1947	Reclaimed Tinker
42-6397	MILLION DOLLAR BABY	2BLURUDSTRP	3/9/1944	468BG	6/27/1949	Reclaimed Davis-Monthan
42-6398			3/10/1944	231BU	2/22/1945	Surveyed at Alamogordo
42-6399	THE AGITATOR	DIAM #	3/9/1944	444BG	6/27/1949	Reclaimed Davis-Monthan
42-6400			3/20/1944	236BU	12/21/1949	Reclaimed Pyote
42-6401			3/11/1944	499BG	12/21/1949	Reclaimed Pyote
42-6402			3/10/1944	243BU	5/15/1947	Reclaimed Tinker
42-6403			3/20/1944	498BG	12/21/1949	Reclaimed Pyote
42-6404			3/20/1944	498BG	12/21/1949	Reclaimed Pyote

I. Master List

Serial #	Name	Identification	Delv	Assign	Off Inv	Circumstances
42-6405			3/28/1944	234BU	3/30/1945	Surveyed at Clovis
42-6406			3/16/1944	497BG	1/14/1949	Reclaimed Kelly
42-6407	LUCKY SEVEN	2YLWRUDSTRP	3/23/1944	468BG	12/21/1949	Reclaimed Pyote
42-6408	REDDY TEDDY	2WHTRUDSTRP	3/29/1944	468BG	8/20/1944	Direct flak hit
42-6409	THE UNINVITED	2YLWRUDSTRP	3/17/1944	468BG	12/21/1949	Reclaimed Pyote
42-6410			3/20/1944	498BG	12/21/1949	Reclaimed Pyote
42-6411		2WHTRUDSTRP	3/30/1944	468BG	12/21/1949	Reclaimed Pyote
42-6412			3/27/1944	234BU	12/21/1949	Reclaimed Pyote
42-6413			3/27/1944	611EXGP	7/31/1946	Reclaimed March
42-6414			3/27/1944	234BU	12/21/1949	Reclaimed Pyote
42-6415			3/23/1944	234BU	12/21/1949	Reclaimed Pyote
42-6416			3/28/1944	234BU	12/21/1949	Reclaimed Pyote
42-6417			3/22/1944	499BG	7/9/1944	Class 26 Smoky Hill
42-6418		FOURYLWSTRP	3/23/1944	40BG	10/26/1944	Chakulia (B-4)
42-6419			4/3/1944	234BU	12/21/1949	Reclaimed Pyote
42-6420			4/1/1944	499BG	10/2/1946	Reclaimed Seattle
42-6421			4/1/1944	499BG	12/21/1949	Reclaimed Pyote
42-6422			4/8/1944	499BG	3/2/1949	Reclaimed Pyote
42-6423	URGIN' VIRGIN	DIAM #	4/15/1944	444BG	12/21/1949	Reclaimed Pyote
42-6424			4/15/1944	499BG	6/27/1949	Reclaimed Davis-Monthan
42-6425	B-SWEET	FOURREDSTRP	4/20/1944	40BG	8/19/1944	Abandoned over Hump enroute Hsinching
42-6426			5/5/1944	499BG	6/27/1949	Reclaimed Davis-Monthan
42-6427			5/18/1944	500BG	5/15/1947	Reclaimed Tinker
42-6428			3/31/1944	29BG	1/13/1945	Lost in training at Pratt
42-6429			4/1/1944	234BU	12/21/1949	Reclaimed Pyote
42-6430			3/31/1944	234BU	6/14/1945	Surveyed at Kirtland
42-6431			3/29/1944	4136BU	5/15/1947	Reclaimed Tinker
42-6432			4/2/1944	238BU	4/1/1947	Reclaimed Pyote
42-6433			4/3/1944	497BG	3/31/1950	Reclaimed Las Vegas
42-6434			4/4/1944	500BG	3/31/1950	Reclaimed Seattle
42-6435			4/7/1944	500BG	8/30/1944	Surveyed at Walker
42-6436			4/2/1944	234BU	12/21/1949	Reclaimed Pyote
42-6437			4/1/1944	498BG	12/21/1949	Reclaimed Pyote
42-6438			3/31/1944	231BU	11/8/1944	Surveyed at Alamogordo
42-6439			4/3/1944	19BG	11/30/1948	Reclaimed Davis-Monthan
42-6440			4/2/1944	500BG	5/15/1947	Reclaimed Tinker
42-6441			4/2/1944	243BU	12/21/1949	Reclaimed Pyote
42-6442			4/2/1944	234BU	1/7/1949	Reclaimed Pyote
42-6443			4/4/1944	500BG	12/21/1949	Reclaimed Pyote
42-6444	HUMPIN' HONEY	SOLGRNRUD	4/12/1944	462BG	11/9/1944	Crashed on T/O 11/5, Salvaged
42-6445			4/4/1944	498BG	5/31/1946	Reclaimed Harlingen
42-6446			4/3/1944	500BG	12/21/1949	Reclaimed Pyote
42-6447			4/12/1944	231BU	7/14/1949	Reclaimed Kelly
42-6448			4/11/1944	500BG	12/21/1949	Reclaimed Pyote
42-6449			4/8/1944	499BG	5/27/1948	Reclaimed Tinker
42-6450			4/11/1944	497BG	5/8/1949	Reclaimed Kelly
42-6451			4/12/1944	234BU	5/15/1947	Reclaimed Tinker
42-6452		2RUDSTRP	4/13/1944	468BG	8/7/1944	Abandoned enroute China for Nagasaki
42-6453			4/20/1944	231BU	12/21/1949	Reclaimed Pyote
42-6454	TOTIN' TO TOKYO	2BLURUDSTRP	4/15/1944	468BG	6/27/1949	Reclaimed Davis-Monthan
42-24420		DIAM #	4/28/1944	444BG	8/11/1944	Ditched, out of fuel from Palembang
42-24421			4/28/1944	500BG	11/4/1948	Reclaimed Pyote
42-24422			4/29/1944	497BG	12/21/1949	Reclaimed Pyote
42-24423			4/29/1944		12/21/1949	Reclaimed Pyote
42-24424			4/30/1944	499BG	12/21/1949	Reclaimed Pyote
42-24425			4/30/1944	499BG	5/10/1949	Reclaimed Tinker
42-24426			4/30/1944	498BG	12/21/1949	Reclaimed Pyote
42-24427			4/30/1944	231BU	6/16/1945	Surveyed at Alamogordo
42-24428			4/30/1944	231BU	5/8/1945	Surveyed Great Bend
42-24429	BLIND DATE	2YLWRUDSTRP	4/30/1944	468BG	4/30/1946	Class O-1Z Chatham
42-24430			4/30/1944	231BU	12/21/1949	Reclaimed Pyote
42-24431			4/30/1944	234BU	12/21/1949	Reclaimed Pyote
42-24432			4/30/1944	497BG	6/27/1949	Reclaimed Davis-Monthan
42-24433			4/30/1944	29BG	12/21/1949	Reclaimed Pyote
42-24434			4/30/1944	231BU	2/13/1945	Surveyed Roswell

16   The B-29 Superfortress

Serial #	Name	Identification	Delv	Assign	Off Inv	Circumstances
42-24435			5/3/1944	234BU	10/19/1944	Surveyed Pyote, crashed on takeoff
42-24436	OLD IRONSIDE		5/3/1944	500BG	12/21/1949	Reclaimed Pyote
42-24437			5/6/1944	505BG	4/1/1945	Surveyed at Pyote
42-24438			5/8/1944	500BG	12/27/1944	Surveyed at Walker
42-24439			5/3/1944	497BG	12/21/1949	Reclaimed Pyote
42-24440			5/3/1944	499BG	12/21/1949	Reclaimed Pyote
42-24441			5/4/1944	4000EXGP	12/21/1949	Reclaimed Pyote
42-24442	POWER PLAY	2BLURUDSTRP	5/8/1944	468BG	6/30/1946	Class 01Z Amarillo
42-24442	WICHITA WITCH	2BLURUDSTRP	5/8/1944	468BG	6/30/1946	Class 01Z Amarillo
42-24443			5/6/1944	499BG	12/21/1949	Reclaimed Pyote
42-24444			5/9/1944	231BU	5/24/1949	Reclaimed Tinker
42-24445			5/16/1944	231BU	5/24/1949	Reclaimed Tinker
42-24446		2WHTRUDSTRP	5/9/1944	468BG	10/22/1944	Crashed into Mt, surveyed, crew abnd
42-24447			5/13/1944	498BG	5/15/1945	Reclaimed Tinker
42-24448			5/10/1944	505BG	12/21/1949	Reclaimed Pyote
42-24449			5/12/1944	499BG	6/27/1949	Reclaimed Davis-Monthan
42-24450		V SQ	5/13/1944	499BG	2/8/1945	Surveyed
42-24451			5/9/1944	499BG	7/14/1949	Reclaimed Kelly
42-24452		FOURYLWSTRP A	5/9/1944	40BG	11/27/1944	Rammed by fighter at Bangkok
42-24453			5/13/1944	247BU	12/21/1949	Reclaimed Pyote
42-24454			5/17/1944	3PRS	6/27/1949	Reclaimed Davis-Monthan
42-24455			5/15/1944	231BU	6/27/1949	Reclaimed Davis-Monthan
42-24456	THE SHRIKE	SOLREDRUD	5/18/1944	462BG	6/27/1949	Reclaimed Davis-Monthan
42-24457	BATTLIN' BEAUTY	FOURREDSTRP T	5/17/1944	40BG	12/14/1944	Bombs exploded prematurely
42-24458			5/16/1944	58BW	7/18/1944	Crashed Calcutta enroute to join
42-24459			5/16/1944	231BU	7/14/1949	Reclaimed Kelly
42-24460			5/17/1944	497BG	7/14/1949	Reclaimed Kelly
42-24461	HAULINAS	SOLYLWRUD	5/18/1944	462BG	2/8/1945	Low Fuel, crash landed Dum Dum, Calcutta
42-24461		SOLGRNRUD	5/18/1944	462BG	2/8/1945	Low Fuel, crash landed Dum Dum, Calcutta
42-24462	PRINCESS EILEEN II	DIAM #	5/18/1944	444BG	5/11/1945	Group transfer to Tinian
42-24462	PRINCESS EILEEN II	TRI N 37 (1)	5/11/1945	444BG	6/30/1946	Reclaimed Amarillo
42-24463	HUMPIN' HONEY	SOLYLWRUD	5/20/1944	462BG	12/21/1949	Reclaimed Pyote
42-24464	FLYING STUD II	TRI N 22 (1)	5/20/1944	444BG	5/11/1945	Group transfer to Tinian
42-24464	FLYING STUD II	DIAM #	5/11/1945	444BG	5/24/1949	Reclaimed Tinker
42-24465			5/18/1944	242BU	12/21/1949	Reclaimed Pyote
42-24466	THE ABLE FOX	FOURBLKSTRP W	5/17/1944	40BG	12/18/1944	Crew bailed over Ankang, China
42-24467			5/19/1944	234BU	6/27/1949	Reclaimed Davis-Monthan
42-24468		T SQ	5/22/1944	498BG	2/10/1945	Ditched from Ota
42-24469	WHAM BAM	2BLURUDSTRP	5/19/1944	468BG	3/2/1945	Abandoned Andaman Sea from Singapore
42-24470			5/22/1944	231BU	2/28/1945	Class 01Z Pueblo
42-24471	CHAT'NOOGA CHOO CHOO	2WHTRUDSTRP	5/28/1944	468BG	5/11/1945	Group transfer to Tinian
42-24471	CHAT'NOOGA CHOO CHOO	TRI I 14	5/11/45	468BG	8/24/1949	Reclaimed Keesler
42-24471	RAMBLIN' WRECK	2BLURUDSTRP	5/28/1944	468BG	5/11/1945	Group transfer to Tinian
42-24471	RAMBLIN' WRECK	TRI I 14	5/11/1945	468BG	8/24/1949	Reclaimed Keesler
42-24472	SKY CHIEF	DIAM #	5/22/1944	444BG	5/11/1945	Group transfer to Tinian
42-24472	SKY CHIEF	TRI N 64	5/11/1945	444BG	6/30/1946	Reclaimed Amarillo
42-24473			5/20/1944	231BU	12/21/1949	Reclaimed Pyote
42-24474		SOLBLURUD	5/23/1944	462BG	8/20/1944	Shot down over Yawata
42-24475	HOODLUM HOUSE II	SOLREDRUD B1	5/23/1944	462BG	5/2/1945	Reclaimed Kelly
42-24476			5/23/1944	236BU	12/21/1949	Reclaimed Pyote
42-24477			5/23/1944	505BG	6/27/1949	Reclaimed Davis-Monthan
42-24478			5/23/1944	611BEXPG	1/27/1947	Salvaged at Eglin
42-24479		SOLREDRUD B1	5/24/1944	462BG	2/24/1945	Ditched Haria Bhangar River near Calcutta
42-2447X	HOLLY HAWK	CIRC E 31	5/24/1944	504BG	12/21/1949	Reclaimed
42-24480			6/27/1944	231BU	10/15/1944	Surveyed Alamogordo, crashed on T/O
42-24481			5/24/1944	231BU	12/21/1949	Reclaimed Pyote
42-24482			5/24/1944		8/10/1944	Crashed on takeoff at Natal, Africa
42-24483			5/24/1944	231BU	6/27/1949	Reclaimed Davis-Monthan
42-24484	OUR GAL	SOLBLURUD	5/24/1944	462BG	6/27/1949	Reclaimed Davis-Monthan
42-24485	LADY MARGE	DIAM #	5/25/1944	444BG	5/10/1945	Group transfer to Tinian
42-24485	LADY MARGE	TRI N 19	5/10/1945	444BG	6/27/1949	Reclaimed Davis-Monthan
42-24486	WINDY CITY II	2YLWRUDSTRP	5/26/1944	468BG	5/11/1945	Group transfer to Tinian
42-24486	WINDY CITY II	TRI I 47	5/11/1945	468BG	2/10/1948	Class 01Z Keesler
42-24487	BENGAL LANCER	2BLURUDSTRP	5/26/1944	468BG	5/11/1945	Group transfer to Tinian
42-24487	BENGAL LANCER	TRI I 32	5/11/1945	468BG	5/2/1949	Reclaimed Pyote

I. Master List

Serial #	Name	Identification	Delv	Assign	Off Inv	Circumstances
42-24488			5/25/1944	231BU	12/21/1949	Reclaimed Pyote
42-24489			5/26/1944	504BG	7/14/1949	Reclaimed Robins
42-24490			5/26/1944	231BU	7/14/1949	Reclaimed Kelly
42-24491			5/30/1944	504BG	12/21/1949	Reclaimed Pyote
42-24492	DEACON'S DISCIPLES II	FOURRUDSTRP	5/30/1945	40BG	5/10/1945	Group transfer to Tinian
42-24492	DEACON'S DISCIPLES II	TRI S 13	5/10/1945	40BG	5/14/1945	Abandoned south of Iwo #2 engine fire
42-24493			5/30/1944	504BG	6/4/1948	Reclaimed Davis-Monthan
42-24494	MARY ANN	2WHTRUDSTRP V31	5/31/1944	468BG	1/14/1945	Crashed returning from Formosa
42-24495			5/31/1944	505BG	6/27/1949	Reclaimed Davis-Monthan
42-24496			5/31/1944	244BU	12/21/1949	Reclaimed Pyote
42-24497			5/31/1944	231BU	7/14/1949	Reclaimed Kelly
42-24498			5/31/1944	231BU	11/9/1944	Surveyed Alamogordo — engine failed
42-24499			6/1/1944	231BU	5/24/1949	Reclaimed Tinker
42-24500			6/1/1944	462BG	3/20/1945	Surveyed at Harvard
42-24501			6/2/1944	504BG	12/21/1949	Reclaimed Pyote
42-24502			6/2/1944	505BG	6/27/1949	Reclaimed Davis-Monthan
42-24503	NIPPON NIPPER II	FOURBLUSTRP A/I	6/2/1944	40BG	6/30/1946	Reclaimed Amarillo
42-24504	GUNGA DIN	2WHTRUDSTRP	6/3/1944	468BG	10/25/1944	Crashed takeoff from A-7 to Omura
42-24505	WILD HAIR	SOLYLWRUD J2	6/6/1944	462BG	12/21/1944	Shot down at Mukden
42-24506	UNTOUCHABLE	SOLREDRUD P1	6/5/1944	462BG	9/25/1944	Crashed on test flight
42-24507	BACHELOR QUARTERS	DIAM # 38	6/7/1944	444BG	5/10/1945	Group transfer to Tinian
42-24507	BACHELOR QUARTERS	TRI N 38	5/10/1945	444BG	4/25/1949	Reclaimed Chanute
42-24508	OUIJA BIRD	FOURREDSTRP S	6/3/1944	40BG	5/15/1947	Reclaimed Tinker
42-24509			6/6/1944	499BG	12/21/1949	Reclaimed Pyote
42-24510		DIAM #	6/3/1944	444BG	11/23/1944	Surveyed, crash landing Chungking
42-24511			6/12/1944	245BU	12/21/1949	Reclaimed Pyote
42-24512			6/6/1944	504BG	11/1/1948	Reclaimed Davis-Monthan
42-24513	SWEATER OUT	FOURREDSTRP	6/8/1944	40BG	10/14/1944	Abandoned China
42-24514			6/6/1944	231BU	8/7/1944	Crashed landing Alamogordo 8/6
42-24515			6/7/1944	236BU	5/2/1949	Reclaimed Pyote
42-24516			6/14/1944	241BU	12/29/1948	Reclaimed Davis-Monthan
42-24517			6/10/1944	9BG	12/21/1949	Reclaimed Pyote
42-24518		DIAM #	6/9/1944	444BG	11/3/1944	Fire, ditched from Singapore
42-24519			6/10/1944	504BG	12/21/1949	Reclaimed Pyote
42-24520			6/10/1944	247BU	7/14/1944	Crashed Atlanta
42-24521			6/12/1944	236BU	8/25/1944	Surveyed
42-24522	B-SWEET II	FOURREDSTRP S	6/14/1944	40BG	12/31/1945	Class 26 Kirtland
42-24523			6/12/1944	236BU	8/20/1944	Surveyed at Pyote
42-24524	SUPER MOUSE	DIAM# 17	6/13/1944	444BG	5/10/1945	Group transfer to Tinian
42-24524	SUPER MOUSE	TRI N 17 (1)	5/10/1945	444BG	6/1/1945	Exploded over Osaka
42-24525	THE MARY K	2BLURUDSTRP	6/15/1944	468BG	5/11/1945	Group transfer to Tinian
42-24525	THE MARY K	TRI I 24	5/11/1945	468BG	2/2/1950	Reclaimed Keesler
42-24526			6/15/1944	AMC	9/27/1949	Unknown
42-24527			6/15/1944	610EXGP	8/21/1946	Surveyed at Eglin
42-24528	ANDY GUMP		6/15/1944	2750ABG	5/4/1949	Reclaimed Wright-Patterson
42-24529			6/14/1944	9BG	9/11/1950	Reclaimed Pyote
42-24530			6/16/1944	9BG	11/21/1947	Class 01Z Smoky Hill
42-24531		SOLYLWRUD	6/17/1944	462BG	11/19/1944	Abandoned India, fire in radar position
42-24532	GREAT SPECKLED BIRD		6/16/1944	505BG	9/11/1950	Reclaimed Pyote
42-24533						Unknown
42-24534			6/16/1944	505BG	9/11/1950	Reclaimed Pyote
42-24535			6/17/1944	500BG	2/24/1945	Surveyed
42-24536			6/17/1944	505BG	10/24/1944	Surveyed at Harvard, crashed takeoff
42-24537			6/17/1944	500BG	2/24/1945	Surveyed at Walker
42-24538	BETER 'N' NUTIN	DIAM #		444BG	5/10/1945	Group transfer to Tinian
42-24538	BETER 'N' NUTIN	TRI N 24	5/10/1945	444BG	7/31/1946	Class 01Z Lowry
42-24539			6/19/1944	231BU	3/14/1946	Class 01Z Fort Worth
42-24540		L TRI	6/19/1944	6BG	9/11/1950	Reclaimed Pyote
42-24541	BOMBIN' BUGGY II	FOURBLUSTRP H	6/20/1944	40BG	5/5/1945	Group transfer to Tinian
42-24541	BOMBIN' BUGGY II	TRI S 25 (1)	5/5/1945	40BG	12/6/1948	Class 01Z Chanute
42-24542	LADY HAMILTON II	2BLURUDSTRP	6/22/1944	468BG	5/11/1945	Group transfer to Tinian
42-24542	LADY HAMILTON II	TRI I 51	5/11/1945	468BG	6/1/1945	Missing from Osaka
42-24543			6/21/1944	236BU	5/12/1945	Surveyed at Pyote
42-24544	LONG DISTANCE	T SQ 49	6/21/1944	498BG	5/31/1950	Reclaimed Keesler
42-24545			6/22/1944		7/31/1946	Reclaimed

Serial #	Name	Identification	Delv	Assign	Off Inv	Circumstances
42-24546		2WHTRUDSTRP	6/22/1944	468BG	5/31/1946	Class 01Z Barksdale
42-24547			6/22/1944	6BG	12/31/1944	Surveyed — crashed at Grand Isle
42-24548			6/26/1944	9BG	9/11/1950	Reclaimed Pyote
42-24549			7/1/1944	236BU	6/22/1950	Reclaimed Davis-Monthan
42-24550	MARY ANN	V SQ 27 (3)	7/1/1944	499BG		Unknown
42-24551			6/23/1944	504BG	9/11/1950	Reclaimed Pyote
42-24552	GREAT SPECKLED BIRD		6/26/1944	234BU	9/11/1950	Reclaimed Pyote
42-24553			6/26/1944	504BG	9/12/1950	Reclaimed Pyote
42-24554			6/28/1944	29BG	9/12/1950	Reclaimed Pyote
42-24555			6/26/1944	504BG	9/11/1950	Reclaimed Pyote
42-24556						Unknown
42-24557			6/29/1944	504BG	8/25/1950	Reclaimed McClellan
42-24558			6/28/1944	247BU	3/21/1950	Reclaimed
42-24559		D SQ	7/1/1944	500BG	9/11/1950	Reclaimed Pyote
42-24560			6/28/1944	499BG	4/21/1950	Reclaimed Kelly
42-24561			7/3/1944	500BG	4/12/1950	Reclaimed Kelly
42-24562			7/3/1944		9/11/1950	Reclaimed
42-24563			7/3/1944	9BG	9/11/1950	Reclaimed Tinker
42-24564			6/30/1944	234BU	9/21/1949	Salvaged at Carswell
42-24565			7/1/1944	236BU	9/11/1950	Reclaimed Pyote
42-24566			7/5/1944	498BG	9/11/1950	Reclaimed Pyote
42-24567		2YLWRUDSTRP	7/5/1944	468BG	9/11/1950	Reclaimed Pyote
42-24568			7/5/1944	243BU	9/11/1950	Reclaimed Pyote
42-24569			7/5/1944	246BU	9/11/1950	Reclaimed Pyote
42-24570			7/8/1944	499BG	12/29/1944	Surveyed Smoky Hill
42-24571			7/7/1944	29BG	9/11/1950	Reclaimed Pyote
42-24572			7/7/1944	29BG	9/11/1950	Reclaimed Pyote
42-24573			7/7/1944	500BG	7/11/1950	Reclaimed McClellan
42-24574	293	FOURYLWSTRP K	7/11/1944	40BG	12/14/1944	Exploded Rangoon
42-24575			7/8/1944	500BG	5/28/1948	Salvaged at Borinquin
42-24576			7/8/1944	242BU	9/11/1950	Reclaimed Pyote
42-24577			7/10/1944	4141BU	9/11/1950	Reclaimed Pyote
42-24578			7/10/1944	39BG	11/29/1944	Class 26 Smoky Hill
42-24579	EDDIE ALLEN	FOURBLUSTRP M	7/10/1944	40BG	5/5/1945	Group transfer to Tinian
42-24579	EDDIE ALLEN	TRI S 36 (1)	5/5/1945	40BG	6/1/1945	Surveyed, battle damage
42-24580	FLYING JACKASS	DIAM #	7/13/1944	444BG	5/10/1945	Group transfer to Tinian
42-24580	FLYING JACKASS	TRI N 18	5/10/1945	444BG	9/11/1950	Unknown
42-24580	UNDECIDED	DIAM #	7/14/1944	444BG	9/11/1950	Reclaimed Pyote
42-24581		FOURBLUSTRP	7/13/1944	40BG	9/11/1950	Reclaimed Pyote
42-24582	LITTLE CLAMBERT	FOURBLUSTRP S	7/14/1944	40BG	1/14/1945	Bomb explosions, fire, at B-4
42-24583		A SQ 7 (1)		497BG	6/8/1945	Crashed Agana, Guam — forced landing
42-24584	LUCKY LADY	DIAM #	7/15/1944	444BG	5/10/1945	Group transfer to Tinian
42-24584	LUCKY LADY	TRI N 44	5/10/1945	444BG	6/10/1945	Lost on training flight
42-24585		F	7/15/1944	3PRS	11/22/1943	MIA
42-24585	POISON IVY	F	7/15/1944	6CMG	6/8/1945	Lost on test hop-all 4 engines quit
42-24586		F	7/15/1944	6CMG	11/21/1944	MIA Nagoya
42-24587	SAN ANTONIO ROSE	FOURBLUSTRP J	7/15/1944	40BG	6/30/1946	Class 01Z Amarillo
42-24588			7/15/1944	234BU	11/9/1950	Reclaimed Tinker
42-24589	CALAMITY JANE	FOURREDSTRP R	7/15/1944	40BG	2/1/1945	Flak at Singapore
42-24590	CELESTIAL PRINCESS	SOLGRNRUD T3	7/17/1944	462BG	5/5/1945	Group transfer to Tinian
42-24590	CELESTIAL PRINCESS	TRI U 40	5/5/1945	462BG	7/14/1949	Reclaimed Chanute
42-24591	LUCKY LYNN	A SQ 3 (1)	7/18/1944	497BG	9/11/1950	Unknown
42-24592	DAUNTLESS DOTTY	A SQ 1 (1)	7/17/1944	497BG	6/6/1945	Crashed takeoff at Kwajalein
42-24593	AMERICAN MAID	A SQ 7 (1)	7/18/1944	497BG	9/11/1950	Reclaimed Tinker
42-24594	BAD BREW	A SQ 6 (1)	7/21/1944	497BG	1/23/1945	Ditched on takeoff, lost #3 & 4
42-24595	PACIFIC UNION	A SQ 2 (1)	7/22/1944	497BG	1/14/1945	Rammed; ditched off Pagan Island
42-24596	LITTLE GEM	A SQ 4 (1)	7/22/1944	497BG	5/10/1949	Reclaimed Tinker
42-24597	OUR BABY	A SQ 10 (1)	7/22/1944	497BG	11/27/1944	Destroyed enemy raid on Saipan
42-24598	WADDY'S WAGON	A SQ 5 (1)	7/22/1944	497BG	1/9/1945	Ditched from Tokyo
42-24599	SKYSCRAPPER	A SQ 9 (1)	7/22/1944	497BG	11/27/1944	Destroyed on Ground — Saipan
42-24600	ADAMS EVE	Z SQ 47 (1)	8/31/1944	500BG	4/7/1945	Rammed; ditched from Tokyo
42-24601	POCOHANTAS PROUD PIGEON	T SQ 23 (1)	7/23/1944	498BG	5/24/1949	Reclaimed Tinker
42-24602			7/22/1944	498BG	9/24/1944	Crashed on takeoff
42-24603		T SQ 7 (1)	7/23/1944	498BG	12/18/1944	Surveyed — destroyed in enemy raid
42-24604	WHEEL'N'DEAL (LT SIDE)	A SQ 24 (1)	7/23/1944	497BG	4/14/1945	MIA Weather Strike

I. Master List

Serial #	Name	Identification	Delv	Assign	Off Inv	Circumstances
42-24605	THE HEAT'S ON	T SQ 2 (1)	7/25/1944	498BG	12/27/1944	Ditched near Pagan Island, out of fuel
42-24606	BATTLIN' BETTY	T SQ 41 (1)	7/25/1944	498BG	1/6/1945	Salvaged after 12/25/44 Jap raid
42-24607	FORBIDDEN FRUIT	T SQ 50	7/25/1944	498BG	9/15/1949	Reclaimed Sheppard
42-24608		T SQ 3 (1)	7/25/1944	498BG	2/4/1945	Surveyed, crashed Saipan
42-24609	LASSIE COME HOME	T SQ 02 (1)	7/26/1944	498BG	3/21/1945	Salvaged from 1/14/45 damage
42-24609	LASSIE COME HOME	T SQ 21 (1)	7/28/1944	498BG	3/21/1945	Surveyed after 1/14/45 mission
42-24610	BEDROOM EYES	T SQ 22 (1)	7/31/1944	498BG	3/13/1945	Per MACR, Historian says survived war
42-24611	LITTLE JO	T SQ 4 (1)	7/28/1944	498BG	4/29/1945	Abandoned over Miyazaki, crashed in sea
42-24612	KLONDIKE KUTEY	12	7/31/1944	236BU	9/11/1950	Reclaimed Pyote
42-24613		T SQ 45 (1)	7/28/1944	498BG	12/27/1944	Ditched after takeoff, lost #'s 2 & 4
42-24614	JOLTIN' JOSIE	T SQ 5 (1)	7/28/1944	498BG	4/1/1945	Crashed Magicienne Bay on T/O to Tokyo
42-24615	CORAL QUEEN	A SQ 8 (1)	7/31/1944	497BG	4/18/1945	Ditched vicinity of Iwo
42-24616	HALEY'S COMET	A SQ 22 (1)	7/29/1944	497BG	1/27/1945	Shot down Tokyo by JNAF Irving
42-24617			7/31/1944	499BG	4/24/1945	Surveyed at Tinker
42-24618			7/31/1944	611EXGP	6/22/1950	Reclaimed Davis-Monthan
42-24619	SHADY LADY	A SQ 23 (1)	7/29/1944	497BG	1/27/1945	Rammed over Tokyo
42-24620	SLEEPY TIME GAL	FOURBLUSTRP C	7/31/1944	40BG	5/5/1945	Reclaimed Pyote
42-24620	SLEEPY TIME GAL	TRI S 21	5/5/1945	40BG	9/11/1950	Reclaimed Pyote
42-24621	YOKOHAMA YOYO	F		6CMG		Unknown
42-24622	LUCKY IRISH	A SQ 26 (1)	7/31/1944	497BG	11/24/1944	Rammed over Tokyo
42-24623	THUMPER	A SQ 21 (1)	7/31/1944	497BG	9/11/1950	Reclaimed Pyote
42-24624	PATCHES	T SQ 44 (2)	7/31/1944	498BG	9/11/1950	Reclaimed Pyote
42-24625	(MAJORETTE)	T SQ 24 (1)	7/31/1944	498BG	9/11/1950	Reclaimed Pyote
42-24625	LADY MARY ANNA	T SQ 24 (1)/T 48	7/31/1944	498BG	9/11/1950	Reclaimed Tinker
42-24626	JOKER'S WILD	A SQ 42 (1)	8/15/1944	497BG	1/3/1945	Rammed Nagoya
42-24627	TEXAS DOLL	A SQ 27 (1)	7/31/1944	497BG	9/11/1950	Reclaimed Pyote
42-24628	SPECIAL DELIVERY	A SQ 29 (1)	8/9/1944	497BG	12/18/1944	Ditched on Weather Strike
42-24629	DEVIL'S DARLIN'	T SQ 9 (1)	8/9/1944	498BG	2/4/1945	Ditched off Saipan returning
42-24630			8/11/1944	4141BU	9/11/1950	Reclaimed Pyote
42-24631			8/12/1944	232BU	9/11/1950	Reclaimed Pyote
42-24632			8/12/1944	504BG	9/11/1950	Reclaimed Pyote
42-24633	WAR WEARY	V SQ 6 (1)	8/13/1944	499BG	9/11/1950	Reclaimed Pyote
42-24634			8/13/1944	505BG	9/11/1950	Reclaimed AMC
42-24635			8/10/1944	611EXGP	6/22/1950	Reclaimed Davis-Monthan
42-24636			8/14/1944	29BG	9/11/1950	Reclaimed Pyote
42-24637			8/14/1944	4141BU	9/11/1950	Reclaimed Pyote
42-24638		V SQ 29 (1)	8/13/1944	499BG	12/13/1944	Ditched from Nagoya
42-24639			8/14/1944	4141BU	9/11/1950	Reclaimed Pyote
42-24640			8/14/1944	231BU	6/22/1950	Reclaimed Davis-Monthan
42-24640	HELLO NATURAL III		8/14/1944	206BU	6/22/1950	Unknown
42-24641	THUNDERHEAD	A SQ 43 (1)	8/15/1944	497BG	11/10/1948	Reclaimed Hickam
42-24642	UNCLE TOM'S CABIN	T SQ 25 (1)	8/15/1944	498BG	12/27/1944	Rammed twice over Tokyo
42-24643	THERE'LL ALWAYS BEAXMAS	Z SQ 5 (1)	8/17/1944	500BG	9/11/1950	Reclaimed Tinker
42-24644		V SQ 23 (1)	8/19/1944	499BG	4/13/1945	Shot up Tokyo, crashed Chiba
42-24645		T SQ 8 (1)	8/20/1944	498BG	11/8/1944	Ditched from Iwo Jima
42-24646	TORCHY	T SQ 27 (1)		498BG	1/5/1950	Reclaimed McClellan
42-24647	HASTA LUEGO	V SQ 22 (1)	8/17/1944	499BG	1/13/1945	Ditched from Nagoya, fuel system malf.
42-24648	HUMP'S HONEY	A SQ 48 (1)	8/19/1944	497BG	9/11/1950	Reclaimed Pyote
42-24649	MISS TITTYMOUSE	T SQ 47 (1)	8/17/1944	498BG	12/18/1944	Crash landing in Saipan ATC area
42-24650	JUG HAID II	V SQ 2 (1)	8/20/1944	499BG	4/2/1945	Lost over Tokyo, flak
42-24651	YANKEE MADE	V SQ 41 (2)	8/18/1944	499BG	9/11/1950	Reclaimed Tinker
42-24652	DEVIL'S DELIGHT	Z SQ 21 (1)	8/20/1944	500BG	6/24/1946	Reclaimed Amarillo
42-24653	SUPINE SUE	Z SQ 42 (1)	8/21/1944	500BG	8/25/1950	Reclaimed McClellan
42-24654	THE WICHITA WITCH	T SQ 30 (1)	8/20/1944	498BG	12/25/1944	Destroyed on Ground — Saipan
42-24655	MISS BEHAVIN'	A SQ 46 (1)	8/19/1944	497BG	1/9/1945	Rammed in #2 engine over Tokyo
42-24656	ROSALIA ROCKET	Z SQ 1 (1)	8/24/1944	500BG	12/3/1944	Rammed — abandoned, fire in left wing
42-24657	MUSN TOUCH	Z SQ 45 (1)	8/21/1944	500BG	1/9/1945	Ditched from Tokyo
42-24658	WUGGED WASCAL	V SQ 3 (1)	8/21/1944	499BG	1/9/1945	Ditched from Tokyo
42-24659		V SQ 1 (1)	8/22/1944	499BG	11/27/1944	Destroyed on Ground — Saipan
42-24660	MILLION DOLLAR BABY	Z SQ 48	8/22/1944	500BG	1/3/1945	Ditched
42-24661	THE BIG STICK	V SQ 44 (1)	8/22/1944	499BG	8/20/1950	Reclaimed Pyote
42-24662		Z SQ 2 (1)	8/22/1944	500BG	11/27/1944	Ditched from Tokyo
42-24663	LADY EVE II	T SQ 43 (1)	8/22/1944	498BG	9/11/1950	Reclaimed Pyote
42-24663	WILLIE MAE	T SQ 43 (1)	8/22/1944	498BG	9/11/1950	Reclaimed Pyote
42-24664	RAMBLIN' ROSCOE	Z SQ 23 (1)	8/23/1944	500BG	4/15/1945	Crash landed Iwo

Serial #	Name	Identification	Delv	Assign	Off Inv	Circumstances
42-24665	SATAN'S SISTER	V SQ 4 (1)	8/24/1944	499BG	1/9/1945	Ditched
42-24666		V SQ 24 (1)	8/26/1944	499BG	12/18/1944	Collision with 29BG aircraft
42-24667		V SQ 21		499BG		Unknown
42-24668	THE CANNUCK	Z SQ 27 (1)		500BG		Unknown
42-24669	HONEY	V SQ 42 (2)	8/25/1944	499BG	9/11/1950	Reclaimed Tinker
42-24670	HAM'S EGGS	V SQ 45 (2)		499BG		Unknown
42-24671	THREE FEATHERS	Z SQ 49 (1)Z 57	8/25/1944	500BG	10/4/1954	Reclaimed Tinker
42-24672	BLACK MAGIC	Z SQ 4 (1)		500BG		Unknown, survived war
42-24673	DREAM GIRL	V SQ 43 (1)		499BG		Reclaimed
42-24674		V SQ 25 (1)	8/27/1944	499BG	4/7/1945	Hit by phosphorus bomb over Tokyo
42-24675	THE BARRONESS	Z SQ 41 (1)	8/30/1944	500BG	10/26/1946	Reclaimed Spokane
42-24676	PRIDE OF THE YANKEES	Z SQ 24		500BG		Transferred to Depot, combat damage
42-24677	DORIS ANN	V SQ 46 (1)		499BG		Unknown
42-24678	KICKAPOO LOU	2WHTRUDSTRP	8/31/1944	468BG	3/2/1945	Flak at Singapore
42-24679		V SQ 48 (1)	8/28/1944	499BG	11/24/1944	Ditched, out of fuel
42-24680	HELL'S BELLE	Z SQ 7 (1)		500BG		Battle damage, repaired, became Z SQ 37
42-24680	BELLE RUTH	Z SQ 37		500BG		Survived war; 31 msns
42-24681		T SQ 29 (1)	8/29/1944	498BG	12/3/1944	Ditched
42-24682	TOKYO TWISTER	V SQ 5 (1)	8/31/1944	499BG	1/27/1945	Hard landing, salvaged
42-24683		V SQ (1)		499BG		Circumstances Unknown
42-24684		V SQ 7 (1)	8/31/1944	499BG	12/22/1944	Ditched, damaged over target
42-24685	THE OUTLAW	FOURBLUSTRP M	8/31/1944	40BG	5/5/1945	Group transfer to Tinian
42-24685	THE OUTLAW	TRI S 19 (1)	5/5/1945	40BG	8/7/1945	Crash landing, surveyed
42-24686	AMERICAN BEAUTY	Z SQ 25 (1)	8/30/1944	500BG	12/29/1944	Ditched on weather strike — out of fuel
42-24687	TOKYO LOCAL	Z SQ 26 (1)	9/1/1944	500BG	12/13/1944	Ditched from Nagoya — flak damage
42-24688		Z SQ (1)		500BG		Shot up, ditched
42-24688		V SQ 29 (2)		499BG		Unknown
42-24689	NINA ROSS	Z SQ 9 (1)	9/1/1944	500BG	9/11/1950	Reclaimed Tinker
42-24690			9/18/1944	4100BU	11/7/1950	Reclaimed Tinker
42-24691	FAST COMPANY	2WHTRUDSTRP		468BG	5/11/1945	Group transfer to Tinian
42-24691	FAST COMPANY	TRI I 7	5/11/1945	468BG		Unknown
42-24692	WABASH CANNONBALL	Z SQ 8 (1)Z 12	9/2/1944	500BG	2/19/1945	Rammed at Tokyo as Z Square 12
42-24693	LI'L LASSIE	V SQ 27 (1)	9/5/1944	499BG	12/7/1944	Destroyed in Jap raid
42-24693	MARY ANNE	V SQ 27 (1)	9/5/1944	499BG	12/7/1944	Destroyed in Jap raid
42-24694	DRAGGIN' LADY	Z SQ 6 (1)	9/5/1944	500BG	2/27/1945	Ditched Saipan on Slow timing flight
42-24695	LUCKY 'LEVEN	T SQ 6 (1)	9/2/1944	498BG	5/10/1949	Reclaimed Chanute
42-24696	FANCY DETAIL	Z SQ 50 (1)		500BG		Unknown
42-24697			9/7/1944	248BU	4/12/1950	Reclaimed Kelly
42-24698	ABROAD WITH 11 YANKS	V SQ 8 (1)		499BG	4/27/1945	MIA
42-24699	SALVO SALLY	V SQ 9 (1)	9/6/1944	499BG	4/28/1945	Ditched off Miyazaki
42-24700	SLICK DICK	V SQ 23 (1)	9/8/1944	499BG	11/00/44	Transferred to 500BG
42-24700	SLICK DICK	Z SQ 33	11/00/44	500BG	12/6/1950	Reclaimed McClellan
42-24701			9/7/1944	236BU	9/11/1950	Reclaimed Pyote
42-24702						Unknown
42-24703	AMERICAN BEAUTY	2WHTRUDSTRP	9/9/1944	468BG	5/11/1945	Group transfer to Tinian
42-24703	AMERICAN BEAUTY	TRI I 4	5/11/1945	468BG	6/5/1945	Surveyed — crashed landing at Iwo
42-24704	THE GEAR BOX	2BLURUDSTRP	9/8/1944	468BG	1/11/1945	Missing from Singapore
42-24705			9/12/1944	234BU	5/16/1945	Abandoned Texas
42-24706		2YLWRUDSTRP	9/12/1944	468BG	11/28/1944	Lost Halliday Island, collision, eng fire
42-24707						Unknown
42-24708						Unknown
42-24709						Unknown
42-24710			9/13/1944	234BU	5/15/1947	Reclaimed Tinker
42-24711		SOLYLWRUD	9/19/1944	462BG	5/5/1945	Group transfer to Tinian
42-24711		TRI U 27	5/5/1945	462BG	11/30/1945	Salvaged, off end of runway 8/7/45
42-24712						Unknown
42-24713						Unknown
42-24714	LUCKY ELEVEN	Z SQ 11 (4)	7/1/1945	500BG		Unknown
42-24714	ROBERT J WILSON	2YLWRUDSTRP	12/23/1944	468BG	5/11/1945	Group transfer to Tinian
42-24714	ROBERT J WILSON	TRI I 46	5/11/1945	468BG		Unknown — transfer to 500BG?
42-24714	ROBERT J WILSON	TRI S		40BG	12/23/1944	Transferred to 468BG 12/23/44
42-24715		2YLWRUDSTRP	9/14/1944	468BG	12/21/1944	Rammed at Mukden
42-24716						Unknown
42-24717	SOUTHERN BELLE	A SQ 11 (1)	9/13/1944	497BG	7/15/1945	Ditched
42-24718	BLACK MAGIC II	FOURYLWSTRP A	9/14/1944	40BG	5/5/1945	Group transfer to Tinian

I. Master List

Serial #	Name	Identification	Delv	Assign	Off Inv	Circumstances
42-24718	BLACK MAGIC II	TRI S 34	5/5/1945	40BG	9/11/1950	Reclaimed Chanute
42-24719	THE UN-INVITED II	2YLWRUDSTRP		468BG	5/5/1945	Group transfer to Tinian
42-24719	THE UN-INVITED II	TRI I 43	5/5/1945	468BG		Unknown
42-24720	FU-KEMAL-TU	DIAM #	9/18/1944	444BG	5/10/1945	Group transfer to Tinian
42-24720	FU-KEMAL-TU	TRI N 11	5/10/1945	444BG	8/30/1945	Ditched 2 miles off Iwo on POW flight
42-24721	SU SU BABY	Z SQ 46 (1)	9/18/1944	500BG	3/7/1945	Ditched on Weather Strike — out of fuel
42-24722			9/22/1944	4006BU	9/22/1953	Reclaimed Pyote
42-24723	SHANGHAI LIL RIDES AGAIN	DIAM #	9/22/1944	444BG	5/11/1945	Group transfer to Tinian
42-24723	SHANGHAI LIL RIDES AGAIN	TRI N 16 (1)	5/11/1945	444BG	5/26/1945	Abandoned over Iwo-flak damage
42-24724	HOLLYWOOD COMMANDO	DIAM #	9/22/1944	444BG	5/11/1945	Group transfer to Tinian
42-24724	HOLLYWOOD COMMANDO	TRI N 55 (1)	5/11/1945	444BG	5/25/1945	Crashed Ibaragi from Tokyo
42-24725			12/31/1944		10/00/54	Reclaimed
42-24726		FOURREDSTRP	9/19/1944	40BG	12/14/1944	MIA Rangoon, hit by falling bomb
42-24727		T SQ 1 (2)	9/26/1944	498BG	3/31/1945	Ditched inbound, fire #3
42-24728		SOLYLWRUD		462BG	5/5/1945	Group transfer to Tinian
42-24728		TRI U 28	5/5/1945	462BG	5/20/1945	Lost on Tokyo mission
42-24729	NIPPON NIPPER III	FOURBLUSTRP X	9/22/1944	40BG	5/5/1945	Group transfer to Tinian
42-24729	NIPPON NIPPER III	TRI S 22	5/5/1945	40BG	5/10/1954	Reclaimed Davis-Monthan
42-24730	HIGH AND MIGHTY	DIAM #	9/26/1944	444BG	5/10/1945	Group transfer to Tinian
42-24730	HIGH AND MIGHTY	TRI N 53 (1)	5/10/1945	444BG	5/10/1954	Reclaimed Davis-Monthan
42-24731	VICTORY GIRL	DIAM# 65	9/23/1944	444BG	5/10/1945	Group transfer to Tinian
42-24731	VICTORY GIRL	TRI N 63	5/10/1945	444BG	5/10/1954	Reclaimed Davis-Monthan
42-24732	HORE-ZONTAL DREAM	DIAM# 5	9/27/1944	444BG	5/10/1945	Group transfer to Tinian
42-24732	HORE-ZONTAL DREAM	TRI N 33	5/10/1945	444BG	3/31/1950	Reclaimed Keesler
42-24733		A SQ 31	9/27/1944	497BG	12/22/1944	Ditched on Weather Strike
42-24734	MISS LEAD	2YLWRUDSTRP	9/24/1944	468BG	5/11/1945	Group transfer to Tinian
42-24734	MISS LEAD	TRI I 45	5/11/1945	468BG	8/6/1945	Crashed landing Tinian
42-24735		T SQ 10 (2)	9/27/1944	498BG	12/3/1944	Ditched
42-24736		DIAM #	9/24/1944	444BG	2/5/1945	Cr ld with #4 engine feathered
42-24737	PIONEER II	2BLURUDSTRP	9/27/1944	468BS	10/12/1953	Reclaimed Pyote
42-24737	KAGU TSUCHI	FOURSTRP	9/27/1944	40BG	10/12/1953	Unknown
42-24738	HONEYWELL HONEY	FOURYLWSTRP B/J	9/26/1944	40BG	5/5/1945	Group transfer to Tinian
42-24738	HONEYWELL HONEY	TRI S 37	5/5/1945	40BG	5/10/1954	Reclaimed Davis-Monthan
42-24739		FOURYLWSTRP C	9/26/1944	40BG	5/5/1945	Group transfer to Tinian
42-24739		TRI S 27	5/5/1945	40BG	3/25/1953	Reclaimed Pyote
42-24740	HARRY MILLER	FOURYLWSTRP Z	9/27/1944	40BG	5/5/1945	Group transfer to Tinian
42-24740	HARRY MILLER	TRI S 38	5/5/1945	40BG	6/7/1945	Salvaged from Tokyo 5/26/45
42-24741		A SQ	9/26/1944	497BG	6/1/1945	Ditched; Jap. say shot down Osaka
42-24742	THE ROCKET	T SQ 31(1)	9/24/1944	498BG	6/5/1945	Ditched from Kobe after ramming
42-24743	WABASH CANNONBALL	Z SQ 25 (2)	9/30/1944	500BG	5/10/1954	Reclaimed Davis-Monthan
42-24743	WABASH CANNONBALL	Z SQ 56	9/30/1944	500BG	5/10/1954	Reclaimed Davis-Monthan
42-24744		Z SQ 2 (2)		500BG	1/10/1945	Ditched from Tokyo
42-24745		A SQ 26 (2)	9/29/1944	497BG	12/7/1944	Surveyed — damaged in Japanese raid
42-24746			9/29/1944	4141BU	9/11/1950	Reclaimed Pyote
42-24747			9/29/1944	4105BU	5/10/1954	Reclaimed Davis-Monthan
42-24748		T SQ 42	10/2/1944	498BG	1/3/1945	Crashed on Anahatan Island
42-24749	TANAKA TERMITE	T SQ 32	10/6/1944	498BG	6/24/1954	Reclaimed Pyote
42-24750	EAGER BEAVER III	T SQ 71?	9/30/1944	498BG	7/26/1946	Reclaimed Amarillo
42-24750	EAGER BEAVER III	DIAM #		444BG		Unknown
42-24751	ANTOINETTE	T SQ 48 (1)	9/30/1944	498BG	5/24/1945	Crashed in Chiba from Tokyo
42-24752	WICHITA WITCH	FOURBLUSTRP T	10/2/1944	40BG	5/5/1945	Group transfer to Tinian
42-24752	WICHITA WITCH	TRI S 15	5/5/1945	40BG	7/31/1946	Reclaimed Lowry
42-24753		V SQ 45 (1)	9/30/1944	499BG	3/11/1945	Ditched
42-24754		V SQ 30	10/7/1944	499BG	3/13/1945	Shot down over Osaka — flak
42-24755	TOMMY HAWK	T SQ 8 (2)	9/30/1944	498BG	10/12/1953	Reclaimed Pyote
42-24756	NEW GLORY	A SQ 31	10/13/1944	497BG	9/15/1953	Reclaimed Pyote
42-24757		FOURREDSTRP J	10/7/1944	40BG	7/28/1948	Reclaimed Chanute
42-24757		TRI S 05 (1)	10/7/1944	40BG	7/28/1948	Reclaimed Chanute
42-24758		V SQ 25 (2)	10/7/1944	499BG	11/18/1947	Reclaimed Smoky Hill
42-24759	BLIND DATE	L TRI	10/7/1944	6BG	5/23/1945	Crashed Yamatogama, Chiba
42-24760	BATTLIN' BETTY II	T SQ 41 (2)	10/6/1944	498BG	2/10/1945	Ditched — out of fuel
42-24760	SMALL FRY	T SQ 41 (2)	10/6/1944	498BG	2/10/1945	Ditched — out of fuel
42-24761	TAIL WIND	Z SQ 51	10/8/1944	500BG	9/15/1953	Reclaimed Pyote
42-24761		Z SQ 1 (3)	10/8/1944	500BG	9/15/1953	Reclaimed Pyote
42-24762	PEE WEE	Z SQ 1 (2)	10/10/1944	500BG	12/18/1944	Ditched — fuel transfer system fasiled

Serial #	Name	Identification	Delv	Assign	Off Inv	Circumstances
42-24763	GEISHA GERTIE	T SQ 34 (1)	10/9/1944	498BG	1/14/1945	Ditched
42-24763	UNCLE TOM'S CABIN II	T SQ 34 (1)	10/9/1944	498BG	1/14/1945	Ditched off Nagoya
42-24764	PACIFIC PLAYBOYS	K TRI 1	10/7/1944	505BG	9/19/1950	Reclaimed Chanute
42-24765	UPPER BERTH	V SQ 1(2)	10/12/1944	499BG	9/15/1953	Reclaimed Pyote
42-24766	GEORGIA ANN	Z SQ 22 (1)	10/9/1944	500BG	1/3/1945	Rammed; crashed SE of Nagoya
42-24766	LEADING LADY	Z SQ 22 (1)	10/9/1944	500BG	1/3/1945	Rammed; crashed SE of Nagoya
42-24767		T SQ 10 (3)	10/9/1944	498BG	1/27/1945	Ditched -out of fuel
42-24768			10/9/1944	619BU	5/10/1954	Reclaimed Davis-Monthan
42-24769	ROVER BOYS EXPRESS	V SQ 27 (2)	10/10/1944	499BG	1/27/1945	Shot down by fighters
42-24769	LASSIE II	V SQ 27 (2)	10/10/1944	499BG	1/27/1945	Shot down, tried to ditch in lake
42-24770		Z SQ 26 (2)	11/13/1944	500BG	5/10/1954	Reclaimed Davis-Monthan
42-24771	PASSION WAGON	T SQ 52	10/14/1944	498BG	5/10/1954	Reclaimed Davis-Monthan
42-24772		A SQ 14	10/13/1944	497BG	1/9/1945	Rammed Tokyo, on fire when last seen
42-24773		V SQ 24	10/10/1944	499BG	12/18/1944	Hit by bomb from 498BG plane
42-24774	MISS HAP	A SQ 37	10/13/1944	497BG	7/7/1945	Crashed Kochi Park, Kochi
42-24775		V SQ 26 (2)	10/18/1944	499BG	5/10/1954	Reclaimed Davis-Monthan
42-24776	WHITE MISTRESS/HUNTRESS	L TRI 57 (1)	10/11/1944	6BG	5/5/1954	Reclaimed Tinker
42-24777		T SQ 7 (2)	10/10/1944	498BG	2/27/1945	Ditched on Weather Strike #238
42-24778	DRAGON LADY	K TRI 8	10/12/1944	505BG	6/17/1954	Class 26 Sheppard
42-24779	SATAN'S LADY	E TRI (1)	10/13/1944	504BG	4/25/1945	Flak over Tachikawa
42-24780	29 USN CONSTRUCTION REG	E TRI (1)	10/13/1944	504BG	9/15/1953	Reclaimed Pyote
42-24780	DOC'S DEADLY DOSE	E TRI (1)	10/13/1944	504BG	9/15/1953	Reclaimed Pyote
42-24781		K TRI	10/11/1944	505BG	1/5/1945	Ditched
42-24781	TWENTY-NINTH USNCB	E TRI	10/11/1944	504BG		Transfer to 505BG
42-24782	STAR DUSTER	V SQ 31	10/13/1944	499BG	6/17/1954	Class 26 Sheppard
42-24783	CONNECTICUT YANKEE	L TRI 41/61	10/14/1944	6BG	11/17/1953	Reclaimed Hill
42-24784	92 NAVAL CONST BATT	K TRI	10/16/1944	505BG	2/10/1945	Collided with 42-24815 at Gumma-Ken
42-24784	SLICK'S CHICKS	K TRI	10/16/1944	505BG	2/10/1945	Collided with 42-24815 at Gumma-Ken
42-24785	HOMING DE-VICE	Z SQ 31/25	10/16/1944	500BG	1/23/1945	Right wing, #3 engine afire, broke off
42-24786	ASSID TEST	SOLGRNRUD	1/28/1945	462BG	5/5/1945	Group Transfer to Tinian
42-24786	ASSID TEST	TRI U 49	5/5/1945	462BG	6/5/1945	Shot down after bombing Kobe — flak
42-24787	THE WOLF PACK	K TRI 4	10/18/1944	505BG	2/10/1945	Last reported twenty minutes from Saipan
42-24788	KRO'S KIDS	CIRC E 26		504BG	6/26/1945	Shot down Nagoya
42-24788	KRO'S KIDS	L TRI		6BG	6/25/1945	Shot down Nagoya
42-24789		X TRI 36 (1)	10/17/1944	9BG	4/7/1945	Surveyed; mine load salvo dmg 3/30/45
42-24790		E TRI	10/18/1944	504BG	2/10/1945	Ditched
42-24791	BIG TIME OPERATOR, THE	X TRI 4	10/17/1944	9BG	12/6/1945	Reclaimed Robins
42-24792	20TH CENTURY LIMITED	Z SQ 2 (3)	10/19/1944	500BG	5/10/1954	Reclaimed Davis-Monthan
42-24792		TRI I		468BG	5/10/1954	Reclaimed Davis-Monthan
42-24793	COUNTRY GENTLEMAN	K TRI 10 (1)	10/19/1944	505BG	5/14/1945	Ditched off Iwo insufficient fuel
42-24794	HOMER'S ROAMERS	T SQ (4)	10/19/1944	498BG	5/19/1950	Reclaimed
42-24794	LITTLE PRINCESS	K TRI 5 (1)	10/19/1944	505BG	5/19/1950	Reclaimed
42-24795		FOURYLWSTRP W	10/17/1944	40BG	5/5/1945	Group transfer to Tinian
42-24795		TRI S 32	5/5/1945	40BG	9/15/1953	Reclaimed Pyote
42-24796	DOTTIE'S DILEMMA	X TRI 16/31	10/25/1944	9BG	7/6/1945	Surveyed off end of runway Tinian 6/1/45
42-24797	JACKPOT	K TRI		505BG	3/19/1945	Ditched 150 miles off Japan-flak damage
42-24798	OLE GAS EATER	FOURYLWSTRP ER	10/23/1944	40BG	1/3/1945	Combat loss?
42-24799		E TRI	10/19/1944	504BG	7/25/1949	Reclaimed
42-24800		SOLYLWRUD	10/24/1944	462BG	5/5/1945	Group transfer to Tinian
42-24800		TRI U 21	5/5/1945	462BG	6/15/1945	Crashed, burned on takeoff
42-24801	PHONY EXPRESS	SOLGRNRUD	10/21/1944	462BG	5/5/1945	Group transfer to Tinian
42-24801	PHONY EXPRESS	TRI U 43	5/5/1945	462BG	5/16/1945	Abandoned, fire #3 engine after takeoff
42-24802	BLACK CAT	K TRI 2	10/25/1944	505BG	12/6/1950	Reclaimed
42-24802	PURPLE SHAFT	K TRI 2	10/25/1944	505BG	12/6/1950	Reclaimed
42-24803			10/20/1944	234BU	1/31/1945	Class 01Z Clovis
42-24804		FOURBLUSTRP S	10/24/1944	40BG	2/26/1945	Abandoned Bay of Bengal, photo mission
42-24805			10/20/1944	4141BU	10/21/1953	Reclaimed Robins
42-24806	INDIAN MAID	E TRI	10/26/1944	504BG	11/7/1950	Reclaimed
42-24806	SATAN'S LADY	E TRI	10/26/1944	504BG	11/7/1950	Reclaimed
42-24807		A SQ	10/23/1944	497BG	1/14/1945	Ditched, flak damage
42-24808		A SQ	10/23/1944	497BG	2/25/1945	Ditched, mid-air collision with 42-63431
42-24809	INDIAN MAID	K TRI 03 (1)	10/26/1944	505BG	6/5/1945	Hit by flak, crashed in Kobe harbor
42-24809	67 SEA BEE	K TRI 03 (1)	10/26/1944	505BG	6/5/1945	Hit by flak, crashed in Kobe harbor
42-24810			10/27/1944	4141BU	10/21/1953	Reclaimed Robins
42-24811		K TRI	10/28/1944	505BG	10/11/1946	Salvaged

I. Master List

Serial #	Name	Identification	Delv	Assign	Off Inv	Circumstances
42-24812	MISS SU-SU	E TRI 19	11/1/1944	504BG	5/15/1945	Landing collision
42-24813		Z SQ 38/18	10/27/1944	500BG	9/22/1947	Reclaimed Far East
42-24814	421ST EMBLEM	E TRI 01	10/26/1944	504BG	9/15/1953	Reclaimed Pyote
42-24814	SITTING PRETTY	E TRI 01	10/26/1944	504BG	9/15/1953	Reclaimed Pyote
42-24815	121 SEA-BEES	K TRI	11/2/1944	505BG	2/10/1945	Collided with 42-24784 at Gumma-Ken
42-24815	DEANER BOY	K TRI	11/2/1944	505BG	2/10/1945	Collided with 42-24784 at Gumma-Ken
42-24816		E TRI	10/27/1944	504BG	7/18/1945	Salvaged
42-24817			10/28/1944		2/13/1945	Surveyed at Herington
42-24818	THE DEACON'S DELIGHT	K TRI	10/26/1944	505BG	2/10/1945	Ditched
42-24819						Unknown
42-24820	DARING DONNA III	X TRI 46	10/31/1944	9BG	9/15/1953	Reclaimed Pyote
42-24821	APHRODITE	E TRI 60	10/31/1944	504BG	4/15/1945	Ditced Kanagawa
42-24822	PATCHES	X TRI 18/33	10/31/1944	9BG	5/10/1954	Reclaimed Davis-Monthan
42-24823	27TH N.C.B. SPECIAL	K TRI 25	11/1/1944	505BG	9/15/1953	Reclaimed
42-24823	DANGEROUS LADY	K TRI 25	11/1/1944	505BG	9/15/1953	Reclaimed Pyote
42-24823	HIGH, EH DOC?	K TRI 25	11/1/1944	505BG	9/15/1953	Reclaimed Pyote
42-24824	HOMING BIRD	K TRI 20	10/31/1944	505BG	2/10/1945	Ditched 150 miles south of Iwo Jima
42-24825	SNOOKY	L TRI 14	11/1/1944	6BG	11/17/1953	Reclaimed Hill
42-24826	GENERAL CONFUSION [1]	E TRI	11/1/1944	504BG	5/25/1945	Crashed Saitama from Tokyo
42-24826	IN THE MOOD [1]	L TRI	11/1/1944	6BG	5/25/1945	Crashed Saitama from Tokyo
42-24827		K TRI 38 (1)	11/2/1944	505BG	5/5/1954	Reclaimed Tinker
42-24828		K TRI ( )	11/2/1944	505BG	5/26/1945	Crashed Kojimachi, Tokyo
42-24829	WHAT HAPPENED	FOURBLKSTRP	3/24/1944	40BG	10/21/1953	Reclaimed Robins
42-24830	IRISH LULLABY	L TRI 06	11/2/1944	6BG	11/22/1973	Reclaimed Robins
42-24831	GOD'S WILL	X TRI 2 (1)	11/3/1944	9BG	9/15/1953	Reclaimed Pyote
42-24832			11/3/1944	4141BU	9/22/1953	Reclaimed Pyote
42-24833		V SQ	11/3/1944	499BG	2/13/1945	MIA
42-24834	VISITING FIREMEN	CIRC E 66/52	11/6/1944	504BG	6/7/1945	Crashed in sea
42-24835		L TRI	11/6/1944	6BG	9/15/1953	Reclaimed Pyote
42-24835		X TRI 47/35 (1)	11/6/1944	9BG	9/15/1953	Reclaimed Pyote
42-24836		L TRI 17 (1)	11/6/1944	6BG	11/17/1953	Reclaimed Hill
42-24837		E TRI	11/6/1944	504BG	10/4/1954	Reclaimed Tinker
42-24838		SOLREDRUD	11/7/1944	462BG	5/5/1945	Group Transfer to Tinian
42-24838		TRI U 02	5/5/1945	462BG	5/7/1954	Reclaimed Robins
42-24839	PADRE AND HIS ANGELS	K TRI ( )	11/8/1944	505BG	7/6/1945	Lost Akashi flying with 73rdBW
42-24840	QUEEN BEE	X TRI 3 (1)	11/6/1944	9BG	4/28/1945	Abandoned north of Iwo, engine failure
42-24841			11/7/1944		5/7/1945	Reclaimed Robins
42-24842		L TRI	11/8/1944	6BG	2/10/1945	Engine fire, explosion, ditched
42-24843		K TRI	11/10/1944	6BG	10/7/1944	Crashed on T/O, Grand Island, surveyed
42-24844		K TRI	11/10/1944	505BG	12/6/1950	Reclaimed
42-24845		A SQ		497BG	5/10/1954	Reclaimed Davis-Monthan
42-24846	MONSOON II	FOURREDSTRP K	11/9/1944	40BG	5/5/1945	Group transfer to Tinian
42-24846	MONSOON II	TRI S 08	5/5/1945	40BG	9/11/1950	Reclaimed
42-24847		X TRI 19	11/10/1944	9BG	9/22/1953	Reclaimed Pyote
42-24848	MISS ROSEMARY	K TRI 13	11/10/1944	505BG	5/10/1954	Reclaimed Davis-Monthan
42-24848		TRI U	11/10/1944	462BG	5/10/1954	Reclaimed Davis-Monthan
42-24849	MISSION TO ALBUQUERQUE	Z SQ 8 (3)	11/9/1944	500BG	3/17/1945	Rammed at Kobe
42-24850		Z SQ	11/10/1944	500BG	4/12/1945	Surveyed
42-24850	BAD MEDICINE	K TRI 15	11/10/1944	505BG	4/12/1945	Surveyed
42-24851	THE MOOSE IS LOOSE	E TRI 66	11/10/1944	504BG	4/12/1945	Surveyed
42-24851	OMAHA ONE MORE TIME	E TRI	11/10/1944	504BG	4/12/1945	Surveyed
42-24851	THE ONE YOU LOVE	E TRI	11/10/1944	504BG	4/12/1945	Surveyed
42-24852	GOOD DEAL	E TRI 27	11/13/1944	504BG	7/7/1948	Reclaimed
42-24852	SANDMAN	CIRC E 27	11/13/1944	504BG	7/7/1948	Reclaimed
42-24853	LIVE WIRE	X TRI 10 (1)	11/15/1944	9BG	4/25/1945	Crashed Iwo Jima
42-24854		E TRI	11/11/1944	504BG	6/17/1954	Class 26 Sheppard
42-24855	[2]	TRI I	11/15/1944	468BG	7/13/1945	Ditched — inoperative fuel transfer system
42-24855	JUMBO II [2]	A SQ 50 (2)	11/15/1944	497BG	5/10/1954	Reclaimed Davis-Monthan
42-24856	GOIN' JESSIE	X TRI 30 (1)	11/11/1944	9BG	5/10/1954	Reclaimed Davis-Monthan
42-24857	JOKERS WILD II	A SQ	11/11/1944	497BG	2/19/1945	Ditched — flak damage
42-24858		2WHTRUDSTRP	11/11/1944	468BG	5/10/1945	Group transfer to Tinian
42-24858		TRI I 3	5/10/1945	468BG	7/13/1945	Ditched near Iwo Jima
42-24859	TOKYO-KO	X TRI 17(1)/32	11/14/1944	9BG	5/7/1954	Reclaimed Robins
42-24860		K TRI 10 (2)	11/17/1945	505BG	4/00/45	Tail code change
42-24860		CIRC W 10 (2)	4/00/45	505BG	7/24/1945	Transferred to 58BW/40BG

Serial #	Name	Identification	Delv	Assign	Off Inv	Circumstances
42-24860		TRI S 51	7/24/1945	40BG	7/14/1954	Salvaged at Davis-Monthan
42-24861		DIAM #	11/15/1944	444BG	5/10/1945	Group transfer to Tinian
42-24861		TRI N 54	5/10/1945	444BG	5/10/1954	Reclaimed Davis-Monthan
42-24862		K TRI (1)	11/15/1944	505BG	9/15/1953	Reclaimed Pyote
42-24863	LUCKY LADY	E TRI 12	11/16/1944	504BG	12/6/1950	Reclaimed
42-24864	STORK CLUB BOYS	K TRI (1)	11/15/1944	505BG	3/27/1945	Shot down, crashed in Onga River
42-24865	BIG BOOTS	E TRI 30	11/17/1944	504BG	9/00/45	Transferred to 58BW462BG
42-24865		TRI U	9/00/45	462BG	4/1/1947	Surveyed
42-24866	EARTHQUAKE MCGOON	CIRC R 51 (1)	11/18/1944	6BG	9/15/1953	Reclaimed Pyote
42-24867	SASSY LASSY	K TRI	11/16/1944	505BG	2/10/1945	Shot Down
42-24868	RIP VAN WINKLE	L TRI 30		6BG		Exploded over Tokyo
42-24869		? TRI	11/17/1944		10/21/1953	Reclaimed Robins
42-24870		CIRC R	11/18/1944	6BG	5/23/1945	MIA
42-24871		? TRI	11/18/1944		10/21/1953	Reclaimed Robins
42-24872	JOLLY ROGER	L TRI 56	11/21/1944	6BG	9/15/1953	Reclaimed Pyote
42-24873	SNUFFY	DIAM #	11/18/1944	444BG	5/10/1945	Group transfer to Tinian
42-24873	SNUFFY	TRI N 10	5/10/1945	444BG	5/10/1954	Reclaimed Davis-Monthan
42-24873	SNAFU, THEM SHIF'LES	TRI N	5/10/1945	444BG	5/10/1954	Reclaimed Davis-Monthan
42-24874	EL PAJARO DE LA GUERRA	L TRI 04	11/21/1944	6BG	9/22/1953	Reclaimed Pyote
42-24875	LIL' IODINE	X TRI 7 (1)	11/18/1944	9BG	3/10/1945	Ditched, out of fuel from Tokyo
42-24876	HON. SPY REPORT	X TRI 52 (1)	11/21/1944	9BG	12/6/1950	Reclaimed Robins
42-24877	DOUBLE EXPOSURE	F		6CMG		Unknown
42-24877	OVER EXPOSED			6CMG		Unknown
42-24878	FLAK ALLEY SALLY	L TRI 33 (1)	11/20/1944	6BG	11/17/1953	Reclaimed
42-24878	HELL'S BELLE	L TRI 53	11/20/1944	6BG	10/31/1953	Reclaimed Hill
42-24879	READY BETTY	2WHTRUDSTRP	11/18/1944	468BG	5/11/1945	Group transfer to Tinian
42-24879	READY BETTY	TRI I 10	5/11/1945	468BG	9/14/1954	Reclaimed
42-24880		L TRI 62	11/22/1944	6BG	9/23/1953	Reclaimed Pyote
42-24881	SITTING PRETTY	E TRI	11/22/1944	504BG	10/21/1953	Reclaimed Salina
42-24882	1919TH COMPANY	CIRC E	11/23/1944	504BG	6/22/1945	Shot down by fighters
42-24882	PAPPY'S PULLMAN	CIRC E	11/23/1944	504BG	6/22/1945	Shot down by fighters
42-24883	CITY OF DALLAS	BLK SQ M 07 (1)	12/2/1945	19BG	5/3/1945	Ditched — flak cut fuel transfer lines
42-24883	JOE'S JUNK	BLK SQ M 07 (1)	12/2/1945	19BG	5/3/1945	Ditched — flak cut fuel transfer lines
42-24884	GRIDER GAL	L TRI 36	11/23/1944	6BG	8/2/1954	Reclaimed Tinker
42-24885	BIG JOE	L TRI 05		6BG	3/25/1949	Reclaimed McClellan
42-24886		FOURREDSTRP V		40BG	5/5/1945	Group transfer to Tinian
42-24886		TRI S 06	5/5/1945	40BG		Unknown
42-24887		V SQ 03 (2)		499BG	9/14/1954	Reclaimed Griffis
42-24888	SMILIN' JACK	FOURREDSTRP S	11/28/1944	40BG	5/5/1945	Group transfer to Tinian
42-24888	SMILIN' JACK	TRI S 12	5/5/1945	40BG	9/23/1953	Reclaimed
42-24889		K TRI (1)	11/24/1944	505BG	4/22/1945	Crashed
42-24890		CIRC W	11/24/1944	505BG		Unknown
42-24891	MONSOON GOON II	DIAM #	11/28/1944	444BG	5/10/1945	Group transfer to Tinian
42-24891	MONSOON GOON II	TRI N 14	5/10/1945	444BG	8/7/1945	Ditched from Yawata
42-24892	LI'L YUTZ	2YLWRUDSTRP	11/28/1944	468BG	5/11/1945	Group transfer to Tinian
42-24892	LI'L YUTZ	TRI I 44	5/11/1945	468BG	11/17/1953	Reclaimed
42-24893	LITTLE ORGAN ANNIE	2YLWRUDSTRP	11/30/1944	468BG	5/11/1945	Group transfer to Tinian
42-24893	LITTLE ORGAN ANNIE	TRI I 41	5/11/1945	468BG	11/12/1953	Reclaimed
42-24894		FOURBLUSTRP D	11/30/1944	40BG	5/5/1945	Group transfer to Tinian
42-24894		TRI S 26 (1)	5/5/1945	40BG	5/29/1945	Rammed Yokohama area
42-24895	CITY OF PITTSBURG	2WHTRUDSTRP	11/30/1944	468BG	5/11/1945	Group transfer to Tinian
42-24895	CITY OF PITTSBURG	TRI I 2	5/11/1945	468BG	9/11/1950	Reclaimed
42-24896	MIS-CHIEF-MAK-ER	SOLYLWRUD	12/1/1944	462BG	5/5/1945	Group Ttransfer to Tinian
42-24896	MIS-CHIEF-MAK-ER	TRI U 25	5/5/1945	462BG	9/22/1953	Reclaimed
42-24897	JOKERS WILD II	DIAM #	11/28/1944	444BG	5/10/1945	Group transfer to Tinian
42-24897	JOKERS WILD II	TRI N 57 (1)	11/28/1944	444BG	5/10/1954	Reclaimed
42-24898		SOLREDRUD S 1	12/1/1944	462BG	5/5/1945	Group transfer to Tinian
42-24898		TRI U 13	5/5/1945	462BG	9/11/1950	Reclaimed
42-24898		K TRI 50	12/1/1944	505BG	9/11/1950	Unknown
42-24899	THE AGITATOR II	DIAM #	12/1/1944	444BG	5/10/1945	Group transfer to Tinian
42-24899	THE AGITATOR II	TRI N 34	5/10/1945	444BG	5/10/1954	Reclaimed
42-24900	OLD 900	X TRI 37(1)48	11/29/1944	9BG	10/12/1953	Reclaimed Pyote
42-24901	THECULTURED VULTURE	L TRI 25 (1)/35		6BG	7/15/1952	Reclaimed
42-24902			12/2/1944	236BU	10/12/1953	Reclaimed Pyote
42-24903		BLK SQ M	12/2/1944	19BG	4/15/1945	MIA Tokyo

I. Master List

Serial #	Name	Identification	Delv	Assign	Off Inv	Circumstances
42-24904	RAMP TRAMP II	K TRI 51	4/00/45	505BG	5/30/1945	Returned to 468BG
42-24904	RAMP TRAMP II	SOLGRNRUD	12/7/1944	462BG	5/5/1945	Group transfer to Tinian
42-24904	RAMP TRAMP II	TRI U 52	5/5/1945	462BG		Transfer to 505BG
42-24905			12/2/1944	4141BU	10/12/1953	Reclaimed Pyote
42-24906	SLICK'S CHICKS	BLK SQ M 42 (1)	12/6/1944	19BG	8/8/1954	Reclaimed Tinker
42-24907	BATTLIN' BONNIE	X TRI 15 (1)		9BG	5/10/1949	Reclaimed
42-24908		TRI S 03	12/12/1944	40BG	9/22/1953	Reclaimed
42-24908		FOURREDSTRP F	12/12/1944	40BG	10/12/1953	Reclaimed Pyote
42-24909	FLYIN' HOME	2BLURUDSTRP	12/8/1944	468BG	5/11/1945	Group transfer to Tinian
42-24909	BLITZ BUGGY	2BLURUDSTRP	12/8/1944	468BG	10/12/1953	Reclaimed Pyote
42-24909	FLYIN' HOME	TRI I 30	5/11/1945	468BG	9/22/1953	Reclaimed
42-24910			3/7/1945	247BU	11/12/1953	Reclaimed Robins
42-24911		BLK SQ Q	12/7/1944	231BU	10/12/1953	Reclaimed Pyote
42-24912	OILY BOID	BLK SQ O (1)	12/13/1944	29BG	10/21/1947	Reclaimed
42-24913	THUNDERIN' LORETTA	X TRI 13	12/12/1944	9BG	4/00/45	Tail code change
42-24913	THUNDERIN' LORETTA	CIRC X 13 (1)	4/00/45	9BG	5/19/1945	Crashed, exploded on takeoff
42-24914		FOURREDSTRP B	12/11/1944	40BG	5/5/1945	Group transfer to Tinian
42-24914		TRI S 07 (1)	5/5/1945	40BG	7/3/1945	Surveyed — crashed off end of runway
42-24915	MISS DONNA LEE	FOURYLWSTRP T	12/9/1944	40BG	5/5/1945	Group transfer to Tinian
42-24915	MISS DONNA LEE	TRI S 33	5/5/1945	40BG	9/11/1950	Reclaimed Pyote
42-24916		L TRI	12/8/1944	6BG	3/28/1945	Flak damage #4, burned off, abandoned
42-24916	THE PEACEMAKER	L TRI	12/8/1944	6BG	3/27/1945	Lost on mining mission, crashed Miyazaki
42-24917	BABY GAIL	BLK SQ O 50 (1)	12/15/1944	29BG	11/4/1945	Salvaged; no data on transfer sequence
42-24917	CITY OF MILWAUKEE	BLK SQ K 66	12/15/1944	330BG	11/4/1945	Salvaged; no data on transfer sequence
42-24917	CITY OF OKLAHOMA CITY	BLK SQ K 66	12/15/1944	330BG	11/4/1945	Salvaged; no data on transfer sequence
42-24917	CITY OF OKLAHOMA CITY	BLK SQ O 08	12/15/1944	29BG	11/4/1945	Salvaged; no data on transfer sequence
42-24917	NIP ON ESE NIPPER	BLK SQ O 50 (1)	12/15/1944	29BG	11/4/1945	Salvaged; no data on transfer sequence
42-24917	NIPP ON ESE	BLK SQ O 08	12/15/1944	29BG	11/4/1945	Salvaged; no data on transfer sequence
42-24918		CIRC E 58	12/11/1944	504BG	7/26/1945	Crash landed
42-24919		SOLREDRUD	12/9/1944	462BG	5/5/1945	Group transfer to Tinian
42-24919		TRI U 03	5/5/1945	462BG	11/12/1953	Reclaimed
42-63352	HULL'S ANGEL	SOLBLURUD	12/30/1943	462BG	5/24/1944	Surveyed, out of fuel, abandoned
42-63353	THE JUKE BOX	2WHTRUDSTRP	12/31/1944	468BG	12/12/1949	Reclaimed
42-63354		2BLURUDSTRP	1/31/1944	468BG	1/7/1949	Reclaimed
42-63355	BELLA BORTION	2BLURUDSTRP	1/31/1944	468BG	12/12/1949	Reclaimed
42-63356	GEORGIA PEACH	2BLURUDSTRP	2/13/1944	468BG	12/21/1949	Reclaimed
42-63356	LASSIE	2BLURUDSTRP	2/13/1944	468BG	12/21/1949	Reclaimed Pyote
42-63356	VALKYRIE QUEEN	2BLURUDSTRP	2/13/1944	468BG	12/21/1949	Reclaimed Pyote
42-63357			2/16/1944	468BG	4/21/1944	Crashed Karachi enroute to CBI
42-63358			2/29/1944	4105BU	6/27/1949	Reclaimed Davis-Monthan
42-63359			2/29/1944	4121BU	3/16/1945	Surveyed Kelly
42-63360	SLOW FREIGHT	DIAM #	2/29/1944	444BG	12/21/1949	Reclaimed Pyote
42-63360	TROJAN SPIRIT	DIAM #	2/29/1944	444BG	12/21/1949	Reclaimed Pyote
42-63361			5/31/1944	4141BU	12/21/1949	Reclaimed Pyote
42-63362	HULL'S ANGEL	SOLGRNRUD	3/14/1944	462BG	6/27/1949	Reclaimed
42-63363	MARIETTA MISFIT	FOURBLKSTRP V	3/18/1944	40BG	12/7/1944	Abandoned China
42-63364			3/31/1944	4141BU	12/21/1949	Reclaimed Pyote
42-63364	MISS CARRIAGE [3]	CIRC E	3/31/1944	504BG		Shot Down
42-63365			3/31/1944	4000BU	12/21/1949	Reclaimed Pyote
42-63366		Z SQ	3/31/1944	500BG	12/21/1949	Reclaimed Pyote
42-63367			4/3/1944	245BU	11/1/1948	Reclaimed McClellan
42-63368			4/10/1944	4136BU	7/8/1944	Class 26 Tinker
42-63369			4/10/1944	499BG	5/10/1949	Reclaimed Tinker
42-63370			4/20/1944	234BU	4/30/1946	Class 01Z Clovis
42-63371			4/20/1944	4141BU	12/21/1949	Reclaimed Pyote
42-63372			4/20/1944	4136BU	12/21/1949	Reclaimed Pyote
42-63373			4/24/1944	505BG	10/24/1944	Surveyed at Harvard
42-63374	TABOOMA II	FOURREDSTRP	4/20/1944	40BG	5/31/1946	Class 01Z Amarillo
42-63375	HER MAJESTY	DIAM #	4/24/1944	444BG	7/31/1946	Class 01Z, Biggs
42-63376	MISS N.C.	DIAM# 36	4/24/1944	444BG	6/30/1946	Class 01Z Amarillo
42-63377			4/30/1944	4141BU	1/7/1949	Reclaimed Pyote
42-63378	FUBAR	DIAM #	4/30/1944	444BG	2/00/45	Returned to ZI as war weary
42-63379			5/4/1944	231BU	12/21/1949	Reclaimed Pyote
42-63380			5/4/1944	4121BU	12/16/1948	Reclaimed Kelly
42-63380	KITTEN		5/2/1944	98BG	12/16/1948	Unknown

Serial #	Name	Identification	Delv	Assign	Off Inv	Circumstances
42-63380	SPITTIN KITTIN		5/2/1944	98BG	12/16/1948	Unknown
42-63381			5/8/1944	499BG	12/12/1949	Reclaimed Pyote
42-63382			5/15/1944	3PRS	2/10/1945	Class 26 Smoky Hill
42-63383			5/15/1944		12/3/1944	Crashed, surveyed at Pyote
42-63384			5/15/1944	504BG	12/8/1944	Surveyed Grand Isle
42-63385			5/29/1944	504BG	7/14/1949	Reclaimed Robins
42-63386	GLOBE GIRDLE MYRTLE	SOLGRNRUD	5/19/1944	462BG	5/15/1947	Reclaimed Tinker
42-63386	GLOBE GIRTLE MURTLE	SOLGRNRUD	5/19/1944	462BG	5/15/1947	Reclaimed Tinker
42-63387			5/29/1944	504BG	7/6/1949	Reclaimed Robins
42-63388			5/2/1944	231BU	2/28/1945	Surveyed Pueblo
42-63389			5/23/1944	19BG	6/6/1949	Reclaimed Kelly
42-63390			5/25/1944	19BG	12/21/1949	Reclaimed Pyote
42-63391			5/27/1944	505BG	3/13/1945	Class 26 Harvard
42-63392			5/27/1944	3030BU	12/12/1949	Reclaimed Pyote
42-63393	RUSH ORDER	SOLREDRUD I 1	5/30/1944	462BG	3/20/1946	Reclaimed Gulfport
42-63394	LAST RESORT	FOURBLUSTRP R	3/7/1944	40BG	1/14/1945	Destroyed in 42-24582 bomb explosion
42-63395		2WHTRUDSTRP	5/31/1944	468BG	12/7/1944	Crashed at A-7 returning
42-63396	MARIETTA BELLE	FOURREDSTRP M/H	5/31/1944	40BG	5/5/1945	Group transfer to Tinian
42-63396	MARIETTA BELLE	TRI S 01	5/5/1945	40BG	6/19/1945	Crashed from Toyohashi — Surveyed
42-63396	PRETTY BABY	FOURREDSTRP M	5/31/1944	40BG	5/5/1945	Group transfer to Tinian
42-63396	PRETTY BABY	TRI S 01	5/5/1945	40BG	6/19/1945	Crashed from Toyohashi — Surveyed
42-63397			6/3/1944	236BU	12/12/1949	Reclaimed Pyote
42-63398			6/7/1944	19BG	6/27/1949	Reclaimed Davis-Monthan
42-63399	LADY MARGE II	DIAM #	6/7/1944	444BG	9/19/1944	Surveyed at Alamogordo
42-63400			6/7/1944	236BU	2/11/1945	Surveyed at Pyote
42-63401			6/8/1944	500BG	9/14/1948	Reclaimed Victorville
42-63402			6/12/1944	610EXGP	5/29/1945	Surveyed at Eglin
42-63403		DIAM# 39	6/13/1944	444BG	5/31/1946	Class 01Z Amarillo
42-63404		FOURYLWSTRP X	6/12/1944	40BG	4/30/1946	Class 01Z Davis-Monthan
42-63405			6/15/1944	4141BU	12/21/1949	Reclaimed Pyote
42-63406			6/20/1944	19BG	1/17/1945	Surveyed Great Bend
42-63407	SHOOT YOU'RE FADED	FOURYLWSTRP S	7/22/1944	40BG	5/15/1947	Reclaimed Tinker
42-63408				4105BU	6/27/1949	Reclaimed Davis-Monthan
42-63409			8/7/1944	499BG	12/21/1949	Reclaimed Pyote
42-63410			8/3/1944	6BG	2/23/1945	Surveyed — crashed T/O with engine fire
42-63411	DUCHESS	FOURREDSTRP	9/16/1944	40BG		Transferred to 444BG in CBI
42-63411	DUCHESS	DIAM 61		444BG	5/5/1945	Group transfer to Tinian
42-63411	DUCHESS	TRI N 61 (1)	5/5/1945	444BG	6/30/1946	Reclaimed
42-63412	PEACE ON EARTH	A SQ 25 (1)	8/24/1944	497BG	3/4/1945	Ditched — out of fuel
42-63413	DIXIE DARLIN'	A SQ 45 (1)	8/30/1944	497BG	12/18/1944	Ditched from Nagoya
42-63414	THE JUMPING STUD	A SQ 49 (1)	9/6/1944	497BG	7/4/1945	Ditched from Minashima
42-63415	JOLLY ROGER	2BLURUDSTRP	9/9/1944	468BG	5/11/1945	Group transfer to Tinian
42-63415	JOLLY ROGER	TRI I 22	5/11/1945	468BG	6/27/1949	Reclaimed Davis-Monthan
42-63416		T SQ 46 (1)	9/7/1944	498BG	5/5/1945	Ditched, engine fire
42-63417	RUSHIN' ROTASHUN	2WHTRUDSTRP	9/22/1944	468BG	5/10/1949	Group transfer to Tinian
42-63417	RUSHIN' ROTASHUN	TRI I 5	9/22/1944	468BG	5/10/1949	Reclaimed Tinker
42-63418	JUMBO, KING OF THE SHOW	A SQ 50 (1)	8/25/1944	497BG	1/3/1945	Ditched — out of fuel
42-63419		DIAM #	9/19/1944	444BG	11/11/1944	Abandoned, possible icing
42-63420	RANKLESS WRECK	FOURREDSTRP D/P	9/16/1944	40BG	5/5/1945	Group transfer to Tinian
42-63420	RANKLESS WRECK	TRI S 09 (1)	5/5/1945	40BG	6/5/1945	Combat damage-salvaged Iwo Jima
42-63421			8/17/1944	245BU	12/21/1949	Reclaimed Pyote
42-63422	LITTLE MIKE	DIAM #	9/7/1944	444BG	5/10/1945	Group transfer to Tinian
42-63422	LITTLE MIKE	TRI N 52	5/10/1945	444BG	11/30/1945	Declared Excess to Inventory
42-63423	WERE WOLF	A SQ 28 (1)	8/25/1944	497BG	1/27/1945	Bombs exploded Bomb Bay-fighter attack
42-63424	HAP'S CHARACTERS	2WHTRUDSTRP	8/29/1944	468BG	5/11/1945	Group transfer to Tinian
42-63424	HAP'S CHARACTERS	TRI I 9	5/11/1945	468BG	3/23/1949	Reclaimed Tinker
42-63425	THE DRAGON LADY	A SQ 41 (1)	8/26/1944	497BG	11/1/1948	Reclaimed Davis-Monthan
42-63425	TERRIBLE TERRY	A SQ 41 (1)	8/26/1944	497BG	11/1/1948	Reclaimed Davis-Monthan
42-63426	FICKLE FINGER	A SQ 47	9/6/1944	497BG	12/1/1944	Crashed Tinian — surveyed
42-63427	MISS MARGARET	A SQ	9/4/1944	497BG	2/26/1945	Ditched. Weather strike?
42-63428		T SQ 10 (1)		498BG	11/27/1944	Surveyed — destroyed in enemy raid
42-63429	PACIFIC QUEEN	Z SQ 35 (1)	9/9/1944	500BG	7/14/1949	Reclaimed Robins
42-63430		T SQ 1 (1)	9/4/1944	498BG	12/13/1944	Ditched from Nagoya
42-63431	PONDEROUS PEG	A SQ 44 (1)	8/31/1944	497BG	2/25/1945	Mid-air collision with 42-24808
42-63432	LUCKY IRISH	T SQ 28 (1)	9/22/1944	498BG	12/3/1944	Ditched, out of fuel

I. Master List

Serial #	Name	Identification	Delv	Assign	Off Inv	Circumstances
42-63433			9/4/1944	498BG	9/28/1944	Surveyed Kearney after collision
42-63434			9/18/1944	462BG	10/7/1949	Class 01Z Chanute
42-63435	SNAFU PER FORT	Z SQ 3 (1) Z19	9/18/1944	500BG	5/31/1946	Salvaged
42-63435	SHARON SUE	Z SQ 19	9/18/1944	500BG	5/31/1946	Salvaged at Memphis
42-63436	OLD IRONSIDES	Z SQ 28 (1)	4/30/1944	500BG	5/10/1949	Reclaimed Pyote
42-63437			9/18/1944	4196BU	7/14/1949	Reclaimed Robins
42-63438	HELL'S BELL	V SQ 42 (1)	9/18/1944	499BG	11/27/1944	Blown up in Jap raid on Saipan
42-63438		V SQ 26 (1)	9/18/1944	499BG	11/27/1944	Surveyed — destroyed in enemy raid
42-63439		V SQ 47 (1)	9/22/1944	499BG	12/13/1944	Crashed in sea off Japan
42-63440	INSPIRATION	V SQ 10 (1)	9/25/1944	499BG	6/4/1945	Lost on test flight
42-63441		Z SQ 46	9/25/1944	500BG	5/10/1949	Reclaimed Tinker
42-63442	BEAUBOMBER II	V SQ 47	9/28/1944	499BG	10/15/1948	Reclaimed Weaver
42-63443	SHIRLEY DEE	T SQ 12 (2)	9/27/1944	498BG	9/11/1950	Reclaimed Pyote
42-63443		T SQ 25 (3)	9/27/1944	498BG	9/11/1950	Reclaimed Pyote
42-63443		T SQ 37	9/27/1944	498BG	9/11/1950	Reclaimed Pyote
42-63444	HONSHU HAWK	T SQ 45(2)	9/30/1944	498BG	9/11/1950	Reclaimed Pyote
42-63444	JOLLY ROGER	L TRI	9/30/1944	6BG	9/11/1950	Reclaimed Pyote
42-63445	CRAIG COMET, THE	TRI I 52	10/1/1944	468BG	5/24/1945	Crashed landing home base
42-63445	[4]	T SQ 56	10/1/1944	498BG		Unknown
42-63446	TROJAN SPIRIT	DIAM #	10/9/1944	444BG	5/10/1945	Group transfer to Tinian
42-63446	TROJAN SPIRIT	TRI N 42	5/10/1945	444BG	9/11/1950	Reclaimed Pyote
42-63447	UMBIAGO III, DAT'S MY BOY	V SQ 50 (1)	10/1/1944	499BG	12/13/1944	Ditched 90 miles off Saipan - out of fuel
42-63448		SOLYLWRUD	10/4/1944	462BG	5/5/1945	Group transfer to Tinian
42-63448		TRI U 23	5/5/1945	462BG	9/11/1950	Reclaimed Pyote
42-63449			10/6/1944	4141BU	9/11/1950	Reclaimed Pyote
42-63450		SOLREDRUD X1		462BG		Reclaimed
42-63450		TRI U 12		462BG		Reclaimed
42-63451	BLACK JACK TOO	DIAM #	10/13/1944	444BG	5/10/1945	Group transfer to Tinian
42-63451	BLACK JACK TOO	TRI N 40 (1)	5/10/1945	444BG	6/5/1945	Wing broke off at #3, Engine fire
42-63452		SOLGRNRUD	10/9/1944	462BG	12/14/1944	Abandoned, fuel trasnsfer system out
42-63453	SATAN'S SISTER	V SQ 4 (2)	10/12/1944	499BG	6/30/1946	Reclaimed Pinecastle
42-63454	THUNDER BIRD	SOLGRNRUD X3	10/16/1944	462BG	5/5/1945	Group transfer to Tinian
42-63454	THUNDER BIRD	TRI U 42	5/5/1945	462BG	9/11/1950	Reclaimed Pyote
42-63455	GENIE	FOURREDSTRP C/L		40BG	5/5/1945	Group transfer to Tinian
42-63455	GENIE	TRI S 10	5/5/1945	40BG		Unknown
42-63456		2BLURUDSTRP	10/15/1944	468BG	5/11/1945	Group transfer to Tinian
42-63456		TRI I 21	5/11/1945	468BG	12/6/1950	Reclaimed
42-63457	BATTLIN' BEAUTY	SOLYLWRUD T1	10/15/1944	462BG	5/5/1945	Group transfer to Tinian
42-63457	BATTLIN' BEAUTY	TRI U 07	5/5/1945	462BG	6/17/1954	Class 26 Sheppard
42-63457	OLD ACQUAINTANCE	SOLREDRUD T 1	10/15/1944	462BG	6/17/1954	Class 26 Sheppard
42-63458	WING DING	DIAM #	10/20/1944	444BG	12/24/1944	Crashed after takeoff
42-63459		SOLREDRUD Z1	10/20/1944	462BG	5/5/1945	Group transfer to Tinian
42-63459		TRI U 8	5/5/1945	462BG	10/1/1954	Reclaimed Tinker
42-63460	LASSIE TOO	2BLURUDSTRP	12/13/1944	468BG	5/11/1945	Group transfer to Tinian
42-63460	LASSIE TOO	TRI I 28	5/11/1945	468BG		Unknown
42-63461		A SQ 30 (1)	10/25/1944	497BG	12/3/1944	Ditched — out of fuel
42-63462	THE LEMON	FOURBLUSTRP K	10/23/1944	40BG	5/5/1945	Group transfer to Tinian
42-63462	THE LEMON	TRI S 20 (1)	5/5/1945	40BG	5/13/1945	Surveyed — broke in two landing
42-63463	SKYSCRAPPER II	A SQ 9 (2)	10/19/1944	497BG	2/19/1945	Surveyed
42-63464	BELLE RINGER	2YLWRUDSTRP	10/20/1944	468BG	5/11/1945	Group transfer to Tinian
42-63464	BELLE RINGER	TRI I 53 (2)	5/11/1945	468BG	9/11/1950	Reclaimed Tinker
42-63465		V SQ 28	10/22/1944	499BG	6/6/1947	Reclaimed Sioux City
42-63466	STRIPPED FOR ACTION	A SQ 32	11/3/1944	497BG	8/25/1950	Reclaimed McClellan
42-63467		A SQ	10/30/1944	497BG	9/11/1950	Reclaimed Tinker
42-63468		T SQ 02 (3)	10/25/1944	498BG	2/10/1945	#1 engine out, disappeared
42-63469	SLEEPY TIME GAL	T SQ 33		498BG		Reclaimed
42-63470			10/30/1944	497BG	11/16/1944	Surveyed at Herington
42-63471	SWEAT'R OUT	A SQ 47 (2)	10/31/1944	497BG	9/11/1950	Reclaimed Pyote
42-63472	MYASAS DRAGGIN	SOLGRNRUD	10/28/1944	462BG	5/5/1945	Group transfer to Tinian
42-63472	MYASAS DRAGGIN	TRI U 45	5/5/1945	462BG	6/22/1950	Reclaimed Davis-Monthan
42-63473		SOLGRNRUD K3	10/29/1944	462BG	5/5/1945	Group transfer to Tinian
42-63473		TRI U 46	5/5/1945	462BG	9/11/1950	Reclaimed Tinker
42-63474		SOLYLWRUD	10/31/1944	462BG	5/5/1945	Group transfer to Tinian
42-63474		TRI U 26	5/5/1945	462BG	12/6/1950	Reclaimed McClellan
42-63475	HOUSTON HONEY	T SQ 51	11/11/1944	498BG	9/11/1950	Reclaimed Tinker

Serial #	Name	Identification	Delv	Assign	Off Inv	Circumstances
42-63476		SOLYLWRUD	11/11/1944	462BG	5/5/1945	Group transfer to Tinian
42-63476		TRI U 24	5/5/1945	462BG	9/11/1950	Reclaimed Tinker
42-63477		V SQ		58BW		Ditched
42-63478	SOUTHERN BELLE	T SQ 35	11/15/1944	498BG	9/11/1950	Reclaimed Tinker
42-63479			11/10/1944	462BG	2/24/1945	Ditched
42-63480	DREAM GIRL	SOLREDRUD H1		462BG	5/5/1945	Group transfer to Tinian
42-63480	DREAM GIRL	TRI U 14	5/5/1945	462BG		Unknown
42-63481	(VARGA GIRL)	E TRI	11/13/1944	504BG	9/11/1950	Reclaimed Tinker
42-63481	PAPPY'S PULLMAN II	E TRI	11/13/1944	504BG	9/11/1950	Reclaimed Tinker
42-63481	(VARGA GIRL)	V SQ 50	11/13/1944	499BG		Transferred to 504BG
42-63481	HONSHU HURRICANE	V SQ 50	11/13/1944	499BG		Transferred to 504BG
42-63482		K TRI	11/11/1944	505BG	3/9/1945	Ditched — out of fuel
42-63483		V SQ 11	11/15/1944	499BG	4/25/1945	Crashed 20 miles off coast of Japan
42-63484	HONORABLE TNT WAGON	K TRI	11/13/1944	505BG	5/19/1945	MIA
42-63485	HOT PANTS	A SQ 12	11/17/1944	497BG	12/11/1944	Unknown
42-63486	HELL'S BELLE	Z SQ 7 (2)	11/17/1944	500BG	5/25/1945	Crashed Tokyo
42-63486	NAUGHTY NANCY	Z 7	11/17/1944	500BG	5/23/1945	Shot down in sea near Tokyo
42-63487	CONSTANT NYMPH	Z SQ 30 (1)	11/22/1944	500BG	1/5/1950	Reclaimed Victorville
42-63488		SOLGRNRUD		462BG	5/5/1945	Group transfer to Tinian
42-63488		TRI U 44	5/5/1945	462BG		Unknown
42-63489	HOLY JOE	Z SQ 11 (2)	11/22/1944	500BG	2/21/1945	Surveyed from crash landing 2/15/45
42-63489	ROUND ROBIN	Z SQ 11 (2)	11/22/1944	500BG	2/21/1945	Surveyed from crash landing 2/15/45
42-63490		Z SQ 45 (2)		500BG		Survived war
42-63490		SPEEDWING GRN		307BG		Unknown
42-63491		V SQ	11/22/1944	499BG	6/17/1954	Class 26 Sheppard
42-63492	STAR DUSTER			499BG	4/24/1945	Rammed by fighter
42-63492	LUCKY IRISH II	A SQ 26 (2)	11/22/1944	497BG	9/11/1950	Reclaimed Pyote
42-63493	STAR DUSTER	V SQ 36	11/22/1944	499BG	3/24/1945	Exploded over target
42-63494		Z SQ 31 (2)	11/22/1944	500BG	2/19/1945	Shot down Tokyo
42-63495	FAST COMPANY	V SQ 12	11/24/1944	499BG	10/30/1946	Reclaimed Victorville
42-63496	LITTLE MISS	DIAM# 50	11/29/1944	444BG	5/10/1945	Group transfer to Tinian
42-63496	NAUGHTY NANCY	DIAM# 50	11/29/1944	444BG	5/10/1945	Group transfer to Tinian
42-63496	NAUGHTY NANCY	TRI N 50 (1)	5/10/1945	444BG	6/1/1945	MIA after collision with 42-65270
42-63497	FEVER FROM THE SOUTH	Z SQ 32		500BG	5/23/1945	
42-63498	B-SWEET III	FOURREDSTRP B	11/24/1944	40BG	5/5/1945	Group transfer to Tinian
42-63498	B-SWEET III	TRI S 02 (1)	5/5/1945	40BG	5/24/1945	Abandoned from Tokyo, fire #3 Engine
42-63498	LAZY BABY	FOURREDSTRP B	11/24/1944	40BG	5/5/1945	Group transfer to Tinian
42-63498	LAZY BABY	TRI S 02 (2)	5/5/1945	40BG	5/24/1945	Abandoned from Tokyo, fire #3 Engine
42-63499	CORAL QUEEN	E TRI	11/30/1944	504BG	2/16/1945	Ditched north of Marianas
42-63499	38 SEA BEES	E TRI	11/30/1944	504BG	2/16/1945	Ditched north of Marianas
42-63500	GRAVEL GERTIE	2WHTRUDSTRP	11/30/1944	468BG	5/11/1945	Group transfer to Tinian
42-63500	GRAVEL GERTIE	TRI I 12	5/11/1945	468BG	7/3/1945	Crashed offshore from Kagawa
42-63501		T SQ 12/2 (2)	11/26/1944	498BG	1/27/1945	Shot down leaving Tokyo
42-63502	LONG JOHN SILVER	SOLGRNRUD	11/30/1944	462BG	5/5/1945	Group transfer to Tinian
42-63502	LONG JOHN SILVER	TRI U 47	5/5/1945	462BG	7/31/1946	Reclaimed Lowry
42-63503	SKY SCRAPPER	SOLYLWRUD		462BG	5/5/1945	Group transfer to Tinian
42-63503	SKY SCRAPPER	TRI U 30	5/5/1945	462BG	6/30/1946	Reclaimed Amarillo
42-63504	FLAG SHIP	E TRI 2	11/28/1944	504BG	9/11/1950	Reclaimed Tinker
42-63505	DRAGGIN' LADY	FOURREDSTRP D	11/27/1944	40BG	5/5/1945	Group transfer to Tinian
42-63505	DRAGGIN' LADY	TRI S 04	5/5/1945	40BG	6/17/1954	Class 26 Sheppard
42-63506	UNTOUCHABLE	SOLREDRUD	11/30/1945	462BG	9/25/1944	#2 engine fire on T/O, hit ground, exploded
42-63507		E TRI	11/30/1944	504BG	9/11/1950	Reclaimed Pyote
42-63508	PEACHY	K TRI 9 (1)	12/7/1944	505BG	5/26/1945	Lost Tokyo, crashed Chiba
42-63509	LONG WINDED	X TRI 38/50	12/5/1944	9BG	5/22/1945	Abandoned,#2 engine shot out, #4 quit
42-6351?	ANN GARRY III	L TRI 37		6BG	1/1/1945	Shot down by P-61 after abnd over Iwo
42-63510		K TRI	11/30/1944	505BG	1/29/1945	Xfr to 498BG. To Guam Air Depot 4/10/45
42-63510	HEAVENLY BODY	T SQ 38	1/29/1945	498BG	5/1/1945	Surveyed Guam Air Depot 4/1/45 damage
42-63511	MAN O' WAR II	X TRI 1/17	12/4/1944	9BG	4/00/45	Tail code change
42-63511	MAN O' WAR II	CIRC X 1	4/00/45	9BG	7/31/1946	Reclaimed Lowry
42-63512	NIP CLIPPER	CIRC X 23	12/5/1944	9BG	8/8/1945	Shot down by flak at Yawata
42-63513	RAMP QUEEN	K TRI 7(1)	11/30/1944	505BG	1/00/45	Transfer to 498BG
42-63513	RAMP QUEEN	V SQ 5(2)	1/00/45	499BG	5/25/1945	Shot down Tokyo
42-63514	CAPT CLAY	L TRI	12/4/1944	6BG	5/25/1945	Shot down Tokyo
42-63515		L TRI	12/7/1944	6BG	9/11/1950	Reclaimed Tinker
42-63516		L TRI	12/5/1944	6BG	10/17/1946	Reclaimed Amarillo

I. Master List

Serial #	Name	Identification	Delv	Assign	Off Inv	Circumstances
42-63517	CITY OF SAN JOSE	BLK SQ K 14(1)	12/9/1944	330BG	1/29/1945	Transferred to 505BG
42-63517	CITY OF BEL AIR	BLK SQ K 42	12/9/1944	330BG	1/29/1945	Transferred to 505BG
42-63517	ROUND ROBIN	BLK SQ K 42	12/9/1944	330BG	1/29/1945	Transferred to 505BG
42-63517	PRINCESS POKEY	K TRI 14 (1)	1/29/1945	505BG	4/13/1945	MIA Koriyame, Saitame Prefecture
42-63517	POKAHUNTAS	K TRI 14 (1)	1/29/1945	505BG	4/13/1945	MIA Koriyame, Saitame Prefecture
42-63518		L TRI	12/7/1944	6BG	12/6/1950	Reclaimed McClellan
42-63519	IRON SHILLALAH	A SQ	12/8/1944	497BG	4/18/1945	Ditched from Tachiari
42-63520		K TRI	12/9/1944	505BG	9/11/1950	Reclaimed Tinker
42-63521		SOLYLWRUD	12/14/1944	462BG	5/5/1945	Group transfer to Tinian
42-63521		TRI U 32 (1)	5/5/1945	462BG	5/25/1945	MIA Tokyo
42-63522	RAIDIN MAIDEN III	T SQ 11 (1)	12/12/1944	498BG	9/11/1950	Reclaimed Pyote
42-63522		E TRI	12/12/1944	504BG	9/11/1950	Reclaimed Pyote
42-63523	MISS BEHAVIN' II	A SQ 46 (3)		497BG		Unknown
42-63524	PASSION WAGON	T SQ 52 (2)	12/10/1944	498BG	9/11/1950	Reclaimed Pyote
42-63524	PASSION WAGON	K TRI	12/10/1944	505BG		Transferred to 498BG
42-63525	BAINBRIDGE BELLE	K TRI 12 (1)	12/9/1944	504BG	4/15/1945	Crashed landing from Kyushu Airfields
42-63525	DRAGON LADY	K TRI 12 (1)	12/9/1944	505BG	4/15/1945	Crashed landing from Kyushu Airfields
42-63526	TEASER	A SQ 52 (2)	12/15/1944	497BG	3/25/1945	Hit by flak, crashed Nagoya
42-63527	DEVILISH SNOOKS	FOURBLUSTRP B	12/29/1944	40BG	5/5/1945	Group transfer to Tinian
42-63527	DEVILISH SNOOKS	TRI S 18	5/5/1945	40BG	9/11/1950	Reclaimed Tinker
42-63528			12/20/1944	244BU	6/29/1945	Surveyed at Biggs
42-63529	CITY OF BURLINGTON	2YLWRUDSTRP	12/15/1944	468BG	5/11/1945	Group transfer to Tinian
42-63529	CITY OF BURLINGTON	TRI I 50	5/11/1945	468BG	5/25/1945	Flak, crashed Saitami Prefecture
42-63530	TOTIN' TO TOKYO II	2BLURUDSTRP	12/12/1944	468BG	5/11/1945	Group transfer to Tinian
42-63530	TOTIN' TO TOKYO II	TRI I 26	5/11/1945	468BG		Unknown
42-63531		SOLREDRUD E1	12/15/1944	462BG	5/5/1945	Group transfer to Tinian
42-63531		TRI U 06	5/5/1945	462BG	7/9/1945	Off end of runway Sendai mission
42-63532	MILLION DOLLAR BABY II	2BLURUDSTRP		468BG	5/11/1945	Group transfer to Tinian
42-63532	MILLION DOLLAR BABY II	TRI I 33	5/11/1945	468BG		Unknown
42-63533	FAITHFUL FAYE	DIAM #		444BG	5/10/1945	Group transfer to Tinian
42-63533		TRI N 32	5/19/1945	444BG		Unknown
42-63533		L TRI		6BG		Unknown
42-63534	PIONEER III	2BLURUDSTRP	12/13/1944	468BG	5/11/1945	Group transfer to Tinian
42-63534	PIONEER III	TRI I 31	5/11/1945	468BG	9/11/1950	Reclaimed Pyote
42-63535		L TRI 01	12/20/1944	6BG	9/11/1950	Reclaimed Tinker
42-63536	MAMMY YOKUM	2WHTRUDSTRP	12/18/1944	468BG	5/11/1945	Group transfer to Tinian
42-63536	MAMMY YOKUM	TRI I 6	5/11/1945	468BG	5/25/1945	Shot down Tokyo
42-63536	THUMPER	E TRI		504BG		Unknown
42-63537	MALE CALL	DIAM# 13	12/26/1944	444BG	5/10/1945	Group transfer to Tinian
42-63537	MALE CALL	TRI N 13	5/10/1945	444BG	5/25/1945	Crashed from Tokyo
42-63538	WINGED VICTORY II	FOURBLUSTRP A 24	12/19/1944	40BG	5/25/1945	Group transfer to Tinian
42-63538	WINGED VICTORY II	TRI S 24 (1)	5/25/1945	40BG	5/26/1945	Crashed Tokyo — flak
42-63539	CITY OF HIGHLAND FALLS	BLK SQ K 67	12/20/1944	330BG		Unknown
42-63539	INFANT OF PRAGUE	BLK SQ K 67	12/20/1944	330BG		Unknown
42-63540		SOLREDRUD N1	12/18/1944	462BG	5/5/1945	Group transfer to Tinian
42-63540		TRI U 11	5/5/1945	462BG	9/11/1950	Reclaimed Tinker
42-63541	THE GHASTLY GOOSE	A SQ 46 (2)	12/23/1944	497BG	1/27/1945	Rammed; ditched from Tokyo
42-63542	KRITZER BLITZER	FOURBLUSTRP R	12/19/1944	40BG	5/5/1945	Group transfer to Tinian
42-63542	KRITZER BLITZER	TRI S 14	5/5/1945	40BG	9/11/1950	Reclaimed Tinker
42-63543			12/19/1944	4141BU	9/11/1950	Reclaimed Pyote
42-63544	COX'S ARMY	X TRI 26	12/28/1944	9BG	1/4/1951	Reclaimed Hill
42-63545	UMBRIAGO	X TRI 28	12/23/1944	9BG	4/16/1945	Crashed Tokyo from Kawasaki
42-63546	INDIANA II	X TRI 44 (1)		9BG	3/17/1945	Crashed Oakibe
42-63547		BLK SQ M		19BG		Unknown
42-63548	MAIDEN USA	DIAM# 36		444BG	5/10/1945	Group transfer to Tinian
42-63548	MAIDEN USA	TRI N 36	5/10/1945	444BG		Unknown
42-63548		L TRI		6BG		Unknown
42-63549	EMPIRE EXPRESS	K TRI	12/31/1944	505BG	5/7/1945	Rammed by a Ki-45-MSgt Murata
42-63550	MARY ANN	V SQ 27 (3)	12/31/1944	499BG	9/11/1950	Reclaimed Pyote
42-63551	BANANA BOAT	L TRI 64	1/4/1945	6BG	4/7/1949	Reclaimed McClellan
42-63552	LUCKY STRIKE	L TRI 27	12/31/1944	6BG	7/21/1946	Crashed in sea
42-63552		TRI I	12/31/1945	468BG	7/21/1946	Crashed in sea
42-63553		L TRI 26	1/5/1945	6BG	3/13/1945	Crashed takeoff to Osaka
42-63554	MISS LACE	T SQ 25 (2)	1/31/1945	498BG	9/11/1950	Reclaimed Pyote
42-63555	DARK EYES	FOURYLWSTRP K	1/5/1945	40BG	5/5/1945	Group transfer to Tinian

Serial #	Name	Identification	Delv	Assign	Off Inv	Circumstances
42-63555	DARK EYES	TRI S 29 (1)	5/5/1945	40BG	9/11/1950	Reclaimed Tinker
42-63556	EARLY BIRD	X TRI 25	1/9/1945	9BG	9/11/1950	Reclaimed Tinker
42-63557	MISSOURI BELLE	DIAM #		444BG	5/10/1945	Group transfer to Tinian
42-63557	MISSOURI BELLE	TRI N 35	5/10/1945	444BG		Unknown
42-63557	SURE THING	TRI N 45		444BG		Unknown
42-63557	SURE THING	BLKSTRP		19BG		Unknown
42-63558		L TRI	1/14/1945	6BG	5/25/1945	Shot down, crashed Tokyo
42-63559	PRINCESS EILEEN III	DIAM #		444BG	5/10/1945	Group transfer to Tinian
42-63559	PRINCESS EILEEN III	TRI N 43	5/10/1945	444BG		Unknown
42-63560		SOLREDRUD A1	1/6/1945	462BG	5/5/1945	Group transfer to Tinian
42-63560		TRI U 04	5/5/1945	462BG	9/11/1950	Reclaimed Tinker
42-63561	READY TEDDY	X TRI 29 (1)	1/11/1945	9BG	5/26/1946	Reclaimed Fairchild
42-63562		BLK SQ O	1/14/1945	29BG	9/11/1950	Reclaimed Tinker
42-63563	MAXIMUM LOAD	BLK SQ M 23 (1)	1/12/1945	19BG	6/13/1945	Crashed on takeoff
42-63564	CHERRY-HORIZONTAL CAT	BLK SQ O	1/11/1945	29BG	3/9/1945	Crashed Ubo Mt in weather per Japanese
42-63565	CITY OF SAN ANTONIO	BLK SQ O 46 (1)	1/13/1945	29BG	6/17/1954	Class 26 Sheppard
42-63565	WHERE'S KILROY	BLK SQ O 46 (1)	1/13/1945	29BG	6/17/1954	Class 26 Sheppard
42-63566	FIRE BUG	BLK SQ O 43(1)	1/14/1945	29BG	8/25/1950	Reclaimed McClellan
42-63567	CITY OF PROVIDENCE	BLK SQ M 28 (1)	1/14/1945	19BG	6/10/1945	Ditched from Kure
42-63568		BLK SQ O (1)		29BG		Reclaimed
42-63569	ZERO AVENGER	BLK SQ M 11 (1)	1/18/1945	19BG	3/9/1945	MIA Tokyo, crashed Kawaguchi
42-63570	THUNDERBIRD	BLK SQ O 07(1)	1/18/1945	29BG		Reclaimed
42-63571		BLK SQ O (1)	1/14/1945	29BG	5/11/1945	Crashed on takeoff
42-63572			1/18/1945	4105BU	6/22/1950	Reclaimed Davis-Monthan
42-63573		BLK SQ M 30 (1)	1/14/1945	19BG	7/18/1949	Reclaimed Chanute
42-63574	OLD 574	X TRI 41 (1)	1/15/1945	9BG	1/4/1951	Reclaimed Hill
42-63575			1/22/1945	4141BU	9/11/1950	Reclaimed Pyote
42-63576			1/19/1945	4141BU	9/11/1950	Reclaimed Pyote
42-63577						Unknown
42-63578			1/17/1945	233BU	1/16/1946	Surveyed at Pyote
42-63579		BLK SQ O	2/19/1945	29BG	6/22/1950	Reclaimed Davis-Monthan
42-63580		FOURYLWSTRP V	1/27/1945	40BG	5/5/1945	Group Transfer to Tinian
42-63580		TRI S 28	5/5/1945	40BG	9/11/1950	Reclaimed Tinker
42-63581			1/21/1945	4105BU	6/22/1950	Reclaimed Davis-Monthan
42-63582			1/26/1945	242BU	9/11/1950	Reclaimed Pyote
42-63583			1/21/1945	611EXGP	9/11/1950	Reclaimed Pyote
42-63584		FA				Unknown
42-63585			1/23/1945	4141BU	9/11/1950	Reclaimed Pyote
42-63586						Unknown
42-63587						Unknown
42-63588			1/26/1945	4141BU	9/11/1950	Reclaimed Pyote
42-63589		DIAM B	2/6/1945	16BG	8/23/1950	Reclaimed Robins
42-63590			1/31/1945	4141BU	9/11/1950	Reclaimed Pyote
42-63591						Unknown
42-63592			1/26/1945	4141BU	9/11/1950	Reclaimed Pyote
42-63593	RAGGED BUT RIGHT	DIAM L 28	1/31/1945	331BG	9/11/1950	Reclaimed Pyote
42-63594		A SQ		497BG		Unknown
42-63595			1/27/1945	4141BU	9/11/1950	Reclaimed Pyote
42-63596			1/27/1945	4141BU	9/11/1950	Reclaimed Pyote
42-63597			1/31/1945	4141BU	9/11/1950	Reclaimed Pyote
42-63598			1/31/1945	4000BU	3/5/1945	Surveyed at Wright-Patterson
42-63599		DIAM Y	2/8/1945	501BG	9/11/1950	Reclaimed Tinker
42-63600	ROAD APPLE	DIAM Y	2/1/1945	501BG	5/9/1945	Missing Kobe with 314th Bomb Wing
42-63601	BELLE OF MARTINEZ	DIAM Y 45	2/8/1945	501BG	8/23/1950	Reclaimed Robins
42-63602			1/29/1945	4141BU	9/11/1950	Reclaimed Pyote
42-63603		DIAM B	2/8/1945	16BG	7/12/1945	Abandoned, prop malfunction
42-63604						Unknown
42-63605	ELLIE BARBARA & HER ORP	DIAM B		16BG		Reclaimed
42-63606		DIAM L 21 (1)	2/3/1945	331BG	8/25/1950	Reclaimed McClellan
42-63607		DIAM Y	2/6/1945	501BG	9/11/1950	Reclaimed Tinker
42-63608		DIAM B	2/7/1945	16BG	8/25/1950	Reclaimed Pyote
42-63609		DIAM L 41 (1)	2/12/1945	331BG	10/3/1950	Reclaimed
42-63610	BUGGER	DIAM L 1 (1)		331BG		Reclaimed
42-63611		DIAM L 47 (1)	2/12/1945	331BG	8/25/1950	Reclaimed McClellan
42-63612		DIAM L 42 (1)		331BG		Reclaimed

I. Master List   31

Serial #	Name	Identification	Delv	Assign	Off Inv	Circumstances
42-63613		DIAM B		16BG		Reclaimed
42-63614		DIAM H		502BG		Reclaimed
42-63614	ISLAND GIRL	CIRC W		505BG		Unknown
42-63615		DIAM Y	2/10/1945	501BG	9/11/1950	Reclaimed Pyote
42-63616		DIAM H	2/12/1945	502BG	8/25/1950	Reclaimed McClellan
42-63617		DIAM B	2/19/1945	16BG	9/14/1951	Reclaimed Hill
42-63618		DIAM L		331BG		Reclaimed
42-63619		DIAM H		502BG		Reclaimed
42-63619	BIG MIKE	K TRI		505BG		Unknown
42-63620		DIAM H	2/19/1945	502BG	8/25/1950	Reclaimed McClellan
42-63621		DIAM B 33	2/20/1945	16BG	9/11/1950	Reclaimed Tinker
42-63622			2/26/1945	4141BU	9/11/1950	Reclaimed Pyote
42-63623		DIAM H		502BG		Reclaimed
42-63624		DIAM H	3/1/1945	502BG	11/8/1950	Reclaimed Hill
42-63625			2/28/1945	245BU	6/21/1945	Class 01Z McCook
42-63626		DIAM B	2/21/1945	16BG	6/22/1950	Reclaimed Davis-Monthan
42-63627		DIAM Y 31	2/21/1945	501BG		Reclaimed
42-63628		DIAM B	2/20/1945	16BG	9/11/1950	Reclaimed Tinker
42-63629			2/25/1945	235BU	9/11/1950	Reclaimed Pyote
42-63630		DIAM B	2/17/1945	16BG	9/11/1950	Reclaimed Tinker
42-63631		DIAM L 22		331BG		Reclaimed
42-63632						Reclaimed
42-63633			2/26/1945	4141BU	9/11/1950	Reclaimed Pyote
42-63634			2/24/1945	4141BU	9/11/1950	Reclaimed Pyote
42-63635			2/24/1945	4141BU	9/11/1950	Reclaimed Pyote
42-63636			2/26/1945	4141BU	9/11/1950	Reclaimed Pyote
42-63637			2/28/1945	4141BU	9/11/1950	Reclaimed Pyote
42-63638			2/26/1945	4141BU	9/11/1950	Reclaimed Pyote
42-63639		DIAM Y	2/24/1945	501BG	9/11/1950	Reclaimed Pyote
42-63640	THE BOOMERANG	DIAM Y	2/24/1945	501BG	6/1/1951	Reclaimed Hill
42-63641		DIAM Y	2/28/1945	501BG	3/31/1951	Reclaimed Tinker
42-63642			2/27/1945	235BU	8/6/1945	Surveyed at Biggs
42-63643				235BU		
42-63644		DIAM B	2/26/1945	16BG	1/4/1951	Reclaimed Hill
42-63645			2/28/1945	4141BU	9/11/1950	Reclaimed Pyote
42-63646		DIAM B	3/2/1945	16BG	9/11/1950	Reclaimed Tinker
42-63647				235BU		Reclaimed
42-63648			2/28/1945	4141BU	9/11/1950	Reclaimed Pyote
42-63649		DIAM B	3/5/1945	16BG	9/11/1950	Reclaimed Tinker
42-63650	FLEET ADMIRAL NIMITZ	DIAM Y	3/9/1945	501BG	3/31/1951	Reclaimed Tinker
42-63651		DIAM B	3/6/1945	16BG	8/25/1950	Reclaimed McClellan
42-63652		DIAM Y		501BG		Reclaimed
42-63653			3/18/1945	16BG	7/12/1945	Ditched from Kawasaki
42-63654		DIAM Y		501BG		Unknown
42-63655		DIAM B		16BG		Reclaimed
42-63656		DIAM H	3/12/1945	502BG	4/25/1949	Reclaimed
42-63657	PINK LADY	DIAM B		16BG		Reclaimed
42-63658		DIAM L 26		331BG	4/20/1949	Reclaimed
42-63659		DIAM B		16BG		Reclaimed
42-63660			3/10/1945	4136BU	9/30/1946	Reclaimed Tinker
42-63661		DIAM B		16BG		Reclaimed
42-63662		DIAM H		502BG		Reclaimed
42-63663		DIAM Y	3/9/1945	501BG	5/5/1954	Reclaimed Tinker
42-63664		DIAM Y		501BG		Reclaimed
42-63665		DIAM Y		501BG		Reclaimed
42-63666		DIAM B	3/12/1945	16BG	6/23/1954	Reclaimed Pyote
42-63667		DIAM B	3/17/1945	16BG	5/5/1954	Reclaimed Tinker
42-63668		DIAM Y		501BG		Reclaimed
42-63669		DIAM Y	3/17/1945	501BG	12/4/1953	Reclaimed Robins
42-63670		DIAM Y 48	3/14/1945	501BG	12/4/1953	Reclaimed Robins
42-63671		DIAM Y		501BG		Reclaimed
42-63672			3/16/1945	4000BU	12/4/1953	Reclaimed Robins
42-63673		DIAM B	3/24/1945	16BG	12/4/1953	Reclaimed Robins
42-63674		DIAM Y	3/17/1945	501BG	12/4/1953	Reclaimed Robins
42-63675		DIAM Y	3/19/1945	501BG	6/23/1954	Reclaimed Pyote

Serial #	Name	Identification	Delv	Assign	Off Inv	Circumstances
42-63676		DIAM L 24	3/12/1945	331BG	12/4/1953	Reclaimed Robins
42-63677		DIAM Y		501BG		Reclaimed
42-63678	MANIUWA	DIAM H	3/19/1945	502BG	7/25/1945	Flak Kawasaki
42-63679		DIAM L 3		331BG		Reclaimed
42-63680		DIAM Y		501BG		Reclaimed
42-63681		DIAM L 22		331BG		Reclaimed
42-63682		DIAM Y	3/20/1945	501BG	12/4/1953	Reclaimed Robins
42-63683		DIAM Y	3/20/1945	501BG	12/4/1953	Reclaimed Tinker
42-63684		DIAM Y	3/21/1945	501BG	3/20/1954	Reclaimed McClellan
42-63685		DIAM Y	3/20/1945	501BG	12/4/1953	Reclaimed Robins
42-63686		DIAM Y	3/19/1945	501BG	12/4/1953	Reclaimed Robins
42-63687		DIAM Y		501BG		Reclaimed
42-63688	LOADED DICE	DIAM B 7	3/24/1945	16BG	12/4/1953	Reclaimed Robins
42-63689		DIAM Y	3/28/1945	501BG	8/19/1954	Reclaimed Hill
42-63690		DIAM L 2	3/22/1945	331BG	3/20/1954	Reclaimed McClellan
42-63691		DIAM B	3/22/1945	16BG	8/19/1954	Reclaimed Hill
42-63692		DIAM B	3/31/1945	16BG	5/27/1948	Reclaimed Tinker
42-63693			3/28/1945	611EXGP	9/14/1953	Reclaimed Tinker
42-63694		DIAM Y		501BG		Reclaimed
42-63695		DIAM H	3/28/1945	502BG	6/23/1954	Reclaimed Pyote
42-63696		DIAM Y	3/28/1945	501BG	12/4/1953	Reclaimed Robins
42-63697		DIAM B	3/30/1945	16BG	9/14/1953	Reclaimed Tinker
42-63698		DIAM L 43	3/29/1945	331BG	12/4/1953	Reclaimed Robins
42-63699		DIAM B 04	3/30/1945	16BG	6/29/1945	Burned
42-63700		DIAM L 57	3/30/1945	331BG	6/23/1954	Reclaimed Pyote
42-63701		DIAM H	3/29/1945	502BG	6/24/1954	Reclaimed Pyote
42-63702		DIAM Y	3/30/1945	501BG	8/19/1954	Reclaimed Hill
42-63703		DIAM B		16BG		Reclaimed
42-63704		DIAM B	4/2/1945	16BG	6/24/1954	Reclaimed Pyote
42-63705		DIAM Y	4/7/1945	501BG	9/14/1953	Reclaimed Tinker
42-63706			4/18/1945	235BU	6/24/1954	Reclaimed Pyote
42-63707		DIAM B	4/3/1945	16BG	12/4/1954	Reclaimed Robins
42-63708		DIAM B		16BG		Reclaimed
42-63709		DIAM Y	4/6/1945	501BG	9/9/1954	Reclaimed Davis-Monthan
42-63710		DIAM Y	4/6/1945	501BG	6/24/1954	Reclaimed Pyote
42-63711		DIAM H	4/9/1945	502BG		Reclaimed
42-63712		DIAM B	4/9/1945	16BG	9/14/1953	Reclaimed Tinker
42-63713		DIAM H	4/9/1945	502BG	9/9/1954	Reclaimed Davis-Monthan
42-63714		DIAM Y	4/9/1945	501BG	8/19/1954	Reclaimed Hill
42-63715		DIAM B	4/9/1945	16BG	6/24/1954	Reclaimed Pyote
42-63716			4/13/1945	245BU	3/20/1954	Reclaimed McClellan
42-63717		DIAM Y	4/6/1945	501BG	9/14/1953	Reclaimed Tinker
42-63718		DIAM Y	4/11/1945	501BG	12/4/1953	Reclaimed Robins
42-63719		DIAM Y 65	4/10/1945	501BG	3/20/1954	Reclaimed McClellan
42-63720		DIAM Y	4/13/1945	501BG	9/14/1953	Reclaimed Tinker
42-63721		DIAM Y	4/13/1945	501BG	12/4/1953	Reclaimed Robins
42-63722		DIAM H	4/10/1945	502BG	9/9/1954	Reclaimed Davis-Monthan
42-63723		DIAM Y	4/13/1945	501BG	5/10/1949	Reclaimed Tinker
42-63724		DIAM Y	4/12/1945	501BG	9/14/1953	Reclaimed Tinker
42-63725	MY NAKED -	DIAM Y	4/11/1945	501BG	9/14/1953	Reclaimed Tinker
42-63726		DIAM Y	4/12/1945	501BG	8/19/1954	Reclaimed Hill
42-63727			4/19/1945	3705BU	9/9/1954	Reclaimed Davis-Monthan
42-63728			4/16/1945	235BU	5/5/1954	Reclaimed Davis-Monthan
42-63729			4/18/1945	235BU	9/9/1954	Reclaimed Davis-Monthan
42-63730			4/14/1945	242BU	6/24/1954	Reclaimed Pyote
42-63731	THE CHALLENGER		4/18/1946	648BU	5/24/1949	Reclaimed Tinker
42-63732			4/16/1945	235BU	6/24/1954	Reclaimed Pyote
42-63733			4/19/1945	235BU	5/5/1954	Reclaimed Davis-Monthan
42-63734		DIAM B	4/23/1945	16BG	3/20/1954	Reclaimed McClellan
42-63735			4/19/1945	4141BU	9/11/1950	Reclaimed Pyote
42-63736		DIAM B		16BG		Reclaimed
42-63737			5/28/1945	233BU	8/19/1954	Reclaimed Pyote
42-63738		DIAM B	4/21/1945	16BG	9/14/1953	Reclaimed Tinker
42-63739		DIAM B	4/23/1945	16BG	12/4/1953	Reclaimed Robins
42-63740		DIAM B	4/27/1945	16BG	8/19/1954	Reclaimed Hill

I. Master List

Serial #	Name	Identification	Delv	Assign	Off Inv	Circumstances
42-63741		DIAM B	4/24/1945	16BG	9/14/1953	Reclaimed Tinker
42-63742		DIAM B	4/30/1945	16BG	6/24/1954	Reclaimed Pyote
42-63743		DIAM B	4/21/1945	16BG	5/10/1949	Reclaimed Chanute
42-63744			5/9/1945	247BU	8/19/1954	Reclaimed Pyote
42-63745		DIAM B	4/27/1945	16BG	12/4/1953	Reclaimed Robins
42-63746		DIAM B	4/28/1945	16BG	9/14/1953	Reclaimed Tinker
42-63747		DIAM H	4/28/1945	502BG	7/9/1945	Exploded on takeoff, crashed in sea
42-63748		DIAM H	4/28/1945	502BG	6/24/1954	Reclaimed Pyote
42-63749	CITY OF MARTINEZ	DIAM Y	4/25/1945	501BG	8/19/1954	Reclaimed Hill
42-63750				22BG	8/31/1954	Reclaimed Tinker
42-63750			5/28/1945	19BG	8/31/1954	Reclaimed Tinker
42-63751		DIAM	4/28/1945		9/14/1953	Reclaimed Tinker
42-65202	SATAN'S ANGEL	DIAM# 15	5/26/1944	444BG	3/25/1945	Collided Bay of Bengal with 42-24507
42-65203			6/2/1944	58BW	9/18/1944	Missing Natal to Accra en route to join
42-65204		DIAM #	6/14/1944	444BG	11/23/1944	Surveyed Ankang, hit by plane
42-65205			6/24/1944	594BU	5/10/1949	Reclaimed Tinker
42-65206			7/31/1944	4141BU	9/11/1950	Reclaimed Pyote
42-65207			7/31/1944	6BG	9/11/1950	Reclaimed Pyote
42-65208	ANDY'S DANDY'S	2YLWRUDSTRP	8/1/1944	468BG	5/15/1947	Reclaimed Tinker
42-65209						Unknown
42-65210	FAY	T SQ 26 (1)	8/9/1944	498BG	3/24/1945	Flak hits, fire, exploded in air
42-65210	FILTHY FAY	T SQ 26 (1)	8/9/1944	498BG	3/24/1945	Flak hits, fire, exploded in air
42-65211	LADY EVE	T SQ 48/58/43	8/29/1944	498BG	5/26/1945	Abandoned, crashed in sea off Iwo
42-65211	LADY EVE	T SQ 43	8/29/1944	498BG	12/13/1944	Threw prop, cut fuselage
42-65212	MRS TITTYMOUSE	T SQ 42 (2)	8/30/1944	498BG	4/7/1945	Tokyo flak blew off left wing
42-65213	CARLA LANI-BATTLE BABY	SOLGRNRUD	8/29/1944	462BG	12/7/1944	Broke up mid-air Lashan, China — ice
42-65214			8/31/1944		1/4/1951	Reclaimed Hill
42-65215			9/9/1944	498BG	9/28/1944	Class 26 at Kearney
42-65216			9/15/1944	504BG	9/11/1950	Reclaimed Pyote
42-65217			9/16/1944	504BG	9/11/1950	Reclaimed Pyote
42-65218		Z SQ 44 (1)	9/20/1944	500BG	11/29/1944	MIA Nuisance raid on Tokyo
42-65219	PUNCHIN' JUDY	Z SQ 10 (1)	9/26/1944	500BG	3/31/1951	Reclaimed Tinker
42-65220	SUPER WABBIT	V SQ 28 (1)	9/26/1944	499BG	11/27/1944	MIA Tokyo
42-65221	GRAVEL GERTIE	Z SQ 29 (1)		500BG	8/6/1945	Crashed on takeoff, surveyed
42-65222	SUPER WABBIT	V SQ 49 (1)	9/29/1944	499BG	2/19/1945	Rammed over Tokyo
42-65223			9/29/1944	2750ABG	2/18/1954	Reclaimed Niagara
42-65224		V SQ 30 (3)		499BG		Unknown
42-65225		DIAM #	10/11/1944	444BG	12/31/1944	Crashed, two engine failure
42-65225		FOURREDSTRP	10/11/1944	40BG	12/31/1944	Crashed, two engine failure
42-65226		DIAM# 54	10/18/1944	444BG	1/11/1945	MIA Singapore
42-65227	LADY BE GOOD	2BLURUDSTRP		468BG	5/11/1945	Group transfer to Tinian
42-65227	LADY BE GOOD	TRI I 15	5/11/1945	468BG		Transferred to 444BG
42-65228	EARTHQUAKE MCGOON	DIAM #		444BG	5/10/1945	Group transfer to Tinian
42-65228	EARTHQUAKE MCGOON	TRI N 59 (1)	5/10/1945	444BG	6/26/1945	Surveyed Iwo
42-65229	DO IT AGAIN	L TRI	10/24/1944	6BG	5/5/1954	Reclaimed Tinker
42-65230	NAMED, NAME NOT KNOWN	SOLREDRUD	10/24/1944	462BG	5/5/1945	Group transfer to Tinian
42-65230	NAMED, NAME UNKNOWN	TRI U 62 (1)	5/5/1945	462BG	5/25/1945	MIA Tokyo
42-65230		TRI U 09	10/24/1944	462BG	5/10/1954	Reclaimed Davis-Monthan
42-65231	GONNA MAK'ER	A SQ 51	10/27/1944	497BG	4/18/1945	Rammed at Tachiara Airfield, Kyushu
42-65232		SOLGRNRUD	10/28/1944	462BG	5/5/1945	Group transfer to Tinian
42-65232		TRI U 48	5/5/1945	462BG	5/10/1949	Reclaimed Tinker
42-65232		T SQ	10/28/1944	498BG	5/10/1949	Reclaimed Tinker
42-65233	SUPERSTITIOUS ALOYSIOUS	FOUREDSTRP W		40BG	5/5/1945	Group transfer to Tinian
42-65233	SUPERSTITIOUS ALOYSIOUS	TRI S 11	5/5/1945	40BG		Reclaimed
42-65233	TABOOMA III	TRI S 11	5/5/1945	40BG		Reclaimed
42-65234			10/18/1944	97BG	9/15/1954	Yokota
42-65235			10/18/1944	504BG	5/10/1954	Reclaimed Davis-Monthan
42-65236						Reclaimed
42-65237						Reclaimed
42-65238						Reclaimed
42-65239			10/26/1945	3502BU	5/10/1954	Reclaimed Davis-Monthan
42-65240						Reclaimed
42-65241	LIFE OF RILEY	E TRI	11/14/1944	504BG	3/25/1945	Ditched from Nagoya
42-65242		E TRI	11/15/1944	504BG	3/17/1945	Flak at kobe — engine fire
42-65243	STEVEADORABLE	K TRI	11/25/1944	505BG	5/10/1954	Reclaimed Davis-Monthan

Serial #	Name	Identification	Delv	Assign	Off Inv	Circumstances
42-65244		E TRI	11/18/1944	504BG	5/10/1949	Reclaimed Tinker
42-65245		Z SQ		500BG		Reclaimed
42-65245		V SQ 51		499BG		Unknown
42-65246	IRISH LASSIE	A SQ 52 (1)	11/27/1944	497BG	1/27/1945	Surveyed after crash landing Saipan
42-65247	MILLION DOLLAR BABY	Z SQ 48	11/27/1944	500BG	10/4/1954	Reclaimed Tinker
42-65248		T SQ 36 (1)	11/27/1944	498BG	6/6/1945	Surveyed Iwo Jima — flak damge 6/5/45
42-65249	ANN DEE	Z SQ 3 (2)	11/27/1944	500BG	9/11/1950	Reclaimed Tinker
42-65250		K TRI	11/30/1944	505BG	11/21/1947	Reclaimed Smoky Hill
42-65251	20TH CENTURY SWEETHEART	Z SQ 52 (1)	11/27/1944	500BG	5/10/1954	Reclaimed Davis-Monthan
42-65252		SOLYLWRUD N2	11/30/1944	462BG	5/5/1945	Group transfer to Tinian
42-65252		TRI U 22	5/5/1945	462BG	6/5/1945	Crashed from Kobe
42-65253	MARY ANNA	K TRI 6 (1)	11/30/1944	505BG	5/7/1945	Ditched from Oita
42-65254		SOLREDRUD K1	11/30/1944	462BG	1/6/1945	Ditched from Omura
42-65255	JOOK GIRL	K TRI	11/29/1944	505BG	2/10/1945	Crashed takeoff for Ota
42-65256		V SQ		499BG		Unknown
42-65257		K TRI	11/30/1944	505BG	7/14/1954	Salvaged at Davis-Monthan
42-65258			11/13/1944	444BG	8/19/1954	Reclaimed Pyote
42-65259			11/11/1944	4141BU	8/19/1954	Reclaimed Pyote
42-65260			11/14/1944	4105BU	5/10/1954	Reclaimed Davis-Monthan
42-65261			11/14/1944	245BU	8/19/1954	Reclaimed Pyote
42-65262			11/14/1944	4117BU	5/7/1954	Reclaimed Robins
42-65263			11/15/1944	4141BU	9/11/1950	Reclaimed Pyote
42-65264						Unknown
42-65265		K TRI	12/12/1944	505BG	4/24/1946	Circumstances Unknown
42-65266	DOC SAID ALL I NEEDED —	E TRI	12/13/1944	504BG	5/23/1945	MIA Morzuri Harbor
42-65266	THE GAMECOCK	E TRI	12/13/1944	504BG	5/24/1945	Shot down Tokyo
42-65267		FOURBLUSTRP A	12/18/1944	40BG	1/22/1945	Ditched returning from Photo mission
42-65268	AIRBORN	DIAM #	12/15/1944	444BG	5/10/1945	Group transfer to Tinian
42-65268	AIRBORN	TRI N 60 (1)	5/10/1945	444BG	9/11/1950	Reclaimed Pyote
42-65269	WINGED VICTORY II	FOURBLUSTRP E	12/18/1944	40BG	5/5/1945	Group transfer to Tinian
42-65269	WINGED VICTORY II	TRI S 17 (1)	5/5/1945	40BG	5/26/1945	Crashed in Tokyo
42-65270	BIG POISON — 2ND DOSE	TRI N 30 (1)	12/15/1944	444BG	6/1/1945	Collided, exploded at Assembly Point
42-65271		FOURYLWSTRP E	12/18/1944	40BG	5/5/1945	Group transfer to Tinian
42-65271		TRI S 39 (1)	5/5/1945	40BG	5/19/1945	Crew bailed at Iwo Jima
42-65272	MISS SHORTY	2BLURUDSTRP	12/19/1944	468BG	5/11/1945	Group transfer to Tinian
42-65272	LUCKY LADY	TRI I 27	12/19/1944	468BG	8/16/1954	Reclaimed Davis-Monthan
42-65272	MISS SHORTY	TRI I 27	5/11/1945	468BG	8/16/1954	Reclaimed Davis-Monthan
42-65272	MISS SHORTY	K TRI	12/19/1944	505BG	8/16/1954	Recd from 468BG. Reclaimed Davis-Monthan
42-65272	BLUETAILFLY	BLKSTRPBLU	12/19/1944	19BG	8/16/1954	Reclaimed Davis-Monthan
42-65273	BIG! AIN'T IT	DIAM #	12/19/1944	444BG	5/10/1945	Group transfer to Tinian
42-65273	BIG! AIN'T IT	TRI N 62 (1)	5/10/1945	444BG	5/25/1945	MIA Tokyo
42-65274	BAD PENNY	TRI S 31	12/18/1944	40BG	9/14/1953	Reclaimed Davis-Monthan
42-65274			12/18/1944	RAF	5/27/1950	Transferred to RAF
42-65274			5/27/1950	USAF	7/7/1953	Returned to USAF
42-65274			7/7/1953		9/14/1953	Reclaimed Davis-Monthan
42-65274	BAD PENNY	FOURYLWSTRP G	12/18/1944	40BG	5/27/1950	Transferred to RAF
42-65275	HI STEPPER	2YLWRUDSTRP	12/19/1944	468BG	5/11/1945	Group transfer to Tinian
42-65275	HI STEPPER	TRI I 48	5/11/1945	468BG	7/15/1954	Reclaimed McClellan
42-65275	HI STEPPER	CIRCLE E	12/19/1944	22BG	7/14/1954	Recd from 468BG. Reclaimed McClellan
42-65276	RAIDEN MAIDEN II	2BLURUDSTRP		468BG	5/11/1945	Group transfer to Tinian
42-65276	RAIDEN MAIDEN II	TRI I 23	5/11/1945	468BG	12/21/1949	Reclaimed
42-65277	DRAGON LADY	DIAM #	12/20/1944	444BG	5/10/1945	Group transfer to Tinian
42-65277	DRAGON LADY	TRI N 20 (1)	5/10/1945	444BG	5/29/1945	Ditched from Tokyo
42-65278	T-N-TEENY	X TRI 5	12/23/1944	9BG	5/16/1945	Surveyed Tinian — battle damage
42-65279	MY BUDDY	2YLWRUDSTRP	12/20/1944	468BG	5/11/1945	Group transfer to Tinian
42-65279	MY BUDDY	TRI I 49	5/11/1945	468BG	7/14/1954	Reclaimed Davis-Monthan
42-65280	DINA MIGHT	E TRI 29	12/22/1944	504BG	5/23/1945	Crashed Iwo from Tokyo
42-65281	MISS AMERICA '62	L TRI 11	12/22/1944	6BG		Reclaimed Tinker post 10/54
42-65282		A SQ 2 (2)	12/23/1944	497BG	5/10/1954	Reclaimed Davis-Monthan
42-65283	THE BIG WHEEL	X TRI 33	12/22/1944	9BG	3/31/1945	Crashed on approach from aborted mission
42-65284	DESTINY'S TOT	X TRI 34	12/22/1944	9BG	9/14/1954	Reclaimed Hill
42-65285	SAD TOMATO	X TRI 22	12/22/1944	9BG	5/10/1954	Reclaimed Davis-Monthan
42-65286	DINAH MIGHT	X TRI 9	12/22/1944	9BG	5/26/1947	Reclaimed Aberdeen Proving Grounds
42-65287			12/22/1944	246BU	5/31/1945	Ditched off Borinquen, PR
42-65288			12/27/1944	246BU	8/19/1954	Reclaimed Pyote

I. Master List   35

Serial #	Name	Identification	Delv	Assign	Off Inv	Circumstances
42-65289			12/28/1944	301BG	5/31/1950	Reclaimed March
42-65290			12/28/1944	246BU	12/4/1953	Reclaimed Robins
42-65291			12/27/1944	4105BU	5/5/1954	Reclaimed Davis-Monthan
42-65292	BAD BREW TOO	A SQ 6 (2)	12/29/1944	497BG	9/11/1950	Reclaimed Pyote
42-65293	DESTINY'S TOTS	A SQ 49 (2)	12/29/1944	497BG	11/17/1953	Reclaimed Hill
42-65294			12/29/1944	248BU	5/6/1945	Surveyed at Walker
42-65295		T SQ 12 (3)	12/28/1944	498BG	4/29/1945	Crashed target due to air-to-air bomb
42-65296	THE ANCIENT MARINER	Z SQ 53	12/29/1944	500BG	5/10/1954	Reclaimed Davis-Monthan
42-65297		BLK SQ M	1/16/1945	19BG	12/4/1953	Reclaimed Robins
42-65298		X TRI 44 (1)	1/17/1945	9BG	3/10/1945	Ditched from Tokyo
42-65299		SOLYLWRUD		462BG	5/5/1945	Group transfer to Tinian
42-65299		TRI U 31	5/5/1945	462BG		Reclaimed
42-65299		SQUARE YLW		2BG		Reclaimed
42-65300	KATIE ANN	BLK SQ M 22 (1)	1/15/1945	19BG	4/13/1945	Ditched off Guam
42-65301		BLK SQ O	1/11/1945	29BG	3/9/1945	Flak shoot down
42-65302	CITY OF LOS ANGELES	BLK SQ O 37 (1)	1/11/1945	29BG	11/17/1953	Reclaimed Hill
42-65302	SNATCH BLACH	BLK SQ O 37 (1)	1/11/1945	29BG	11/17/1953	Reclaimed Hill
42-65303	CITY OF RICHMOND	BLK SQ M 43 (1)	1/15/1945	19BG	5/31/1945	Salvaged
42-65303	CITY OF SANTA MONICA	BLK SQ M 24 (1)	1/12/1945	19BG	5/31/1945	Salvaged
42-65304	CITY OF BURLINGTON	BLK SQ M 27 (1)	1/12/1945	19BG	5/27/1945	Crash landed on test flight
42-65305		BLK SQ O (1)	1/19/1945	29BG	5/5/1945	Abandoned Kyushu, fire #1 engine & tank
42-65306	THE OUTLAW	BLKSTRPGRN		19BG	10/2/1951	Crashed on takeoff— test flight
42-65306		BLK SQ O (1)		29BG		Survived war
42-65307	CITY OF OGDEN	BLK SQ M 21 (1)	1/14/1945	19BG	11/17/1953	Reclaimed Hill
42-65307	MISS OGDEN	BLK SQ M 21 (1)	1/14/1945	19BG	11/17/1953	Reclaimed Hill
42-65307	SALT CENSORED RESISTOR	BLK SQ M 21 (1)	1/14/1945	19BG	11/17/1953	Reclaimed Hill
42-65308	CITY OF CLIFTON	BLK SQ M 45 (1)	1/23/1945	19BG	11/17/1953	Reclaimed Hill
42-65309	CITY OF RICHMOND [5]	BLK SQ M-43 (1)	1/15/1945	19BG	5/31/1945	Salvaged
42-65309	[5]	A SQ	1/15/1945	497BG		Possible transfer to 19BG prior to 3/9/45
42-65310		BLK SQ M	1/17/1945	19BG	3/9/1945	Crashed Fubo Mountain per Japanese
42-65311		BLK SQ O	1/16/1945	29BG	3/9/1945	MIA Tokyo
42-65312	NAMED, NAME UNKNOWN	BLK SQ O (1)	1/29/1945	29BG	4/13/1945	Crashed Chiba from Tokyo
42-65313			1/24/1945	248BU	5/5/1954	Reclaimed Davis-Monthan
42-65315	LIMBER DUGAN II	2WHTRUDSTRP	1/25/1945	468BG	5/11/1945	Group transfer to Tinian
42-65315	LIMBER DUGAN II	TRI I 8	5/11/1945	468BG	5/10/1954	Reclaimed Davis-Monthan
42-65316			1/27/1945	248BU	5/5/1954	Reclaimed Davis-Monthan
42-65317			1/24/1945	248BU	7/25/1945	Reclaimed Tinker
42-65318			1/24/1945	243BU	11/12/1953	Reclaimed Robins
42-65319						Unknown
42-65320			1/25/1945	246BU	8/19/1954	Reclaimed Pyote
42-65321			1/30/1945	4141BU	8/19/1954	Reclaimed Pyote
42-65322			1/26/1945	236BU	12/21/1949	Reclaimed Pyote
42-65323			1/31/1945	4141BU	8/19/1954	Reclaimed Pyote
42-65324			1/27/1945	236BU	8/19/1954	Reclaimed Pyote
42-65325			1/25/1945	236BU	8/19/1954	Reclaimed Pyote
42-65326			1/25/1945	236BU	8/19/1954	Reclaimed Pyote
42-65327	PRINCESS EILEEN IV	DIAM #	1/29/1945	444BG	5/10/1945	Group transfer to Tinian
42-65327	PRINCESS EILEEN IV	TRI N 39	5/10/1945	444BG	5/26/1945	Exploded over Tokyo
42-65328	EIGHT BALL CHARLIE	FOURBLUSTRP W	1/26/1945	40BG	5/5/1945	Group transfer to Tinian
42-65328	EIGHT BALL CHARLIE	TRI S 16	5/5/1945	40BG	7/14/1954	Reclaimed Davis-Monthan
42-65329		SOLGRNRUD		462BG	5/5/1945	Group transfer to Tinian
42-65329		TRI U 50	5/5/1945	462BG		Reclaimed
42-65330		V SQ 14	1/27/1945	499BG	5/10/1954	Reclaimed Davis-Monthan
42-65331	ARKANSAS TRAVELLER	T SQ 11 (2)	1/30/1945	498BG	11/17/1953	Reclaimed Hill
42-65331	ARKANSAS TRAVELLER	T SQ 17	1/30/1945	498BG	11/17/1953	Reclaimed Hill
42-65332	THE HEAT'S ON	T SQ 02 (3)		498BG		Unknown
42-65333		V SQ	1/28/1945	499BG	9/14/1954	Reclaimed Hamilton
42-65333		BLKSTRP	1/28/1945	19BG	9/14/1954	Reclaimed Hamilton
42-65334		K TRI 7 (2)	1/30/1945	505BG	6/3/1945	Crashed Iwo, reason unknown
42-65335	BETTY BEE	V SQ 37	1/28/1945	499BG	11/17/1953	Reclaimed Hill
42-65336	ASSID TEST II	SOLGRNRUD	10/17/1944	462BG	5/5/1945	Group transfer to Tinian
42-65336	ASSID TEST II (6)	TRI U 49	5/5/1945	462BG	6/5/1945	Flak — crashed Iseda-cho, Uji City, Kyoto
42-65337	ELEANOR	DIAM# 29	1/30/1945	444BG	5/26/1945	Surveyed Iwo after crash landing
42-65337	JO	DIAM #	1/30/1945	444BG	5/10/1945	Group transfer to Tinian
42-65337	JO	TRI N 21 (1)	5/10/1945	444BG	5/26/1945	Surveyed Iwo after crash landing

Serial #	Name	Identification	Delv	Assign	Off Inv	Circumstances
42-65338	RED HOT RIDER	A SQ	2/6/1945	497BG	5/10/1954	Reclaimed Davis-Monthan
42-65339		L TRI	2/9/1945	6BG	5/10/1954	Reclaimed Davis-Monthan
42-65340		V SQ 15	2/8/1945	499BG	11/17/1953	Reclaimed Hill
42-65341		Z SQ 25	2/9/1945	500BG	5/10/1954	Reclaimed Davis-Monthan
42-65342		BLK SQ M 31 (1)	2/10/1945	19BG	5/16/1945	Crashed at Iwo from Nagoya
42-65343		A SQ 44 (2)		497BG	9/5/1945	Circumstances Unknown
42-65344	BALL OF FIRE	V SQ 52	2/10/1945	499BG	4/14/1945	Shot down Tokyo Arsenal
42-65345		T SQ 21(2)	2/14/1945	498BG	9/4/1945	Missing-circumstances unknown
42-65346		Z SQ 14	2/13/1945	500BG	7/14/1953	Reclaimed Robins
42-65347		L TRI	2/10/1945	6BG	4/8/1945	Exploded 2 miles off Tinian
42-65348		A SQ 16	2/12/1945	497BG	6/1/1945	Shot down, crashed Omine Mt, Nara
42-65349	PRETTY BABY	CIRC E 15	2/12/1945	504BG	5/5/1954	Reclaimed Tinker
42-65350		BLK SQ O (1)	2/12/1945	29BG	4/7/1945	Rammed at Nagoya
42-65351			2/13/1945	4105BU	5/5/1954	Reclaimed Davis-Monthan
42-65352	HOT TO GO	BLKSTRPGRN	2/13/1945	19BG	10/1/1954	Reclaimed Davis-Monthan
42-65352	HOT TO GO	SQUARE H	2/13/1945	98BG	10/1/1954	Reclaimed Davis-Monthan
42-65352		TRI U	2/13/1945	462BG		Survived war
42-65353		SQUARE H	2/15/1945	98BG	10/5/1950	Engine fire, 10 mi W of Takamatsu
42-65354			2/17/1945	233BU	10/27/1950	Reclaimed Kelly
42-65355		SQUARE H	2/19/1945	98BG	5/15/1950	Salvaged at Spokane
42-65356			2/20/1945	4141BU	8/19/1954	Reclaimed Pyote
42-65357	SHADY LADY	BLKSTRP	2/20/1945	19BG	1/31/1953	Surveyed at Kadena
42-65357	SHADY LADY	SQUARE H	2/20/1945	98BG		Transfer to 19BG
42-65358			2/17/1945	234BU	8/19/1954	Reclaimed Pyote
42-65359			2/20/1945	236BU	8/19/1954	Reclaimed Pyote
42-65360			2/20/1945	234BU	8/19/1954	Reclaimed Pyote
42-65361	CITY OF JEWETT CITY	BLK SQ P 47	2/17/1945	39BG		Survived war
42-65361	THE KICK-A-POO-JOY II	BLK SQ P 47	2/17/1945	39BG		Survived War
42-65361	PURPLE SHAFT	BLKSTRPBLU	2/17/1945	19BG	5/16/1954	Salvaged at Davis-Monthan
42-65362	TIAN LONG (SKY DRAGON)	BLK SQ P 52 (1)	2/23/1945	39BG	5/23/1945	Crashed near Tokyo
42-65363	CITY OF AKRON	BLK SQ K 56	2/25/1945	330BG	11/17/1953	Reclaimed Hill
42-65363	LADY JANE	BLK SQ K 56	2/25/1945	330BG	11/17/1953	Reclaimed Hill
42-65364	CITY OF ROSWELL	BLK SQ P 30 (1)	2/27/1945	39BG	6/1/1945	Lost prop-abandoned Sofu-Gan Island
42-65364	SKYSCRAPPER	BLK SQ P 30 (1)	2/27/1945	39BG	6/1/1945	Lost prop-abandoned Sofu-Gan Island
42-65365	CITY OF SANTA FE	BLK SQ P 41		39BG		Reclaimed
42-65365	CITY OF WILMINGTON	BLK SQ P 41		39BG		Reclaimed
42-65365	HELL'S BELLE	BLK SQ P 41		39BG		Reclaimed
42-65365	ONE WEAKNESS	BLK SQ P 41		39BG		Unknown
42-65365	CITY OF LAS VEGAS	BLK SQ P 55		39BG		Unknown
42-65365	ROUND TRIP TICKET	BLK SQ P 55		39BG		Unknown
42-65366		BLK SQ P	2/21/1945	39BG	10/27/1945	Crashed, engine fire
42-65367	BLACKJACK	BLK SQ P 02	2/26/1945	39BG		Renamed
42-65367	BATTLIN BITCH II	BLK SQ P 02		39BG	6/00/45	Scrapped Iwo Jima
42-65367	CITY OF MIAMI	BLK SQ P 02		39BG	6/00/45	Scrapped Iwo Jima
42-65368		BLK SQ P 04 (1)	2/26/1945	39BG	4/10/1945	Ditched training mission — lost prop
42-65369	BLACK SHEEP	BLKSTRPRED		19BG	4/12/1951	Crashed, burned off runway at Kadena
42-65369	CITY OF JACKSON	BLK SQ P 34		39BG		Survived war
42-65370	CITY OF MIAMI BEACH (2)	BLK SQ K 55	2/21/1945	330BG		Survived war
42-65370	OLE SMOKER II	BLK SQ K 55	2/21/1945	330BG		Survived war
42-65370		BLKSTRPBLU	2/21/1945	19BG	7/14/1954	Salvaged at Davis-Monthan
42-65371	CITY OF OMAHA	BLK SQ K 28	2/27/1945	330BG	5/10/1954	Reclaimed Davis-Monthan
42-65371	YONKEE DOLL-AH	BLK SQ K 28	2/27/1945	330BG	5/10/1954	Reclaimed Davis-Monthan
42-65372			2/24/1945	235BU	11/12/1953	Reclaimed Robins
42-65373		BLK SQ P 42 (1)	2/27/1945	39BG	6/19/1945	Collision with 499BG aircraft
42-65374			2/27/1945	235BU	5/10/1954	Reclaimed Davis-Monthan
42-65375			2/26/1945	247BU	4/1/1947	Reclaimed Goodfellow
42-65376			2/27/1945	231BU	8/19/1954	Reclaimed Pyote
42-65377			2/26/1945	236BU	8/5/1945	Surveyed at Pyote
42-65378			2/27/1945	236BU	8/19/1954	Reclaimed Pyote
42-65379			2/27/1945	235BU	8/19/1954	Reclaimed Pyote
42-65380			2/28/1945	235BU	5/5/1954	Reclaimed Davis-Monthan
42-65381		L TRI	2/27/1945	6BG	5/5/1954	Reclaimed Davis-Monthan
42-65382			2/28/1945	233BU	5/10/1945	Surveyed at Davis-Monthan
42-65383						Reclaimed
42-65384			2/15/1945	428BU	7/15/1954	Salvaged at Davis-Monthan

I. *Master List*   37

Serial #	Name	Identification	Delv	Assign	Off Inv	Circumstances
42-65385			2/15/1945	237BU	3/31/1947	Salvaged at Kirtland
42-65386			2/15/1945	428BU	10/00/54	Reclaimed post 1954
42-65387			2/15/1945	237BU	3/12/1946	Salvaged at Kirtland
42-65388			3/5/1945	233BU	8/19/1954	Reclaimed Pyote
42-65389		TRIANGLE O	3/8/1945	97BG	7/27/1954	Reclaimed Dow
42-65390	LAGGIN WAGON	SQUARE Y	3/8/1945	307BG	11/16/1954	Reclaimed Davis-Monthan
42-65391		SQUARE A	3/9/1945	301BG	7/14/1954	Salvaged at Davis-Monthan
42-65392		SQUARE Y		307BG		Reclaimed
42-65393			3/9/1945	4105BU	5/10/1954	Reclaimed Davis-Monthan
42-65394						Reclaimed
42-65395			3/9/1945	326BU	11/12/1953	Reclaimed Robins
42-65396			3/10/1945	4141BU	8/19/1954	Reclaimed Pyote
42-65397			3/27/1945	326BU	11/12/1953	Reclaimed Robins
42-65398			3/13/1945	331BU	4/20/1945	Surveyed Barksdale
42-65399			3/14/1945	4141BU	8/19/1954	Reclaimed Pyote
42-65400			3/13/1945	4141BU	8/19/1954	Reclaimed Pyote
42-65401						Reclaimed
42-65402					9/22/1954	Reclaimed Brookley
42-93824			2/8/1944	243BU	9/13/1944	Surveyed at Clovis
42-93825		SOLREDRUD	2/16/1944	462BG	8/31/1944	Surveyed CBI, Hump crash in China
42-93826		2BLURUDSTRP	2/29/1944	468BG	6/15/1944	Crashed into cliff from Yawata
42-93827		SOLREDRUD	3/4/1944	462BG	11/2/1945	Reclaimed Randolph
42-93828	MONSOON GOON	2YLWRUDSTRP	3/21/1944	468BG	11/30/1949	Reclaimed Pyote
42-93829	CAIT PAORNAT	FOURBLKSTRP	3/24/1944	40BG	8/20/1944	Interned Khabarovsk from Yawata
42-93830	TORRID TOBY	SOLREDRUD	3/30/1944	462BG	3/2/1949	Reclaimed Pyote
42-93831	QUEENIE	FOURYLWSTRP W	4/3/1944	40BG	12/14/1944	Abandoned Rangoon
42-93832			4/15/1944	234BU	11/17/1944	Surveyed at Clovis
42-93833			4/21/1944	242BU	12/21/1949	Reclaimed Pyote
42-93834			4/21/1944	241BU	9/26/1945	Reclaimed Mountain Home
42-93835			5/16/1944	502BG	3/7/1945	Class 26 Grand Isle
42-93836			5/18/1944	231BU	6/27/1949	Reclaimed Davis-Monthan
42-93837	CITY OF MIAMI BEACH	BLK SQ K 59 (1)		330BG	4/12/1945	Ditched on first mission
42-93837	OL' SMOKER	BLK SQ K 59 (1)		330BG	4/12/1945	Ditched on first mission
42-93838			6/5/1944	244BU	4/1/1945	Surveyed at Pyote
42-93839			6/10/1944	248BU	3/21/1945	Surveyed at Clovis
42-93840			6/20/1944	245BU	6/27/1945	Reclaimed Davis-Monthan
42-93841			6/27/1944	236BU	12/21/1949	Reclaimed Pyote
42-93842			6/28/1944	6BG	12/7/1944	Surveyed Grand Isle
42-93843			7/8/1944	6BG	10/7/1944	Crashed, engine fire
42-93844			7/17/1944	242BU	6/27/1949	Reclaimed Davis-Monthan
42-93845			7/22/1944	4101BU	5/17/1949	Reclaimed Chanute
42-93846	SPIRIT OF FDR	K TRI 4	7/20/1944	505BG	11/10/1944	Surveyed McCook per ARC
42-93847			7/20/1944	245BU	5/10/1949	Reclaimed Robins
42-93848		SOLGRNRUD	7/29/1944	462BG	11/21/1944	Abandoned, combat damage
42-93849	UNDER-EXPOSED		8/1/1944		4/25/1949	Reclaimed Robins
42-93850		F/V60	8/3/1944	499BG	5/10/1949	Reclaimed Robins
42-93851		F	8/7/1944	6CMG	5/10/1949	Reclaimed Robins
42-93852	TOKYO ROSE	T SQ 12 (4)	8/11/1944	498BG	6/27/1949	Reclaimed Davis-Monthan
42-93853	QUAN YIN CHA ARA					Unknown
42-93854	BROOKLYN BESSIE	F	8/15/1944	40BG	2/4/1945	On fire, crew abandoned North China
42-93855	DOUBLE EXPOSURE	F	8/16/1944	6CMG	5/24/1945	Salvaged overseas
42-93855	DOUBLE EXPOSURE	F	8/16/1944	444BG	5/25/1945	Salvaged overseas
42-93856	WEE MISS AMERICA	F			11/3/1944	Missing from Nagoya
42-93857	HELLON WINGS	DIAM #		444BG	5/10/1945	Group transfer to Tinian
42-93857	HELLON WINGS	TRI N 58	5/10/1945	444BG		Unknown
42-93858	STAR DUSTER	A SQ 30		497BG		Reclaimed
42-93859	SHAG'N HOME	FOURYLWSTRP N		40BG	5/5/1945	Group transfer to Tinian
42-93859	SHAG'N HOME	TRI S 35	5/5/1945	40BG		Reclaimed
42-93860			9/7/1944	BOEING	8/19/1944	Crashed Seattle
42-93861			8/26/1944	4141BU	9/11/1950	Reclaimed Pyote
42-93862			8/30/1944	331BG	4/10/1947	Surveyed McCook
42-93863		F	9/1/1944	6CMG	2/14/1945	Missing on Sea Search
42-93864	SHUTTERBUG	SOLGRNRUD		462BG		Reclaimed
42-93864	SHUTTERBUG			6CMG		Reclaimed
42-93865	SNOOPIN' KID	F		462BG		Reclaimed

Serial #	Name	Identification	Delv	Assign	Off Inv	Circumstances
42-93865	SNOOPIN' KID	SOLGRNRUD		462BG		Reclaimed
42-93866		F	9/13/1944	6CMG	12/7/1944	Destroyed on ground in enemy raid
42-93867		F	9/9/1944	6CMG	12/9/1944	Ditched
42-93868						Unknown
42-93869		F	9/16/1944	6CMG	5/2/1945	Crashed, surveyed
42-93870	VALIANT LADY	F		6CMG		Reclaimed
42-93871		DIAM #	9/21/1944	444BG	1/7/1945	Crashed on takeoff
42-93872		F		6CMG		Reclaimed
42-93873		SOLYLWRUD		462BG	5/5/1945	Group transfer to Tinian
42-93873		TRI U 51	5/5/1945	462BG		Reclaimed
42-93874	SOUTH SEA SINNER	BLKSTRPGRN		19BG		Unknown
42-93875	CORAL QUEEN	CIRC E	9/27/1944	504BG	7/22/1949	Salvaged at Keesler
42-93875	GEORGIA ANN	Z SQ 22 (2)	9/27/1944	500BG	7/22/1949	Salvaged at Keesler
42-93876		Z SQ 13 (1)	9/27/1944	500BG	2/10/1945	Ditched from Ota
42-93877	SNOOKY'S BRATS	2WHTRUDSTRP		468BG	5/5/1945	Group transfer to Tinian
42-93877	SNOOKY'S BRATS	TRI I 1	5/5/1945	468BG		Reclaimed
42-93878	HELLES BELLES	K TRI 11	9/30/1944	505BG	1/00/45	Transfer to 500BG
42-93878		Z SQ 54	1/00/45	500BG	7/14/1954	Salvaged at Davis-Monthan
42-93879						Unknown
42-93880	THE DUCHESS ALMOST READY	SQUARE H	10/5/1944	98BG	9/27/1954	Reclaimed McClellan
42-93880			10/5/1944	5SRC	9/27/1954	Reclaimed McClellan
42-93881		T SQ 10 (5)	10/23/1944	498BG	4/6/1945	Surveyed; crashed returning on two eng
42-93882		K TRI	10/9/1944	505BG	5/13/1954	Reclaimed Randolph
42-93883		A SQ	10/9/1944	497BG	4/1/1945	MIA Tokyo
42-93884	URGIN' VIRGIN II	DIAM #	10/21/1944	444BG	5/10/1945	Group transfer to Tinian
42-93884	URGIN' VIRGIN II	TRI N 12	5/10/1945	444BG	5/29/1949	Reclaimed Keesler
42-93885	SHRIMPER, THE	A SQ 33	10/21/1944	497BG	5/14/1945	Ditched from Nagoya
42-93886	KRISTY ANN	X TRI 6 (1)	10/27/1944	9BG	9/11/1950	Reclaimed Tinker
42-93887		L TRI 63	10/30/1944	6BG	8/24/1949	Reclaimed Keesler
42-93888	MARK OF ZORRO	CIRC E 10	10/29/1944	504BG	5/10/1949	Reclaimed Tinker
42-93888	TAMERLANE	CIRC E 10	10/29/1944	504BG	5/10/1949	Reclaimed Tinker
42-93888		X TRI 21 (1)	10/29/1944	9BG	5/10/1949	Reclaimed Tinker
42-93888		X TRI	10/29/1944	9BG	5/10/1949	Reclaimed Tinker
42-93889	FRISCO NANNY	Z SQ 34		500BG		Reclaimed
42-93889				97BG		Unknown
42-93889		SPEEDWING BLU		307BG		Unknown
42-93890		CIRC W 4 (2)	11/1/1944	505BG	8/8/1945	Crashed into sea on T/O for Yawata
42-93891			11/1/1944	4141BU	9/11/1950	Reclaimed Pyote
42-93892	DRAGON LADY	X TRI 45	11/3/1944	9BG	9/11/1950	Reclaimed Tinker
42-93893		X TRI 12 (1)	11/4/1944	9BG	4/15/1945	Crashed in sea near Yokosuka
42-93894	LASSY TOO	K TRI 41	11/7/1944	505BG	6/23/1945	MIA Fukuoka
42-93894	TIMES-A-WASTIN'	DIAM #	11/7/1944	444BG	6/23/1945	MIA Fukuoka
42-93895			11/7/1944	234BU	8/18/1945	Surveyed at Clovis
42-93896	B. A. BIRD	X TRI 14	11/7/1944	9BG	9/11/1950	Reclaimed Pyote
42-93896	BIG GASS BIRD	SQUARE H	11/7/1944	98BG		Unknown
42-93896	BIG GASS BIRD	CIRC R 14	11/7/1944	9BG		Transfer to 98BG
42-93897		V SQ	11/7/1944	499BG	12/6/1950	Reclaimed McClellan
42-93898		L TRI 31	11/9/1944	6BG	1/4/1951	Reclaimed Hill
42-93899			11/9/1944	233BU	9/11/1945	Reclaimed Pyote
42-93900		BLK SQ O		29BG		Reclaimed
42-93901	BIG FAT MAMA	L TRI 34	11/11/1944	6BG	9/11/1945	Reclaimed Tinker
42-93901	PATRICIA LYNN	L TRI 34	11/11/1944	6BG	9/11/1945	Reclaimed Tinker
42-93902		L TRI	11/13/1944	6BG	3/10/1945	Surveyed Tinian
42-93903			11/14/1944	19BG	8/11/1950	Salvaged at Kadena
42-93903			11/14/1944	31SRP	8/11/1950	Salvaged at Kadena
42-93904			12/8/1944	231BU	9/11/1950	Reclaimed Pyote
42-93905		BLK SQ O (1)	12/7/1944	29BG	3/10/1945	MIA Tokyo
42-93906	UNCLE SAM'S MILK WAGON	L TRI 10	11/21/1944	6BG	5/23/1945	Crashed Iwo Jima
42-93907		CIRC X	11/21/1944	505BG	9/10/1945	Condemned Tinian
42-93908	BATTLIN BULLDOZER	BLK SQ K 31	11/22/1944	330BG	9/11/1950	Reclaimed Pyote
42-93908	CITY OF CEDAR RAPIDS	BLK SQ K 31	11/22/1944	330BG	9/11/1950	Reclaimed Pyote
42-93909	MISSION TO ALBUQUERQUE II	K TRI	11/24/1944	505BG	6/17/1949	Salvaged Victorville
42-93910		BLK SQ K	11/24/1944	330BG	5/10/1949	Reclaimed Chanute
42-93911	358TH AIR SERVICE GROUP	L TRI 28	11/24/1944	6BG	3/13/1950	Reclaimed Keesler
42-93911	TRIGGER MORTIS	L TRI 28	11/24/1944	6BG	3/13/1950	Reclaimed Keesler

I. Master List

Serial #	Name	Identification	Delv	Assign	Off Inv	Circumstances
42-93912		K TRI	11/26/1944	505BG	5/00/45	Returned to Guam — unit unspecified
42-93912	CITY OF GLENDALE [7]	BLK SQ K 62	11/26/1944	330BG	8/31/1945	Crashed Harmon Field, Guam
42-93912	MOTLEY CREW [7]	BLK SQ K 62	11/26/1944	330BG	8/31/1945	Crashed Harmon Field, Guam
42-93912	[7]	BLK SQ O	11/26/1944	29BG	8/31/1945	Crashed Harmon Field, Guam
42-93913	CITY OF DENVER	BLK SQ O 03		29BG	9/11/1950	Reclaimed Pyote
42-93913		BLK SQ M 03 (1)	11/28/1944	19BG	9/11/1950	Reclaimed Pyote
42-93914						Unknown
42-93915		X TRI 40		9BG		Reclaimed Robins
42-93916		BLK SQ O	11/28/1944	29BG	9/11/1950	Reclaimed Pyote
42-93917	BABY GAIL	BLK SQ M 50	11/30/1944	19BG	6/22/1949	Reclaimed McClellan
42-93917	CITY OF MEMPHIS	BLK SQ M 50	11/30/1944	19BG	6/22/1949	Reclaimed McClellan
42-93917	NIP ON ESE NIPPER	BLK SQ M 50	11/30/1944	19BG	6/22/1949	Reclaimed McClellan
42-93917	CHARACTER CARRIAGE	BLK SQ M 50	11/30/1944	19BG	6/22/1949	Reclaimed McClellan
42-93918		BLK SQ O	11/29/1944	29BG	2/2/1950	Reclaimed Keesler
42-93919		BLK SQ M		19BG		Reclaimed, date unknown
42-93920		BLK SQ O		29BG		Reclaimed, date unknown
42-93921			12/4/1944	4000EXG	7/8/1954	Reclaimed Sheppard
42-93922			2/2/1945	326BG	4/25/1945	Surveyed at MacDill
42-93923	CITY OF DETROIT	BLK SQ M 04	12/4/1944	19BG	9/11/1950	Reclaimed Tinker
42-93924			12/14/1944		4/26/1954	Transferred to Navy
42-93925	CITY OF ARCADIA	BLK SQ O 20	12/5/1944	29BG	7/24/1954	Reclaimed Hickam
42-93925	NO BALLS ATOLL	BLK SQ O 20	12/5/1944	29BG	7/24/1954	Reclaimed Hickam
42-93926		X TRI 53	12/6/1944	9BG	6/22/1950	Reclaimed Davis-Monthan
42-93927		BLK SQ O	12/5/1944	29BG	6/12/1945	Structural failure
42-93928	CITY OF HARTFORD	BLK SQ O	12/5/1944	29BG	6/15/1945	Crashed on takeoff
42-93928	CITY OF HARTFORD	BLK SQ O	12/5/1945	29BG	6/15/1945	Crashed on takeoff
42-93929		BLK SQ O	12/4/1944	29BG	8/19/1954	Reclaimed Pyote
42-93930		BLK SQ O	12/10/1944	29BG	5/10/1954	Reclaimed Davis-Monthan
42-93931			12/13/1944	326BU	5/31/1946	Reclaimed Amarillo
42-93932			12/11/1944	4141BU	8/19/1954	Reclaimed Davis-Monthan
42-93933			12/13/1944	234BU	9/6/1945	Surveyed at Clovis
42-93934			12/14/1944	326BU	9/9/1954	Reclaimed Davis-Monthan
42-93935	CITY OF BEDFORD	BLK SQ K 10	12/21/1944	330BG	8/19/1954	Reclaimed Tinker
42-93935	CITY OF HIGHMAN	BLK SQ K 10	12/21/1944	330BG	8/19/1954	Reclaimed Tinker
42-93935	SHILLELAGH	BLK SQ K 10	12/21/1944	330BG	8/19/1954	Reclaimed Tinker
42-93936			12/23/1944	4105BU	5/5/1945	Reclaimed Davis-Monthan
42-93937	CITY OF MIAMI BEACH [10]	BLK SQ K 59 (2)		330BG	4/12/1945	Ditched
42-93937			12/21/1944	247BU	8/19/1954	Reclaimed Pyote
42-93938			12/21/1944	4141BU	8/19/1954	Reclaimed Pyote
42-93939	TAKE IT OFF	L TRI	12/23/1944	6BG	7/9/1945	Crashed in sea off Moji City
42-93940			12/23/1944	4121BU	10/29/1950	Reclaimed Kelly
42-93941	ZZZUNNAMED	V SQ	12/23/1944	499BG	12/4/1953	Reclaimed Robins
42-93942			12/23/1944	233BU	8/19/1954	Reclaimed Pyote
42-93943	BEATS ME	BLK SQ K 30	1/2/1945	330BG	8/19/1954	Reclaimed
42-93943	CITY OF PORTSMOUTH VA	BLK SQ K 30	1/2/1945	330BG	8/19/1954	Reclaimed
42-93943	CITY OF SPRINGFIELD IL	BLK SQ K 30	1/2/1945	330BG	8/19/1954	Reclaimed
42-93943	JANIE	BLK SQ K 30	1/2/1945	330BG	8/19/1954	Reclaimed
42-93944	MIGHTY FINE	T SQ 7 (3)	12/28/1944	498BG	7/28/1945	Taxi accident 7/8/45 — Surveyed
42-93944			12/31/1944		3/31/1945	Transfer to 498BG
42-93945		T SQ 14 (2)	12/31/1944	498BG	1/5/1950	Reclaimed McClellan
42-93945			12/31/1944		2/26/1945	Transfer to 498BG
42-93946		SPEEDWING BLU		307BG	6/17/1954	Class 26
42-93947		Z SQ 55	12/28/1944	500BG	7/14/1954	Reclaimed McClellan
42-93948			12/31/1944	4121BU	10/29/1950	Reclaimed Kelly
42-93949			12/29/1944	4105BU	5/5/1945	Reclaimed Davis-Monthan
42-93950		BLK SQ P	10/3/1945	39BG	6/28/1950	Reclaimed Tinker
42-93951			12/31/1944	56WRS	9/15/1954	Reclaimed Yokota
42-93951	LUCKY LEVEN	L TRI		6BG		Survived war
42-93952			1/1/1945		11/4/1948	Reclaimed Hickam
42-93953		BLK SQ K	1/1/1945	330BG	5/5/1945	Shot down Shikoku
42-93954		BLK SQ K 41	1/4/1945	330BG		Unknown; not listed MACR Summary
42-93954	BEATS ME TOO	BLK SQ K 34	1/4/1945	330BG	3/31/1950	Reclaimed Seattle
42-93954	CITY OF PORTSMOUTH VA	BLK SQ K 34	1/4/1945	330BG	3/31/1950	Reclaimed Seattle
42-93955	COLLEEN	BLK SQ K 32	1/4/1945	330BG	6/22/1945	Crashed landing at North Field Guam
42-93956	LUCKY LEVEN	X TRI 8	1/1/1945	9BG	5/5/1945	Reclaimed Davis-Monthan

40 THE B-29 SUPERFORTRESS

Serial #	Name	Identification	Delv	Assign	Off Inv	Circumstances
42-93957	CITY OF DULUTH MN	BLK SQ K 05	1/4/1945	330BG	12/4/1953	Reclaimed Robins
42-93957	SHE WOLF	BLK SQ K 05	1/4/1945	330BG	12/4/1953	Reclaimed Robins
42-93957	DON'T WORRY ABOUTA THING	BLK SQ K 05	1/4/1945	330BG		12/4/1953 Reclaimed Robins
42-93958		BLK SQ O (1)	1/3/1945	29BG	11/1/1945	Non-combat loss overseas
42-93959	TUMBLING TUMBLEWEEDS	T SQ 55	1/5/1945	498BG	9/11/1950	Reclaimed Pyote
42-93960			1/6/1945	XXIBC	12/4/1953	Reclaimed Robins
42-93961	CITY OF ABERDEEN	BLK SQ K 60	1/7/1945	330BG	8/19/1954	Reclaimed Tinker
42-93962		X TRI 49	1/7/1945	9BG	4/16/1945	Crashed at Yokohama from Kawasaki
42-93963			1/8/1945	4141BU	8/19/1954	Reclaimed Pyote
42-93964	BABY'S BUGGY	BLK SQ K 52	5/8/1945	330BG	11/17/1953	Reclaimed
42-93964	CITY OF ROCK ISLAND	BLK SQ K 52	5/8/1945	330BG	11/17/1953	Reclaimed
42-93965			1/8/1945	XXIBC	5/11/1945	Shot down Kobe
42-93966	EIGHT BALL	BLK SQ P 08 (1)	1/12/1945	39BG	9/10/1945	Salvaged Guam — cause unknown
42-93967	CITY OF LANDSFORD PA	BLK SQ M	1/11/1945	19BG	12/4/1953	Reclaimed Robins
42-93967	WET BULB WILLY	BLK SQ M	1/11/1945	19BG	12/4/1953	Reclaimed Robins
42-93968		K TRI	1/9/1945	505BG	9/1/1954	Reclaimed Tinker
42-93969		BLK SQ K 01 (1)	1/13/1945	330BG	5/24/1945	Shot down over Tokyo Bay
42-93970	CITY OF WEST PALM BEACH	BLK SQ K 26	1/13/1945	330BG	5/5/1945	Survived war
42-93970		SQUARE Y	1/13/1945	307BG	5/5/1945	Reclaimed Davis-Monthan
42-93971	CITY OF COUNCIL BLUFFS	BLK SQ K 06 (1)	1/14/1945	330BG	11/16/1954	Survived war
42-93971	THE GERM	BLK SQ K 06 (1)	1/14/1945	330BG	11/16/1954	Survived war
42-93971		BLKSTRPGRN	1/14/1945	19BG	11/16/1954	Reclaimed Davis-Monthan
42-93971		SQUARE Y	1/14/1945	307BG	11/16/1954	Reclaimed Davis-Monthan
42-93972			1/12/1945	4105BU	3/12/1954	Transferred to Navy
42-93973			1/16/1945	231BU	10/24/1945	Reclaimed Alamogordo
42-93974	CITY OF EL PASO	BLK SQ P 58		39BG		Survived war
42-93974	RAINBOW'S END	BLK SQ P 58		39BG		Survived war
42-93974	THE UNINVITED	BLK SQ P 58		39BG		Survived war
42-93974		SQUARE H		98BG	11/9/1951	Flak at Chongju — Abandoned
42-93975	CITY OF SANTA FE	BLK SQ P 37	1/16/1945	39BG	8/30/1954	Survived war
42-93975	CITY OF VIRGINIA BEACH	BLK SQ P 28	1/16/1945	39BG	8/30/1954	Survived war
42-93975	FOUR ACES & HER MAJESTY	BLK SQ P 37	1/16/1945	39BG	8/30/1954	Survived war
42-93975	HER MAJESTY	BLK SQ P 37	1/16/1945	39BG	8/30/1954	Survived war
42-93975	PIECE OF MEANNESS	BLK SQ P 28	1/16/1945	39BG	8/30/1954	Survived war
42-93975		CIRCLE E	1/16/1945	22BG	8/30/1945	Survived war
42-93975			1/16/1945	54WRS	8/30/1954	Reclaimed Tinker
42-93976		BLK SQ K 01 (2)	1/18/1945	330BG	9/8/1954	Reclaimed Davis-Monthan
42-93976					4/25/1950	Transferred to RAF
42-93976			4/25/1950	RAF	3/22/1954	Returned to USAF
42-93976			3/22/1954	USAF	9/8/1954	Reclaimed Davis-Monthan
42-93977			1/18/1945	4105BU	5/5/1954	Reclaimed Davis-Monthan
42-93978	CITY OF JACKSONVILLE	BLK SQ K 07		330BG		Reclaimed
42-93978	MISS TAKE	BLK SQ K 07		330BG		Reclaimed
42-93979	SLIM	BLK SQ P 44	1/19/1945	39BG		Reclaimed
42-93979		BLK SQ K	1/19/1945	330BG	8/19/1954	Reclaimed Hill
42-93980	CITY OF ST PETERSBURG	BLK SQ K 08 (1)	1/19/1945	330BG	8/30/1954	Survived war
42-93980	CITY OF WILLIAMSPORT	BLK SQ K 08 (1)	1/19/1945	330BG	8/30/1954	Survived war
42-93980	MY GAL	BLK SQ K 08 (1)	1/19/1945	330BG	8/30/1954	Survived war
42-93980	SS ANNABELLE	BLK SQ K 08 (1)	1/19/1945	330BG	8/30/1954	Survived war
42-93980			8/21/1950	514WRS	8/30/1954	Reclaimed Robins
42-93981			1/19/1945	4105BU	5/5/1954	Reclaimed Davis-Monthan
42-93982	CITY OF FORT WORTH	BLK SQ K 03	1/22/1945	330BG	8/19/1954	Reclaimed Pyote
42-93982	ISLAND QUEEN	BLK SQ K 03	1/22/1945	330BG	8/19/1954	Reclaimed Pyote
42-93983			1/27/1945	4105BU	5/5/1945	Reclaimed Davis-Monthan
42-93984		BLK SQ M 32 (1)	1/22/1945	19BG	12/4/1953	Reclaimed Robins
42-93984	LASSIE TOO!	SOLREDRUD L	1/22/1945	462BG		Transferred to 497BG
42-93985		A SQ 45 (2)		497BG	4/8/1954	Transferred to Navy
42-93986			1/26/1945	4105BU	5/10/1954	Reclaimed Davis-Monthan
42-93987		SQUARE I	1/28/1945	91SRP	11/3/1954	Reclaimed Yokota
42-93988			1/28/1945	233BU	6/24/1945	Surveyed at Davis-Monthan
42-93989	CITY OF ASHEVILLE	BLK SQ M 52	1/28/1945	19BG	8/19/1954	Reclaimed Kelly
42-93989	CITY OF CINCINNATI	BLK SQ M 52	1/28/1945	19BG	8/19/1954	Reclaimed Kelly
42-93990			1/28/1945	4105BU	5/10/1954	Reclaimed Davis-Monthan
42-93991			1/28/1945	326BU	12/4/1953	Reclaimed Robins
42-93992		E TRI		504BG		Reclaimed

I. Master List

Serial #	Name	Identification	Delv	Assign	Off Inv	Circumstances
42-93993		SQUARE I	1/29/1945	91SRP	2/17/1955	Reclaimed Eglin
42-93994			1/28/1945	331BU	7/6/1948	Salvaged at Barksdale
42-93995	BEHRENS BROOD	BLK SQ K 37(1)	1/4/1945	330BG	6/22/1945	Abandoned over Iwo-shot down by P-61
42-93995	CITY OF OSCEOLA	BLK SQ K 37 (1)	1/4/1945	330BG	6/22/1945	Abandoned over Iwo-shot down by P-61
42-93996	CITY OF COLLEGE PARK	BLK SQ M	1/29/1945	19BG	6/7/1945	Abandoned 50 mi from Guam- no fuel
42-93996	CITY OF RICHMOND CA	BLK SQ K 37	1/29/1945	330BG		Transfer to 19BG
42-93996	REBEL'S ROOST	BLK SQ K 37	1/29/1945	330BG		Transfer to 19BG
42-93996	VIVACIOUS LADY	BLK SQ K 59 (2)	1/29/1945	330BG		Transfer to 19BG
42-93997		K TRI 11(3)	1/29/1945	505BG	8/19/1954	Reclaimed Hill
42-93998		BLK SQ O	1/31/1945	29BG	8/19/1954	Reclaimed Pyote
42-93999	FILTHY FAY II	T SQ 26 (2)	1/31/1945	498BG	4/1/1945	Flak damage, hit mountain
42-94000	TIGER LIL	SQUARE I	1/31/1945	91SRP	8/26/1954	Reclaimed Yokota
42-94001		Z SQ 56 (2)	2/2/1945	500BG	8/19/1954	Reclaimed Hill
42-94002	504BG INSIGNIA	CIRC E	2/1/1945	504BG	5/25/1945	Crashed off coast of Chiba Prefecture
42-94002	HAMMER OF THOR	CIRC E	2/1/1945	504BG	5/25/1945	Crashed off coast of Chiba prefecture
42-94003		BLK SQ M 12 (1)	2/3/1945	19BG	7/14/1954	Reclaimed Davis-Monthan
42-94004			2/1/1945	233BU	8/19/1954	Reclaimed Pyote
42-94005		V SQ	2/2/1945	499BG	8/19/1954	Reclaimed Hill
42-94006		V SQ	2/1/1945	499BG	6/17/1954	Class 26 Sheppard
42-94007		X TRI 10 (2)	2/2/1945	9BG	9/22/1947	Reclaimed Clark Field
42-94008		L TRI		6BG		Reclaimed
42-94009		BLK SQ O	2/3/1945	29BG		Survived war
42-94009		BLKSTRPBLU	2/3/1945	19BG	7/14/1954	Reclaimed Davis-Monthan
42-94009		SQUARE Y	2/3/1945	307BG	7/14/1954	Reclaimed Davis-Monthan
42-94010	112 SEABEES	CIRC E 26	2/3/1945	9BG	8/19/1954	Reclaimed Tinker
42-94010	INDIANA	X TRI 44+C1174	2/3/1945	9BG	8/19/1954	Reclaimed Tinker
42-94011			2/3/1945	237BU	5/8/1945	Surveyed at Kirtland
42-94012			2/3/1945	233BU	5/10/1954	Reclaimed Davis-Monthan
42-94013		BLK SQ O	2/3/1945	29BG	5/13/1945	Crash landed — T/O engine loss
42-94014	LITTLE BUTCH	T SQ 05	2/8/1945	498BG	9/11/1950	Reclaimed Pyote
42-94015			2/3/1945	4105BU	5/5/1954	Reclaimed Davis-Monthan
42-94016	CITY OF JERSEY CITY	BLK SQ K 13	2/3/1945	330BG	1/5/1950	Reclaimed McClellan
42-94016	MCNAMARA'S BAND	BLK SQ 13	2/3/1945	330BG	1/5/1950	Reclaimed McClellan
42-94016	MCNAMARA'S BAND	BLK SQ K 13	2/3/1945	330BG	1/5/1950	Reclaimed McClellan
42-94017			2/10/1945	237BU	5/5/1954	Reclaimed Davis-Monthan
42-94018			2/7/1945	233BU	4/30/1945	Surveyed at Davis-Monthan
42-94019		K TRI	2/6/1945	505BG	8/19/1954	Reclaimed Hill
42-94020			2/12/1945		9/9/1954	Reclaimed Davis-Monthan
42-94021		BLK SQ P 11	2/8/1945	39BG	4/16/1945	Ditched Pago Bay, Guam, enroute
42-94022	OLD DOUBLE DEUCE	SQUARE I	2/10/1945	91SRP	8/26/1954	Reclaimed Yokota
42-94022	MAYA'S DRAGON	L TRI	2/10/1945	6BG		Unknown
42-94023	CITY OF LINDSAY	BLK SQ O	2/12/1945	29BG	8/19/1954	Reclaimed Hill
42-94024	CITY OF CLAYTON	BLK SQ K 12	2/12/1945	330BG	5/7/1954	Reclaimed McClellan
42-94024	OUR BABY	BLK SQ K 12	2/12/1945	330BG	5/7/1954	Reclaimed McClellan
42-94025	HEAVENLY FLOWER	X TRI 51	2/12/1945	9BG	8/19/1954	Reclaimed Hill
42-94025	THE JUDY ANN	X TRI 51	2/12/1945	9BG	8/19/1954	Reclaimed Hill
42-94025	LITTLE EVIL	X TRI 51	2/12/1945	9BG	8/19/1954	Reclaimed Hill
42-94026		BLK SQ M 15	2/12/1945	19BG	4/24/1945	Lost Tachikawa to flak, crashed Saitame
42-94027	RUTHLESS	T SQ 01	2/12/1945	498BG	8/19/1954	Reclaimed Pyote
42-94028		BLK SQ M 55	2/12/1945	19BG	7/6/1945	Flak at Kofu
42-94029	CITY OF KANKAKEE	BLK SQ K 16	2/13/1945	330BG	1/24/1947	Condemned overseas per ARC
42-94029	LUCKY STRIKE	BLK SQ K 16	2/13/1945	330BG	1/24/1947	Condemned overseas per ARC
42-94030	LUCKY STRIKES	T 10	2/12/1945	498BG	8/19/1954	Reclaimed Pyote
42-94030	LUCKY STRIKES	T SQ 10	2/13/1945	498BG	8/19/1954	Reclaimed Pyote
42-94031		BLK SQ P 07	2/17/1945	39BG		Survived war
42-94031		SQUARE Y	2/17/1945	307BG	9/27/1954	Reclaimed McClellan
42-94032	CITY OF ST PETERSBURG	BLK SQ K 14	2/17/1945	330BG		Survived war
42-94032	MY GAL II	BLK SQ K 14	2/17/1945	330BG		Survived war
42-94032		SQUARE Y	2/17/1945	307BG	6/27/1954	Reclaimed Kadena
42-94033		BLK SQ O	2/15/1945	29BG	5/7/1954	Reclaimed McClellan
42-94034		BLK SQ O 06	2/14/1945	29BG	4/15/1945	Crashed off coast at Enoshima Island
42-94035		BLK SQ O	2/14/1945	29BG	7/14/1954	Salvaged at Davis-Monthan
42-94036			3/17/1945	BOEING	1/31/1945	Reclaimed Boeing Seattle
42-94037	CITY OF HERSHEY	BLK SQ K 39	2/15/1945	330BG	12/4/1953	Reclaimed Robins
42-94037	THE WILFUL WITCH	BLK SQ K 39	2/25/1945	330BG	12/4/1953	Reclaimed Robins

Serial #	Name	Identification	Delv	Assign	Off Inv	Circumstances
42-94038		CIRCLE E	2/15/1945	22BG	9/9/1954	Salvaged at Davis-Monthan
42-94039		BLK SQ M	2/16/1945	19BG	5/29/1945	Ditched from Yokohama
42-94040	CITY OF ROCHESTER	BLK SQ K 64	2/16/1945	330BG		Survived war
42-94040	FEATHER MERCHANTS	BLK SQ K 64	2/16/1945	330BG		Survived war
42-94040		SQUARE A	2/16/1945	301BG	9/2/1954	Reclaimed Robins
42-94041	UMBRIAGO II	E TRI	2/17/1945	505BG	7/27/1945	Ditched — flak damage
42-94041	UMBRIAGO II	CIRC X 28	2/17/1945	9BG	7/27/1945	Ditched with 505BG — flak damage
42-94042	MYA'S DRAGON	L TRI 40	2/17/1945	6BG	8/19/1954	Reclaimed Tinker
42-94043	PASSION WAGON	X TRI 12	2/17/1945	9BG		Survived war
42-94043	PASSION WAGON	BLKSTRPRED	2/17/1945	19BG	4/6/1949	Reclaimed Kadena
42-94044	CITY OF ATHENS	BLK SQ P 48	2/19/1945	39BG	6/17/1954	Class 26 Sheppard
42-94044	OLE FORTY AND EIGHT II	BLK SQ P 48	2/19/1945	39BG	6/17/1954	Class 26 Sheppard
42-94045		SQUARE Y		307BG	10/23/1951	Shot down Namsi, North Korea
42-94046		T SQ 27	2/20/1945	498BG		Survived war
42-94046		SQUARE Y	2/20/1945	97BG	6/17/1954	Class 26 Sheppard
42-94047	CITY OF JAMESTOWN NY	BLK SQ K 63	2/19/1945	330BG	9/1/1954	Reclaimed Yokota
42-94047	THROBBING MONSTER	BLK SQ K 63	2/19/1945	330BG	9/1/1954	Reclaimed Yokota
42-94047		SQUARE A	2/19/1945	301BG	9/1/1954	Reclaimed Yokota
42-94048			2/20/1945		9/9/1954	Reclaimed Davis-Monthan
42-94049	RAMBLIN' ROSCOE II	Z SQ 23 (2)	2/20/1945	500BG	9/9/1954	Salvaged at Davis-Monthan
42-94050		A SQ	2/21/1945	497BG	5/26/1945	Ditched from Tokyo
42-94051			2/19/1945		5/14/1945	Lost on Nagoya mission
42-94052	CITY OF TERRA HAUTE	BLK SQ K 36	2/21/1945	330BG	6/28/1950	Transferred to RAF
42-94052	STAR DUST	BLK SQ K 36	2/21/1945	330BG	6/28/1950	Transferred to RAF
42-94052			6/28/1950	RAF	8/25/1953	Returned to USAF
42-94052			8/25/1953	USAF	3/22/1954	Reclaimed
42-94053	QUEEN CATHY	BLK SQ P 06 (1)	2/20/1945	39BG	5/19/1945	Caught fire on takeoff, crash landed
42-94054						Unknown — cannot locate ARC
42-94055			2/22/1945		8/19/1954	Reclaimed Tinker
42-94056		E TRI	2/22/1945	504BG	5/29/1945	Flak at Yokohama
42-94057			2/23/1945	234BU	8/19/1945	Reclaimed Tinker
42-94058		L TRI	2/21/1945	6BG	5/10/1954	Reclaimed Davis-Monthan
42-94059	CITY OF FARMINGTON	BLK SQ K 61	2/24/1945	330BG	7/11/1950	Reclaimed McClellan
42-94059	LONESOME POLECAT	BLK SQ K 61	2/24/1945	330BG	7/11/1950	Reclaimed McClellan
42-94060			2/24/1945			Unknown
42-94061		BLK SQ O	2/24/1945	29BG	5/10/1954	Reclaimed Davis-Monthan
42-94062	CITY OF CHATTANOOGA	BLK SQ K 09	2/26/1945	330BG		Survived war
42-94062	PLUTO	BLK SQ K 09	2/26/1945	330BG		Survived war
42-94062		SQUARE Y	2/26/1945	307BG	4/20/1954	Reclaimed Kelly
42-94063	THE WOLF PACK	L TRI 12	2/26/1945	6BG	9/15/1953	Reclaimed Pyote
42-94064			2/24/1945	XXIBC	6/17/1954	Class 26 Sheppard
42-94065			3/5/1945	234BU	5/14/1945	Surveyed at Clovis
42-94066		BLK SQ O	2/28/1945	29BG	3/22/1953	Class 26 Kirtland
42-94067	THE STARDUSTER	X TRI 35	2/27/1945	9BG	6/23/1954	Reclaimed Pyote
42-94068			2/27/1945	4141BU	5/15/1953	Reclaimed Pyote
42-94069			3/2/1945	4117BU	12/4/1953	Reclaimed Robins
42-94070			2/28/1945	4105BU	5/10/1954	Reclaimed Davis-Monthan
42-94071	CITY OF EVANSTON	BLK SQ K 65	2/28/1945	330BG	9/9/1954	Salvaged at Davis-Monthan
42-94071	CITY OF GAINSVILLE	BLK SQ K 65	2/28/1945	330BG	9/9/1954	Salvaged at Davis-Monthan
42-94072		SQUARE Y	2/27/1945	307BG	9/8/1950	Crashed E China Sea 5 miles off Okinawa
42-94072			2/27/1945	39BG		Survived war
42-94073			3/1/1945	4141BU	8/19/1954	Reclaimed Pyote
42-94074			3/1/1945	234BU	7/20/1945	Surveyed at Clovis
42-94075			3/3/1945	4117BU	12/4/1953	Reclaimed Robins
42-94076			3/3/1945	4105BU	9/9/1954	Reclaimed Davis-Monthan
42-94077			3/8/1945	326BU	5/12/1945	Surveyed at MacDill
42-94078			3/2/1945	4105BU	5/10/1954	Reclaimed Davis-Monthan
42-94079		BLK SQ P 21	3/5/1945	39BG	5/26/1945	Lost Tokyo to flak
42-94080						Unknown
42-94081			3/5/1945	234BU	6/8/1950	Salvaged Scultho
42-94082			3/6/1945	4105BU	5/10/1954	Reclaimed Davis-Monthan
42-94083			3/6/1945	4141BU	9/15/1953	Reclaimed Pyote
42-94084			3/5/1945	4105BU	5/10/1954	Reclaimed Davis-Monthan
42-94085			3/8/1945	4141BU	9/15/1953	Reclaimed Pyote
42-94086			3/8/1945	4105BU	5/10/1954	Reclaimed Davis-Monthan

I. Master List

Serial #	Name	Identification	Delv	Assign	Off Inv	Circumstances
42-94087			3/9/1945	4141BU	9/15/1953	Reclaimed Pyote
42-94088			3/9/1945	4105BU	5/10/1954	Reclaimed Davis-Monthan
42-94089			3/9/1945	4105BU	9/9/1954	Reclaimed Davis-Monthan
42-94090			3/9/1945	4141BU	9/15/1953	Reclaimed Pyote
42-94091			3/9/1945	4105BU	5/10/1954	Reclaimed Davis-Monthan
42-94092			3/9/1945	4141BU	8/19/1954	Reclaimed Pyote
42-94093			3/14/1945	4141BU	9/12/1950	Reclaimed Pyote
42-94094		Z SQ	3/9/1945	500BG	12/4/1953	Reclaimed Robins
42-94095		T SQ	3/16/1945	498BG	4/27/1954	Transferred to Navy
42-94096			3/10/1945	4141BU	9/15/1953	Reclaimed Pyote
42-94097			3/10/1945	4141BU	9/15/1953	Reclaimed Pyote
42-94098		BLK SQ M	3/12/1945	19BG	7/26/1945	Exploded over Matsuyama
42-94099		BLKSTRP	3/10/1945	19BG	7/21/1954	Reclaimed McClellan
42-94100			3/22/1945	326BU	12/4/1953	Reclaimed Robins
42-94101			3/12/1945	4105BU	5/10/1954	Reclaimed Davis-Monthan
42-94102			3/12/1945	XXIBC	12/4/1953	Reclaimed Robins
42-94103		L TRI	3/14/1945	6BG	9/15/1953	Reclaimed Pyote
42-94104			3/13/1945	4141BU	9/15/1953	Reclaimed Pyote
42-94105			3/13/1945	331BU	10/29/1945	Salvaged at Barksdale
42-94106			3/14/1945	4141BU	9/15/1953	Reclaimed Pyote
42-94107		SQUARE A	3/16/1945	301BG	6/24/1954	Reclaimed Tinker
42-94108		E TRI	3/16/1945	504BG	8/19/1954	Reclaimed Hill
42-94109			3/16/1945	326BU	12/4/1953	Reclaimed Robins
42-94110			3/17/1945	4141BU	9/15/1953	Reclaimed Pyote
42-94111			3/16/1945	326BU	10/29/1950	Reclaimed Kelly
42-94112			3/19/1945	4141BU	9/15/1953	Reclaimed Pyote
42-94113		CIRCLE X	3/19/1945	5RCNGP	8/16/1954	Reclaimed Davis-Monthan
42-94114	WILD WESTY'S WABBITS		3/19/1945	3PRS	6/11/1946	Surveyed overseas, crashed on takeoff
42-94115			3/19/1945	4141BU	9/15/1953	Reclaimed Pyote
42-94116			3/19/1945	326BU	12/4/1953	Reclaimed Robins
42-94117			3/22/1945	4141BU	9/15/1953	Reclaimed Pyote
42-94118			3/19/1945	4141BU	9/15/1953	Reclaimed Pyote
42-94119		X TRI 7 (3)	3/22/1945	9BG	9/15/1953	Reclaimed Pyote
42-94120			3/21/1945	4141BU	8/19/1954	Reclaimed Pyote
42-94121			3/22/1945	4141BU	9/15/1953	Reclaimed Pyote
42-94122			3/20/1945	248BU	8/11/1945	Surveyed at Walker
42-94123			3/20/1945	4141BU	8/19/1954	Reclaimed Pyote
44-27259			3/14/1945	4141BU	9/15/1953	Reclaimed Pyote
44-27260		CIRCLE E	3/19/1945	22BG	1/6/1953	Assigned Storage Sqdn at Davis-Monthan
44-27261				301BG	2/2/1953	Engines 1 & 2 failed on takeoff
44-27262		SQUARE Y		307BG	1/27/1953	Crashed near Tom-ni, Korea — crew bailed
44-27263		CIRCLE E	3/15/1945	22BG	5/20/1954	Class 26 Eglin
44-27264	WILD GOOSE	SQUARE A		301BG	11/21/1952	Circumstances Unknown
44-27265			3/17/1945	4141BU	9/15/1953	Reclaimed Pyote
44-27266			3/16/1945	4141BU	9/15/1953	Reclaimed Pyote
44-27267		SQUARE A	3/19/1945	301BG	7/14/1954	Salvaged at Davis-Monthan
44-27268			3/19/1945	421ARM	9/1/1954	Last reported at Yokota
44-27269			3/20/1945	308WRG	12/31/1954	Last reported at Hickam
44-27270			3/22/1945	4141BU	9/15/1953	Reclaimed Pyote
44-27271			3/19/1945	308WRG	1/5/1955	Last reported at Hickam
44-27272			3/19/1945	4141BU	9/15/1953	Reclaimed Pyote
44-27273		SQUARE P	3/19/1945	306BG	7/14/1954	Reclaimed McClellan
44-27274		SQUARE A	3/20/1945	301BG	7/18/1954	Reclaimed Mountain Home
44-27275		SQUARE A	3/22/1945	301BG	6/16/1953	Reclaimed Davis-Monthan
44-27276		CIRCLE E	3/21/1945	22BG	9/9/1954	Salvaged at Davis-Monthan
44-27277		CIRCLE E	3/21/1945	22BG	7/14/1954	Reclaimed McClellan
44-27278		CIRCLE E	3/22/1945	22BG	6/17/1953	Reclaimed Davis-Monthan
44-27279			3/22/1945	244BU	7/6/1952	Surveyed at Harvard
44-27280			3/22/1945	4105BU	1/14/1949	Surveyed Hill
44-27281		TRIANGLE S	3/26/1945	28BG	7/14/1954	Reclaimed Davis-Monthan
44-27282	TOWN PUMP	CIRCLE K	3/24/1945	43ARS	7/14/1954	Salvaged at Davis-Monthan
44-27283			3/24/1945	4196BU	9/15/1953	Reclaimed Pyote
44-27284			3/24/1945	4141BU	9/15/1953	Reclaimed Pyote
44-27285			3/24/1945	4141BU	9/15/1953	Reclaimed Pyote
44-27286			3/24/1945	4141BU	9/15/1953	Reclaimed Pyote

The B-29 Superfortress

Serial #	Name	Identification	Delv	Assign	Off Inv	Circumstances
44-27287		SQUARE Y	3/28/1945	307BG	7/14/1954	Salvaged at Davis-Monthan
44-27288		BLKSTRPGRN	3/27/1945	19BG	7/14/1954	Reclaimed Davis-Monthan
44-27288		SQUARE H	3/27/1945	98BG	7/14/1954	Unknown
44-27289			3/29/1945	4105BU	6/27/1948	Salvaged at Davis-Monthan
44-27290		CIRCLE ARROW	3/28/1945	509CG	8/19/1954	Reclaimed Hill
44-27291	NECESSARY EVIL	CIRC R 91	5/18/1945	509CG	7/14/1954	Salvaged at Tinker
44-27292		CIRCLE ARROW	3/30/1945	509CG	7/14/1954	Reclaimed McClellan
44-27292		CIRCLE E BLU	3/30/1945	22BG	7/14/1954	Reclaimed McClellan
44-27293		CIRCLE ARROW	3/28/1945	509CG	8/19/1954	Reclaimed Hill
44-27294		CIRC X	3/28/1945	9BG	8/19/1954	Reclaimed Hill
44-27295			3/19/1945	428BU	1/30/1955	Last reported in Germany
44-27296	SOME PUNKINS	A 84*	3/19/1945	509CG	4/9/1948	Reclaimed Kirtland
44-27296	SOME PUMPKINS	L TRI 13	3/19/1945	6BG		Unknown
44-27297	BOCKSCAR	TRI N 77*	3/19/1945	509CG		On display USAF Museum
44-27298	FULL HOUSE	BLK SQ O 83*	3/20/1945	509CG	9/22/1954	Last reported at Tinker
44-27299	NEXT OBJECTIVE	TRI N 86*	3/20/1945	509CG	5/26/1949	Salbaged at Biggs
44-27300	STRANGE CARGO	A 73*	4/2/1945	509CG	8/25/1954	Last reported at Tinker
44-27301	STRAIGHT FLUSH	TRI N 85*	4/2/1945	509CG	7/14/1954	Salvaged at Davis-Monthan
44-27302	TOP SECRET	A 72*	4/2/1945	509CG	7/14/1954	Salvaged at Davis-Monthan
44-27303	JABBITT III	A 71*	4/3/1945	509CG	2/18/1946	Salvaged at Chicago Municipal
44-27304	UP AN' ATOM	TRI N 88*	4/3/1945	509CG	9/14/1954	Reclaimed Hamilton
44-27305			4/10/1945		8/5/1948	Salvaged at Harmon
44-27306		TRI N	4/7/1945	4196BU	5/7/1954	Reclaimed McClellan
44-27307	SLAVE GIRL	TRI N	4/9/1945	444BG	8/5/1948	Reclaimed Harmon
44-27308	HONGCHOW	2YLWBANDS	4/9/1945	2ARS	12/9/1954	Last reported at Andersen
44-27309			4/16/1945	4135BU	8/19/1954	Reclaimed Hill
44-27310			4/10/1945	XXIBC	8/19/1954	Reclaimed Hill
44-27311			4/12/1945		12/30/1955	Last reported at Yuma
44-27312		2YLWBANDS	4/14/1945	2ARS	5/7/1954	Reclaimed Robins
44-27313			4/12/1945		12/22/1948	Salvaged at Harmon
44-27314		SQUARE Y		307BG	12/4/1951	Crashed 2 miles east of Kadena
44-27314		CIRC X		9BG		Survived war
44-27315			4/13/1945		11/12/1945	Salvaged overseas
44-27316		CIRC R	4/21/1945	6BG	9/15/1953	Reclaimed Pyote
44-27317		CIRC X	4/14/1945	9BG	12/4/1953	Reclaimed Robins
44-27318		CIRC X	4/16/1945	9BG	6/25/1946	Salvaged at Tinker
44-27319			4/16/1945		8/19/1954	Reclaimed Hill
44-27320		CIRCLE W		92BG		Unknown
44-27321			4/19/1945		4/4/1954	Last reported at Eielsen
44-27322			4/20/1945	328BU	7/24/1945	Salvaged at Gulfport
44-27323			4/19/1945	4141BU	9/15/1953	Reclaimed Pyote
44-27324			4/19/1945	308WRG	5/19/1954	Last reported at McClellan
44-27325			4/20/1945		8/16/1954	Reclaimed Davis-Monthan
44-27326		SQUARE Y		307BG	9/13/1951	Crashed on mountain 5 miles east of Taegu
44-27326	UNITED NOTIONS	CIRCLE W		92BG		Transferred to 98BG when 92nd rotated
44-27326	UNITED NOTIONS	SQUARE H		98BG	10/25/1951	Salvaged
44-27326	DARK SLIDE	F		6CMG		Unknown
44-27327		CIRCLE E	4/20/1945	22BG	9/2/1953	Reclaimed Davis-Monthan
44-27328			4/21/1945	4141BU	9/15/1953	Reclaimed Pyote
44-27329		SQUARE A	4/20/1945	301BG	2/9/1955	Last reported at Kadena
44-27330		SQUARE I	4/21/1945	301BG	9/8/1954	Last reported at Biggs
44-27331			4/23/1945	331BU	6/6/1945	Surveyed Barksdale
44-27332	MISS MINOOKI	SQUARE H	4/23/1945	98BG	10/19/1951	Crashed in Sea of Japan
44-27332	MISS SPOKANE	CIRCLE W	4/23/1945	92BG	6/19/1955	Reclaimed Davis-Monthan
44-27332	MISS SPOKANE	SQUARE H	4/23/1945	98BG	6/19/1955	Reclaimed Davis-Monthan

*These planes were assigned to the 509th Composite Group, 393rd Bomb Squadron, the unit that dropped the atomic bombs on Hiroshima and Nagasaki. The official tail code for this group was a broad, forward-facing arrow in a circle. The tail codes of all other bomb groups were letters within various geometric shapes, so the 509th's arrow was distinctive. Perhaps fearing that this code would make the 509th planes more recognizable to the Japanese, someone in the upper chain of command decided to assign them tail codes that would make them look more like other B29s, but using I.D. numbers outside the usual range of 1 to 60. Those "disguise" codes are listed here.

The planes in the 509th were equipped with only tail guns. It seems likely that this was at least one reson to avoid giving them a distinctive tail code that might have drawn closer inspection from the Japanese.

I. Master List

Serial #	Name	Identification	Delv	Assign	Off Inv	Circumstances
44-27332	MISS SPOKANE	SQUARE Y	4/23/1945	307BG	6/19/1955	Reclaimed Davis-Monthan
44-27333		CIRCLE R	4/23/1945	9BG	8/16/1954	Reclaimed Davis-Monthan
44-27334	NOAH'S ARK		4/23/1945	2759EXGP	8/23/1954	Last reported at Wright Field
44-27335			4/25/1945	308WRG	11/2/1954	Last reported at Tinker
44-27336		SQUARE A	4/24/1945	301BG	7/14/1954	Salvaged at Davis-Monthan
44-27337			4/24/1945	308WRG	9/27/1954	Last reported at Laurel
44-27338		SQUARE A	4/25/1945	301BG	8/16/1954	Reclaimed Davis-Monthan
44-27339			4/25/1945	308WRG	8/16/1954	Last reported at Laurel
44-27340		SQUARE V	4/25/1945	55RCN	7/14/1954	Reclaimed Davis-Monthan
44-27340	IRON GEORGE	TRI U 50	4/25/1945	462BG	7/14/1954	Salvaged at Davis-Monthan
44-27341		SQUARE H	4/30/1945	98BG	9/27/1954	Reclaimed McClellan
44-27342			4/30/1945		5/20/1950	Transferred to RAF
44-27342			5/20/1950	RAF	3/16/1954	Returnerd to USAF
44-27342			3/16/1954	USAF		Unknown
44-27343			4/28/1945		5/16/1954	Last reported at Hickam
44-27344			4/28/1945	308WRG	9/1/1954	Reclaimed Hickam
44-27345		SQUARE T	4/27/1945	2BG	5/25/1954	Reclaimed Randolph
44-27346		BLK SQ Q 94	4/27/1945	39BG		Unknown
44-27347		SQUARE Y	4/28/1945	307BG	9/15/1954	Reclaimed Davis-Monthan
44-27348		SQUARE T	4/27/1945	2BG	8/11/1954	Last reported at Alexandria
44-27349			4/28/1945		9/27/1954	Reclaimed
44-27350		SQUARE P	4/28/1945	306BG	9/27/1954	Reclaimed McClellan
44-27351			4/28/1945	3902ABG	6/17/1953	Reclaimed Davis-Monthan
44-27352			4/28/1945	3902ABG	1/7/1955	Last reported at Hill
44-27353	THE GREAT ARTISTE	CIRC R 89*	4/20/1945	509CG	9/27/1949	Salvaged at Goose Bay
44-27354	DAVE'S DREAM	CIRC X 90*	4/20/1945	509CG	10/10/1954	Reclaimed
44-27354	BIG STINK	CIRC X 90*	4/20/1945	509CG	10/10/1954	Reclaimed Tinker
44-27355			5/7/1945	28BG	9/14/1953	Reclaimed Tinker
44-27356			5/4/1945	22BG	9/14/1953	Reclaimed Tinker
44-27357			5/4/1945	303BG	7/14/1954	Salvaged at Davis-Monthan
44-27358			5/15/1945	4105BU	5/10/1954	Reclaimed Davis-Monthan
44-61509				2759EXGP		
44-61510		SQUARE A	3/24/1945	301BG	5/11/1948	In-flight explosion Suadi Arabia
44-61511			3/24/1945		9/15/1953	Reclaimed Pyote
44-61512			3/22/1945	4141BU	9/15/1953	Reclaimed Pyote
44-61513			3/24/1945	4117BU	11/12/1953	Reclaimed Robins
44-61514			3/27/1945	4141BU	9/15/1953	Reclaimed Pyote
44-61515		TRI U	3/24/1945	462BG	4/27/1954	Transferred to Navy
44-61516		TRI I 51 (2)	3/24/1945	468BG	9/27/1954	Reclaimed McClellan
44-61517		TRI I 16	3/28/1945	468BG	8/3/1950	Salvaged
44-61518			3/24/1945	4141BU	9/15/1953	Reclaimed
44-61519			3/24/1945	4141BU	9/15/1953	Reclaimed
44-61520		CIRC E 14	3/28/1945	504BG	6/23/1954	Reclaimed Pyote
44-61521	CONVINCER	TRI I 25	3/29/1945	468BG	5/10/1954	Reclaimed Davis-Monthan
44-61522			3/22/1945	4141BU	9/15/1953	Reclaimed Pyote
44-61523			3/29/1945	328BG	10/25/1945	Salvaged
44-61524	CITY OF YOUNGSTOWN	BLK SQ P 09 (2)	3/30/1945	39BG	11/12/1953	Reclaimed Robins
44-61524	CONFEDERATE SOLDIER	BLK SQ P 09 (2)	3/30/1945	39BG	11/12/1953	Reclaimed Robins
44-61525		TRI N 56	3/31/1945	444BG	5/10/1954	Reclaimed Davis-Monthan
44-61526		TRI N	3/28/1945	444BG	9/15/1953	Reclaimed
44-61527		C C2	3/29/1945	4141BU	9/15/1953	Reclaimed Pyote
44-61528		F	3/31/1945	6CMG	8/5/1948	Salvaged
44-61529	VIRGINIA TECH	TRI S 29 (2nd)/38		40BG	4/8/1951	Salvaged
44-61530		Z 39	3/30/1945	500BG	6/15/1945	Non-combat crash, Saipan
44-61531		CIRCLE X	4/2/1945	5RCN	8/16/1954	Reclaimed Davis-Monthan
44-61532	CITY OF SANTA BARBARA	BLK SQ P 41	3/30/1945	39BG		Survived war
44-61533	SALEM WITCH	SQUARE I	3/30/1945	91RCN	8/26/1954	Reclaimed Yokota
44-61533			3/30/1945	5SRG	10/21/1954	Reclaimed Yokota
44-61534			3/31/1945	4141BU	9/15/1953	Reclaimed Pyote
44-61535		TRI U 43	3/31/1945	462BG		Survived war
44-61535		BLKSTRP	3/31/1945	19BG	6/27/1954	Reclaimed
44-61535		SQUARE Y	3/31/1945	307BG	6/27/1954	Reclaimed
44-61536			6/30/1945	4141BU	9/15/1953	Reclaimed Pyote
44-61537		BLK SQ K	3/30/1945	330BG	10/4/1953	Reclaimed
44-61537		SQUARE H	3/30/1945	98BG	9/13/1954	Reclaimed Davis-Monthan

The B-29 Superfortress

Serial #	Name	Identification	Delv	Assign	Off Inv	Circumstances
44-61538		CIRC R		6BG	5/1/1952	Salvaged
44-61539		BLK SQ K	3/31/1945	330BG		Survived war
44-61540			4/3/1945	59RCN	3/17/1947	Salvaged
44-61541			4/2/1945	4141BU	9/15/1953	Reclaimed Pyote
44-61542		TRI S 02 (3)	4/4/1945	40BG	7/7/1948	Salvaged
44-61543	INCHCLIFFE CASTLE	TRI N 30 (2)	4/4/1945	444BG	9/27/1954	Reclaimed McClellan
44-61544			4/4/1945	611BU	7/14/1954	Salvaged at Davis-Monthan
44-61545			4/3/1945	4141BU	9/15/1953	Reclaimed Pyote
44-61546	HUN-DA-GEE	TRI N 20 (2)	4/6/1945	444BG	9/27/1954	Reclaimed McClellan
44-61547			4/7/1945		5/27/1948	Salvaged
44-61548	MUS'N TOUCH IT	TRI S 24 (2)	4/4/1945	40BG	7/14/1954	Salvaged at Davis-Monthan
44-61549	SHASTA	CIRC R 02	4/9/1945	6BG	9/15/1953	Reclaimed Pyote
44-61550	BIG BOOTS II	CIRC E 48	4/9/1945	504BG	9/15/1953	Reclaimed Pyote
44-61551	ACE OF THE BASE	CIRC E 60	4/6/1945	504BG	6/23/1954	Reclaimed Pyote
44-61552	RENEGADE	BLK SQ P 21	4/7/1945	39BG	9/27/1954	Reclaimed McClellan
44-61553		TRI N	4/6/1945	444BG	9/27/1954	Reclaimed McClellan
44-61554		TRI S 36 (3)	4/6/1945	40BG	8/30/1945	Crashed on POW Supply flight to Fukuoka
44-61555	MISS JUDY	TRI U 53	4/6/1945	462BG	9/28/1954	Reclaimed McClellan
44-61556	USS COMFORT'S REVENGE	TRI S 39 (2)	4/6/1945	40BG		Survived war
44-61556		SQUARE EMPTY	4/6/1945	2BG	7/11/1954	Reclaimed
44-61556			4/6/1945	54WRS	7/11/1954	Reclaimed
44-61557		SPEEDWING	4/6/1945	307BG	5/5/1949	Salvaged
44-61558	PORCUPINE	CIRC R 13	4/10/1945	6BG	1/24/1947	Condemned
44-61559		TRI U	4/7/1945	462BG	9/15/1953	Reclaimed Davis-Monthan
44-61560		TRI U	4/7/1945	462BG	9/15/1953	Reclaimed Davis-Monthan
44-61561		CIRC W 3 (2)	4/9/1945	505BG	8/7/1948	Salvaged
44-61562	CHICAGO SAL	TRI I 17	4/9/1945	468BG		Survived war
44-61562	NEVER HOPPEN	BLKSTRPGRN	4/9/1945	19BG	8/25/1954	Reclaimed Davis-Monthan
44-61562		SQUARE H	4/9/1945	98BG	8/25/1954	Reclaimed Davis-Monthan
44-61563		CIRC W 7 (3)	4/9/1945	505BG	3/25/1953	Reclaimed Pyote
44-61564	SENTIMENTAL JOURNEY	TRI N 37 (2)	4/9/1954	444BG	9/28/1954	Reclaimed
44-61565	DRUNKARD — STAGGER INN	TRI S 45	4/10/1945	40BG	9/15/1953	Reclaimed Pyote
44-61566	JACK'S HACK	TRI I 54	4/14/1945	468BG	9/15/1953	Reclaimed Pyote
44-61567			4/10/1945	4141BU	7/14/1954	Reclaimed McClellan
44-61568			4/10/1945	4141BU	9/15/1953	Reclaimed Pyote
44-61569		TRI U 7	4/12/1945	462BG	7/19/1945	Crashed, burned Tinian
44-61570			4/14/1945	308BG	1/6/1953	Reclaimed Davis-Monthan
44-61571			4/12/1954	3415TTW	9/9/1954	Salvaged at Davis-Monthan
44-61572		CIRC W	4/14/1945	505BG	8/5/1948	Salvaged
44-61573		TRI I 6	4/12/1945	468BG	6/28/1945	MIA Fushiki
44-61574			4/17/1945	324BU	6/20/1945	Surveyed
44-61575			4/14/1945	4141BU	9/15/1953	Reclaimed Pyote
44-61576			4/17/1945	4141BU	8/19/1954	Reclaimed Pyote
44-61577	SUELLA J	BLK SQ F	4/16/1945	16PRS	8/19/1954	Reclaimed
44-61578	SWEET'N LOLA	BLK SQ F	4/16/1945	4136BU	8/19/1954	Reclaimed
44-61579			4/16/1945	4141BU	9/15/1953	Reclaimed Pyote
44-61580			4/18/1945	4141BU	9/15/1953	Reclaimed Pyote
44-61581			4/18/1945	4141BU	9/15/1953	Reclaimed Pyote
44-61582			4/18/1945	4141BU	7/28/1953	Reclaimed Pyote
44-61583	KAMRA-KAZE	BLK SQ F	4/17/1945	16PRS	8/19/1954	Reclaimed Kelly
44-61584		SQUARE A	4/18/1945	301BG	8/19/1954	Reclaimed
44-61584			4/18/1945	215BU	9/14/1953	Reclaimed Davis-Monthan
44-61585			4/19/1945	AMC	8/16/1954	Transferred to Navy
44-61586			4/15/1945	4141BU	6/23/1954	Reclaimed Pyote
44-61587	YUCATAN KIDS	TRI S 26 (2)	4/15/1945	40BG	9/15/1953	Reclaimed Pyote
44-61588			4/18/1945	3510FTW	9/15/1953	Reclaimed Randolph
44-61589			4/18/1945	4141BU	9/15/1953	Reclaimed
44-61589			4/18/1945	4141BU	9/15/1953	Reclaimed Pyote
44-61590			4/18/1945	3510FTW	7/29/1954	Salvaged at Davis-Monthan
44-61591		SQUARE Y	4/20/1945	307BG	9/15/1953	Reclaimed Pyote
44-61592			4/18/1945	4141BU	9/15/1953	Reclaimed Pyote
44-61593			4/19/1945	4141BU	6/23/1954	Reclaimed Pyote
44-61594			4/19/1945	4117BU	11/12/1953	Reclaimed Robins
44-61595		SQUARE A	4/20/1948	301BG	6/17/1954	Class 26 Sheppard
44-61596			4/19/1945	4141BU	6/23/1954	Reclaimed Pyote

I. Master List

Serial #	Name	Identification	Delv	Assign	Off Inv	Circumstances
44-61597			4/21/1945	310BG	3/28/1954	Reclaimed Davis-Monthan
44-61598			4/20/1945	4141BU	9/15/1953	Reclaimed Pyote
44-61599			4/20/1945	324BU	3/22/1950	Delivered to RAF
44-61599			3/22/1950	RAF	7/5/1956	Broken up, scrapped England 8/8/57
44-615XX	MISS IRENE				9/9/1954	Reclaimed Burtonwood
44-61600			4/20/1945	301BG	9/9/1954	Reclaimed
44-61600	HAULIN' ASS		4/20/1945	512WRS	9/9/1954	Reclaimed Burtonwood
44-61601			4/21/1945		5/30/1948	Salvaged
44-61602			4/21/1945	4105BU	5/5/1954	Reclaimed Davis-Monthan
44-61603					5/5/1954	Reclaimed
44-61604			4/20/1945	4105BU	5/5/1954	Reclaimed Davis-Monthan
44-61605			4/21/1945	310BG	3/29/1954	Reclaimed Davis-Monthan
44-61606			4/21/1945	4105BU	5/5/1954	Reclaimed Davis-Monthan
44-61607			4/21/1945	72SRW	8/16/1954	Reclaimed Davis-Monthan
44-61608			4/21/1945	308RCN	9/1/1954	Reclaimed Wright-Patterson
44-61609			4/24/1945	4105BU	9/28/1954	Reclaimed McClellan
44-61610			4/23/1945	4105BU	5/10/1954	Reclaimed Davis-Monthan
44-61611			4/26/1945	4105BU	5/5/1945	Reclaimed Davis-Monthan
44-61612			4/25/1945	4105BU	5/5/1945	Reclaimed Davis-Monthan
44-61613			4/24/1945	247BU	6/13/1945	Surveyed
44-61614			4/24/1945	4105BU	5/5/1945	Reclaimed Davis-Monthan
44-61615			4/24/1945	4105BU	5/5/1954	Reclaimed Davis-Monthan
44-61616			4/24/1945	235BU	9/23/1945	Surveyed
44-61617	NIPP-ON-NEES	CIRCLE W	4/24/1945	92BG	10/26/1950	Unknown
44-61617		DIAMOND EMPTY	4/24/1945	92BG	10/26/1950	Unknown
44-61618	BABY'S BUGGY	CIRCLE W	5/8/1945	92BG	11/17/1953	Reclaimed Hill
44-61619			4/25/1945	XXIBC	11/17/1953	Salvaged
44-61620			4/24/1945	4105BU	5/10/1954	Reclaimed
44-61621			4/26/1945	4105BU	5/10/1954	Reclaimed Davis-Monthan
44-61622			4/25/1945	4105BU	5/10/1954	Reclaimed Davis-Monthan
44-61623		SQUARE Y	4/27/1945	307BG	9/28/1954	Reclaimed McClellan
44-61623	SALLY DELLE	T SQ 16/12	4/27/1945	498BG		Tail code change, date unknown
44-61623		SQUARE Y		307BG	9/28/1954	Reclaimed
44-61624			4/26/1945	308BG	3/15/1954	Reclaimed Davis-Monthan
44-61625			4/26/1945	233BU	9/22/1953	Reclaimed Pyote
44-61626	JANKE'S JINX		4/27/1945	242BU	8/19/1954	Reclaimed Hill
44-61627			4/27/1945	4105BU	5/10/1954	Reclaimed Davis-Monthan
44-61628		TRI U	4/27/1945	462BG	5/10/1954	Reclaimed Davis-Monthan
44-61629			4/30/1945	XXIBC	9/15/1953	Reclaimed Pyote
44-61630			4/28/1945	XXIBC	5/10/1954	Reclaimed Davis-Monthan
44-61631			4/28/1945	4105BU	5/10/1954	Reclaimed Davis-Monthan
44-61632			4/28/1945	234BU	9/22/1953	Reclaimed Pyote
44-61633		SQUARE P	4/30/1945	306BG	9/28/1954	Reclaimed McClellan
44-61634	FLAK MAGNET	TRI S 25	4/28/1945	40BG	5/10/1950	Transferred to RAF
44-61634	SENTIMENTAL JOURNEY	TRI S 25	4/28/1945	40BG	5/10/1950	Transferred to RAF
44-61635		CIRC R 08	5/9/1945	6BG	9/15/1953	Reclaimed Pyote
44-61636		CIRCLE E YLW	5/3/1945	9SRC	9/28/1954	Reclaimed McClellan
44-61637			5/2/1945	611BU	5/10/1954	Reclaimed Davis-Monthan
44-61638	BUG'S (BALL) BUSTER	BLKSTRPGRN	4/30/1945	19BG	6/27/1954	Reclaimed Kadena
44-61638	GOING MY WAY	BLK SQ M	4/30/1945	19BG		Survived war
44-61639	HELLBIRD INSIGNIA	TRI U	5/1/1945	462BG	7/14/1954	Salvaged at Davis-Monthan
44-61640		TRI I 18		468BG		Survived war
44-61640		SQUARE A		301BG		Converted to weather configuration
44-61640				54WRS	2/27/1952	Lost on weather recon mission
44-61641			5/1/1945	611BU	11/12/1953	Reclaimed Robins
44-61642		BLKSTRPBLU	4/30/1945	19BG	10/22/1951	Reclaimed Davis-Monthan
44-61642		BLK SQ M	4/30/1945	19BG	11/27/1950	Transferred to RAF
44-61643	YONKEE DOLLAR	BLK SQ P	5/5/1945	39BG	9/22/1953	Reclaimed Pyote
44-61644			5/2/1945	4141BU	9/22/1953	Reclaimed Pyote
44-61645			5/2/1945	320BG	3/29/1954	Reclaimed Davis-Monthan
44-61646	THIS IS IT!	TRI N 61 (2)	5/3/1945	444BG	5/10/1954	Reclaimed Davis-Monthan
44-61647		BLK SQ M 26 (?)	5/2/1945	19BG	9/22/1953	Reclaimed Pyote
44-61648		CIRC X 18	5/2/1945	9BG	9/22/1953	Reclaimed Pyote
44-61649			5/3/1945		9/8/1954	Reclaimed Davis-Monthan
44-61650		TRI U	5/3/1945	462BG	5/10/1954	Reclaimed Davis-Monthan

Serial #	Name	Identification	Delv	Assign	Off Inv	Circumstances
44-61651	TAGALONG	TRI S 01(2)	5/3/1945	40BG	1/3/1953	Reclaimed Davis-Monthan
44-61651			5/3/1945		8/3/1953	Reclaimed Seattle
44-61652		TRI U	5/4/1945	462BG	9/22/1953	Reclaimed Pyote
44-61653	FIRE BELLE	TRI N 22 (2)	5/4/1945	444BG	5/10/1954	Reclaimed Davis-Monthan
44-61654			5/4/1945		9/23/1953	Reclaimed Davis-Monthan
44-61655	MIGHTY FINE II	T SQ 07 (3)	5/9/1945	498BG	5/10/1954	Reclaimed Davis-Monthan
44-61656		TRI S 50		40BG	11/23/1951	Salvaged
44-61656		BLKSTRPBLU		19BG	10/22/1951	Shot down by Mig's — abandoned
44-61657	CREAM OF THE CROP	BLKSTRPBLU	5/4/1945	19BG		Transferred to 98BG
44-61657	CREAM OF THE CROP	SQUARE H		98BG	9/13/1954	Reclaimed
44-61657		TRI U		462BG		Survived War
44-61657		SQUARE Y	5/4/1945	307BG	9/13/1954	Reclaimed
44-61658	BARBARA ANN	Z SQ 21 (2)	8/20/1944	500BG	5/1/1954	Reclaimed Davis-Monthan
44-61658		BLK SQ O	5/5/1945	29BG	5/24/1948	Salvaged at Tinker
44-61659		CIRCLE X	5/5/1945	5RCN	6/5/1953	Reclaimed Randolph
44-61660	STORMY PETREL	E TRI 03	5/5/1945	504BG	9/22/1953	Reclaimed Pyote
44-61661		T 26 (4)	5/7/1945	498BG		Survived War
44-61661		CIRCLE E	5/7/1945	22BG	9/22/1953	Reclaimed Pyote
44-61661						Not known if left in FEAF 10/50
44-61662			5/5/1945		11/17/1953	Reclaimed Hill
44-61663		A	5/5/1945	497BG	12/3/1945	Salvaged
44-61664	CITY OF LYNN	BLK SQ K	5/7/1945	330BG	11/17/1953	Reclaimed
44-61664		CIRC W 18	5/7/1945	505BG	11/17/1953	Reclaimed Hill
44-61665		CIRC W	5/7/1945	505BG	5/6/1949	Salvaged
44-61666	SWEET SIXTEEN	T SQ 16 (1)	5/10/1945	498BG	6/1/1945	Abandoned over Iwo
44-61667		CIRC W 18 (1)	5/10/1945	505BG	3/29/1954	Reclaimed Davis-Monthan
44-61669	FLAG SHIP 500	Z 49	5/8/1945	500BG	5/10/1954	Reclaimed Davis-Monthan
44-61669	FLAG SHIP 500	Z SQ 49 (2)	5/8/1945	500BG	5/10/1954	Reclaimed Davis-Monthan
44-61669	MISSION INN	CIRCLE W		92BW	10/19/1950	Transferred to 19bg when 92nd rotated
44-61669	MISSION INN	BLKSTRP		19BG		Unknown
44-61669	MISSION INN	CIRCLE E		22BG		Unknown
44-61670	LADY FRANCIS	TRI N 25	5/8/1945	444BG	5/10/1954	Reclaimed
44-61670		BLK SQ K	5/8/1945	330BG	5/10/1954	Reclaimed Davis-Monthan
44-61671		CIRC R		6BG		Unknown
44-61672		TRI I 37	5/7/1945	468BG	9/27/1954	Reclaimed
44-61673			5/9/1945		5/10/1954	Reclaimed Davis-Monthan
44-61674	JOHNNY REBEL	TRI I 53	5/7/1945	468BG	3/28/1954	Reclaimed Davis-Monthan
44-61675		CIRC R	5/11/1945	6BG	9/22/1953	Reclaimed Pyote
44-61676	CITY OF FLINT	BLK SQ O 49	5/8/1945	29BG	6/17/1953	Reclaimed
44-61676	SAD SAC	SQUARE H	5/8/1945	98BG	6/17/1953	Reclaimed
44-61676		SPEEDWING	5/8/1945	307BG	6/17/1953	Reclaimed
44-61676		SQUARE Y	5/8/1945	307BG	6/17/1953	Reclaimed
44-61677	SOUTHERN BELL	TRI I 35	5/9/1945	468BG	6/23/1954	Reclaimed
44-61677			5/9/1945	581ARC	6/23/1954	Reclaimed Clark Field
44-61678	MAIDEN'S PRAYER	T 28 (3)		498BG		Survived war
44-61678		SQUARE EMPTY		2BG	7/25/1951	Salvaged
44-61679			5/10/1945	6BG	6/9/1948	Reclaimed
44-61680		BLK SQ M	5/12/1945	19BG	7/9/1948	Reclaimed
44-61681		TRI I 57	5/11/1945	468BG	3/19/1954	Reclaimed
44-61681			5/11/1945	581ARC	3/19/1954	Salvaged at Wheelus
44-61682			5/14/1945		5/10/1954	Reclaimed Davis-Monthan
44-61683			5/10/1945		5/10/1954	Reclaimed Davis-Monthan
44-61684	SILVER THUNDER	Z SQ 22	9/27/1944	500BG		Tail code change
44-61684	SILVER THUNDER	Z 22	5/10/1945	500BG		Survived war
44-61684		SQUARE Y	5/10/1945	307BG	6/13/1952	Salvaged
44-61685	BARBARA ANN	Z 21 (2)	5/11/1945	500BG	5/10/1954	Reclaimed Davis-Monthan
44-61686		CIRC R 38	5/11/1945	6BG	9/22/1953	Reclaimed Pyote
44-61687			5/11/1945		9/22/1953	Reclaimed
44-61688		CIRC R 10	5/11/1945	6BG	12/3/1950	Transferred to RAF
44-61688			12/3/1950	RAF	7/22/1953	Returned to USAF
44-61688			7/22/1953	USAF		Unknown
44-61689		CIRC X 28 (3)	5/14/1945	9BG	12/4/1945	Condemned
44-61690		V ?	5/12/1945	499BG		Survived war
44-61690		CIRC E		22BG	6/17/1953	Reclaimed
44-61691		CIRC X 50	5/12/1945	9BG	9/21/1949	Transferred to Army

I. Master List

Serial #	Name	Identification	Delv	Assign	Off Inv	Circumstances
44-61692	SHARON SUE	Z SQ 19	9/18/1944	500BG	4/00/45	Tail code change
44-61692	SHARON SUE	Z 19	5/14/1945	500BG	5/10/1954	Reclaimed Davis-Monthan
44-61693	CONSTANT NYMPH	Z SQ 30	5/14/1945	500BG		Survived War
44-61693		BLKSTRPBLU	5/14/1945	19BG	5/5/1952	Condemned
44-61693		SQUARE Y	5/14/1945	307BG	5/5/1952	Condemned
44-61694		TRI S 07 (2)	5/14/1954	40BG		Survived war
44-61694		CIRCLE E	5/14/1945	22BG	10/00/50	Transfer to 98BG when 22BG rotated to ZI
44-61694		SQUARE H	5/14/1954	98BG	9/13/1954	Reclaimed
44-61694		SQUARE Y	5/14/1954	307BG	9/13/1954	Reclaimed
44-61695		TRI I 47	5/14/1954	468BG	11/17/1950	Transferred to RAF
44-61695			11/17/1950	RAF	3/22/1954	Returned to USAF
44-61695			3/22/1954	USAF		Unknown
44-61696		BLK SQ K	5/14/1945	330BG	5/28/1946	Salvaged
44-61697	LITTLE FELLOW		5/15/1945		9/22/1953	Reclaimed Pyote
44-61698		SQUARE Y	5/15/1945	307BG	9/28/1954	Reclaimed McClellan
44-61699	STING SHIFT	Z SQ 30 (2)	5/15/1945	500BG	5/24/1945	MIA Tokyo
44-61700		BLK SQ M	5/15/1945	19BG	10/1/1948	Salvaged
44-61701			5/15/1945	306BG	7/14/1954	Reclaimed Davis-Monthan
44-61702	HERO HEATER	TRI I 32	5/15/1945	468BG	9/28/1954	Reclaimed
44-61703	O'REILLYS DAUGHTER II	TRI I 14	5/16/1945	468BG	9/8/1954	Reclaimed Davis-Monthan
44-61704			5/18/1945	4105BU	5/10/1954	Reclaimed Davis-Monthan
44-61705		CIRC R 18	5/19/1945	6BG		Survived war
44-61705		BLKSTRPRED	5/19/1945	19BG	7/14/1954	Reclaimed McClellan
44-61706		BLK SQ O	5/16/1945	29BG		Survived war
44-61706			5/16/1945		1/18/1954	Reclaimed
44-61707						unknown
44-61708		TRI I 32	5/16/1945	468BG	9/14/1954	Reclaimed
44-61709		A	5/17/1945	497BG	8/12/1945	Salvaged Marianas
44-61710			5/18/1945	53WRS	11/30/1953	Reclaimed
44-61711		T SQ 52	5/21/1945	498BG	9/22/1953	Reclaimed
44-61712			5/18/1945		8/20/1945	Circumstances Unknown
44-61713			5/18/1945		10/7/1948	Salvaged at Smoky Hill
44-61714			5/19/1945	326BU	5/16/1950	Transferred to RAF
44-61714			5/16/1950	RAF	9/16/1958	Scrapped in England
44-61715		CIRC W 16	5/18/1945	497BG	9/22/1953	Reclaimed
44-61716		A	5/18/1945		1/24/1947	Lost returning to ZI
44-61717			5/23/1945	303BG	3/1/1953	Reclaimed
44-61718		BLKSTRP	5/18/1945	19BG	7/14/1954	Salvaged at Davis-Monthan
44-61719			5/18/1945		6/17/1953	Reclaimed
44-61720			5/19/1945	29BG	9/14/1954	Reclaimed
44-61721	MASON'S HONEY	SQUARE H	5/22/1945	98BG	7/18/1945	Surveyed
44-61722			5/23/1945	326BU	9/8/1954	Reclaimed
44-61723			5/19/1945	3510TTW	11/12/1953	Reclaimed Robins
44-61724			5/21/1945	4105BU	7/14/1954	Salvaged at Davis-Monthan
44-61725			5/24/1945	307BG	5/5/1954	Reclaimed Davis-Monthan
44-61726		SQUARE Y	5/19/1945	91RCN	6/27/1954	Reclaimed Kadena
44-61727	SO TIRED, 7 TO 7			505BG	7/4/1952	Shot down by Mig's
44-61728			5/21/1945	4105BU	12/3/1950	Transferred to RAF
44-61728			12/3/1950	RAF	1/5/1954	Returned to USAF
44-61728			1/5/1954	USAF	7/14/1954	Reclaimed Davis-Monthan
44-61729			5/19/1945	241BU	5/10/1954	Reclaimed Davis-Monthan
44-61730			5/19/1945	43BG	11/12/1953	Reclaimed Robins
44-61731		CIRCLE K	5/22/1945	4105BU	5/5/1945	Reclaimed Davis-Monthan
44-61732			5/21/1945	3510FTW	9/9/1954	Salvaged at Davis-Monthan
44-61733				498BG		Unknown
44-61734	LUCKY SEVEN	T SQ 06	5/21/1945	498BG	5/19/1946	Converted to weather plane
44-61734			1/30/1951	54WRS		Unknown
44-61735			5/21/1945	4117BU	11/12/1953	Reclaimed Robins
44-61736			5/23/1945	500BG	9/22/1953	Reclaimed Davis-Monthan
44-61737		Z SQ 1			1/3/1945	Lost on Nagoya mission
44-61738			5/22/1945	513WRS	3/30/1954	Reclaimed
44-61739			5/24/1945	581ARCW	5/31/1954	Reclaimed
44-61740			5/23/1945	4141BU	9/22/1953	Reclaimed Davis-Monthan
44-61741			5/24/1945	68BW3	7/15/1953	Reclaimed Davis-Monthan
44-61742			5/23/1945	3121ERE	8/16/1954	Reclaimed

Serial #	Name	Identification	Delv	Assign	Off Inv	Circumstances
44-61743		TRIANGLE O	5/25/1945	97BG	6/29/1950	Transferred to RAF
44-61743			5/25/1945	326BU	6/29/1950	Transferred to RAF
44-61743			6/29/1950	RAF	10/20/1953	Returned to USAF
44-61743			10/20/1953	USAF	7/14/1954	Reclaimed Davis-Monthan
44-61744			5/24/1954	462BG	11/12/1953	Reclaimed Robins
44-61745		TRI U	5/26/1945	40BG	8/5/1948	Salvaged at Harmon
44-61746		TRI S	5/25/1945	40BG	12/22/1948	Salvaged at Harmon
44-61747		TRI S	6/19/1945	31SR	3/31/1953	Surveyed at Kadena
44-61748	HAWG WILD	SQUARE Y	5/26/1945	307BG	6/27/1954	Salvaged at Kadena
44-61748	IT'S HAWG WILD	SQUARE Y	5/26/1945	307BG	6/27/1954	Salvaged at Kadena
44-61749	SOUTHERN COMFORT	BLKSTRPBLU	5/25/1945	19BG	12/13/1950	Salvaged at Itazuke
44-61750			5/26/1945	19BG	5/5/1954	Reclaimed Davis-Monthan
44-61751		BLK SQ M		19BG		Survived war
44-61751	LUBRICATING LADY	BLKSTRPRED		19BG	10/31/1952	Ditched off Kadena
44-61752			5/28/1945	19BG	5/10/1954	Reclaimed Davis-Monthan
44-61753			5/28/1945	68BG	7/14/1953	Reclaimed Davis-Monthan
44-61754			5/26/1945	582ARC	3/7/1954	Reclaimed
44-61755			5/26/1945	4141BU	9/22/1953	Reclaimed Pyote
44-61756	USS PINTADO	BLK SQ K 13	5/28/1945	330BG	6/8/1948	Salvaged Guam
44-61757		T 24 (2)	5/28/1945	498BG		Survived war
44-61757		TRIANGLE O	5/28/1945	97BG		Unknown
44-61757		SQUARE Y	5/28/1945	307BG	7/14/1954	Salvaged at Davis-Monthan
44-61758					8/5/1945	Unknown
44-61759			5/29/1945	4105BU	5/5/1954	Reclaimed Davis-Monthan
44-61760		CIRC X 17(3)	5/29/1945	9BG	7/25/1949	Salvaged at Kelly
44-61761			5/29/1945	68BG	7/28/1953	Reclaimed Davis-Monthan
44-61762			5/28/1945	4105BU	5/10/1954	Reclaimed Davis-Monthan
44-61763			5/29/1945	308BG	11/9/1953	Reclaimed Randolph
44-61764			5/30/1945	303BG	9/28/1954	Reclaimed McClellan
44-61765			5/30/1945	308BG	9/22/1953	Reclaimed Davis-Monthan
44-61766		T ?	5/31/1945	498BG	9/17/1953	Reclaimed Davis-Monthan
44-61767			5/30/1945	19BG	12/31/1950	Reclaimed Kelly
44-61768		BLK SQ M	5/31/1945	19BG	9/22/1953	Reclaimed Pyote
44-61769			5/30/1945	581ARC	9/8/1954	Reclaimed Davis-Monthan
44-61770			5/30/1945	4141BU	9/22/1953	Reclaimed Pyote
44-61771		SQUARE Y	5/30/1945	307BG	6/27/1954	Reclaimed
44-61772			5/31/1945	4141BU	9/22/1953	Reclaimed
44-61773			5/31/1945	303BG	3/1/1953	Reclaimed Davis-Monthan
44-61774			5/31/1945	300ABU	10/1/1954	Reclaimed
44-61775		T SQ 34 (3)	5/31/1945	498BG		Survived war
44-61775		TRIANGLE O	5/31/1945	97BG	6/28/1949	Salvaged at Marham
44-61776	RAPID RABBIT [7]	SQUARE H		98BG	3/31/1952	Engine failure; abnd 3 miles from Kansong
44-61776	[7]	DIAM L 56		331BG		Converted to WB-29
44-61776		TRI N 59 (2)		444BG	9/00/45	Transfer to 331BG
44-61777			6/1/1945		12/31/1945	Surveyed overseas
44-61778		CIRC W	6/1/1945	505BG	9/1/1954	Reclaimed
44-61779			6/9/1945	3171EREG	8/16/1954	Reclaimed Griffis
44-61780			6/1/1945	582ARC	3/10/1954	Reclaimed Molesworth
44-61781		T 57	6/4/1945	498BG	9/22/1953	Reclaimed Pyote
44-61782	LITTLE FELLOW	Z 09 (2)	6/2/1945	500BG	9/9/1953	Reclaimed Clark Field
44-61783			6/2/1945		7/14/1954	Reclaimed McClellan
44-61784		CIRC R 07	6/4/1945	6BG	9/22/1953	Reclaimed Pyote
44-61785		A	6/5/1945	497BG	9/13/1954	Reclaimed Davis-Monthan
44-61786		TRI U	6/5/1945	462BG	9/4/1945	Crashed on takeoff
44-61787		TRIANGLE O	6/4/1945	97BW	3/16/1950	Transferred to RAF
44-61787			3/16/1950	RAF	7/5/1958	Broken up at Shoeburyness
44-61788		CIRC W 10 (3)		505BG	11/20/1951	Salvaged
44-61789		Z SQ 38 (2)	6/6/1945	500BG	8/31/1948	Salvaged at North Field, Guam
44-61790	PEACE ON EARTH	BLKSTRPRED	6/4/1945	19BG		Ditched returning to ZI
44-61790	PEACE ON EARTH	CIRCLE W	6/4/1945	92BG	10/00/50	Transferred to 19BG when Group rotated
44-61790	PEACE ON EARTH		6/4/1945	10/00/50	7/14/1954	Reclaimed McClellan
44-61791		BLK SQ M 16 (2)	6/7/1945	19BG		Survived the war
44-61792		CIRC E 63	6/5/1945	504BG	3/15/1950	Transferred to RAF
44-61792		TRIANGLE O	6/5/1945	97BG	3/15/1950	Transferred to RAF
44-61793			6/7/1945	59RCN	10/16/1946	Salvaged at Eglin

I. Master List   51

Serial #	Name	Identification	Delv	Assign	Off Inv	Circumstances
44-61794	CITY OF LAREDO		6/5/1945	19BG	4/19/1949	Salvaged at Kelly
44-61795	LUELLA JEAN	BLK SQ P 07 (2)	6/5/1945	39BG	11/17/1953	Reclaimed Hill
44-61796			6/8/1945	59RCN	9/22/1953	Reclaimed Pyote
44-61797	TWENTIETH CENTURY LIM.	Z SQ		500BG		Transferred to 9BG
44-61797	20TH CENTURY LIMITED	CIRC X 36 (3)		9BG	12/4/1952	Salvaged at Smoky Hill
44-61798			6/7/1945	59RCN	9/16/1946	Salvaged at Eglin
44-61799			6/11/1945	4105BU	5/10/1954	Reclaimed Davis-Monthan
44-61800		CIRC X 65	6/7/1945	9BG	9/3/1953	Reclaimed Davis-Monthan
44-61801			6/7/1945		9/10/1947	Surveyed overseas
44-61802		SQUARE Y		307BG	1/10/1953	Mig damage at Yalu, abandoned
44-61802		CIRCLE W	6/7/1945	92BG	10/00/50	Transferred to 307BG when 92nd rotated
44-61803	RATTLE N'ROLL	CIRC R 55	6/8/1945	6BG	6/30/1949	Salvaged at Kelly
44-61804			6/12/1945	4117BU	11/12/1953	Reclaimed Robins
44-61805		BLK SQ O	6/8/1945	29BG	7/20/1953	Reclaimed Davis-Monthan
44-61806					2/19/1948	Class 01Z
44-61807		TRI I 3	6/9/1945	468BG	11/9/1953	Reclaimed Randolph
44-61808			6/12/1945		8/19/1954	Reclaimed Hill
44-61809	SIC EM!	SQUARE H	6/8/1945	98BG	9/13/1953	Reclaimed Davis-Monthan
44-61809		A	6/8/1945	497BG	9/13/1953	Reclaimed Davis-Monthan
44-61810	ICHIBAN	CIRCLE X		91RCN	6/13/1952	Missing-possible USSR shoot-down
44-61811			6/14/1945	4141BU	9/22/1953	Reclaimed Pyote
44-61812	GENIE II	TRI S 09 (2)	6/13/1945	40BG	9/22/1953	Reclaimed Davis-Monthan
44-61813	OVEREXPOSED	SQUARE V	6/14/1945	55RCN	12/19/1950	Reclaimed Johnson AB, Japan
44-61813	PACIFIC PRINCESS	TRI N	6/14/1945	444BG	12/19/1950	Reclaimed Johnson AB, Japan
44-61814			6/14/1945	308BG	9/1/1954	Salvaged Randolph
44-61815	MOONSHINE RAIDERS	DIAMOND L	6/15/1945	331BG		Survived war
44-61815	GAY TIMES	DIAMOND L	6/15/1945	331BG		Survived war
44-61815	MOON'S MOONBEAM	CIRCLE X	6/15/1945	5RCN		Transferred to 19BG or 98BG
44-61815	DIAJOBU	SQUARE H		98BG	7/10/1952	Salvaged
44-61815	BUB	BLKSTRPBLU		19BG	7/10/1952	Salvaged
44-61816		TRI I 45	6/15/1945	468BG	6/27/1954	Reclaimed Kadena
44-61816		SQUARE EMPTY	6/15/1945	2BG	6/27/1954	Reclaimed Kadena
44-61816		SQUARE Y	6/15/1945	307BG	6/27/1954	Reclaimed Kadena
44-61817	AH SOOOOOOOO	CIRC X	6/15/1945	91RCN	7/14/1954	Salvaged at Davis-Monthan
44-61817	SHUTTERBUG		6/15/1945	91SRP	7/14/1954	Salvaged at Davis-Monthan
44-61818	BOOMERANG		6/16/1945	31PHRCN	8/26/1954	Reclaimed Yokota
44-61819		SQUARE I	6/18/1945	91RCN	12/3/1955	Reclaimed Eglin
44-61820		CIRCLE Z	6/18/1945		3/1/1953	Reclaimed Davis-Monthan
44-61821	JOY-OUS VENTURE	TRI N 45	6/14/1945	444BG	7/14/1954	Salvaged at Davis-Monthan
44-61822	THE BELLE OF BIKINI	SQUARE H	6/14/1945	98BG	8/19/1954	Crashed south of Yokota
44-61822	DESTINATION KNOWN	SQUARE H	6/14/1945	98BG	8/19/1954	Crashed south of Yokota
44-61822	HEAVENLY LADEN	SQUARE H	6/14/1945	98BG	8/19/1954	Crashed south of Yokota
44-61823	DADDY COME HOME	TRI U	6/15/1945	462BG	10/28/1945	Salvaged at Lowry
44-61824		SQUARE Y	6/15/1945	307BG	6/27/1954	Reclaimed Kadena
44-61825			6/27/1945	582ARC	7/20/1945	Reclaimed Tinker
44-61827			6/15/1945	4141BU	9/22/1953	Reclaimed Pyote
44-61828		828	6/22/1945	320BG	9/28/1954	Reclaimed McClellan
44-61829			6/18/1945	68BW	6/17/1953	Reclaimed Davis-Monthan
44-61830	EVERY MAN A TIGER	SQUARE H	6/25/1945	98BG	7/14/1954	Reclaimed Davis-Monthan
44-61830	MYASIS DRAGON	BLKSTRP	6/25/1945	19BG	7/14/1954	Reclaimed Davis-Monthan
44-61830	MYASIS DRAGON	CIRCLE W	6/25/1945	92BG	10/00/50	Transferred to 19bg when 92nd rotated
44-61830		BLKSTRP	10/00/50	19BG	7/14/1954	Received when 92BG rotated
44-61831		SQUARE H	6/15/1945	98BG	8/16/1954	Reclaimed Davis-Monthan
44-61832			6/18/1945	24BG	8/5/1948	Salvaged at Harmon
44-61833			6/16/1945		7/20/1953	Reclaimed Davis-Monthan
44-61834	SLOW FREIGHT IV	SQUARE H	6/28/1945	98BG	6/17/1954	Class 26
44-61835	DRAGON LADY	BLKSTRPBLU		19BG	10/31/1951	Exploded, crashed 15 minutes after T/O
44-61836		CIRC R 41	6/15/1945	6BG		Survived War
44-61836		TRI U	6/15/1945	462BG	3/16/1954	Reclaimed
44-61837			6/16/1945	4141BU	9/22/1953	Reclaimed
44-61838		SQUARE A	6/16/1945	301BG	6/2/1953	Reclaimed Davis-Monthan
44-61839			6/18/1945	4141BU	9/22/1953	Reclaimed Pyote
44-61840		CIRC X 57	6/15/1945	9BG	5/6/1949	Salvaged at Kelly
44-61841			6/16/1945	308??	11/9/1953	Reclaimed Randolph
44-61842			6/19/1945	19BG	6/30/1949	Salvaged at Kelly

Serial #	Name	Identification	Delv	Assign	Off Inv	Circumstances
44-61843	MARGIE'S MAD GREEK III		6/19/1945	71RG	8/26/1954	Reclaimed
44-61843			6/19/1945	31SRP	8/26/1954	Reclaimed
44-61844			6/16/1945	4141BU	9/22/1953	Reclaimed
44-61845	PACIFIC PRINCESS (?)	TRI N	6/19/1945	444BG		Survived war
44-61845			6/19/1945	90SRC	7/13/1953	Reclaimed
44-61846			7/24/1945	4141BU	9/22/1953	Reclaimed Pyote
44-61847					7/13/1950	Surveyed
44-61848		TRI S 19	6/18/1945	40BG		Survived war
44-61848		CIRCLE E	6/18/1945	22BG	3/18/1949	Reclaimed Petersen
44-61849			6/22/1945	4135BU	2/10/1948	Reclaimed Hill
44-61850			6/20/1945	3510TTW10/8/53	10/8/1953	Reclaimed
44-61851	LA BOHEME		6/21/1945	68BG	11/9/1953	Reclaimed Randolph
44-61852			6/19/1945	581ASL	11/30/1953	Reclaimed Clark Field
44-61853			6/26/1945	3415TTW	3/17/1954	Reclaimed Davis-Monthan
44-61854	BUTTERFLY BABY		6/20/1945	31SRP	7/14/1954	Reclaimed Davis-Monthan
44-61855			6/21/1945	31SRP		Unknown
44-61856			6/22/1945	308??	9/9/1954	Reclaimed Randolph
44-61857			6/21/1945	4105BU	8/19/1954	Reclaimed Kelly
44-61858			6/20/1945	247BU	9/17/1946	Surveyed McChord
44-61859			6/21/1945	68BG	9/9/1954	Salvaged at Davis-Monthan
44-61860			6/22/1945	55SRC	5/18/1953	Reclaimed Robins
44-61861		SQUARE Y	6/23/1945	307BG	6/27/1954	Reclaimed Kadena
44-61862			6/22/1945	FEAF	4/29/1948	Salvaged overseas
44-61863			6/22/1945	247BU	3/31/1949	Reclaimed Tinker
44-61864			6/23/1945	303BG	3/1/1953	Reclaimed Davis-Monthan
44-61865		CIRCLE T	6/22/1945	44BG	6/17/1953	Reclaimed Davis-Monthan
44-61866			6/22/1945	4105BU	11/17/1953	Reclaimed Kelly
44-61867					5/20/1952	Salvaged
44-61868			6/23/1945	3415TTW	8/27/1953	Reclaimed Davis-Monthan
44-61869			6/25/1945	376BG	1/5/1953	Reclaimed Davis-Monthan
44-61870			6/29/1945	3415TTW	3/17/1954	Reclaimed Davis-Monthan
44-61871			6/25/1945	3510FTW	7/14/1954	Salvaged at Davis-Monthan
44-61872	ACE IN THE HOLE	SQUARE H	6/28/1945	98BG	9/13/1954	Reclaimed Davis-Monthan
44-61872	SAC'S APPEAL	SQUARE H	6/28/1945	98BG	9/13/1954	Reclaimed Davis-Monthan
44-61872			6/28/1945	307BG	9/13/1954	Reclaimed Davis-Monthan
44-61873			6/25/1945	581ARC	1/5/1954	Reclaimed
44-61874		SQUARE H	6/25/1945	98BG	10/4/1953	Reclaimed Davis-Monthan
44-61875			6/23/1945	308BG	7/16/1953	Reclaimed Davis-Monthan
44-61876			6/25/1945	4127BU	9/27/1954	Reclaimed McClellan
44-61877			6/25/1945	376BG	1/5/1953	Reclaimed Davis-Monthan
44-61878		SQUARE H	6/25/1945	98BG	7/22/1953	Reclaimed
44-61879			6/27/1945	68BG	9/29/1953	Reclaimed Davis-Monthan
44-61880			6/27/1945	3415TTW	7/14/1954	Salvaged at Davis-Monthan
44-61881			6/26/1945	308BG	9/13/1954	Reclaimed Davis-Monthan
44-61882			6/26/1945	54WRS	8/19/1954	Reclaimed Hill
44-61883			6/30/1945	301BG	5/16/1950	Transferred to RAF
44-61883			5/16/1950	RAF	7/22/1953	Returned to USAF
44-61883			7/22/1953	USAF	9/8/1954	Reclaimed
44-61884			6/27/1945	580ARC	6/20/1954	Reclaimed
44-61885		SQUARE A	6/28/1945	301BG		Converted to WB-29
44-61885			6/28/1945	56WRS	9/1/1954	Salvaged Yokota
44-61886			6/25/1945	310BW	11/9/1953	Reclaimed Randolph
44-61887			6/25/1945	303BW	3/1/1953	Reclaimed Davis-Monthan
44-61888			6/25/1945	54WRS	3/9/1954	Reclaimed
44-61889			6/27/1945	429BU	11/21/1950	Transferred to RAF
44-61889			11/21/1950	RAF	2/22/1954	Returned to USAF
44-61889			2/22/1954	USAF	8/18/1954	Transferred to Navy
44-61890		SQUARE A	6/28/1945	301BG	12/6/1948	Salvaged at Peterson
44-61891			6/26/1945	373WRS	9/22/1953	Reclaimed Pyote
44-61892			6/26/1945	247BU	9/22/1953	Reclaimed Pyote
44-61893		SQUARE A	6/27/1945	301BG		Converted to WB-29
44-61893			6/27/1945	56WRS	9/1/1954	Reclaimed
44-61894		SQUARE H	6/28/1945	98BG	2/11/1951	Transferred to RAF
44-61894			2/11/1951	RAF	1/8/1953	Crashed England
44-61895			6/26/1945	427BU	12/1/1950	Transferred to RAF

I. Master List

Serial #	Name	Identification	Delv	Assign	Off Inv	Circumstances
44-61895			12/1/1950	RAF	8/1/1953	Returned to USAF
44-61895			8/1/1953	USAF	7/14/1954	Reclaimed Davis-Monthan
44-61896		SQUARE H	6/28/1945	98BG	8/16/1954	Reclaimed Davis-Monthan
44-61897		SQUARE A	6/27/1945	301BG	3/15/1950	Transferred to RAF
44-61897			3/15/1950	RAF	8/8/1957	Broken up Shoeburyness
44-61898		376BW	6/28/1945	376BW	5/15/1951	Transferred to RAF
44-61898			5/15/1951	RAF	1/5/1954	Returnerd to USAF
44-61898			1/5/1954	USAF	7/14/1954	Salvaged
44-61899			6/28/1945	68BW	8/16/1954	Reclaimed Davis-Monthan
44-61900			6/28/1945	303BW	7/14/1954	Reclaimed McClellan
44-61901			6/29/1945	56WRS	3/9/1954	Reclaimed
44-61902		BLKSTRPGRN		19BG		Unknown
44-61903			6/29/1945	68BW	11/9/1953	Reclaimed Randolph
44-61904			6/30/1945	376BW	7/22/1954	Reclaimed Wright-Patterson
44-61905		CIRC R 18	6/29/1945	6BG	5/9/1954	Reclaimed
44-61906			6/29/1945	320BW	9/28/1954	Reclaimed McClellan
44-61907			6/29/1945	308BW	3/27/1954	Reclaimed Tinker
44-61908		SQUARE Y		307BG	2/1/1951	Mid-air with 42-65392, crashed in sea
44-61909			6/29/1945	301BG		Converted to WB-29
44-61909			6/29/1945	54WRS	9/22/1954	Reclaimed Tinker
44-61910			7/17/1945	2759EXPG	7/15/1954	Reclaimed Tinker
44-61911			6/30/1945	4141BU	9/22/1953	Reclaimed Pyote
44-61912			6/30/1945	68BW	7/22/1952	Surveyed
44-61913			6/30/1945		9/13/1954	Reclaimed Davis-Monthan
44-61914			7/2/1945	376BG	1/5/1953	Reclaimed Davis-Monthan
44-61915			6/30/1945	376BG	1/4/1953	Reclaimed Davis-Monthan
44-61916			7/6/1945	54RCN	9/22/1953	Reclaimed Pyote
44-61917			6/30/1945	376BG	1/6/1953	Reclaimed Davis-Monthan
44-61918		SQUARE I	6/30/1945	91RCN	9/9/1954	Salvaged at Davis-Monthan
44-61919			6/30/1945	310BG	6/17/1953	Reclaimed Davis-Monthan
44-61920		BLKSTRP	6/30/1945	19BG	9/15/1954	Reclaimed Davis-Monthan
44-61921			7/2/1945	236BU	9/22/1953	Reclaimed Pyote
44-61922			7/2/1945	11RDC	7/14/1954	Salvaged at Davis-Monthan
44-61923		CIRCLE W	7/2/1945	92BG	7/13/1950	Crashed near Gogo Island
44-61924					6/2/1952	Surveyed
44-61925	CHIEF SPOKANE	SQUARE H	9/4/1945	98BG		Unknown
44-61925		CIRCLE W		92BG		Unknown
44-61926			7/3/1945		9/9/1954	Salvaged at Davis-Monthan
44-61927		SQUARE H	7/2/1945	98BG	8/16/1954	Reclaimed Davis-Monthan
44-61928		SQUARE Y	7/2/1945	307BG	10/5/1954	Reclaimed Davis-Monthan
44-61929	HONEY BUCKET HONSHOS	SQUARE I	7/3/1945	91RCN	10/5/1954	Reclaimed McClellan
44-61930			7/5/1945	4136BU	5/27/1948	Salvaged at Tinker
44-61931		CIRCLE X	7/5/1945	5RCN	7/14/1954	Salvaged at Davis-Monthan
44-61932	OUR GIRL	SQUARE H		98BG	10/24/1951	Shot down Wonson Harbor, Korea
44-61932	OUR GAL	BLKSTRP		19BG	11/23/1951	Surveyed with 98th BG
44-61933					11/28/1951	Salvaged
44-61934			7/3/1945	5RCN	7/28/1954	Salvaged
44-61935		CIRCLE T	7/5/1945	44BG	7/14/1954	Salvaged at Davis-Monthan
44-61936		SQUARE H	8/6/1945	98BG	9/13/1954	Reclaimed
44-61937					7/31/1951	Transferred to RAF
44-61937			7/13/1951	RAF	4/17/1958	Broken up; scrapped
44-61938			7/9/1945	98BG	2/11/1951	Transferred to RAF
44-61938			2/11/1951	RAF	8/11/1953	Returned to USAF
44-61938			8/11/1953	USAF	7/14/1954	Salvaged
44-61939				55RCN	5/5/1952	Surveyed
44-61940	MISS NORTH CAROLINA	SQUARE Y		307BG	10/23/1951	Lost Namsi Airfield raid
44-61940				97BG	11/23/1951	Salvaged
44-61941			7/5/1945	3510FTW	4/7/1953	Salvaged Randolph
44-61942			7/5/1945	4105BU	6/17/1953	Reclaimed Davis-Monthan
44-61943			7/5/1945	4141BU	9/22/1953	Reclaimed Pyote
44-61944		CIRCLE Z	7/7/1945	90RCN	9/8/1953	Reclaimed Davis-Monthan
44-61945		SQUARE V		55RCN	4/12/1951	Surveyed
44-61946			7/6/1945	4136BU	11/17/1953	Surveyed Hill
44-61947		SQUARE I	7/6/1945	91SRC	8/19/1954	Reclaimed Kelly
44-61948	THE GYPSY		7/6/1945	91SRP	7/14/1954	Reclaimed

Serial #	Name	Identification	Delv	Assign	Off Inv	Circumstances
44-61948	SHEER MADNESS	BLKSTRP	7/6/1945	19BG	7/14/1954	Reclaimed McClellan
44-61949			7/6/1945	4127BU	3/25/1953	Reclaimed McClellan
44-61950		CIRCLE E BLU	7/6/1945	22BG	7/14/1954	Reclaimed
44-61951	OUR L'LASS	CIRCLE W	7/6/1945	92BG	8/26/1954	Reclaimed
44-61951	OUR L'LASS	SQUARE I	7/6/1945	91RCN	10/00/50	Transferred to 91RCN when 92nd rotated
44-61951		CIRCLE X	7/6/1945	91RCN	8/26/1954	Reclaimed
44-61952					7/20/1951	Transferred to RAF
44-61952			7/20/1951	RAF	2/27/1954	Returned to USAF
44-61952			2/27/1954	USAF	9/8/1954	Reclaimed Davis-Monthan
44-61953		SQUARE H	7/6/1945	98BG	6/17/1954	Class 26 Sheppard
44-61954		CIRCLE E	7/7/1945	22BG	9/9/1954	Salvaged at Davis-Monthan
44-61955			7/7/1945	4141BU	9/22/1953	Reclaimed Pyote
44-61956		SQUARE V	7/9/1945	55RCN	9/22/1953	Reclaimed Pyote
44-61957		BLKSTRPGRN	7/7/1945	19BG	9/28/1954	Reclaimed McClellan
44-61958			7/10/1945	310BW	6/17/1953	Reclaimed Davis-Monthan
44-61959			7/11/1945	308BW	11/9/1953	Reclaimed Randolph
44-61960	MARY LOU	BLK SQ F	7/9/1945	611BEXP	8/19/1954	Reclaimed Kelly
44-61961		SQUARE I	7/10/1945	91RCN	8/19/1954	Reclaimed Kelly
44-61962		CIRCLE M	7/10/1945	93BG	5/1/1950	Salvaged at Kelly
44-61963					7/3/1951	Transferred to RAF
44-61963			7/3/1951	RAF	10/1/1957	Scrapped england
44-61964			7/9/1945	54WRS	5/24/1954	Reclaimed McClellan
44-61965		CIRCLE Z	7/16/1945	90RCN	7/27/1953	Reclaimed Davis-Monthan
44-61966	RAIDEN MAIDEN	47	7/10/1945	468BG	6/17/1953	Reclaimed Davis-Monthan
44-61967		BLKSTRPBLU		19BG	6/10/1952	Missing over North Korea
44-61968					7/22/1951	Transferred to RAF
44-61968			7/22/1951	RAF	7/29/1955	Scrapped England
44-61969		T SQ 11 (3)		498BG	6/4/1951	Transferred to RAF
44-61969			6/4/1951	RAF	1/9/1954	Returned to USAF
44-61969			1/9/1954	USAF	7/14/1954	Salvaged Davis-Monthan
44-61970			7/11/1945	9BG	6/17/1953	Reclaimed Davis-Monthan
44-61971			7/11/1945	580ARC	6/23/1953	Reclaimed
44-61972			7/12/1945	93BG	9/14/1954	Reclaimed Hamilton
44-61973			7/12/1945	376BW	6/14/1954	Reclaimed
44-61974			7/11/1945	53WRS		Unknown
44-61974		CIRCLE E	7/11/1945	22BG		Converted to WB-29
44-61975			7/13/1945	582ARC	9/12/1954	Reclaimed Molesworth
44-61976		CIRCLE Z	7/13/1945	90RCN	9/9/1954	Salvaged at Davis-Monthan
44-61977			7/12/1945	376BW	9/9/1954	Salvaged at Davis-Monthan
44-61978					3/31/1951	Transferred to RAF
44-61978			3/31/1951	RAF	1/5/1954	Returned to USAF
44-61978			1/5/1954	USAF	8/22/1954	Transferred to Navy
44-61979			7/14/1945	3415tTTN	7/14/1954	Salvaged at Davis-Monthan
44-61980			7/13/1945	308BW	9/30/1953	Reclaimed Davis-Monthan
44-61981		SQUARE A	7/13/1945	301BG	10/4/1954	Reclaimed
44-61981	CHEECHAKO	SQUARE I	7/13/1945	91RCN		Transferred to 91RCN
44-61982		TRIANGLE O	7/16/1945	97BG	10/15/1950	Transferred to RAF
44-61982			10/15/1950	RAF	8/11/1953	Returned to USAF
44-61982			8/11/1953	USAF	7/14/1954	Reclaimed
44-61983					7/14/1954	Salvaged at Davis-Monthan
44-61984		CIRCLE Z	7/16/1945	90RCN	9/8/1953	Reclaimed Davis-Monthan
44-61985			7/14/1945	7RDC	7/14/1954	Salvaged at Davis-Monthan
44-61986		SQUARE EMPTY	7/14/1945	2BG		Unknown
44-61987		SQUARE A	7/17/1945	301BG	7/31/1953	Reclaimed
44-61988						Unknown
44-61989			7/14/1945	16PHRCN	8/19/1954	Reclaimed Kelly
44-61990	QUANTRELL'S RAIDERS		7/17/1945	320BW	9/28/1954	Reclaimed McClellan
44-61991	THE ANGELLIC PIG	BLK SQ F	7/14/1945	40BG	8/19/1954	Reclaimed Kelly
44-61991			7/14/1945	40BG	8/19/1954	Reclaimed Kelly
44-61992			7/14/1945	320BW	9/28/1954	Reclaimed McClellan
44-61993			7/19/1945	4141BU	6/17/1953	Reclaimed Davis-Monthan
44-61994			7/16/1945	68BW	9/14/1954	Reclaimed Davis-Monthan
44-61995			7/17/1945	4141BU	9/22/1953	Reclaimed Pyote
44-61996			7/17/1945	320BW	7/14/1954	Reclaimed Davis-Monthan
44-61997		CIRCLE Z	7/20/1945	90RCN	9/8/1953	Reclaimed

I. Master List

Serial #	Name	Identification	Delv	Assign	Off Inv	Circumstances
44-61998			7/19/1945	4105BU	9/14/1953	Reclaimed Tinker
44-61999	OVER EXPOSED	SQUARE EMPTY	7/19/1945	2BG	11/3/1948	Crashed from Scampton to Burtonwood
44-61999	KAMODE HEAD	BLK SQ F	7/19/1945	16PRS		Transfer to 2 BG
44-62000			7/17/1945	91SRC	8/19/1954	Reclaimed Kelly
44-62001			7/20/1945	4141BU	2/11/1951	Transferred to RAF
44-62001			2/11/1951	RAF	11/3/1953	Returned to USAF
44-62001			11/3/1953	USAF	9/8/1954	Reclaimed
44-62002		BLKSTRP	7/19/1945	19BG	7/14/1954	Reclaimed McClellan
44-62003			7/28/1945	4141BU	3/18/1951	Transferred to RAF
44-62003			3/18/1951	RAF	1/22/1954	Returned to USAF
44-62003			1/22/1954	USAF	8/8/1954	Reclaimed
44-62004			7/20/1945	3415TTN	8/31/1953	Reclaimed
44-62005			7/21/1945	4141BU	3/18/1951	Transferred to RAF
44-62005			3/8/1951	RAF	1/22/1954	Returned to USAF
44-62005			1/22/1954	USAF	9/8/1954	Reclaimed
44-62006					5/3/1951	Transferred to RAF
44-62006			5/3/1951	RAF	2/25/1954	Returned to USAF
44-62006			2/25/1954	USAF	5/17/1954	Transferred to Navy
44-62007			7/21/1945	43BG	7/14/1954	Reclaimed McClellan
44-62008		BLKSTRP	7/20/1945	19BG	7/14/1954	Salvaged at Davis-Monthan
44-62009		SQUARE H	7/20/1945	98BG	8/16/1954	Assigned Storage Sqdn at Davis-Monthan
44-62010	LUCKY STRIKE II	SQUARE H	9/10/1945	98BG	9/8/1954	Reclaimed Davis-Monthan
44-62010	LUCKY STRIKE II	CIRCLE W	9/10/1945	92BG	9/8/1954	Reclaimed Davis-Monthan
44-62011		BLKSTRPGRN		19BG	12/30/1952	Shot down 25 miles N of Pyongyang
44-62012			7/23/1945	4121BU	12/3/1950	Transferred to RAF
44-62012			12/3/1950	RAF	11/3/1953	Returned to RAF
44-62012			11/3/1953	USAF	9/8/1954	Reclaimed
44-62013					5/3/1951	Transferred to RAF
44-62013			5/3/1951	RAF	1/19/1954	Returned to USAF
44-62013			1/19/1954	USAF	7/14/1954	Reclaimed Davis-Monthan
44-62014					6/4/1951	Transferred to RAF
44-62014			6/4/1951	RAF	1/5/1954	Returned to USAF
44-62014			1/5/1954	USAF	8/8/1954	Reclaimed Davis-Monthan
44-62015			7/18/1945		7/14/1954	Salvaged at Davis-Monthan
44-62016		SQUARE A	7/18/1945	301BG	3/16/1951	Transferred to RAF
44-62016			3/16/1951	RAF	6/18/1953	Returned to USAF
44-62016			6/18/1953	USAF	7/14/1954	Reclaimed Davis-Monthan
44-62017			7/20/1945	4141BU	9/22/1953	Reclaimed Pyote
44-62018			7/21/1945		6/17/1953	Reclaimed Davis-Monthan
44-62019			7/23/1945		6/4/1951	Transferred to RAF
44-62019			6/4/1951	RAF	12/1/1953	Returned to USAF
44-62019			12/1/1953	USAF	3/14/1954	Reclaimed Davis-Monthan
44-62020			7/21/1945	4000EXG		Reclaimed Griffis
44-62021			7/25/1945		3/29/1954	Reclaimed Davis-Monthan
44-62022	OLD DOUBLE DEUCE	CIRCLE X	7/21/1945	5RCN	12/7/1953	Reclaimed Tinker
44-62022	PEACHY		7/21/1945	91SRS		Unknown
44-62023		SQUARE Y	7/27/1945	307BG	9/28/1954	Reclaimed McClellan
44-62024			7/23/1945		12/10/1953	Reclaimed Davis-Monthan
44-62025		BLKSTRPRED		19BG		Unknown
44-62026			7/23/1945		5/25/1954	Assigned Storage Sqdn at Davis-Monthan
44-62027			7/24/1945		10/21/1954	Reclaimed Tinker
44-62028			7/24/1945	68BG		Unknown
44-62029			7/24/1945		7/19/1953	Assigned Storage Sqdn at Davis-Monthan
44-62030					6/20/1951	Transferred to RAF
44-62030			6/20/1951	RAF	1/15/1954	Returned to USAF
44-62030			1/15/1954	USAF	7/14/1954	Reclaimed Davis-Monthan
44-62031					3/14/1951	Transferred to RAF
44-62031			3/14/1951	RAF	2/2/1953	Crashed England, written off
44-62032					3/14/1951	Transferred to RAF
44-62032			3/14/1951	RAF	11/17/1953	Returned to USAF
44-62032			11/17/1953	USAF	7/14/1954	Reclaimed Davis-Monthan
44-62033			7/25/1945	376BG	1/15/1953	Assigned Storage Sqdn at Davis-Monthan
44-62034	PADDY DADDY		7/25/1945		9/1/1954	Reclaimed Yokota
44-62035			7/27/1945	310BG	5/25/1954	Assigned Storage Sqdn at Davis-Monthan
44-62036			7/25/1945		9/9/1954	Salvaged at Davis-Monthan

Serial #	Name	Identification	Delv	Assign	Off Inv	Circumstances
44-62037			7/27/1945	4141BU	3/20/1951	Transferred to RAF
44-62037			3/20/1951	RAF	2/15/1954	Returned to USAF
44-62037			2/15/1954	USAF	7/14/1954	Reclaimed Davis-Monthan
44-62038		TRIANGLE	7/27/1945	6BG	7/14/1954	Reclaimed McClellan
44-62039					2/19/1952	Salvaged
44-62040			7/27/1945	4141BU	7/23/1953	Assigned Storage Sqdn at Davis-Monthan
44-62041		SQUARE H	7/27/1945	98BG	3/14/1954	Assigned Storage Sqdn at Davis-Monthan
44-62042		SQUARE H		98BG		Unknown
44-62043			7/27/1945	4121BU	12/5/1950	Transferred to RAF
44-62043			12/5/1950	RAF	7/7/1953	Returned to USAF
44-62043			7/7/1953	USAF	9/14/1953	Reclaimed Davis-Monthan
44-62044		CIRCLE K	7/30/1945	9BG	7/14/1954	Salvaged at Davis-Monthan
44-62045			7/30/1945		9/9/1954	Salvaged at Davis-Monthan
44-62046					5/9/1951	Transferred to RAF
44-62046			5/9/1951	RAF	2/22/1954	Returned to USAF
44-62046			2/22/1954	USAF	6/8/1954	Trasnsferred to Navy
44-62047			7/30/1945	320BG	7/13/1953	Reclaimed Davis-Monthan
44-62048			7/30/1945		6/20/1954	Reclaimed Davis-Monthan
44-62049						Unknown
44-62050					3/13/1951	Transferred to RAF
44-62050			3/31/1951	RAF	4/4/1954	Returned to USAF
44-62050			4/4/1954	USAF	10/4/1954	Reclaimed Davis-Monthan
44-62051		CIRCLE Z	7/28/1945	90RCN	9/17/1953	Assigned Storage Sqdn at Davis-Monthan
44-62052	CAT GIRL	BLKSTRPGRN	6/17/1950	19BG	7/14/1954	Salvaged at Davis-Monthan
44-62053	ROCK HAPPY	BLKSTRPRED	7/28/1945	19BG	7/14/1954	Salvaged at Davis-Monthan
44-62054		CIRCLE T	7/30/1945	44BG	9/28/1954	Reclaimed McClellan
44-62055		CIRCLE T	7/30/1945	44BG	7/14/1954	Reclaimed McClellan
44-62056		CIRCLE T	7/30/1945	44BG	7/14/1954	Reclaimed McClellan
44-62057			7/30/1945	4121BU	12/6/1954	Reclaimed Tinker
44-62058					7/19/1951	Transferred to RAF
44-62058			7/19/1951	RAF	1/5/1954	Returned to USAF
44-62058			1/5/1954	USAF	7/14/1954	Salvaged Davis-Monthan
44-62059		CIRCLE Z	7/31/1945	90RCN	9/9/1954	Salvaged at Davis-Monthan
44-62060	SPIRIT OF FREEPORT	CIRCLE E	7/31/1945	22BG	10/00/50	Transferred to 307BG when 22nd rotated
44-62060	SPIRIT OF FREEPORT	SQUARE Y	10/00/50	307BG	9/28/1954	Reclaimed McClellan
44-62061			7/31/1945		9/22/1954	Reclaimed Tinker
44-62062		SQUARE A	7/31/1945	301BG	6/29/1950	Transferred to RAF
44-62062			6/29/1950	RAF	7/7/1953	Returned to USAF
44-62062			7/7/1953	USAF	9/8/1954	Reclaimed Davis-Monthan
44-62063	BIG SHMOO	BLKSTRPRED		19BG		Unknown
44-62064		SQUARE A	7/31/1945	301BG	12/16/1948	Surveyed Germany
44-62065		CIRCLE W	7/31/1945	92BG	7/21/1953	Reclaimed Hill
44-62066	TOWNSWICK'S TERRORS	SQUARE H	7/31/1945	98BG	11/17/1953	Reclaimed Hill
44-62067			7/31/1945	68BG	7/26/1953	In storage Davis-Monthan
44-62068		SQUARE A	7/31/1945	301BG	7/14/1954	Salvaged at Davis-Monthan
44-62069			8/7/1945	523ABG	6/22/1949	Salvaged at Lagens
44-62070	FIFI					Still flying with CAF
44-62070	LUCKY STRIKE			310BG		Still flying with CAF
44-62071		BLKSTRPBLU		19BG	10/27/1951	Crash landed K-14, Korea, salvaged
44-62072		SQUARE A	7/31/1945	301BG	7/14/1954	Salvaged at Davis-Monthan
44-62073		SQUARE Y		307BG	11/8/1952	Shot down on night mission
44-62074			8/1/1945	97BG	12/9/1950	Transferred to RAF
44-62074			12/9/1950	RAF	10/20/1953	Returned to USAF
44-62074			10/20/1953	USAF	7/14/1954	Reclaimed Davis-Monthan
44-62075		CIRCLE T	8/1/1945	44BG	11/7/1954	Assigned Storage Sqdn at Davis-Monthan
44-62076		SQUARE H	8/1/1945	98BG	3/8/1949	Surveyed at kadena
44-62077			8/2/1945		12/31/1954	Reclaimed Hickam
44-62078					12/29/1951	Surveyed
44-62079			8/3/1945	22BG	3/14/1954	Reclaimed Robins
44-62080			8/2/1945			Unknown
44-62081			8/1/1945	68BG		Unknown
44-62082		CIRCLE W	8/1/1945	92BG	7/14/1954	Reclaimed McClellan
44-62083		SQUARE Y		307BG	1/31/1952	Operational loss, missing from Kadena
44-62084		CIRCLE W	8/2/1945	92BG	9/9/1950	Flak at Wolbang-ni, exploded
44-62085		CIRCLE T	8/2/1945	44BG	9/28/1954	Reclaimed McClellan

I. Master List

Serial #	Name	Identification	Delv	Assign	Off Inv	Circumstances
44-62086		CIRCLE T	8/2/1945	44BG	9/13/1954	Reclaimed Davis-Monthan
44-62087			8/2/1945	68BG	7/22/1953	Assigned Storage Sqdn at Davis-Monthan
44-62088			8/2/1945	320BG	9/28/1954	Reclaimed McClellan
44-62089	SUKEBE GIRL		8/3/1945	56WRS	12/31/1954	Reclaimed Kindley
44-62090		SQUARE A	8/3/1945	301BG		Converted to weather plane
44-62090			8/3/1945		9/22/1954	Reclaimed Tinker
44-62091			8/6/1945	68BG	7/14/1953	Storage at Davis-Monthan
44-62092		SQUARE A	8/3/1945	301BG	7/14/1954	Salvaged at Davis-Monthan
44-62093			8/11/1945		11/20/1951	Crashed
44-62094			8/5/1945	54WRS	8/24/1954	Reclaimed Robins
44-62095			8/6/1945	310BG	7/21/1953	Assigned Storage Sqdn at Davis-Monthan
44-62096		SQUARE Y	8/8/1945	307BG	11/16/1954	Assigned Storage Sqdn at Davis-Monthan
44-62097			8/9/1945	611EXPGP	2/2/1955	Reclaimed Tinker
44-62098		SQUARE A	8/9/1945	301BG	4/7/1949	Surveyed Great Falls
44-62098		SQUARE H	8/9/1945	98BG	4/7/1949	Surveyed Great Falls
44-62099	NIPPONESE BABY	BLKSTRPRED	8/9/1945	19BG	7/14/1954	Salvaged at Davis-Monthan
44-62100		CIRCLE W	8/9/1945	92BG	3/18/1949	Salvaged England
44-62100		SQUARE EMPTY		2BG		Unknown
44-62101		SQUARE Y		307BG	4/3/1951	Transferred to RAF
44-62101			4/3/1951	RAF	7/22/1953	Returned to USAF
44-62101			7/22/1953	USAF	9/21/1954	Reclaimed Davis-Monthan
44-62102	WRIGHT'S DELIGHT	SQUARE H	10/00/50	98BG	11/19/1952	Ch'o-Do Island, North Korea
44-62102		CIRCLE W	8/9/1945	92BG	10/00/50	Transferred to 19BG or 307BG Oct 1950
44-62102		SQUARE Y	8/9/1945	307BG	11/19/1952	Ch'o-Do Island, North Korea
44-62103	HAULIN' ASS	SQUARE H	8/9/1945	98BG	9/28/1954	Reclaimed McClellan
44-62104			8/8/1945	326BU	12/4/1953	Reclaimed Robins
44-62105					5/25/1951	Transferred to RAF
44-62105			5/25/1951	RAF	2/15/1954	Returned to USAF
44-62105			2/15/1954	USAF	7/14/1954	Reclaimed Davis-Monthan
44-62106	READY WILLIN', WANTON	SQUARE H	8/8/1945	98BG	5/19/1953	Assigned Storage Sqdn at Davis-Monthan
44-62106	VICIOUS ROOMER	SQUARE H	8/8/1945	98BG	5/19/1953	Assigned Storage Sqdn at Davis-Monthan
44-62107		CIRCLE T	8/11/1945	44BG	7/14/1954	Reclaimed McClellan
44-62108	MYAKINAS	SQUARE H		98BG	4/10/1951	Crashed Taegu — two engines out
44-62109		CIRCLE T	8/11/1945	44BG	9/8/1954	Assigned Lake Charles
44-62110		BLKSTRP	10/2/1945	19BG	9/28/1954	Reclaimed McClellan
44-62111		CIRCLE W	9/14/1945	92BG	12/18/1950	Surveyed at Yokota
44-62112		CIRCLE Z	8/22/1945	90RCN		Unknown
44-62113		CIRCLE Z	8/25/1945	90RCN	9/9/1954	Salvaged at Davis-Monthan
44-62114		CIRCLE W	8/11/1945	92BG	6/9/1948	Surveyed Spokane
44-62115			8/11/1945		9/12/1954	Reclaimed Davis-Monthan
44-62116		SQUARE Y	8/20/1945	307BG	11/16/1954	Reclaimed Davis-Monthan
44-62117					5/17/1951	Transferred to RAF
44-62117			5/17/1951	RAF	1/5/1954	Returned to USAF
44-62117			1/5/1954	USAF	9/8/1954	Reclaimed Davis-Monthan
44-62118		CIRCLE Z	8/11/1945	90RCN	2/2/1954	Assigned Storage Sqdn at Davis-Monthan
44-62119					5/18/1952	Salvaged
44-62120			8/14/1945	68BG	9/13/1954	Reclaimed Davis-Monthan
44-62121		SQUARE Y	8/14/1945	307BG	1/7/1953	Assigned Storage Sqdn at Davis-Monthan
44-62122		CIRCLE Z	8/27/1945	90RCN	9/7/1953	Assigned Storage Sqdn at Davis-Monthan
44-62123			8/14/1945	308WG	11/17/1953	Reclaimed Hill
44-62124			8/18/1945	303BG	7/14/1954	Reclaimed McClellan
44-62125			8/14/1945	513WRS		Last at Tinker 10/54
44-62126			8/14/1945	308WRG	10/25/1949	Salvaged at Muroc
44-62127					4/1/1950	Surveyed
44-62128	WARM FRONT		8/14/1945	308WRG	11/27/1950	Transferred to RAF
44-62128	WARM FRONT		11/27/1950	RAF	1/27/1954	Crashed England
44-62129					3/30/1951	Transferred to RAF
44-62129			3/30/1951	RAF	1/4/1954	Returned to USAF
44-62129			1/4/1954	USAF	4/8/1954	Transferred to Navy
44-62130			8/14/1945	376BG	1/6/1955	Assigned Storage Sqdn at Davis-Monthan
44-62131			8/18/1945	376BG	1/5/1953	Assigned Storage Sqdn at Davis-Monthan
44-62132		CIRCLE T	8/18/1945	44BG	9/28/1954	Reclaimed McClellan
44-62133			8/24/1945	308BG	7/15/1953	Assigned Storage Sqdn at Davis-Monthan
44-62134		CIRCLE T	8/14/1945	44BG	7/14/1954	Reclaimed
44-62135		CIRCLE T	8/14/1945	44BG	6/15/1951	Transferred to RAF

Serial #	Name	Identification	Delv	Assign	Off Inv	Circumstances
44-62135			6/15/1951	RAF	3/16/1954	Returned to USAF
44-62135			3/16/1954	USAF	7/14/1954	Salvaged Davis-Monthan
44-62136			8/18/1945	320BG	9/28/1954	Reclaimed McClellan
44-62137		CIRCLE T	8/20/1945	44BG	7/14/1954	Reclaimed McClellan
44-62138			8/20/1945	310BW	3/28/1954	Assigned Storage Sqdn at Davis-Monthan
44-62139			8/22/1945	4000BU	7/22/1946	Surveyed at Wright-Patterson
44-62140			8/20/1945	320BG	7/14/1954	Reclaimed McClellan
44-62141		SQUARE H	8/21/1945	98BG	8/23/1949	Salvaged Scultho
44-62142		SQUARE A	8/21/1945	301BG	6/17/1953	Assigned Storage Sqdn at Davis-Monthan
44-62143		CIRCLE R	8/21/1945	28BG	5/11/1953	Assigned Storage Sqdn at Davis-Monthan
44-62144			8/21/1945			Unknown
44-62145		SQUARE P	8/21/1945	306BW	9/13/1954	Reclaimed Davis-Monthan
44-62146			8/22/1945		2/7/1955	Reclaimed Tinker
44-62147			8/23/1945	4121BU	10/20/1950	Surveyed at Randolph
44-62148			8/22/1945	4141BU	9/22/1953	Reclaimed Pyote
44-62149		CIRCLE T	8/21/1945	44BG	6/17/1953	Reclaimed Davis-Monthan
44-62150		SQUARE Y	8/21/1945	307BG	11/16/1954	Assigned Storage Sqdn at Davis-Monthan
44-62151	LONESUM POLL CAT		8/22/1945	308WG		Unknown
44-62151	LONESOME POLECAT		8/22/1945	308WG		Unknown — at Eielsen 12/31/54
44-62152	STATESIDE REJECT	BLKSTRPRED	8/24/1950	19BG	12/19/1950	Surveyed Kadena, 11/15/50 T/O crash
44-62153			8/23/1945	2753BU	3/10/1951	Transferred to RAF
44-62153			3/10/1951	RAF	3/28/1954	Returned to USAF
44-62153			3/28/1954	USAF	5/12/1954	Reclaimed Davis-Monthan
44-62154					3/13/1951	Transferred to RAF
44-62154			3/13/1951	RAF	7/28/1953	Returned to USAF
44-62154			7/28/1953	USAF	9/8/1954	Reclaimed Davis-Monthan
44-62155			8/22/1945	4121BU	11/20/1950	Transferred to RAF
44-62155			11/20/1950	RAF	8/18/1953	Returned to USAF
44-62155			8/18/1953	USAF	9/8/1954	Reclaimed Davis-Monthan
44-62156			8/23/1945	308WG		At Hickam 12/31/54
44-62157	POLAR QUEEN		8/23/1945	308WG	12/16/1948	Salvaged at Ladd
44-62158		SQUARE A	8/28/1945	301BG	7/14/1954	Salvaged at Davis-Monthan
44-62159		SQUARE A	8/23/1945	301BG	7/7/1950	Transferred to RAF
44-62159			7/7/1950	RAF	11/17/1953	Returned to USAF
44-62159			11/17/1953	USAF	7/14/1954	Reclaimed Davis-Monthan
44-62160		CIRCLE E	8/23/1945	22BG	7/14/1954	Reclaimed McClellan
44-62161		CIRCLE Z	10/31/1945	90RCN	5/11/1953	Assigned Storage Sqdn at Davis-Monthan
44-62162			8/24/1945	90BG	12/12/1953	Assigned Storage Sqdn at Davis-Monthan
44-62163	POLAR QUEEN		8/25/1945	375WRS		At Tinker 1/7/55
44-62164					6/25/1952	Surveyed
44-62165		CIRCLE Z	8/24/1945	90RCN	9/18/1953	Assigned Storage Sqdn at Davis-Monthan
44-62166	FUJIGMO	BLKSTRPRED		19BG	7/22/1952	Engine fire, exploded
44-62167		SQUARE H		98BG	8/30/1952	Crashed 4 miles east of Taegu (K-2)
44-62168			8/27/1945		3/28/1954	Assigned Storage Sqdn at Davis-Monthan
44-62169			8/27/1945	4141BU	2/28/1953	Reclaimed Pyote
44-62170		BLKSTRP	8/28/1945	19BG	7/14/1954	Salvaged at Davis-Monthan
44-62171			8/28/1945		3/28/1954	Assigned Storage Sqdn at Davis-Monthan
44-62172			9/10/1945	4105BU	6/29/1953	Assigned Storage Sqdn at Davis-Monthan
44-62173		SQUARE H	9/10/1945	98BG	9/28/1954	Reclaimed McClellan
44-62174			8/29/1945	376BG	1/5/1953	Assigned Storage Sqdn at Davis-Monthan
44-62175			8/29/1945	68BG	3/4/1954	Assigned Storage Sqdn at Davis-Monthan
44-62176			8/28/1945	4141BU	9/22/1953	Reclaimed Pyote
44-62177		CIRCLE Z	9/1/1945	90RCN	6/11/1953	Assigned Storage Sqdn at Davis-Monthan
44-62178			8/27/1945	308BG		Unknown
44-62179			8/28/1945	308BG		Unknown
44-62180						Unknown
44-62181		SQUARE P	8/28/1945	306BG	7/14/1954	Reclaimed McClellan
44-62182		SQUARE Y	8/29/1945	307BG	11/16/1954	Reclaimed Davis-Monthan
44-62183	HOT TO GO	BLKSTRPGRN		19BG	6/10/1952	Mig Rocket attack
44-62184		SQUARE Y	8/29/1945	307BG	1/27/1955	Assigned AMC
44-62185			8/30/1945	68BG	10/31/1954	Assigned Storage Sqdn at Davis-Monthan
44-62186	CHIEF MAC'S 10 LITTLE INDIANS	SQUARE H		98BG		Unknown
44-62187			9/10/1945	68BG		Unknown
44-62188	TREMLIN GREMLINS	CIRCLE W	8/30/1945	92BG	7/14/1954	Salvaged at Davis-Monthan
44-62189			8/29/1945	4105BU	5/10/1954	Reclaimed Davis-Monthan

I. Master List

Serial #	Name	Identification	Delv	Assign	Off Inv	Circumstances
44-62190		YLW BANDS	9/10/1945		9/22/1954	Reclaimed Tinker
44-62191			9/10/1945	310BG	3/28/1954	Assigned Storage Sqdn at Davis-Monthan
44-62192		SQUARE Y	8/30/1945	307BG	11/16/1954	Reclaimed Davis-Monthan
44-62193			8/30/1945	4121BU	7/14/1954	Salvaged at Davis-Monthan
44-62194		YLW BANDS	9/10/1945	3RESGP	10/1/1954	Reclaimed Davis-Monthan
44-62195			9/4/1945	375RCNW	9/22/1954	Reclaimed Tinker
44-62196	NEVER HOPPEN	CIRCLE E	8/30/1945	22BG	9/28/1954	Reclaimed McClellan
44-62197			9/4/1945	308WG	1/3/1955	At Tinker
44-62198		SQUARE EMPTY		2BG	10/31/1950	Transferred to RAF
44-62198		SQUARE Y	8/30/1945	307BG	10/31/1950	Transferred to RAF
44-62198			10/31/1950	RAF	8/25/1953	Returned to USAF
44-62198			8/25/1953	USAF	9/14/1953	Reclaimed Tinker
44-62199		CIRCLE E	8/30/1945	22BG	5/18/1953	Assigned Storage Sqdn at Davis-Monthan
44-62200			8/30/1945	308WRG	6/18/1948	Surveyed at Fairfield
44-62201		BLKSTRP		19BG	7/14/1954	Reclaimed Davis-Monthan
44-62201						Unknown; WB-29
44-62202			9/4/1945		11/25/1954	Reclaimed Tinker
44-62203			9/4/1945	580ASL		Unknown-Record card page missing
44-62204			9/4/1945	581ASL		Unknown
44-62205			9/4/1945	4000BU	7/14/1954	Salvaged at Davis-Monthan
44-62206			9/4/1945	581ASL		At Kadena 10/24/54
44-62207	TO EACH HIS OWN	SQUARE H		98BG		Unknown
44-62208		CIRCLE W	9/4/1945	92BG	10/00/50	Transfer to 307BG when 92nd rotated
44-62208		SQUARE Y	10/00/50	307BG	2/1/1955	Assigned Storage Sqdn at Davis-Monthan
44-62209		SQUARE Y	9/20/1945	307BG	11/16/1954	Assigned Storage Sqdn at Davis-Monthan
44-62210			9/30/1945		7/14/1954	Salvaged at Davis-Monthan
44-62211			9/19/1945	581ASL		At Kadena 10/24/54
44-62212			9/20/1945		7/14/1954	Salvaged at Davis-Monthan
44-62213	FRIENDLY UNDERTAKER	SQUARE H	9/20/1945	98BG	7/14/1954	Salvaged at Davis-Monthan
44-62213		CIRCLE W	9/20/1945	92BG	10/00/50	Transfer to 307BG when 92nd rotated
44-62214			9/5/1945	308WG		At Eielson 12/31/54
44-62215		SQUARE EMPTY		2BG	2/14/1952	Surveyed
44-62216	BLIZZARD WIZARD		9/1/1945	375WRS	10/12/1950	Surveyed Shemya
44-62216	DUFFY'S TAVERN		9/1/1945	375WRS	7/23/1949	Surveyed Shemya
44-62217			9/5/1945		2/18/1954	Reclaimed Far East
44-62218		BLKSTRP	9/6/1945	19BG	7/14/1954	Salvaged at Davis-Monthan
44-62218		SQUARE H	9/6/1945	98BG	7/14/1954	Salvaged at Davis-Monthan
44-62219			9/26/1945	308WG		Unknown
44-6221X	SHINPAINAI	SQUARE H		98BG		Unknown
44-62220			9/5/1945	57WRS		Unknown
44-62221		CIRCLE T	9/30/1945	44BG	6/7/1953	Assigned Storage Sqdn at Davis-Monthan
44-62222			9/6/1945	581ASL		Unknown
44-62223					3/12/1951	Salvaged
44-62224	GUARDIANS OF PEACE		9/7/1945		7/14/1954	Salvaged at Davis-Monthan
44-62224	THE WANDERER	BLKSTRP	9/7/1945	19BG	7/14/1954	Salvaged at Davis-Monthan
44-62224	THE WANDERER	CIRCLE W	9/7/1945	92BG		Transfer to 19BG or 98BG when rotated
44-62224	THE WANDERER	SQUARE H	9/7/1945	98BG	7/14/1954	Salvaged at Davis-Monthan
44-62225			9/6/1945		5/6/1954	Reclaimed McClellan
44-62226		CIRCLE Z	9/20/1945	90RCN	5/28/1951	Transferred to RAF
44-62226			5/28/1951	RAF	2/22/1954	Returned to USAF
44-62226			2/22/1954	USAF	9/9/1954	Salvaged Davis Monthan
44-62227					8/8/1951	Transferred to RAF
44-62227			8/8/1951	RAF	2/22/1954	Returned to USAF
44-62227			2/22/1954	USAF	7/14/1954	Salvaged at Davis-Monthan
44-62228			9/19/1945	2622BU	8/9/1947	Salvaged at Davis-Monthan
44-62229			9/7/1945		12/31/1954	Last reported at Eielsen
44-62230		CIRCLE Z	9/30/1945		9/9/1954	Salvaged at Davis-Monthan
44-62231	FOREVER AMBLING	SQUARE EMPTY	9/19/1945	2BG	2/11/1951	Transferred to RAF
44-62231			2/11/1951	RAF	7/28/1953	Returned to USAF
44-62231			7/28/1953	USAF	9/14/1953	Salvaged Davis Monthan
44-62232			9/19/1945			Last reported at Eielsen
44-62233			9/19/1945	308WG	4/6/1950	Reclaimed Davis-Monthan
44-62234			9/19/1945		3/18/1951	Transferred to RAF
44-62234			3/18/1951	RAF	8/18/1953	Returned to USAF
44-62234			8/18/1953	USAF	9/8/1954	Reclaimed Davis-Monthan

THE B-29 SUPERFORTRESS

Serial #	Name	Identification	Delv	Assign	Off Inv	Circumstances
44-62235					6/19/1951	Transferred to RAF
44-62235			6/19/1951	RAF	7/7/1953	Returned to USAF
44-62235			7/7/1953	USAF	9/8/1954	Reclaimed Davis-Monthan
44-62236					5/24/1951	Transferred to RAF
44-62236			5/24/1951	RAF	1/19/1954	Returned to USAF
44-62236			1/19/1954	USAF	7/14/1954	Reclaimed Davis-Monthan\
44-62237	(EIGHT BALL)	SQUARE H		98BG		Unknown
44-62238		CIRCLE Z	9/20/1945	90RCN	4/7/1951	Transferred to RAF
44-62238			4/7/1951	RAF	10/20/1953	Returned to USAF
44-62238			10/20/1953	USAF	9/8/1954	Reclaimed Davis-Monthan
44-62239					7/12/1950	Transferred to RAF
44-62239			7/12/1950	RAF	10/1/1957	Broken up; scrapped
44-62240		CIRCLE Z	9/20/1945	90RCN	9/9/1954	Salvaged at Davis-Monthan
44-62241					6/12/1951	Transferred to RAF
44-62241			6/12/1951	RAF	1/8/1953	Crashed, scrapped England
44-62242					6/5/1951	Transferred to RAF
44-62242			6/5/1951	RAF	11/3/1953	Returned to USAF
44-62242			11/3/1953	USAF	9/8/1954	Reclaimed Davis-Monthan
44-62243					5/25/1951	Transferred to RAF
44-62243			5/25/1951	RAF	2/15/1954	Returned to USAF
44-62243			2/15/1954	USAF	9/8/1954	Reclaimed Davis-Monthan
44-62244					7/4/1951	Transferred to RAF
44-62244			7/4/1951	RAF	11/17/1953	Returned to USAF
44-62244			11/17/1953	USAF	9/8/1954	Reclaimed Davis-Monthan
44-62245		CIRCLE M	9/30/1945	93BG	7/14/1954	Reclaimed McClellan
44-62246		CIRCLE Z	9/20/1945	90RCN	9/9/1954	Salvaged at Davis-Monthan
44-62247					8/5/1951	Surveyed
44-62248		CIRCLE Z	9/20/1945	90RCN	9/9/1954	Salvaged at Davis-Monthan
44-62249			9/20/1945	320BG	9/28/1954	Reclaimed McClellan
44-62250					5/8/1951	Transferred to RAF
44-62250			5/8/1951	RAF	1/4/1954	Returned to USAF
44-62250			1/4/1954	USAF	7/14/1954	Reclaimed Davis-Monthan
44-62251		CIRCLE Z	9/20/1945	90RCN	9/9/1954	Salvaged at Davis-Monthan
44-62252		SQUARE Y		307BG	5/14/1951	Salvaged, crashed Suwon 4/12/51
44-62253	(DRAGON AND LADY)	SQUARE H	9/20/1945	98BG	7/14/1954	Salvaged at Davis-Monthan
44-62253	RELUCTANT DRAG'ON	SQUARE H	9/20/1945	98BG	7/14/1954	Salvaged at Davis-Monthan
44-62253	RELUCTANT DRAG'ON	BLKSTRP	9/20/1945	19BG	7/14/1954	Salvaged at Davis-Monthan
44-62254					5/8/1951	Transferred to RAF
44-62254			5/8/1951	RAF	1/3/1952	Written off, scrapped England
44-62255					8/8/1951	Transferred to RAF
44-62255			8/8/1951	RAF	2/25/1954	Returned to USAF
44-62255			2/25/1954	USAF	5/10/1954	Reclaimed Davis-Monthan
44-62256					5/24/1951	Transferred to RAF
44-62256			5/24/1951	RAF	3/16/1954	Returned to USAF
44-62256			3/16/1954	USAF	8/16/1954	Reclaimed Davis-Monthan
44-62257					5/28/1951	Transferred to RAF
44-62257			5/28/1951	RAF	8/11/1953	Returned to USAF
44-62257			8/11/1953	USAF	7/14/1954	Reclaimed Davis-Monthan
44-62258						Transferred to RAF
44-62258				RAF	1/15/1954	Returned to USAF
44-62258			1/15/1954	USAF	7/14/1954	Salvaged Davis-Monthan
44-62259					6/5/1951	Transferred to RAF
44-62259			6/5/1951	RAF	2/17/1954	Returned to USAF
44-62259			2/17/1954	USAF	7/14/1954	Salvaged Davis-Monthan
44-62260			11/23/1945	581ASL		Last reported Kadena 10/24/54
44-62261	NIP ON NEES	SQUARE H	11/13/1945	98BG	8/16/1954	Assigned Storage Sqdn at Davis-Monthan
44-62262			11/13/1945	4141BU	9/22/1953	Reclaimed Pyote
44-62263			2/21/1946	3200PTS	7/14/1954	Salvaged Davis-Monthan
44-62264			12/14/1945	581ASL	5/10/1954	Reclaimed Davis-Monthan
44-62265					6/27/1951	Transferred to RAF
44-62265			6/27/1951	RAF	1/15/1954	Returned to USAF
44-62265			1/15/1954	USAF	5/10/1954	Reclaimed Davis-Monthan
44-62266					5/28/1951	Transferred to RAF
44-62266			5/28/1951	RAF	2/22/1954	Returned to USAF
44-62266			2/22/1954	USAF	6/27/1954	Reclaimed Davis-Monthan

I. Master List

Serial #	Name	Identification	Delv	Assign	Off Inv	Circumstances
44-62267			1/8/1946	4127BU	5/7/1954	Reclaimed McClellan
44-62268			1/17/1946	580ASL		Unknown
44-62269			1/24/1946	580ASL		Unknown
44-62270		SQUARE H	1/15/1946	98BG	9/13/1954	Reclaimed Davis-Monthan
44-62270		SQUARE Y	1/15/1946	307BG	9/13/1954	Reclaimed Davis-Monthan
44-62271			12/28/1945	4105BU	9/16/1948	Salvaged at Ladd
44-62272			1/11/1946	3203MSU		Unknown
44-62273			2/4/1946			Unknown
44-62274	DEE-FENCE BUSTER	SQUARE A	2/4/1946	301BG		Unknown
44-62274		SQUARE V	2/4/1946	55RCN		Unknown
44-62275			1/25/1946	4136BU	3/16/1948	Surveyed at Tinker
44-62276		SQUARE A	2/15/1946	301BG	3/31/1949	Crashed Scotland — severe icing
44-62277			2/11/1946		11/23/1953	Reclaimed Robins
44-62278			2/15/1946	4117BU	11/12/1953	Reclaimed Robins
44-62279		CIRCLE E YLW	2/19/1946	22BG	10/19/1950	Engine fire on T/O, crashed in sea
44-62280					4/3/1951	Transferred to RAF
44-62280			4/3/1951	RAF	4/17/1958	Broken up, scrapped England
44-62281	FIRE BALL	SQUARE H	2/28/1946	98BG	9/8/1953	Assigned Storage Sqdn at Davis-Monthan
44-62282					3/11/1952	Salvaged
44-62282			3/11/1952	RAF	4/17/1958	Broken up, scrapped England
44-62283		55, A			4/19/1952	Transferred to RAF
44-62283			4/19/1952	RAF	4/17/1958	Broken up; scrapped England
44-62284		SQUARE I	3/14/1946	91RCN		Unknown
44-62285		SQUARE I	3/12/1946	91RCN		Unknown
44-62286					2/7/1951	Surveyed
44-62287		SQUARE Y	3/19/1946	307BG		Unknown
44-62288		SQUARE I	3/31/1946	91RCN		Unknown
44-62289		SQUARE I	3/26/1946	91RCN	9/14/1953	Reclaimed Tinker
44-62290		SQUARE I	4/16/1946	91RCN		Unknown
44-62291			4/22/1946			Unknown
44-62292			4/22/1946	3171ERE	9/8/1954	Reclaimed Davis-Monthan
44-62293			4/22/1946	3203MSU		Unknown
44-62294			4/22/1946	3203MSU		Unknown
44-62295	THE IRON BIRD		4/24/1946	4135BU	11/17/1953	Reclaimed Hill
44-62296		SQUARE I	4/29/1946	91RCN	8/21/1950	Transferred to RAF
44-62296		57, C	8/21/1950	RAF	4/17/1958	Scrapped England
44-62297			4/26/1946	308WG	11/17/1953	Reclaimed Kelly
44-62298		CIRCLE V	4/26/1946	55RCN	11/17/1953	Reclaimed Hill
44-62299						Unknown
44-62300			5/1/1946	3203MSU		Unknown
44-62301		TRIANGLE C	5/1/1946	509BG	9/9/1954	Salvaged at Davis-Monthan
44-62302		BLKSTRP	5/3/1946	19BG	7/13/1953	Assigned Storage Sqdn at Davis-Monthan
44-62303		BLKHORIZSTRP	5/3/1946	19BG		Unknown
44-62304	LUCKY LADY	DIAM		43BG		Unknown
44-62304	MISS LACE	DIAM B		16BG		Unknown
44-62305						Unknown
44-62306				308WG	3/17/1952	Class 26
44-62307		TRIANGLE C	5/31/1946	509BG	5/20/1953	Assigned Storage Sqdn at Davis-Monthan
44-62308					3/17/1952	Class 26
44-62309			5/20/1946	248BU	11/10/1948	Surveyed at Davis-Monthan
44-62310		CIRCLE K		43BG		Unknown
44-62311		TRIANGLE C	5/21/1946	509BG	5/20/1953	Assigned Storage Sqdn at Davis-Monthan
44-62312		TRIANGLE C	5/20/1946	509BG	9/9/1954	Reclaimed Davis-Monthan
44-62313		TRIANGLE C	5/22/1946	509BG	5/24/1953	Assigned Storage Sqdn at Davis-Monthan
44-62314	GAS GOBBLER	CIRCLE K		43BG		Unknown
44-62315	SATAN'S MATE	SQUARE EMPTY	5/22/1946	2BG	9/9/1954	Reclaimed Davis-Monthan
44-62315		TRIANGLE C	5/22/1946	509BG	9/9/1954	Reclaimed Davis-Monthan
44-62316		TRIANGLE C	5/29/1946	509BG	5/24/1953	Assigned Storage Sqdn at Davis-Monthan
44-62316		SQUARE EMPTY	5/29/1946	2BG	5/24/1953	Reclaimed Davis-Monthan
44-62317					3/17/1952	Class 26
44-62318						Unknown
44-62319		CIRCLE R	5/29/1946	9BG	3/12/1951	Salvaged Fairfield
44-62320						Unknown
44-62321						Unknown
44-62322		TRIANGLE C	5/29/1946	509BG	6/10/1953	Assigned Storage Sqdn at Davis-Monthan

Serial #	Name	Identification	Delv	Assign	Off Inv	Circumstances
44-62323		TRIANGLE C	6/7/1946	509BG	9/9/1954	Salvaged at Davis-Monthan
44-62324			5/29/1946	4121BU	3/1/1950	Surveyed Kelly
44-62325						Unknown
44-62326		SQUARE EMPTY		2BG	4/25/1951	Transferred to RAF
44-62326			4/25/1951	RAF	2/22/1954	Returned to USAF
44-62326			2/22/1954	USAF	7/14/1954	Reclaimed Davis-Monthan
44-62327					3/20/1951	Salvaged
44-62328		SQUARE T	6/11/1946		4/7/1951	Transferred to RAF
44-62328			4/7/1951	RAF	7/22/1953	Returned to USAF
44-62328			7/22/1953	USAF		Unknown
44-69655		V SQ 11	3/5/1944	499BG	6/26/1945	Shot down from Osaka Aresenal
44-69656		SQUARE H	12/14/1944	98BG	9/13/1954	Reclaimed Davis-Monthan
44-69656		SQUARE Y	12/14/1944	307BG	9/13/1954	Reclaimed Davis-Monthan
44-69657	LI'L ABNER	Z SQ 36	12/13/1944	500BG		Unknown
44-69658			12/11/1944	IRSBU	5/12/1954	Reclaimed
44-69659		FOURBLUSTRP F	12/15/1944	40BG	5/5/1945	Group transfer to Tinian
44-69659		TRI S 23	12/15/1944	40BG	8/7/1945	Surveyed-landed long & hot at Tinian
44-69660	MY GAL SAL	2YLWRUDSTRP	12/14/1944	468BG	5/11/1945	Group Transfer to Tinian
44-69660	MY GAL SAL	TRI I 42	12/14/1944	468BG	12/19/1954	Tinker
44-69660			12/14/1944	580ARC	12/19/1954	Tinker
44-69661		SOLYLWRUD	12/13/1944	462BG	5/5/1945	Group transfer to Tinian
44-69661		TRI U 20	12/13/1944	462BG	6/17/1954	Class 26 Sheppard
44-69661		CIRCLE E	12/13/1944	22BG	6/17/1954	Class 26 Sheppard
44-69662		K TRI	12/11/1944	505BG	7/28/1948	Salvaged
44-69663	DRAGON LADY	2BLURUDSTRP	12/13/1944	468BG	5/11/1945	Group transfer to Tinian
44-69663	DRAGON LADY	TRI I 29	12/13/1944	468BG	3/18/1954	Reclaimed Robins
44-69664		K TRI 05	12/11/1944	505BG	6/22/1945	Salvaged for 4/28 damage
44-69665		2BLURUDSTRP	12/12/1945	468BG	6/5/1945	Group transfer to Tinian
44-69665		TRI I 25	12/12/1944	468BG	6/5/1945	Shot down by ftrs ar Shoren-ji
44-69666		Z SQ 11	1/4/1945	500BG	4/2/1945	Crashed Tokyo
44-69667	SNUGGLEBUNNY	L TRI 32	12/12/1944	6BG		Survived war
44-69667	SNUGGLEBUNNY	SQUARE H		98BG	3/10/1954	Reclaimed Tinker
44-69668	THE OUTLAW	FOURYLWSTRP H	12/14/1944	40BG	5/5/1945	Group transfer to Tinian
44-69668	THE OUTLAW	TRI S 30 (1)	12/14/1944	40BG		Survived war
44-69668	THE WILD GOOSE	SQUARE H	12/14/1944	98BG	9/13/1954	Reclaimed Davis-Monthan
44-69669		BLK SQ O	12/15/1944	29BG	4/3/1945	Flak, crashed Nagoya
44-69670		BLK SQ 0	12/16/1944	29BG		Unknown
44-69671			12/19/1944	4105BU	5/10/1954	Reclaimed Davis-Monthan
44-69672	REAMATROID	L TRI 33	12/14/1944	6BG		Survived war
44-69672		BLKSTRPRED	12/14/1944	19BG	6/13/1954	Reclaimed Tinker
44-69673		BLK SQ M 44	12/16/1944	19BG	4/15/1945	Lost Kawasaki
44-69674		K TRI	12/13/1944	505BG	8/4/1954	Reclaimed
44-69675	BAD PENNY	L TRI	12/13/1944	6BG	3/28/1945	Crashed Fukuoka
44-69676	JUG HAID III	BLK SQ O 04	12/15/1945	29BG		Survived war
44-69676		CIRCLE E	12/15/1945	22BG		Converted to WB-29
44-69676			12/15/1945	53WRS	11/30/1953	Reclaimed Burtonwood
44-69677		BLK SQ O	12/16/1944	29BG	5/7/1954	Reclaimed Tinker
44-69678	CITY OF UNIVERSITY PARK	BLK SQ M 05 (1)	12/17/1944	19BG	8/8/1954	Reclaimed Tinker
44-69678	SOUND AND FURY	BLK SQ M 05 (1)	12/17/1944	19BG	8/8/1954	Reclaimed Tinker
44-69679		BLK SQ M	12/17/1944	19BG	5/5/1954	Reclaimed Davis-Monthan
44-69680	CITY OF BAKERSFIELD	BLK SQ M 02 (1)	12/18/1945	19BG		Survived war
44-69680	CITY OF TRENTON	BLK SQ M 02 (1)	12/18/1945	19BG		Survived war
44-69680	PRINCESS PAT	BLK SQ M 02 (1)	12/18/1945	19BG		Survived war
44-69680		SQUARE EMPTY	12/18/1945	2BG	3/10/1950	Transferred to RAF
44-69680			3/10/1950	RAF	10/20/1953	Returned to USAF
44-69680			10/20/1953	USAF		Unknown
44-69681	CITY OF AUSTIN	BLK SQ M 06	12/17/1944	19BG	8/8/1954	Reclaimed Tinker
44-69681	BLACK SHEEP	BLK SQ P 34		39BG	8/8/1954	Reclaimed Tinker
44-69681	CITY OF AUSTIN	BLK SQ P 34		39BG	8/8/1954	Reclaimed Tinker
44-69682	CITY OF FLATBUSH	BLK SQ M 04		19BG		Survived war
44-69682	CITY OF FLATBUSH	BLK SQ M 49		19BG		Survived war
44-69682	ATOMIC TOM	BLKSTRPRED		19BG	5/14/1951	Salvaged, major battle damage 4/12/51
44-69683			12/21/1944	236BU	3/5/1945	Surveyed at Pyote
44-69684		BLK SQ M 08	12/18/1944	19BG	6/23/1954	Reclaimed Pyote
44-69685	MISS BEHAVIN'	BLK SQ M 01	12/19/1944	19BG	8/8/1954	Reclaimed Tinker

I. Master List

Serial #	Name	Identification	Delv	Assign	Off Inv	Circumstances
44-69685	CITY OF BOSTON	BLK SQ M 01	12/19/1944	19BG	8/8/1954	Reclaimed Tinker
44-69686	TALL IN THE SADDLE	BLK SQ M	12/17/1944	19BG	3/10/1945	MIA Tokyo, crashed Ibaraki
44-69687					6/23/1949	Salvaged
44-69688		BLK SQ P 26	12/18/1944	39BG	6/23/1949	Abandoned Smoky Hill, two engines afire
44-69689	CITY OF ORLANDO	BLK SQ M 10	12/17/1944	19BG		Survived war
44-69689		SQUARE Y	12/17/1944	307BG	6/27/1954	Reclaimed Kadena
44-69690			12/17/1944	4105BU	5/5/1954	Reclaimed Davis-Monthan
44-69691	TINY TIM	BLK SQ O	12/19/1945	29BG	3/9/1945	MIA Tokyo
44-69692			12/22/1945	245BU	2/7/1945	Surveyed McCook
44-69693	OLD RUSTY	BLK SQ O 38	12/20/1944	29BG	9/16/1954	Reclaimed Bergstrom
44-69694		SOLREDRUD	12/19/1944	462BG	5/5/1945	Group transfer to Tinian
44-69694		TRI U 05	12/19/1944	462BG	5/10/1954	Reclaimed Davis-Monthan
44-69695			12/20/1944	4105BU	1/10/1954	Reclaimed Tinker
44-69696	HEAVENLY	BLK SQ M 29 (1)	12/20/1944	19BG	5/10/1949	Reclaimed Tinker
44-69696	CITY OF RED BANK	BLK SQ K 02		330BG		Unknown
44-69696	HAPPY SAVAGE	BLK SQ K 02		330BG		Unknown
44-69697			12/19/1945	231BU	4/4/1945	Surveyed at Alamogordo
44-69698	SKY QUEEN	FOURRUDSTRP	12/20/1944	40BG		transfer to 444BG ?
44-69698	SKY QUEEN	DIAM # 42		444BG		Group transfer to Tinian
44-69698	SKY QUEEN	TRI S 42	12/20/1944	40BG	5/25/1954	Reclaimed Davis-Monthan
44-69699		T SQ 29 (2)	12/19/1944	498BG	8/16/1954	Reclaimed Davis-Monthan
44-69700			12/28/1944		8/8/1954	Reclaimed Tinker
44-69701		2WHTRUDSTRP	12/21/1944	468BG	5/11/1945	Group transfer to Tinian
44-69701		TRI I 11	5/11/1945	468BG	9/22/1954	Reclaimed Tinker
44-69702			12/31/1945		10/4/1954	Reclaimed Tinker
44-69703		BLK SQ M 25 (1)	12/22/1945	19BG	3/9/1945	Crashed from Osaka, surveyed Iwo Jima
44-69704			1/2/1945		9/1/1954	Reclaimed Yokota
44-69705		SQUARE A	12/21/1944	301BG	6/17/1954	Class 26 Sheppard
44-69706	BLACK BART'S REVENGE	T SQ 53	12/26/1944	498BG	9/22/1953	Reclaimed Pyote
44-69706	VANISHING RAE	T SQ 08 (3)		498BG	9/22/1953	Reclaimed Pyote
44-69707		BLK SQ M	12/22/1945	19BG	3/20/1945	Lost on Nagoya mission
44-69708		SQUARE H	12/31/1944	98BG	6/17/1954	Class 26 Sheppard
44-69709			12/27/1944		10/4/1954	Reclaimed Davis-Monthan
44-69710	HOMOGENIZED ETHYL	CIRCLE K	12/27/1949	43BG		Unknown
44-69711			12/31/1944	242BU	9/14/1954	Reclaimed Hill
44-69712		Z	12/22/1944	500BG	7/31/1945	Crash landed with engine afire
44-69713			1/2/1945	4105BU	5/5/1954	Reclaimed Davis-Monthan
44-69714			12/28/1944	4105BU	5/10/1954	Reclaimed Davis-Monthan
44-69715			1/2/1945	4105BU	5/10/1954	Reclaimed Davis-Monthan
44-69716	TIMELY REMINDER	BLK SQ P 57	1/3/1945	39BG	9/19/1954	Reclaimed Robins
44-69717			1/2/1945	4105BU	5/5/1954	Reclaimed Davis-Monthan
44-69718			1/2/1945	4105BU	5/10/1945	Reclaimed Davis-Monthan
44-69719			1/3/1945	4105BU	5/10/1945	Reclaimed Davis-Monthan
44-69720			1/3/1945	326BU	4/17/1945	Surveyed at MacDill
44-69721		Z SQ 12 (2)	1/3/1945	500BG	5/10/1954	Reclaimed Davis-Monthan
44-69722	BATTLING BETTY III	T SQ 41 (3)	1/3/1945	498BG	9/22/1953	Reclaimed Pyote
44-69723			1/4/1945	4136BU	8/8/1954	Reclaimed Tinker
44-69724	PANCHITO, THE FIGHTING	A SQ 54	1/8/1945	497BG	5/10/1954	Reclaimed Davis-Monthan
44-69725	BEN'S RAIDERS	Z SQ 43	1/4/1945	500BG	3/19/1953	Reclaimed Tucson
44-69726			1/4/1945		4/10/1945	Salvaged Marianas
44-69727	HOT T' TROT	SQUARE H	1/4/1945	98BG	8/30/1954	Reclaimed Biggs
44-69727	THE ONE YOU LOVE	E TRI 23	1/4/1945	504BG	8/30/1954	Reclaimed Biggs
44-69728		BLK SQ O	1/4/1945	29BG	5/25/1945	Abandoned Tokyo
44-69728		T SQ 28	1/14/1945	498BG		Transfer to 29BG
44-69729		T SQ 54	1/4/1945	498BG	8/30/1954	Reclaimed Biggs
44-69730	BUTCH	SOLREDRUD		462BG	5/5/1945	Group transfer to Tinian
44-69730	MISS HART OF AMERICA	TRI U 10	5/5/1945	462BG	9/14/1954	Reclaimed Hamilton
44-69731		A SQ	1/5/1945	497BG	8/16/1954	Reclaimed Davis-Monthan
44-69732	MARIANNA RAM	A SQ 35	1/5/1945	497BG		Survived war
44-69732		SQUARE Y	1/5/1945	307BG	9/27/1954	Reclaimed McClellan
44-69733	NIP NEMESIS	X TRI 24	1/5/1945	9BG	7/28/1948	Salvaged at Tinker
44-69734		SOLYLWRUD	1/8/1945	462BG	5/5/1945	Group transfer to Tinian
44-69734		TRI U 29	5/5/1945	462BG	5/23/1945	Hit by flak at Tokyo, abandoned
44-69735		V SQ 34	1/8/1945	499BG	11/8/1954	Reclaimed Robins
44-69736	LOOK HOMEWARD, ANGEL	L TRI 29	1/8/1945	6BG	11/30/1945	Condemned Kwajalien

Serial #	Name	Identification	Delv	Assign	Off Inv	Circumstances
44-69737			1/8/1945			Unknown-Record card page missing
44-69738			1/9/1945		4/21/1945	Lost Shikoku Airfield raid
44-69739	(WINGED NUDE WITH DRINK)	E TRI 08	1/10/1945	504BG	2/27/1945	Crashed on takeoff, surveyed
44-69739	890TH AVIATION ENGR CO	K TRI		505BG		Unknown
44-69739	HONORABLE COCK WAGON	CIRC E		504BG	2/26/1945	Unknown
44-69740		V SQ	1/10/1945	499BG	5/10/1954	Reclaimed Davis-Monthan
44-69741	CITY OF COLUMBUS	BLK SQ K 51	1/12/1945	330BG	2/15/1949	Salvaged at Spokane
44-69741	TEN UNDER PARR	BLK SQ K 51	1/12/1945	330BG	2/15/1949	Salvaged at Spokane
44-69741	DEFIANT LASSIE	BLK SQ K 51	1/12/1945	330BG	2/15/1949	Salvaged at Spokane
44-69742	STING SHIFT	Z SQ 30	1/9/1945	500BG	5/23/1945	Shot down over Tokyo
44-69743		A SQ	1/8/1945	497BG	5/10/1954	Reclaimed Davis-Monthan
44-69744	TRIGGER MORTIS II	L TRI 39	1/9/1945	6BG	4/00/45	Tail Code Change
44-69744	TRIGGER MORTIS II	CIRC R	4/00/45	6BG	4/28/1946	Transferred to Army
44-69745	SKYSCRAPPER III	A SQ 9 (3)	1/11/1945	497BG		Survived war
44-69746	BOOZE HOUND	Z SQ 6 (2)	1/12/1945	500BG		Survived war
44-69746	CHARLIE'S WAGON	CIRCLE E	1/12/1945	22BG	10/00/50	Rotated back to ZI; transfer to 98BG
44-69746	IT SHOULDN'T HAPPEN	V SQ	1/12/1945	500BG		Survived war
44-69746	SEPTEMBER SONG	SQUARE H	10/00/45	98BG	9/13/1954	Reclaimed Davis-Monthan
44-69746	SNOOPY DROOPY		1/12/1945	54WRS		Unknown
44-69747		T SQ 15 (1)	1/9/1945	498BG	3/10/1945	Crashed Fubo Mt. in weather
44-69748	LIL IODINE II	X TRI 07	1/10/1945	9BG	3/24/1945	Exploded in air, crashed, burned
44-69749		T SQ 45/55 (2)	1/9/1945	498BG	9/24/1953	Reclaimed Tinker
44-69750		A SQ	1/8/1945	497BG	5/10/1954	Reclaimed Davis-Monthan
44-69751	HOMING DEVICE	Z SQ 31 (3)	1/10/1945	500BG	4/4/1945	MIA
44-69752		T SQ 15 (2)	1/11/1945	498BG	4/2/1945	MIA Tokyo
44-69753	BATAAN AVENGER	L TRI	1/12/1945	6BG	9/6/1949	Salvaged
44-69754	THE UNINVITED	X TRI 42	1/15/1945	9BG	4/19/1946	Condemned — Clark Field
44-69755		Z TRI 13 (2)	1/11/1945	500BG	5/26/1945	Ditched from Tokyo
44-69756	SILVER STREAK	BLK SQ P 23	1/13/1945	39BG	8/8/1954	Reclaimed Tinker
44-69757		L TRI 42	1/11/1945	6BG	7/9/1948	Reclaimed Clark Field
44-69757		BLK SQ P	1/11/1945	39BG	7/9/1948	Reclaimed Clark Field
44-69758		E TRI 37	1/13/1945	504BG	4/20/1954	Reclaimed Robins
44-69759		K TRI	1/12/1945	505BG	3/5/1945	Crashed on takeoff
44-69760	PURPLE HEARTLESS	X TRI 20	1/15/1945	9BG	5/24/1949	Salvaged at Lowry
44-69760		SQUARE A	1/15/1945	301BG	5/24/1949	Salvaged at Lowry
44-69761		Z SQ 1	1/12/1945	500BG	7/27/1954	Salvaged Andersen with 509BW
44-69762	CITY OF FORT GIBSON	BLK SQ O	1/15/1945	29BG	8/8/1954	Reclaimed Tinker
44-69763	CHERIE	BLK SQ P 15	1/16/1945	39BG		Survived war
44-69763	CITY OF BOULDER	BLK SQ P 15	1/16/1945	39BG		Survived war
44-69763	TOP OF THE MARK	BLKSTRPGRN	1/16/1945	19BG	8/16/1954	Reclaimed Davis-Monthan
44-69763	TOP OF THE MARK	SQUARE H	1/16/1945	98BG	8/16/1954	Reclaimed Davis-Monthan
44-69764	MISS MI-NOOKIE	X TRI 43	1/12/1945	9BG	7/27/1948	Salvaged
44-69765		T SQ 16 (2)	1/15/1945	498BG		Unknown
44-69766	CITY OF BURBANK	BLK SQ K 57	1/17/1945	330BG	6/5/1945	Ftr damage, crashed Kyoto Prefecture
44-69766	OLD SOLDIER'S HOME	BLK SQ K 57	1/17/1945	330BG	6/5/1945	Ftr damage, crashed Kyoto Prefecture
44-69767		K TRI		505BG	5/21/1952	Class 26
44-69768	CITY OF CLEVELAND	BLK SQ P 49	1/20/1945	39BG	8/11/1948	Salvaged
44-69768	EXPERIMENT PERILOUS	BLK SQ P 49	1/20/1945	39BG	8/11/1948	Salvaged
44-69769	CITY OF SCOTLAND NECK	BLK SQ P 14	1/16/1945	39BG		Converted to WB-29
44-69769	SOUTHERN DRAWL	BLK SQ P 14	1/16/1945	39BG		Converted to WB-29
44-69769		AIR WEA SERV	1/16/1945	53WRS	2/3/1954	Reclaimed Kindley
44-69770	TYPHOON GOON II	AIR WEA SERV		54WRS	10/26/1952	Missing in typhoon
44-69771	DANNY MITE	CIRC E	1/18/1945	504BG		Survived war
44-69771	OLD WILD GOOSE	SQUARE H	1/18/1945	98BG	7/14/1954	Reclaimed Davis-Monthan
44-69771		BLKSTRP	1/18/1945	19BG	7/14/1954	Reclaimed Davis-Monthan
44-69772	BATTLIN' BETTY III	T SQ 41(3)48	1/18/1945	498BG	9/22/1953	Reclaimed Pyote
44-69772		BLK SQ O	1/18/1945	29BG	9/22/1953	Reclaimed Pyote
44-69773	CITY OF PITTSFIELD	BLK SQ P 21	1/19/1945	39BG	5/15/1945	Ditched from Nagoya
44-69773	TWO PASSES AND A CRAP	BLK SQ P 21	1/19/1945	39BG	5/15/1945	Ditched from Nagoya
44-69774	CITY OF PATTERSON	BLK SQ K 54	1/19/1945	330BG	10/4/1954	Reclaimed Tinker
44-69774	KOEHANE'S KULPRITS	BLK SQ K 54	1/19/1945	330BG	10/4/1954	Reclaimed Tinker
44-69774	WALTZING MATILDA	BLK SQ K 54	1/19/1945	330BG	10/4/1954	Reclaimed Tinker
44-69775		Z SQ 13	1/23/1945	500BG	5/26/1945	Ditched from Tokyo
44-69776		BLK SQ O	1/1/2045	29BG	9/8/1953	Reclaimed Tinker
44-69777	DANNY MITE	T SQ 28	1/20/1945	498BG	5/24/1945	Lost 2 engines over Tokyo

I. Master List

Serial #	Name	Identification	Delv	Assign	Off Inv	Circumstances
44-69778			1/24/1945	235BU	4/10/1945	Surveyed at Biggs
44-69779	CITY OF MONTGOMERY	BLK SQ P 03	1/23/1945	39BG	5/7/1954	Reclaimed Robins
44-69779	WEDDING BELLE	BLK SQ P 03	1/23/1945	39BG	5/7/1954	Reclaimed Robins
44-69780			2/2/1945	4105BU	5/10/1954	Reclaimed Davis-Monthan
44-69781			1/24/1945	4136BU	7/28/1948	Salvaged at Tinker
44-69782			1/23/1945		8/2/1954	Reclaimed Tinker
44-69783		BLK SQ P	1/23/1945	39BG	5/10/1954	Reclaimed Davis-Monthan
44-69784			1/23/1945		9/22/1953	Reclaimed Pyote
44-69785	CITY OF GALVESTON	BLK SQ P 13 (1)	1/25/1945	39BG	6/26/1945	Abandoned flak at Nagoya Arsenal
44-69786	CITY OF RENO	BLK SQ K 53	1/25/1945	330BG		Survived war
44-69786	HERE TO STAY	BLK SQ K 53	1/25/1945	330BG		Survived war
44-69786	HERE TO STAY	CIRCLE E	1/25/1945	22BG		Survived war
44-69786		BLKSTRPBLU	1/25/1945	19BG	7/14/1954	Reclaimed Davis-Monthan
44-69787			1/26/1945		9/24/1953	Reclaimed Tinker
44-69788	CITY OF ATLANTA	BLK SQ P 29	1/26/1945	39BG	7/22/1954	Reclaimed Mountain Home
44-69789			1/29/1945		9/9/1954	Reclaimed Wright-Patterson
44-69790	CITY OF BIRMINGHAM	BLK SQ K 27	1/26/1945	330BG	5/26/1950	Salvaged
44-69790	OLE BOOMERANG	BLK SQ K 27	1/26/1945	330BG	5/26/1950	Salvaged
44-69791	CITY OF CHARLESTON	BLK SQ P 27	1/26/1945	39BG		Converted to WB-29
44-69791	LITTLE GEORGE JR.	BLK SQ P 27	1/26/1945	39BG		Converted to WB-29
44-69791		AIR WEA SERV		53WRS	12/5/1953	Reclaimed Kindley
44-69792	FIFTY-SECOND SEABEES	BLK SQ P 33	1/29/1945	39BG	10/4/1945	Crashed 8 miles west of Aguijian Island
44-69793			1/30/1945	4105BU	5/10/1954	Reclaimed Davis-Monthan
44-69794	OLE FORTY EIGHT	BLK SQ P 48	1/30/1945	39BG	5/13/1945	Scrapped Harmon Field per crew
44-69795		BLK SQ K 43	1/31/1945	330BG	4/12/1945	Crash landed Agana from Koriyama
44-69795	BUSTY BABE BOMBER	BLK SQ K 44		330BG	4/12/1945	Crash landed Agana from Koriyama
44-69795	CITY OF GRASS VALLEY	BLK SQ K 44		330BG	4/12/1945	Crash landed Agana from Koriyama
44-69796	CITY OF ALBUQUERQUE	BLK SQ P 32	1/31/1945	39BG	5/10/1954	Reclaimed Davis-Monthan
44-69796	PILEDRIVER	BLK SQ P 32	1/31/1945	39BG	5/10/1954	Reclaimed Davis-Monthan
44-69797	CITY OF SPRINGFIELD	BLK SQ O	1/31/1945	29BG	9/22/1954	Reclaimed Tinker
44-69797		BLK SQ K	1/31/1945	330BG	9/22/1954	Reclaimed Tinker
44-69798			2/1/1945		8/16/1954	Reclaimed Davis-Monthan
44-69799		BLK SQ K	2/1/1945	330BG	4/13/1945	Crashed from Tokyo Arsenal
44-69799	WEDDIN' BELLE	BLK SQ P 03	2/1/1945	39BG		Transfer to 330BG
44-69800	BEETLE BOMB	SQUARE H	2/1/1945	98BG	11/16/1954	Reclaimed Davis-Monthan
44-69800	BEETLE BOMBER	SQUARE H	2/1/1945	98BG	11/16/1954	Reclaimed Davis-Monthan
44-69800	CITY OF SAN FRANCISCO	BLK SQ K 29	2/1/1945	330BG		Survived war
44-69800	SOMETHING TO FIGHT	BLK SQ K 29	2/1/1945	330BG		Survived war
44-69801	CITY OF MEDFORD	BLK SQ K 58	2/1/1945	330BG	6/3/1954	Reclaimed Davis-Monthan
44-69801	LIGHTNING LADY	BLK SQ K 58	2/1/1945	330BG	6/3/1954	Reclaimed Davis-Monthan
44-69802	BANANA BOAT	L TRI		6BG		Combat loss in Korea
44-69802	BAIT ME?	BLKSTRPGRN		19BG	9/12/1952	Icing, stalled, cr 21 miles SW Kangnung
44-69803	LUCKY ELEVEN	SQUARE H		98BG	2/28/1952	Engine prob, abnd NE corner of Punchbowl
44-69803	LOADED LEVEN	SQUARE H		98BG	2/28/1952	Engine prob, abnd NE corner of Punchbowl
44-69803	LUCKY LEVEN	SQUARE H		98BG	2/28/1952	Engine prob, abnd NE corner of Punchbowl
44-69804		BLK SQ M	2/9/1945	19BG	5/7/1945	Surveyed
44-69805	DEAL ME IN	CIRCLE W	2/3/1945	92BG	8/8/1954	Reclaimed Tinker
44-69805	MISS BEA HAVEN	CIRCLE W	2/3/1945	92BG	8/8/1954	Reclaimed Tinker
44-69806			2/6/1945		8/25/1954	Reclaimed Biggs
44-69807			2/6/1945		8/16/1954	Reclaimed Davis-Monthan
44-69808			2/6/1945		5/26/1954	Reclaimed Davis-Monthan
44-69809	DEUCES WILD	CIRC W 17	2/3/1945	505BG	7/14/1954	Reclaimed Davis-Monthan
44-69810	CENSORED	BLK SQ P 07	3/3/1945	39BG	12/4/1951	Salvaged
44-69810	CENSORED	BLK SQ P 11	3/3/1945	39BG	12/4/51	Salvaged
44-69810	CENSORED LADY	BLK SQ P 11	3/3/1945	39BG	12/4/1951	Salvaged
44-69810	OLD P-7	BLK SQ P 07	3/3/1945	39BG	12/4/1951	Salvaged
44-69811	TINNY ANNE	X TRI 7 (2)	2/9/1945	9BG	5/27/1945	Flak at Shimonoseki
44-69812		SQUARE Y	2/6/1945	307BG	11/16/1954	Reclaimed Davis-Monthan
44-69812	THE FRY'IN PAN	SQUARE H	2/6/1945	98BG	11/16/1954	Reclaimed Davis-Monthan
44-69812	SUCHOSI? NI!	SQUARE H	2/6/1945	98BG	11/16/1954	Reclaimed Davis-Monthan
44-69813		K TRI	2/7/1945	505BG	4/25/1946	Transferred to Army
44-69814	CITY OF INDIANAPOLIS	BLK SQ K 33	2/5/1945	330BG	4/20/1954	Reclaimed Davis-Monthan
44-69814	MARY KATHLEEN	BLK SQ K 33	2/5/1945	330BG	4/20/1954	Reclaimed Davis-Monthan
44-69815	CITY OF TULSA	BLK SQ M 11	2/5/1945	19BG	8/16/1954	Reclaimed Davis-Monthan
44-69816			2/8/1945	4141BU	9/22/1953	Reclaimed Pyote

Serial #	Name	Identification	Delv	Assign	Off Inv	Circumstances
44-69817	CITY OF ROANOKE	BLK SQ K 04		330BG	6/2/1951	Salvaged
44-69817	READY BETTIE	BLK SQ K 04		330BG	6/2/1951	Salvaged
44-69817	HOT TO GO (II)	BLKSTRPGRN		19BG		Unknown
44-69818	STAR DUSTER	BLKSTRPGRN	2/9/1945	19BG	7/7/1953	Crash landed on approach to Pohang — K3
44-69818		BLK SQ M 33 (1)	2/9/1945	19BG		Survived war
44-69819			2/8/1945		9/22/1954	Reclaimed Tinker
44-69820		K TRI	2/7/1945		4/29/1945	Surveyed at Kirtland
44-69821		K TRI	2/14/1945	505BG	6/30/1954	Reclaimed Fairchild
44-69822			2/7/1945		5/10/1954	Reclaimed McClellan
44-69823			2/7/1945	505BG	2/17/1954	Reclaimed Tinker
44-69824		K TRI	2/14/1945	505BG	5/16/1945	Lost on Nagoya mission
44-69825		L TRI 66	2/7/1945	6BG	5/26/1945	Ditched, no fuel, # $ shot out
44-69826			2/8/1945		9/15/1954	Reclaimed Dow
44-69827			2/8/1945		5/18/1953	Reclaimed Tinker
44-69828		TRI S 30 (2)	2/10/1945	40BG	11/9/1954	Reclaimed Dow
44-69829	DUKE OF ALBUQUERQUE	Z SQ 8 (4)	2/9/1945	500BG	8/16/1954	Reclaimed Davis-Monthan
44-69830			2/8/1945	6BG		Reclaimed Kirtland
44-69831		L TRI 32	2/9/1945	6BG	9/22/1953	Reclaimed Pyote
44-69832			2/8/1945		7/14/1954	Reclaimed Tinker
44-69833		V SQ	2/9/1945	499BG	3/22/1954	Salvaged at Wheelus
44-69834		X TRI 21 (2)	2/14/1945	9BG	4/15/1945	Crashed at Yokohama from Kawasaki
44-69835		K TRI	2/10/1945	505BG	5/26/1946	Transferred to Army
44-69836		SQUARE A	2/8/1945	301BG	8/8/1954	Reclaimed Tinker
44-69837		TRIANGLE O	2/10/1945	97BG	7/11/1954	Tinker
44-69838			2/16/1945	6BG	4/14/1945	Surveyed
44-69839	FOREVER AMBER	L TRI 41	2/10/1945	6BG	6/5/1945	Salvaged Iwo, Battle damage
44-69840		L TRI	2/10/1945	6BG	5/16/1945	Lost on Nagoya mission
44-69841		L TRI	2/10/1945	6BG	8/16/1954	Reclaimed Davis-Monthan
44-69842		V SQ	2/10/1945	499BG	5/16/1954	Reclaimed Wheelus
44-69843		BLK SQ M	2/14/1945	19BG	11/15/1947	Salvaged overseas
44-69843		V SQ	2/14/1945	499BG	11/15/1947	Salvaged overseas
44-69844		E TRI 09	2/10/1945	504BG	7/31/1945	Surveyed Iwo, crash landing
44-69844	CITY OF TYLER	BLK SQ P 43		39BG		Unknown
44-69844	LANCER	BLK SQ P 43		39BG		Unknown
44-69845		BLK SQ M 57 (1)	2/15/1945	19BG	6/17/1945	Crashed in sea near Kashiwa Island
44-69846	THE SPIRIT OF FDR	E TRI 04	2/13/1945	504BG	6/9/1954	Reclaimed Alexandria
44-69847	BATTLIN' BETTY	BLK SQ O	2/14/1945	29BG	10/11/1954	Reclaimed Bergstrom
44-69847	BATTLIN' BETTY	CIRC R 56	2/14/1945	6BG	10/11/1954	Reclaimed Bergstrom
44-69848		T 15 (3)	2/13/1945	498BG	8/6/1945	Ditched from Nishinomiya
44-69848		Z SQ	2/13/1945	500BG	8/6/1945	Ditched from Nishinomiya with 498BG
44-69849	WARSAW PIGEON	CIRC X 27	2/14/1945	9BG	10/4/1954	Reclaimed Tinker
44-69850		L TRI	2/13/1945	6BG	9/14/1954	Reclaimed Hamilton
44-69851		BLK SQ O	2/14/1945	29BG	6/20/1945	Crashed in weather
44-69852	FILTHY FAY III	T SQ 26	3/15/1945	498BG	5/23/1945	MIA Tokyo
44-69853		CIRC W	2/16/1945	505BG	8/8/1954	Reclaimed Tinker
44-69854			2/15/1945	6BG	8/8/1954	Reclaimed Tinker
44-69855	LITTLE JEFF	CIRC R 65	2/15/1945	6BG	4/1/1949	Salvaged at Tinker
44-69856	CITY OF BUFFALO	BLK SQ M 53	2/15/1945	19BG		Survived war
44-69856		BLKSTRPRED	2/15/1945	19BG	10/4/1954	Reclaimed Tinker
44-69857	OLE SMOKER	BLK SQ K 59	2/17/1945	330BG	4/12/1945	Ditched from Koriyama
44-69857		CIRC R	2/17/1945	6BG		Unknown
44-69857	MERRY FORTUNE [8]				5/29/1945	
44-69858			2/16/1945		9/12/1954	Reclaimed Great Falls
44-69859		E TRI	2/17/1945	504BG	5/19/1945	Lost on mining mission, Hamamatsu
44-69860	WE DOOD IT	BLK SQ M 14	2/16/1945	19BG	8/16/1954	Reclaimed Davis-Monthan
44-69861		SQUARE A	2/17/1945	301BG	6/30/1950	Salvaged at Lake Charles
44-69862		BLK SQ M 35	2/17/1945	19BG	8/8/1954	Reclaimed Tinker
44-69863		A	2/17/1945	497BG	10/4/1953	Reclaimed Davis-Monthan
44-69864		CIRC R 09	2/16/1945	6BG	5/22/1946	Transferred to Army
44-69865		CIRC R 16	2/16/1945	6BG	7/28/1948	Salvaged
44-69866		BLKSTRPGRN	2/17/1945	19BG	7/12/1950	Attacked by YAK-9, engine fire, abandoned
44-69867	MARY FORTUNE [8]	BLK SQ P 01	2/19/1945	39BG	5/29/1945	Abandoned 50 miles off coast-flak damage
44-69868		Z SQ 41	2/19/1945	500BG	10/4/1953	Reclaimed Davis-Monthan
44-69869		BLK SQ O	2/23/1945	29BG	8/8/1954	Reclaimed Tinker
44-69870	CABOOSE	DIAM	2/19/1945	315BW		Transferred to 39BG ?

I. Master List

Serial #	Name	Identification	Delv	Assign	Off Inv	Circumstances
44-69870	THE CABOOSE	BLK SQ P 35	2/19/1945	39BG	7/9/1945	Engine fire — abandoned over Gifu
44-69870	CITY OF AURORA	BLK SQ P 35	2/19/1945	39BG	7/9/1945	Engine fire — abandoned over Gifu
44-69871		BLK SQ M 34	2/19/1945	19BG	4/15/1945	Lost Kawasaki, crashed Chiba
44-69872	CITY OF OAKLAND	BLK SQ M 36	2/23/1945	19BG	2/22/1954	Salvaged at Molesworth
44-69872	WHITE'S CARGO	BLK SQ M 36	2/23/1945	19BG	2/22/1954	Salvaged at Molesworth
44-69873		BLK SQ M 13	2/19/1945	19BG	6/26/1945	Rammed; exploded at Nagoya
44-69874	LADY JAYNE	CIRC X 21	2/19/1945	9BG	10/21/1953	Reclaimed Pyote
44-69875	CITY OF COVINGTON	BLK SQ O 13	2/19/1945	29BG		Survived war
44-69875		SQUARE A	2/19/1945	301BG	5/17/1950	Surveyed Azores — crashed, burned 5/14
44-69876		BLK SQ O	2/23/1945	29BG	5/10/1954	Reclaimed Davis-Monthan
44-69877		BLK SQ O	2/19/1945	29BG	6/29/1954	Reclaimed Laurel
44-69878		Z SQ 46	2/23/1945	500BG	9/22/1954	Reclaimed Tinker
44-69878		Z SQ 49	2/23/1945	500BG	9/22/1954	Reclaimed Tinker
44-69878			2/23/1945		12/2/1944	Reclaimed Tinker
44-69879		V SQ	2/20/1945	499BG	5/5/1954	Reclaimed Davis-Monthan
44-69880		V SQ	2/20/1945	499BG	8/16/1954	Reclaimed Griffis
44-69881		BLK SQ O	2/23/1945	29BG	6/19/1945	Mid-air with 42-65373 at Shizuoko
44-69882		BLK SQ O	2/24/1945	29BG	4/15/1945	Lost Kawasaki
44-69883	MARIANNA BELLE	X TRI 11	2/24/1945	9BG	4/00/45	Tail Code Change
44-69883	MARIANNA BELLE	CIRC X 11	4/00/45	9BG	8/7/1945	Abandoned Iwo Jima from Toyokawa
44-69884	CITY OF TYLER (TEXAS)	BLK SQ P 43	2/24/1944	39BG	8/8/1954	Reclaimed Tinker
44-69884	LANCER	BLK SQ P 43	2/24/1944	39BG	8/8/1954	Reclaimed Tinker
44-69884		CIRC R 36	2/24/1944	6BG	8/8/1954	Reclaimed Tinker
44-69885		Z 47 (2)	2/24/1945	500BG	4/20/1954	Reclaimed Davis-Monthan
44-69886		BLK SQ O	2/24/1945	29BG	3/31/1954	Reclaimed Davis-Monthan
44-69887		V ?	2/26/1945	499BG	5/5/1945	Crashed, exploded at Utogi
44-69888	GENERAL ANDREWS	BLK SQ P 50	2/24/1945	39BG	4/27/1945	Crashed from Kushira Airfield
44-69889	THE SLIC CHIC	BLK SQ P 46	2/24/1945	39BG	5/29/1945	Ditched 120 miles from Tori shima
44-69889	THE OLD GIRL					Unknown
44-69890	THE ANTAGONIZER	BLK SQ P 25	2/14/1945	39BG	8/8/1954	Reclaimed Tinker
44-69890	BOEING WICHITA 1000	BLK SQ P 25	2/14/1945	39BG	8/8/1945	Reclaimed Tinker
44-69891	CITY OF GRUNDY CENTER	BLK SQ O		29BG		Unknown
44-69892	ARKANSAS TRAVELER	BLK SQ O	2/24/1945	29BG	5/24/1945	Shot down Tokyo
44-69893	MAXIMUM EFFORT #3	BLK SQ M	2/24/1945	19BG	6/22/1945	Shot down offshore from Mizushima
44-69894	GRUMPY	BLK SQ P 20	2/26/1945	504BG		Survived war
44-69894		SQUARE H	2/26/1945	98BG	8/16/1954	Reclaimed Davis-Monthan
44-69895		BLK SQ P 26	2/27/1945	39BG	4/28/1945	Rammed at Miyazaki Airfield, ditched
44-69896		V ?		499BG	11/30/1951	Salvaged
44-69897		BLK SQ K	3/5/1945	330BG	5/25/1945	Shot down by fighters
44-69898		A SQ	3/5/1945	497BG	5/25/1954	Reclaimed Davis-Monthan
44-69898		CIRCLE E	3/5/1945	22BG	5/25/1954	Reclaimed Davis-Monthan
44-69899		A 56	2/27/1945	497BG	5/5/1945	3 & 4 out, wing broke off between them
44-69900		CIRC W 05	3/13/1945	505BG	5/14/1945	Shot up at Nagoya, abandoned Iwo Jima
44-69901	CITY OF MAYWOOD	BLK SQ P 10	3/5/1945	39BG	4/26/1954	Reclaimed Davis-Monthan
44-69901	DOUBLE TROUBLE	BLK SQ P 10	3/5/1945	39BG	4/26/1954	Reclaimed Davis-Monthan
44-69902		BLK SQ P	3/9/1945	39BG	4/20/1954	Reclaimed Robins
44-69903		BLK SQ P	2/28/1945	39BG	8/8/1954	Reclaimed Tinker
44-69904			3/12/1945		3/16/1954	Reclaimed Dow
44-69905			3/10/1945		10/14/1953	Reclaimed Pyote
44-69906	TABOOMA IV	TRI S 46	5/8/1945	40BG	9/24/1953	Reclaimed Tinker
44-69907		BLK SQ P 09	3/10/1945	39BG	4/15/1945	Lost Kawasaki
44-69908	CITY OF LA GRANGE (IL)	BLK SQ P 54		39BG		Unknown
44-69908	LOW & LONELY	BLK SQ P 54		39BG		Unknown
44-69909	DYNA-MITE II	SQUARE H	3/10/1945	98BG	8/16/1954	Reclaimed Davis-Monthan
44-69909		SQUARE Y	3/10/1945	307BG	8/16/1954	Reclaimed Davis-Monthan
44-69910	BETTY MARIAN	BLK SQ P 17	3/10/1945	39BG	8/16/1954	Reclaimed Davis-Monthan
44-69910	CITY OF EUGENE	BLK SQ P 17	3/10/1945	39BG	8/16/1954	Reclaimed Davis-Monthan
44-69910	CITY OF SPOKANE	BLK SQ P 17	3/10/1945	39BG	8/16/1954	Reclaimed Davis-Monthan
44-69911	CITY OF RICHMOND CA	BLK SQ K 59	3/8/1945	330BG	5/7/1954	Reclaimed Davis-Monthan
44-69911	VIVACIOUS LADY	BLK SQ K 59	3/8/1945	330BG	5/7/1954	Reclaimed Davis-Monthan
44-69912		A	3/8/1945	497BG	8/10/1954	Reclaimed Mountain Home
44-69913		A	3/8/1945	497BG	4/28/1954	Reclaimed Davis-Monthan
44-69914	CITY OF EAGLE ROCK	BLK SQ P 05	3/8/1945	39BG	7/18/1954	Reclaimed Dow
44-69914	LORD'S PRAYER	BLK SQ P 05	3/8/1945	39BG	7/18/1954	Reclaimed Dow
44-69915			3/9/1945		9/22/1954	Reclaimed Tinker

Serial #	Name	Identification	Delv	Assign	Off Inv	Circumstances
44-69916			3/8/1945	4141BU	10/14/1953	Reclaimed Pyote
44-69917			3/8/1945	4105BU	5/5/1954	Reclaimed Davis-Monthan
44-69918		CIRC E	3/8/1945	504BG	4/29/1945	Crashed from Shikoku Airfields
44-69919		CIRC E	3/13/1945	504BG	6/8/1945	Surveyed-Flak damage at Osaka
44-69920	T-N-TEENY II	CIRC X 3	3/8/1945	9BG	9/22/1953	Reclaimed Pyote
44-69921			3/8/1945	4141BU	10/14/1953	Reclaimed Pyote
44-69922		TRI N 48	3/8/1945	444BG	9/27/1953	Reclaimed Tinker
44-69923			3/8/1945		8/16/1954	Reclaimed Davis-Monthan
44-69924			3/9/1945	4105BU	5/5/1954	Reclaimed Davis-Monthan
44-69925			3/10/1945	4105BU	5/5/1954	Reclaimed Davis-Monthan
44-69926		V 23	3/12/1945	499BG	5/14/1945	Abandoned Iwo Jima from Nagoya
44-69927			3/14/1945	4141BU	10/14/1953	Reclaimed Pyote
44-69928		BLK SQ K	3/8/1945	330BG	8/8/1954	Reclaimed Tinker
44-69929		A	3/12/1945	497BG	8/24/1953	Reclaimed Tinker
44-69930			3/17/1945	4141BU	9/22/1953	Reclaimed Pyote
44-69931			3/9/1945	4105BU	5/5/1954	Reclaimed Davis-Monthan
44-69932	LITTLE JO	T SQ 4	3/12/1945	498BG		Transferred to 497BG
44-69932	LITTLE JO	A	3/12/1945	497BG	9/24/1953	Reclaimed Tinker
44-69933			3/12/1945		8/8/1954	Reclaimed Tinker
44-69934		CIRC X 49	3/13/1945	9BG	9/25/1953	Reclaimed Pyote
44-69935		BLK SQ K 11	3/12/1945	330BG	7/14/1954	Reclaimed McClellan
44-69936	DINA MIGHT	CIRC E 29	3/13/1945	504BG	6/26/1945	Flak Eitoku
44-69937			3/10/1945		9/27/1953	Reclaimed Tinker
44-69938		CIRC E	3/19/1945	504BG	8/8/1954	Reclaimed Tinker
44-69939		CIRC E 61	3/13/1945	504BG	11/17/1949	Salvaged
44-69940			3/14/1945	4196BU	9/22/1953	Reclaimed Pyote
44-69941		CIRC W 13 (2)	3/13/1945	505BG	8/16/1945	Ditched on test hop
44-69942			3/13/1945		10/4/1954	Reclaimed Tinker
44-69943			3/16/1945	233BU	7/17/1945	Surveyed — hit mountain 6/30/45
44-69944	FIRE BUG	Z SQ 15	3/30/1945	500BG		Survived the war
44-69944	LOADED LADY	SQUARE H	3/30/1945	98BG	9/13/1954	Reclaimed Davis-Monthan
44-69944		SQUARE Y	3/30/1945	307BG	9/13/1954	Reclaimed Davis-Monthan
44-69944		SPEEDWING	3/30/1945	307BG	9/13/1954	Reclaimed Davis-Monthan
44-69945			3/16/1945	4105BU	5/5/1945	Reclaimed Davis-Monthan
44-69946			3/16/1945	4117 BU	4/20/1954	Reclaimed Robins
44-69947			3/16/1945	247BU	4/20/1954	Reclaimed Robins
44-69948			3/13/1945	4105BU	5/5/1954	Reclaimed Davis-Monthan
44-69949		CIRCLE T	3/14/1945	44BW	10/4/1954	Reclaimed Tinker
44-69950			3/17/1945		8/8/1954	Reclaimed Tinker
44-69951			3/17/1945		9/1/1954	Reclaimed Yokota
44-69952		CIRC E	3/17/1945	504BG	10/10/1954	Reclaimed Randolph
44-69953		CIRC E 29	3/19/1945	504BG	10/4/1954	Reclaimed Tinker
44-69954			3/19/1945		3/31/1954	Reclaimed Davis-Monthan
44-69955			3/19/1945		2/8/1954	Reclaimed Randolph
44-69956			3/17/1945		9/22/1953	Reclaimed Pyote
44-69957		V ?	3/19/1945	499BG	3/28/1954	Reclaimed Robins
44-69957		YLW BANDS	3/19/1945		3/24/1954	Reclaimed Robins
44-69958		CIRC W	3/20/1945	505BG	8/16/1954	Reclaimed Davis-Monthan
44-69958			3/20/1945		8/16/1954	Reclaimed Davis-Monthan
44-69959		BLK SQ M 60 (1)		19BG		Unknown
44-69959		BLKSTRPBLU		19BG		Unknown
44-69960			3/19/1945		8/8/1954	Reclaimed Tinker
44-69961		CIRC W	3/21/1945	505BG	5/19/1945	Ditched from Hamamatsu
44-69962	CITY OF MILES CITY	BLK SQ O	3/22/1945	29BG	8/16/1954	Reclaimed Davis-Monthan
44-69963	BIG SHMOO	TRI N 15		444BG	12/8/1951	Salvaged
44-69963	WINNIE II	TRI N 15		444BG	12/8/1951	Salvaged
44-69964	MARY ANNA II	CIRC W 6	3/21/1945	505BG	5/25/1945	Shot up Tokyo, crashed Chiba
44-69965		TRI U 53	3/22/1945	462BG	6/5/1945	Shot down, crashed in sea at Nu Island
44-69966		TRI U	3/23/1945	462BG	5/14/1945	Shot down by Jack II at Nagoya
44-69967		BLK SQ M 72	3/21/1945	19BG	6/8/1945	Surveyed
44-69968		BLK SQ M 41	3/27/1945	19BG	7/28/1948	Salvaged
44-69969	OREGON EXPRESS	T 11	3/23/1945	498BG	4/20/1954	Reclaimed Kelly
44-69970	NIP FINALE	CIRC E 28	3/23/1945	504BG	5/29/1945	Exploded over Yokohama
44-69971			3/23/1945	330BG	2/17/1954	Reclaimed Robins
44-69971		YLW BANDS	3/23/1945		2/7/1954	Reclaimed Robins

I. Master List

Serial #	Name	Identification	Delv	Assign	Off Inv	Circumstances
44-69972	DOC		3/23/1945		10/1/1954	Reclaimed Tinker
44-69973			3/24/1945	236BU	8/9/1945	Surveyed at MacDill
44-69974		BLK SQ P	3/26/1945	39BG		Unknown
44-69975	THE SPEARHEAD	CIRC X 1	3/24/1945	9BG	5/23/1946	Transferred to Army
44-69976		SQUARE H		98BG	4/16/1951	Salvaged
44-69977	MISS PEGGY	TRI I 34	3/24/1945	468BG	6/27/1954	Reclaimed Kadena
44-69977		SQUARE H	3/24/1945	98BG	6/27/1954	Reclaimed Kadena
44-69977		SQUARE Y	3/24/1945	307BG	6/27/1954	Reclaimed Kadena
44-69978		CIRC E	3/25/1945	504BG	5/25/1945	Shot down over Tokyo
44-69979		CIRC R	3/27/1945	6BG	8/19/1954	Reclaimed Kelly
44-69980	LAKE SUCCESS EXPRESS	SQUARE H	3/27/1945	92BG	8/8/1954	Reclaimed Tinker
44-69980		CIRC R 67	3/27/1945	6BG	8/8/1954	Reclaimed Tinker
44-69981	CITY OF ALLENDALE	BLK SQ P 19	3/28/1945	39BG	8/8/1954	Reclaimed Tinker
44-69981	OLD EIGHTY ONE	BLK SQ P 19	3/28/1945	39BG	8/8/1954	Reclaimed Tinker
44-69982	110TH NCB	TRI N 31	3/28/1945	444BG	9/15/1954	Harmon
44-69982	1885 AVIATION ENGINEERS	TRI N 31	3/28/1945	444BG	9/15/1954	Harmon
44-69982	CAJUN QUEEN	TRI N 31	3/28/1945	444BG	9/15/1954	Harmon
44-69983		BLK SQ O 35	3/29/1945	29BG		Unknown
44-69984		BLK SQ M 20	3/27/1945	19BG	2/22/1954	Reclaimed Randolph
44-69984		CIRC X	3/27/1945	9BG	2/22/1954	Reclaimed Randolph
44-69985	JAKE'S JALOPY	X TRI 53	3/27/1945	9BG	9/14/1948	Salvaged at McClellan
44-69986	CITY OF VINCENNES	BLK SQ P 18	3/29/1945	39BG	6/17/1954	Reclaimed Davis-Monthan
44-69986	MANY HAPPY RETURNS	BLK SQ P 18	3/29/1945	39BG	6/17/1954	Reclaimed Davis-Monthan
44-69987	HURRICANE HUNTERS		3/28/1945	53WRS	10/1/1954	Reclaimed Burtonwood
44-69987	KAYO KID	T SQ 14	3/28/1945	498BG		Converted to WB-29
44-69987		CIRCLE E	3/28/1945	22BG	10/1/1954	Reclaimed Burtonwood
44-69988		TRI N 49	3/28/1945	444BG	5/18/1954	Reclaimed Hickam
44-69989		TRI U	3/28/1945	462BG	3/25/1948	Salvaged
44-69990		BLK SQ M 25	3/29/1945	19BG	7/2/1945	Abandoned from Minoshims, #2 engine fire
44-69991		BLK SQ O	3/29/1945	29BG	4/20/1954	Reclaimed Kelly
44-69992		TRI S 40	3/29/1945	40BG	9/14/1954	Reclaimed Griffis
44-69993		T 34 (3)	3/31/1945	498BG	6/22/1945	Lost #3,2 & 4 after T/O, crash landed
44-69994		CIRC X 36	3/30/1945	9BG	9/22/1953	Reclaimed Pyote
44-69995	CITY OF KNOXVILLE	BLK SQ K 38	3/30/1945	330BG	4/20/1954	Reclaimed Hill
44-69995	ERNIE PYLE	BLK SQ K 38	3/30/1945	330BG	4/20/1954	Reclaimed Hill
44-69996		BLK SQ M 58	3/30/1945	19BG	8/20/1945	Surveyed
44-69997	CITY OF SPANISH FORKS	BLK SQ K 35	4/2/1945	330BG	8/8/1954	Reclaimed Tinker
44-69997	HEAVENLY BODY	BLK SQ K 35	4/2/1945	330BG	8/8/1954	Reclaimed Tinker
44-69997		CIRC W	4/2/1945	505BG	8/8/1954	Reclaimed Tinker
44-69998		A	4/2/1945	497BG	8/28/1945	Possible crash dropping POW supplies
44-69998		SQUARE H	4/2/1945	98BG		Unknown
44-69999		CIRC W 12	4/3/1945	505BG		Survived War
44-69999	FOUR OF A KIND	BLKSTRPBLU	4/3/1945	19BG	7/14/1954	Salvaged at Davis-Monthan
44-70000		BLK SQ M 59	4/2/1945	19BG	4/30/1946	Transferred to Army
44-70001	SWEET SUE	T 42	4/2/1945	498BG	7/18/1954	Reclaimed McClellan
44-70002	SKY CHIEF	DIAM #	4/3/1945	444BG	5/10/1945	Group transfer to Tinian
44-70002	SKY CHIEF	TRI N 51	4/3/1945	444BG	5/14/1945	Cr, burned short of runway, engine failed
44-70003	CITY OF CHICAGO	BLK SQ M 12	2/3/1945	19BG	5/6/1948	Reclaimed Tinker
44-70003	PARKER'S VAN	BLK SQ M 12	4/2/1945	19BG	5/6/1948	Reclaimed Tinker
44-70004	CITY OF COOPERSTOWN	BLK SQ P 16	4/3/1945	39BG	5/4/1945	Shot down Oita — 1st mission
44-70005	HERD OF BALD GOATS	CIRC W 14 (2)	4/3/1945	505BG	11/5/1945	Abandoned near Saipan
44-70006		CIRC R 58	4/4/1945	6BG	9/29/1954	Reclaimed Bergstrom
44-70007	TARGET FOR TONIGHT	BLKSTRPRED	4/5/1945	19BG	6/27/1954	Reclaimed Kadena
44-70007		A	4/5/1945	497BG		Survived war
44-70007		SQUARE EMPTY	4/5/1945	2BG		Possible transfer to 19BG
44-70008		BLK SQ O	4/4/1945	29BG	6/5/1945	Flak damage, crashed Hyogo Prefecture
44-70010	CITY OF HATCH	BLK SQ K 41	4/4/1945	330BG	7/18/1954	Reclaimed Mountain Home
44-70010	CITY OF PACIFIC PALISADES	BLK SQ K 41	4/4/1945	330BG	7/18/1954	Reclaimed Mountain Home
44-70010	CUE BALL	BLK SQ K 41	4/4/1945	330BG	7/18/1954	Reclaimed Mountain Home
44-70010	CUE BALL	BLK SQ K 41		505BG		Unknown
44-70011	FRENCH'S KABAZIE WAGON	CIRC X 54	4/4/1945	9BG	10/12/1953	Reclaimed Pyote
44-70012		BLKSTRPGRN	4/5/1945	19BG	10/4/1954	Reclaimed Tinker
44-70013		BLK SQ M 09	4/3/1945	19BG	5/16/1946	Transferred to Army
44-70014		TRI I	4/6/1945	468BG	6/27/1954	Reclaimed Kadena
44-70014		V 9	4/6/1945	499BG		Unknown

Serial #	Name	Identification	Delv	Assign	Off Inv	Circumstances
44-70015	MARIANNA BELLE	TRI S 20		40BG	6/2/1952	Salvaged
44-70016	CITY OF QUAKER CITY	BLK SQ K 40	4/6/1945	330BG	9/15/1954	Reclaimed Griffis
44-70016	DOPEY		4/6/1945		9/14/1954	Reclaimed Griffis
44-70016	QUAKER CITY	BLK SQ K 40	4/6/1945	330BG	9/14/1954	Reclaimed Griffis
44-70016	SENTIMENTAL JOURNEY	BLK SQ K 40	4/6/1945	330BG	9/14/1954	Reclaimed Griffis
44-70017		BLK SQ M	4/6/1945	19BG	5/14/1945	Shot down Nagoya
44-70018	CITY OF SPOKANE	CIRC E 56	4/6/1945	504BG	4/24/1947	Salvaged
44-70018	THE ERNIE PYLE	CIRC E 56	4/6/1945	504BG	4/24/1947	Salvaged
44-70019		CIRCLE K	4/6/1945	43BG	9/1/1954	Reclaimed Yokota
44-70019	WHO'S NEXT	CIRCLE K	4/6/1945	43BG	9/1/1954	Reclaimed Yokota
44-70020		CIRCLE K	4/9/1945	43BG	10/4/1954	Tinker
44-70021		SQUARE Y	4/10/1945	307BG	11/16/1954	Reclaimed Davis-Monthan
44-70021		SQUARE P	4/10/1945	306BG		Transfer to 307BG
44-70022			4/10/1945		5/25/1954	Reclaimed Davis-Monthan
44-70023			4/9/1945		8/8/1954	Reclaimed Tinker
44-70024			4/9/1945		11/16/1950	Salvaged
44-70025			4/9/1945	4105BU	5/5/1954	Reclaimed Davis-Monthan
44-70026			4/9/1945		10/12/1954	Salvaged Turner
44-70027			4/10/1945	4105BU	5/5/1954	Reclaimed Davis-Monthan
44-70028			4/10/1945		11/1/1954	Reclaimed Tinker
44-70029			4/10/1945	4141BU	10/14/1953	Reclaimed Pyote
44-70030			4/11/1945		11/1/1954	Reclaimed Tinker
44-70031			4/11/1945	235BU	5/7/1945	Surveyed at Biggs
44-70032			4/11/1945	4105BU	5/5/1954	Reclaimed Davis-Monthan
44-70033			4/11/1945		11/1/1954	Reclaimed Tinker
44-70034			4/11/1945		8/8/1954	Reclaimed Tinker
44-70035			4/12/1945		11/1/1954	Reclaimed Tinker
44-70036			4/13/1945		8/8/1954	Reclaimed Tinker
44-70037			4/12/1945	4141BU	10/14/1953	Reclaimed Pyote
44-70038			4/12/1945	4196BU	4/20/1954	Reclaimed Robins
44-70039			4/12/1945		11/15/1957	Removed form inventory, Alaska
44-70040			4/20/1945	4141BU	10/14/1953	Reclaimed Pyote
44-70041		BLKSTRP	4/13/1945	19BG	7/14/1954	Salvaged at Davis-Monthan
44-70042	THE LEMON DROP KID	BLKSTRPGRN	4/12/1945	19BG	12/18/1951	Circumstances Unknown
44-70042		TRI I 54	4/12/1945	468BG		Survived the war
44-70042		CIRCLE E	4/12/1945	22BG		Transfer to 19BG
44-70043			4/13/1945		8/8/1954	Reclaimed Tinker
44-70044			4/14/1945		8/16/1954	Reclaimed Davis-Monthan
44-70045			4/14/1945		9/23/1953	Reclaimed Tinker
44-70046			4/14/1945		9/22/1954	Reclaimed Tinker
44-70047			4/16/1945		8/16/1954	Reclaimed Davis-Monthan
44-70048			4/14/1945	326BU	2/10/1954	Reclaimed Randolph
44-70049			4/14/1945	581ASL	3/23/1954	Reclaimed Clark Field
44-70050			4/16/1945	326BU	12/3/1953	Reclaimed Great Falls
44-70051			4/18/1945	4141BU	9/22/1953	Reclaimed Pyote
44-70052			4/17/1945		8/8/1954	Reclaimed Tinker
44-70053			4/17/1945		9/24/1953	Reclaimed Tinker
44-70054			4/17/1945		8/8/1954	Reclaimed Tinker
44-70055		SQUARE Y	4/18/1945	307BG	6/27/1954	Reclaimed Kadena
44-70056			4/17/1945	236BU	5/8/1945	Surveyed at Pyote
44-70057			4/18/1945		9/13/1954	Reclaimed Davis-Monthan
44-70058			4/19/1945		7/14/1954	Salvaged at Davis-Monthan
44-70059			4/19/1945		10/4/1954	Reclaimed Tinker
44-70060			4/20/1945	2750ABG	8/18/1954	Reclaimed Wright-Patterson
44-70061			4/19/1945		11/3/1953	Reclaimed Tinker
44-70062			4/19/1945		7/15/1954	Reclaimed McClellan
44-70063			4/28/1945		10/4/1954	Reclaimed Tinker
44-70064		BLK SQ M 39	4/20/1945	19BG		Unknown
44-70065			4/20/1945			Unknown
44-70066			4/18/1945		9/24/1953	Reclaimed Tinker
44-70067			4/25/1945	326BU	4/20/1954	Reclaimed Robins
44-70068			4/21/1945		11/1/1954	Reclaimed Tinker
44-70069	DEARLY BELOVED	CIRC R 59	4/20/1945	6BG	8/8/1954	Reclaimed Tinker
44-70069	THE RAMP TRAMP	CIRC R 59	4/20/1945	6BG	8/8/1954	Reclaimed Tinker
44-70069	THE RAMP TRAMP	CIRC X		9BG	8/8/1954	Reclaimed Tinker

I. Master List

Serial #	Name	Identification	Delv	Assign	Off Inv	Circumstances
44-70070	THE 8 BALL	CIRC X 16	4/21/1945	9BG	3/31/1948	Salvaged at Tinker
44-70071		CIRC W	4/21/1945	505BG	3/25/1954	Reclaimed Hill
44-70072	LIMBER RICHARD	CIRC X 55	4/23/1945	9BG	7/18/1954	Salvaged at Mountain Home
44-70073	MAC'S EFFORT	CIRCLE W	4/23/1945	92BG	11/8/1954	Reclaimed Smoky Hill
44-70074			4/22/1945	326BU	4/20/1954	Reclaimed Robins
44-70075		T 46 (2)	4/24/1945	498BG	5/24/1945	Crashed near Haneda
44-70076		TRI U 15	4/30/1945	462BG	7/25/1954	Reclaimed Turner
44-70077	CITY OF CLEVELAND	BLK SQ P 31	4/25/1945	39BG		Survived War
44-70077	LITTLE BULLY	BLK SQ P 31	4/25/1945	39BG		Survived War
44-70077		BLKSTRP	4/25/1945	19BG	7/14/1954	Salvaged at Davis-Monthan
44-70078	WILLIAM ALLEN WHITE	CIRC W 05	4/28/1945	505BG	11/8/1950	Reclaimed Hill
44-70079	CITY OF REDFIELD	BLK SQ P 51	4/25/1945	38BG	6/23/1954	Reclaimed Pyote
44-70079	FOREVER AMBER	BLK SQ P 51	4/25/1945	39BG	6/23/1954	Reclaimed Pyote
44-70080	CITY OF GRIFFIN, GA	BLK SQ O 16	4/30/1945	29BG	3/25/1954	Reclaimed Hill
44-70081		V ?	4/30/1945	499BG	8/16/1954	Reclaimed Davis-Monthan
44-70082		T 30	4/25/1945	498BG	10/14/1953	Reclaimed Pyote
44-70083	FLYING FOOL	V SQ 8	4/22/1945	499BG	6/1/1945	Flak, crashed Osaka Bay
44-70083	FLYING FOOL	V SQ 10/8	4/22/1945	499BG	6/1/1945	Flak, crashed Osaka Bay
44-70084		TRI I 35	4/28/1945	468BG	6/22/1945	Flak at Kure — crashed Kochi City
44-70085	PATCHES	TRI S 41	4/25/1945	40BG	7/25/1954	Reclaimed Mountain Home
44-70086		TRI U 33	4/25/1945	462BG	9/29/1953	Reclaimed Tinker
44-70087			4/25/1945		10/4/1954	Reclaimed Tinker
44-70088			5/7/1945		8/8/1954	Reclaimed Tinker
44-70089		YLW BANDS	4/26/1945			Unknown
44-70090			4/27/1945	326BU	5/15/1945	Surveyed at MacDill
44-70091			4/27/1945		1/10/1954	Reclaimed Tinker
44-70092		BLK SQ M 22	4/26/1945	19BG	11/4/1945	Salvaged
44-70093		BLK SQ O	4/27/1945	29BG	4/20/1954	Reclaimed Hill
44-70094	B-SWEET IV	TRI S 43	4/28/1945	40BG		Unknown — Record Card Missing
44-70095		BLK SQ M		19BG	6/10/1951	Salvaged
44-70096		V 16	4/30/1945	499BG	10/4/1954	Reclaimed Tinker
44-70097			5/2/1945	4000EXP	7/2/1954	Reclaimed Wright-Patterson
44-70098		BLK SQ M 19	4/30/1945	19BG	9/24/1949	Salvaged at McClellan
44-70099		V 23	4/30/1945	499BG	9/14/1953	Reclaimed Davis-Monthan
44-70100	MISS YOU	TRI S 44	4/30/1945	40BG	4/20/1954	Reclaimed Hill
44-70101	CITY OF BERKELEY	BLK SQ K 11	4/30/1945	330BG	9/13/1954	Reclaimed Yokota
44-70101	JE REVIENS	BLK SQ K 11	4/30/1945	330BG	9/13/1954	Reclaimed Yokota
44-70101	JE REVIENS	Z 16	4/30/1945	500BG	9/13/1954	Reclaimed Yokota
44-70101		YLW BANDS	4/30/1945		9/13/1954	Reclaimed Yokota
44-70102	HERE'S HOPIN	TRI N 41	4/30/1945	444BG	10/8/1953	Reclaimed Tinker
44-70102		BLK SQ M 40	4/30/1945	19BG	10/8/1953	Reclaimed Tinker
44-70103	CITY OF LINCOLN	BLK SQ M 07	4/30/1945	19BG	11/15/1947	Salvaged at FEAMCOM
44-70103	PRINCESS PAT II	BLK SQ M 07	4/30/1945	19BG	11/15/1947	Salvaged at FEAMCOM
44-70104		TRI U 16	5/2/1945	462BG	3/11/1954	Reclaimed Robins
44-70105		BLK SQ O 06	5/3/1945	29BG	4/20/1954	Reclaimed Hill
44-70106	NIP FINALE II	CIRC E 28	5/2/1945	504BG	8/8/1954	Reclaimed Tinker
44-70107		BLK SQ O	5/2/1945	29BG	11/12/1953	Reclaimed Pyote
44-70108	SWEET THING	TRI N 32/39	5/4/1945	444BG	7/31/1954	Reclaimed Tinker
44-70109	DELILAH	CIRC E 26	5/2/1945	504BG	6/26/1945	Lost Eitoku
44-70110		TRI U 34	5/3/1945	462BG	11/12/1953	Reclaimed Pyote
44-70111	STRAP HANGER	CIRC E 57	5/2/1945	504BG	7/9/1948	Salvaged
44-70111		CIRC W 6	5/2/1945	505BG	7/9/1948	Salvaged at Clark Field with 504th
44-70112	SWEET SUE	CIRC X 05	5/7/1945	9BG	8/19/1954	Reclaimed Kelly
44-70113	ANCIENT MARINER,	Z 58		500BG		Static display, Florence NC
44-70113	THE MARYLIN GAY	Z 58		500BG		Static display, Florence NC
44-70113	THE MARYLN GAY,	Z SQ 58		500BG		Static display, Florence NC
44-70114		CIRC R	5/4/1945	6BG	4/20/1954	Reclaimed Hill
44-70115		CIRC W 9	5/2/1945	505BG	8/24/1953	Reclaimed Tinker
44-70116		CIRC R	5/4/1945	6BG	7/19/1945	Shot down Niigata area
44-70117		Z 07	5/7/1945	500BG		Survived war, converted to SB-29
44-70117		YLW BANDS	5/7/1945		9/1/1954	Reclaimed Yokota
44-70118		CIRC R	5/4/1945	6BG	6/23/1954	Reclaimed Pyote
44-70118		CIRC E	5/4/1945	504BG	6/23/1954	Reclaimed Pyote
44-70118	THE ERNIE PYLE	CIRC E 56	4/6/1945	504BG	4/24/1947	Salvaged
44-70119		V ?	5/7/1945	499BG	2/9/1954	Reclaimed Hickam

Serial #	Name	Identification	Delv	Assign	Off Inv	Circumstances
44-70119		YLW BANDS	5/7/1945		2/9/1954	Reclaimed Hickam
44-70120	MEMPHIS MAID	TRI N 23	5/9/1945	444BG	9/24/1953	Reclaimed Tinker
44-70121	WILLIAM ALLEN WHITE	CIRC X 50	5/3/1945	9BG	9/25/1953	Reclaimed Pyote
44-70122		CIRC W	5/8/1945	505BG	10/11/1945	Missing
44-70123	HO HUM	TRI N 16	5/9/1945	444BG	10/4/1954	Reclaimed Tinker
44-70124	TOJO'S NIGHTMARE	CIRC R 60	5/8/1945	6BG	10/14/1953	Reclaimed Tinker
44-70125		BLKSTRP	7/4/1945	19BG	7/14/1954	Reclaimed Davis-Monthan
44-70126		CIRC R	5/8/1945	6BG	4/20/1954	Reclaimed Kelly
44-70127	HERE'S HOPN	TRI N 40	5/10/1945	444BG		No 444BG Record
44-70127	UNCONDITIONAL SURRENDER	TRI N 40	5/10/1945	444BG	5/12/1954	Reclaimed Robins
44-70128	BIG! AIN'T IT?	TRI N 62	5/9/1945	444BG	7/14/1954	Reclaimed Andersen
44-70129	ARSON INC	TRI N 47	5/9/1945	444BG	5/16/1954	Reclaimed Robins
44-70129	FLAK MAID	TRI N 47	5/9/1945	444BG	5/16/1954	Reclaimed Robins
44-70130		TRI N	5/10/1945	444BG	8/19/1954	Reclaimed Kelly
44-70131	PATIENCE REWARD	TRI N 21	5/11/1945	444BG		Converted to WB-29
44-70131			5/11/1945	53WRS	2/00/55	Kindley
44-70132		TRI N 55	5/11/1945	444BG	7/24/1945	Direct flak hit Handu, crashed Osaka Bay
44-70133		CIRC W 06	5/10/1945	505BG	10/14/1953	Reclaimed Pyote
44-70134	NO SWEAT (II)	BLKSTRPRED	5/12/1945	19BG	7/14/1954	Salvaged at Davis-Monthan
44-70134		V ?	5/12/1945	499BG		Survived war
44-70134		SQUARE EMPTY	5/12/1945	2BG		Survived war
44-70135	ANTOINETTE II	T 48 (3)	5/16/1945	498BG	4/20/1954	Reclaimed Hill
44-70136	BUCKIN' BRONCO	Z 28	6/11/1945	500BG	8/29/1945	Shot down by Russians on POW mission
44-70136	HOG WILD	Z 28	6/11/1945	500BG	8/29/1945	Shot down by Russians on POW mission
44-70137		TRI N 51	5/11/1945	444BG	4/13/1954	Reclaimed Robins
44-70138		CIRC W	5/12/1945	505BG	4/19/1946	Condemned overseas
44-70139	GLOBAL GLAMOR	TRI S 17	5/12/1945	40BG	11/12/1953	Reclaimed Pyote
44-70140		TRI I 52	5/12/1945	468BG	9/22/1954	Reclaimed Tinker
44-70141		T 36	5/14/1945	498BG	3/31/1954	Reclaimed Davis-Monthan
44-70142		TRI U		462BG	12/4/1951	Condemned
44-70143	TALIE HO!	TRI N 17	5/14/1945	444BG	8/8/1954	Reclaimed Tinker
44-70144		Z 39	5/14/1945	500BG	7/14/1954	Salvaged at Davis-Monthan
44-70145		T 46	5/15/1945	498BG	4/11/1954	Prestwick
44-70146		TRI I 50	5/15/1945	468BG	4/20/1954	Reclaimed Kelly
44-70147	SENTIMENTAL JOURNEY	T 31	5/15/1945	498BG	9/7/1949	Salvaged at Smoky Hill
44-70148		TRI U	5/15/1945	462BG	11/12/1953	Reclaimed Pyote
44-70149	LUCKY 13	TRI N 13	5/16/1945	444BG	8/8/1954	Reclaimed Tinker
44-70150		CIRC R 26	5/18/1945	6BG	5/10/1954	Reclaimed patrick
44-70151		SQUARE Y		307BG	10/23/1951	Damaged Namsi, abandoned Inchon area
44-70151	EDDY ALLEN II	TRI S 47		40BG		Survived the war
44-70151		SQUARE A		301BG	12/31/1951	Salvaged
44-70152		V	5/18/1945	499BG	8/8/1954	Reclaimed Tinker
44-70153		V ?		499BG	4/12/1951	Salvaged
44-70154			5/22/1945	4141BU	10/14/1953	Reclaimed Pyote
44-83890		DIAM L 10	4/28/1945	331BG	1/31/1952	Salvaged
44-83891		DIAM L 6	4/28/1945	331BG	5/7/1954	Reclaimed
44-83892		DIAM L 23	4/28/1945	331BG	5/7/1954	Reclaimed McClellan
44-83893	DODE	DIAM L 4	4/28/1945	331BG	6/24/1954	Reclaimed Pyote
44-83893	SALOME, WHERE SHE DANCED	DIAM L 4	4/28/1945	331BG		Renamed DODE for POW rescue flt
44-83894			4/28/1945	233BU	9/22/1953	Reclaimed Pyote
44-83895		DIAM L 46	4/30/1945	331BG	1/6/1955	Reclaimed Bergstrom
44-83896		DIAM L 44	4/28/1945	331BG	10/11/1954	Reclaimed Turner
44-83897	CRAMER'S CRAPPER	DIAM L 25	4/28/1945	331BG	7/8/1954	Reclaimed Tinker
44-83897	NO NAME!	DIAM L 25	4/28/1945	331BG	7/8/1954	Reclaimed Tinker
44-83897	SALTY DOG	DIAM L 25	4/28/1945	331BG	7/8/1954	Reclaimed Tinker
44-83898		DIAM L 45	4/30/1945	331BG	5/7/1954	Reclaimed McClellan
44-83899		DIAM B	4/28/1945	16BG	8/31/1945	Crashed overseas
44-83900			5/23/1945	4105BU	7/14/1954	Reclaimed Davis-Monthan
44-83901		DIAM H	4/23/1945	502BG	7/21/1954	Reclaimed
44-83902		DIAM L 7	4/30/1945	331BG	4/7/1954	Reclaimed Wright-Patterson
44-83903		DIAM L 53	4/30/1945	331BG	12/1/1949	Reclaimed Tinker
44-83904		DIAM L	5/28/1945	331BG	10/12/1953	Reclaimed Pyote
44-83905		DIAM L 9	5/12/1945	331BG	4/7/1956	Crashed in lake, Eielsen AFB
44-83905		TRIANGLE C	5/12/1945	509BW	8/15/1954	Reclaimed Tinker
44-83906	REGINA COELI	DIAM B	5/3/1945	16BG		Unknown

I. Master List

Serial #	Name	Identification	Delv	Assign	Off Inv	Circumstances
44-83907		DIAM L 5	4/30/1945	331BG	5/7/1954	Reclaimed Tinker
44-83908			5/29/1945	233BU	10/12/1953	Reclaimed Pyote
44-83909		DIAM H	5/8/1945	502BG	5/7/1954	Reclaimed McClellan
44-83910		DIAM H	5/10/1945	502BG	10/11/1954	Reclaimed Turner
44-83911			5/28/1945	4117BU	11/12/1953	Reclaimed Robins
44-83912		DIAM H	5/5/1945	502BG	5/5/1954	Reclaimed Davis-Monthan
44-83913		DIAM B	4/30/1945	16BG	5/7/1954	Reclaimed McClellan
44-83914			5/30/1945	233BU	10/12/1953	Reclaimed Pyote
44-83915		DIAM B	5/11/1945	16BG	7/14/1954	Salvaged at Davis-Monthan
44-83916			4/30/1945	4105BU	5/5/1954	Reclaimed Davis-Monthan
44-83917			5/28/1945	233BU	10/12/1953	Reclaimed Pyote
44-83918		DIAM B	5/12/1945	16BG	9/22/1954	Reclaimed Tinker
44-83919	TRIFLIN' GAL	SQUARE A	5/12/1945	301BW		Unknown
44-83919		DIAM L 30	5/12/1945	331BG		Unknown
44-83920			5/31/1945	233BU	10/12/1953	Reclaimed Pyote
44-83921			5/17/1945	XXIBC	4/29/1950	Salvaged at Barksdale
44-83922			5/10/1945	XXIBC	4/8/1954	Reclaimed Tinker
44-83923			5/30/1945	233BU	10/12/1953	Reclaimed Pyote
44-83924		DIAM L 31	5/11/1945	331BG	10/4/1954	Reclaimed Tinker
44-83925		DIAM L 08	5/9/1945	331BG	7/18/1954	Lost at Mt Home
44-83926		SQUARE I	5/31/1945	91RCN	7/14/1954	Salvaged at Davis-Monthan
44-83927			5/9/1945	91SRC	7/5/1954	Reclaimed Davis-Monthan
44-83928		DIAM L	5/31/1945	331BG	7/14/1954	Salvaged at Davis-Monthan
44-83929		DIAM L 27	5/8/1945	331BG	6/24/1954	Reclaimed Pyote
44-83930		DIAM B	5/31/1945	16BG	7/14/1954	Salvaged at Davis-Monthan
44-83931		DIAM B	5/16/1945	16BG	7/14/1954	Salvaged at Davis-Monthan
44-83932		DIAM B	5/31/1945	16BG	5/10/1954	Reclaimed Davis-Monthan
44-83933		DIAM B	5/12/1945	16BG	7/15/1954	Salvaged at Davis-Monthan
44-83934	SHACK RABBIT	SQUARE H	5/31/1945	98BG	9/13/1954	Reclaimed Davis-Monthan
44-83934	UNDECIDED	SQUARE H	5/31/1945	98BG	9/13/1954	Reclaimed Davis-Monthan
44-83935			5/10/1945	XXIBC	2/11/1954	Reclaimed Randolph
44-83936			5/31/1945	4105BU	5/10/1954	Reclaimed Davis-Monthan
44-83937		L DIAM 48	5/15/1945	331BG	7/31/1954	Reclaimed Tinker
44-83938		TRIANGLE C	5/31/1945	509BW	7/14/1954	Reclaimed McClellan
44-83939		DIAM L 50		331BG		Unknown
44-83940			5/31/1945	233BU	10/12/1953	Reclaimed Pyote
44-83941		DIAM L 49	5/19/1945	331BG	3/31/1948	Condemned Tinker
44-83942		DIAM H	5/19/1945	502BG	9/10/1945	Condemned overseas
44-83943		DIAM L 32	5/21/1945	331BG	5/18/1954	Reclaimed Tinker
44-83943		SQUARE I	5/21/1945	91RCN	5/18/1954	Reclaimed Tinker
44-83944		DIAM L 11	5/19/1945	331BG	2/14/1951	Salvaged at Barksdale
44-83945			5/31/1945	233BU	10/12/1953	Reclaimed Pyote
44-83946	VICTORY JEAN	DIAM L 51	5/25/1945	331BG	5/7/1954	Reclaimed McClellan
44-83947			5/31/1945	4105BU	5/5/1954	Reclaimed Davis-Monthan
44-83948		DIAM H	5/24/1945	502BG	11/23/1954	Reclaimed Tinker
44-83949			5/31/1945	233BU	10/12/1953	Reclaimed Pyote
44-83950				6BG	9/12/1951	Salvaged
44-83951			6/2/1945	233BU	8/17/1954	Reclaimed Tinker
44-83952		DIAM L 13	5/28/1945	331BG	11/23/1954	Reclaimed Tinker
44-83953		CIRCLE W	5/31/1945	92BG	12/7/1954	Reclaimed Tinker
44-83953		SQUARE Y	5/31/1945	307BG	12/7/1954	Reclaimed Tinker
44-83954		DIAM L 52	5/25/1945	331BG	7/28/1948	Condemned Tinker
44-83955			5/31/1945	233BU	10/12/1953	Reclaimed Pyote
44-83956		DIAM L 37	5/26/1945	331BG	10/12/1954	Reclaimed Bergstrom
44-83957			6/18/1945	233BU	9/22/1954	Reclaimed Pyote
44-83958		DIAM L 15	5/29/1945	331BG	7/14/1954	Salvaged at Davis-Monthan
44-83959		BLK SQ M	6/6/1945	19BG	9/2/1954	Reclaimed Hunter
44-83960		CIRC X	6/23/1945	9BG	1/2/1947	Condemned returning to ZI
44-83961		DIAM L 33	5/29/1945	331BG	2/16/1954	Reclaimed Randolph
44-83962			6/22/1945		2/28/1946	Salvaged overseas
44-83963			5/18/1945	620BU	8/25/1950	Reclaimed McClellan
44-83964			6/21/1945	333BG	10/7/1945	Lost, conditions unknown
44-83965			5/19/1945			Unknown
44-83966	NEVER BEEN TRIED		6/23/1945		5/7/1954	Reclaimed Robins
44-83966		BLK SQ M	6/23/1945	19BG	5/7/1954	Reclaimed Robins

Serial #	Name	Identification	Delv	Assign	Off Inv	Circumstances
44-83967		BLK SQ M 72	5/12/1945	19BG	1/24/55	Reclaimed Bergstrom
44-83968		CIRC X	6/23/1945	9BG	12/4/1953	Reclaimed Robins
44-83969			5/31/1945	608BEXG	1/9/1955	Reclaimed Tinker
44-83970			6/19/1945		11/12/1953	Reclaimed Robins
44-83971		TRIANGLE O	6/8/1945	97BG	9/22/1954	Reclaimed Tinker
44-83972			6/28/1945		7/28/1948	Reclaimed Tinker
44-83973			6/7/1945	620BU	8/29/1947	Salvaged, crashed Muroc 6/4/46
44-83974	PRINCESS PAT III	BLK SQ M 15		19BG		Survived War
44-83974	PUNCH BOWL QUEEN		6/18/1945	559SR	1/24/1950	Salvaged at Kadena
44-83975			6/14/1945	320PTS	12/31/1954	Reclaimed Kelly
44-83976			6/18/1945	xxibc	11/12/1953	Reclaimed Robins
44-83977			6/9/1945	608BEXG	7/25/1954	Reclaimed McClellan
44-83978		BLK SQ M	6/22/1945	19BG	5/4/1950	Salvaged at North Field, Guam
44-83979		CIRC X	6/11/1945	9BG	5/31/1954	Reclaimed Tinker
44-83980		CIRC X	6/26/1945	9BG	4/6/1953	Reclaimed Robins
44-83981		DIAM L 29	6/8/1945	331BG	7/27/1954	Reclaimed Mountain Home
44-83982			6/27/1945	XXIBC	11/12/1953	Reclaimed Robins
44-83983		BLK SQ M	6/8/1945	19BG	10/13/1954	Reclaimed Turner
44-83984			6/23/1945	314BW	10/17/1950	Salvaged at Robins
44-83985		DIAM L 12	6/11/1945	331BG	7/21/1953	Surveyed Hunter, crashed 7/10/53
44-83986		DIAM Y	6/29/1945	501BG	11/17/1953	Reclaimed Hill
44-83987		DIAM Y	6/9/1945	501BG	9/23/1954	Reclaimed Bergstrom
44-83988		BLK SQ M	7/9/1945	19BG	9/14/1954	Reclaimed Hill
44-83989		BLK SQ M	6/12/1945	19BG	7/5/1954	Reclaimed Tinker
44-83990		CIRCLE E		22BG		Salvaged
44-83991			6/9/1945	4141BU	6/24/1954	Reclaimed Pyote
44-83992	WHERE NEXT?	SQUARE H	7/9/1945	98BG	5/7/1954	Reclaimed Robins
44-83992		DIAM L	7/9/1945	331BG		Survived war
44-83993		DIAM L 54	6/8/1945	331BG	10/12/1954	Reclaimed Turner
44-83994			6/30/1945	XXIBC	11/12/1953	Reclaimed Robins
44-83995		CIRC X	6/11/1945	9BG	3/31/1954	Reclaimed Bergstrom
44-83996		CIRC X	6/28/1945	9BG	8/17/1949	Condemned Kadena
44-83997			6/15/1945	XXIBC	7/14/1954	Salvaged at Davis-Monthan
44-83998			6/28/1945	XXIBC	8/31/1945	Surveyed after crash
44-83999		CIRC X	6/18/1945	9BG	12/3/1954	Reclaimed Tinker
44-84000		CIRC X		9BG	7/24/1951	Salvaged
44-84001		Z ?	6/12/1945	500BG	10/12/1954	Reclaimed Turner
44-84002		L TRI 11	6/12/1945	6BG	5/24/1949	Reclaimed Tinker
44-84003		CIRC R	6/13/1945	6BG	9/22/1954	Reclaimed Tinker
44-84004			7/9/1945		7/14/1954	Salvaged at Davis-Monthan
44-84005			6/19/1945	XXIBC	11/17/1954	Reclaimed Tinker
44-84006		Z ?	6/30/1945	500BG	7/20/1950	Salvaged at Eglin
44-84007			6/21/1945	XXIBC	1/16/1955	Reclaimed Tinker
44-84008		CIRC R		6BG	4/26/1951	Salvaged
44-84009		BLK SQ O	6/19/1945	29BG	9/7/1954	Reclaimed Tinker
44-84010		CIRC X	6/30/1945	9BG	10/17/1950	Reclaimed Robins
44-84011			6/22/1945		3/22/1954	Reclaimed Great Falls
44-84012		BLK SQ M	6/30/1945	19BG	10/17/1950	Reclaimed Robins
44-84013		T	6/21/1945	498BG	12/7/1954	Reclaimed Bergstrom
44-84014		T ?	6/30/1945	498BG	6/25/1950	Condemned North Field
44-84015		BLK SQ P	6/23/1945	39BG	9/9/1954	Reclaimed Bergstrom
44-84016		BLK SQ P	6/30/1945	39BG	1/2/1947	Condemned — unknown reasons
44-84017		SQUARE A	6/25/1945	301BG	9/9/1954	Reclaimed Davis-Monthan
44-84018		BLK SQ M	7/13/1945	19BG	10/17/1950	Salvaged at Robins
44-84019	CITY OF SAM BEE FLA		6/25/1945	333BG	12/22/1945	Crashed on Training flight
44-84020		BLK SQ M	7/10/1945	19BG	10/21/1953	Reclaimed Robins
44-84021			6/25/1945	333BG	2/6/1950	Reclaimed Ladd
44-84022			7/13/1945		7/14/1954	Salvaged at Davis-Monthan
44-84023			6/25/1945	242BU	6/24/1954	Reclaimed Pyote
44-84024		BLK SQ K	7/12/1945	330BG	11/16/1949	Salvaged at Castle
44-84025		TRIANGLE C	6/28/1945	509BW	10/10/1954	Reclaimed Tinker
44-84026			7/13/1945		7/14/1954	Salvaged at Davis-Monthan
44-84027			6/25/1945	233BU	6/24/1954	Reclaimed Pyote
44-84028		BLK SQ P	7/10/1945	39BG	8/31/1949	Scrapped England
44-84029		BLK SQ P	6/26/1945	39BG	1/10/1957	Crash landed Bergstrom AFB

I. Master List

Serial #	Name	Identification	Delv	Assign	Off Inv	Circumstances
44-84030			7/14/1945		7/14/1954	Salvaged at Davis-Monthan
44-84031		DIAM L 55	6/28/1945	331BG	7/5/1954	Reclaimed Tinker
44-84032		CIRCLE W	7/7/1945	92BG	7/14/1954	Salvaged at Davis-Monthan
44-84033		CIRCLE E	6/26/1945	22BG	1/30/1955	Reclaimed Offut
44-84034			7/10/1945	XXIBC	10/12/1953	Reclaimed Pyote
44-84035			6/30/1945	4112BU	9/21/1954	Reclaimed Laurinburg
44-84036		BLK SQ P	7/12/1945	39BG	4/15/1948	Salvaged at Davis-Monthan
44-84037		BLK SQ P	6/29/1945	39BG	7/29/1954	Reclaimed Mountain Home
44-84038			7/24/1945	XXIBC	7/18/1954	Reclaimed Turner
44-84039			6/30/1945	4117BU	12/19/1954	Reclaimed Tinker
44-84040		BLK SQ K	7/17/1945	330BG	8/5/1948	Reclaimed Overseas
44-84041			6/30/1945	4117BU	2/24/1954	Reclaimed George
44-84042		CIRC X	7/14/1945	9BG	4/29/1948	Reclaimed Overseas
44-84043			6/26/1945		5/27/1954	Reclaimed Schenectady
44-84044		CIRC X	7/14/1945	9BG	8/28/1946	Salvaged overseas
44-84045		T	7/11/1945	498BG	10/12/1954	Reclaimed Turner
44-84046		T ?	7/18/1945	498BG	5/5/1946	Transferred to Army
44-84047			7/18/1945		10/11/1954	Reclaimed Turner
44-84048			7/18/1945	4117BU	7/14/1954	Salvaged at Davis-Monthan
44-84049			7/16/1945	XXIBC	9/30/1954	Great Falls
44-84050			7/18/1945	XXIBC	6/8/1948	Reclaimed Harmon
44-84051			7/16/1945	346BG	9/19/1954	Reclaimed Bergstrom
44-84052			7/19/1945	XXIBC	7/14/1954	Salvaged at Davis-Monthan
44-84053	BONNIE LEE	YLW BANDS	7/28/1945	ASR	11/8/1954	Reclaimed Komaki
44-84054	BLACK HILLS BABY	CIRC E BLUE		22BG		Converted to SB-29
44-84054	BLACK HILLS BABY	TRI S		40BG		Survived War
44-84054		YLW BANDS	7/28/1945	asr		Unknown
44-84055		SQUARE T	7/19/1945	2BG	12/19/1954	Reclaimed Turner
44-84056		DIAM L	7/23/1945	331BG	1/7/1949	Reclaimed at Harmon
44-84057		DIAM L 14	7/18/1945	331BG	10/6/1954	Reclaimed Hill
44-84058		CIRC R	7/23/1945	6BG	1/20/1949	Reclaimed Far East
44-84059		BLK SQ P	7/16/1945	39BG	9/9/1954	Reclaimed Davis-Monthan
44-84060		TRI N	7/20/1945	444BG	11/12/1953	Reclaimed Robins
44-84061	PACUSAN DREAMBOAT	BLK SQ P	7/20/1945	39BG	8/8/1954	Reclaimed Tinker
44-84062		BLK SQ P	7/24/1945	39BG	4/29/1948	Salvaged at Hill
44-84063			7/23/1945	4141BU	6/24/1954	Reclaimed Pyote
44-84064			7/31/1945	24BG	8/5/1948	Salvaged Overseas
44-84065	QUEEN OF THE NECHES		7/21/1945	92BG	11/28/1954	Reclaimed Tinker
44-84066		CIRCLE W	7/21/1945	92BG	9/27/1954	Reclaimed McClellan
44-84066		DIAM L	7/21/1945	331BG	9/27/1954	Reclaimed McClellan
44-84067		DIAM L 34	7/25/1945	331BG	9/22/1954	Reclaimed Tinker
44-84068	LADY IN WAITING	TRI U	7/31/1945	462BG	8/17/1949	Reclaimed Far East
44-84068	SHAFT ABSORBER	TRI N	7/31/1945	444BG	8/17/1949	Reclaimed Far East
44-84068	LADY IN WAITING		7/31/1945	333BG		
44-84069		DIAM L 35	7/23/1945	331BG	8/19/1954	Reclaimed Great Falls
44-84070			7/31/1945	303BW	7/14/1954	Salvaged at Davis-Monthan
44-84071			7/26/1945		10/10/1954	Reclaimed Great Falls
44-84072			7/28/1945	346BG	8/14/1945	Surveyed at Herington
44-84073			7/31/1945		1/28/1954	Reclaimed Burbank
44-84074			7/28/1945	4105BU	8/29/1947	Salvaged at Smoky Hill
44-84075		CIRCLE M	7/27/1945	93BW	7/15/1954	Reclaimed Bergstrom
44-84076	PERCUSSION STEAMBOAT	TRIANGLE S	8/4/1945	2BW	10/13/1954	Reclaimed Tinker
44-84077		DIAM B	7/31/1945	16BG	8/30/1945	Condemned non-combat
44-84078		TRIANGLE S	7/31/1945	28BG	4/15/1954	Reclaimed Robins
44-84078		YLW BANDS	7/31/1945	3RES	4/15/1954	Reclaimed Robins
44-84079			7/27/1945		10/11/1954	Reclaimed McClellan
44-84080		SQUARE H	7/31/1945	98BG	7/14/1954	Salvaged at Davis-Monthan
44-84080		TRIANGLE S	7/31/1945	28BG	7/14/1954	Salvaged at Davis-Monthan
44-84081		TRIANGLE EMPTY		6BG	7/15/1952	Salvaged
44-84082			7/31/1945		8/5/1948	Reclaimed Overseas
44-84083		TRIANGLE S	7/31/1945	28BG	11/7/1954	Reclaimed Hanscomb
44-84083		TRIANGLE S	7/31/1945	28BG	9/1/1954	Reclaimed Yokota
44-84084		YLW BANDS	7/31/1945		9/15/1954	Reclaimed Yokota
44-84085		SQUARE A	7/31/1945	301BW	5/7/1953	Reclaimed Davis-Monthan
44-84086		YLW BANDS	8/8/1945		5/5/1953	Reclaimed Komaki

Serial #	Name	Identification	Delv	Assign	Off Inv	Circumstances
44-84087		DIAM L 00	7/31/1945	331BG	5/9/1951	Salvaged
44-84088	IITYWYBAD	TRIANGLE S	7/31/1945	28BG	4/27/1954	Reclaimed
44-84088		YLW BANDS	7/31/1945		4/27/1954	Reclaimed
44-84089			8/3/1945	4141BU	6/23/1954	Reclaimed Pyote
44-84090		BLK SQ M	8/8/1945	19BG	10/17/1950	Reclaimed Robins
44-84091			8/3/1945	4141BU	6/23/1954	Reclaimed Pyote
44-84092			8/6/1945	4141BU	7/14/1954	Salvaged at Davis-Monthan
44-84093			8/6/1945	4141BU	6/23/1954	Reclaimed Pyote
44-84094		TRIANGLE S	8/6/1945	28BG	6/23/1954	Reclaimed Pyote
44-84094		YLW BANDS	8/6/1945	2RESGP	9/14/1954	Reclaimed Yokota
44-84095			8/9/1945	427BU	9/14/1954	Reclaimed Kelly
44-84096		TRIANGLE S	8/10/1945	28BG	2/2/1954	Reclaimed Robins
44-84096		YLW BANDS	8/10/1945	ASR	2/2/1954	Reclaimed Robins
44-84097			8/6/1945		3/25/1953	Reclaimed Pyote
44-84098		BLK SQ M	8/6/1945	19BG	10/21/1953	Reclaimed Robins
44-84099			8/14/1945	4141BU	6/23/1954	Reclaimed Pyote
44-84100		TRIANGLE S	8/14/1945	28BG	9/27/1954	Reclaimed McClellan
44-84101			8/18/1945	1TOWSQ	12/15/1954	Reclaimed Biggs
44-84102		BLK SQ M	8/13/1945	19BG	11/12/1953	Reclaimed Robins
44-84103			8/6/1945	2754ABG	4/20/1954	Reclaimed Kelly
44-84104			8/20/1945	303BW	7/14/1954	Salvaged at Davis-Monthan
44-84105			8/18/1945	4117BU	1/24/1955	Reclaimed Eglin
44-84106			8/22/1945	303BW	7/14/1954	Reclaimed Davis-Monthan
44-84107			8/23/1945		10/11/1954	Reclaimed Turner
44-84108			8/28/1945	4141BU	10/12/1953	Reclaimed Pyote
44-84109			8/18/1945	4127BU	5/7/1954	Reclaimed McClellan
44-84110	WILLY'S ICE WAGON	TRIANGLE S	8/20/1945	28BG	11/25/1952	Surveyed at Randolph
44-84111	MONSTRO		8/18/1945	2759ABG	7/14/1954	Reclaimed Eglin
44-84112		YLW BANDS	8/11/1945	ASR	3/9/1954	Reclaimed Andersen
44-84112		TRIANGLE S	8/11/1945	28BG	3/9/1954	Reclaimed Andersen
44-84113			8/27/1945	2TOWSQ	7/11/1954	Reclaimed Biggs
44-84114	UNPREDICTABLE		8/20/1945	112RDC	7/14/1954	Salvaged at Davis-Monthan
44-84114		TRIANGLE S	8/20/1945	28BG	7/14/1954	Salvaged at Davis-Monthan
44-84115			8/20/1945	4117BU	12/9/1954	Reclaimed Tinker
44-84116		BLK SQ M	8/25/1945	19BG	10/20/1950	Reclaimed Robins
44-84117			8/22/1945	2750ABW	5/12/1954	Reclaimed Wright-Patterson
44-84118	LEGAL EAGLE	SQUARE I	8/24/1945		10/10/1954	Reclaimed Turner
44-84119		SQUARE I	8/24/1945	91RCN	10/10/1954	Reclaimed
44-84120			8/27/1945	559SR	10/4/1954	Reclaimed
44-84121			8/23/1945		7/22/1954	Reclaimed Wright-Patterson
44-84122	THE RED	SQUARE A	8/25/1945	28BG	7/14/1949	Reclaimed Weaver
44-84123		SQUARE I	8/22/1945	91RCN	8/29/1954	Reclaimed Tinker
44-84124		TRIANGLE O	8/27/1945	97BG	1/31/1951	Surveyed at Tinker
44-84125			8/27/1945	4121BAG	8/19/1954	Reclaimed Kelly
44-84126			8/30/1945	303BW	7/14/1954	Salvaged at Davis-Monthan
44-84127		DIAM L 36	8/29/1945	331BG	10/12/1954	Reclaimed Turner
44-84128			8/31/1945	4121BU	8/19/1954	Reclaimed Kelly
44-84129		DIAM L 58	9/4/1945	331BG	3/18/1954	Reclaimed Kelly
44-84130			8/31/1945	303BW	7/14/1954	Salvaged at Davis-Monthan
44-84131		DIAM L 16	9/4/1945	331BG	7/14/1954	Reclaimed McClellan
44-84132		TRIANGLE O	8/29/1945	97BW	5/20/1954	Reclaimed Tinker
44-84133			9/6/1945	4105BU	5/5/1954	Reclaimed Davis-Monthan
44-84134		TRIANGLE O	8/30/1945	97BW	8/13/1954	Cr near Lockbourbe AFB, engine failure
44-84135		SQUARE I	8/30/1945	91RCN	7/18/1954	Reclaimed Turner
44-84136			8/31/1945	4141BU	10/12/1953	Reclaimed Pyote
44-84137		TRIANGLE C	9/5/1945	509BW	9/22/1954	Reclaimed Tinker
44-84138			9/19/1945	4121BU	8/19/1954	Reclaimed Kelly
44-84139		SQUARE T	9/30/1945		10/12/1954	Reclaimed Langley
44-84140						Unknown
44-84141				4121BU	7/24/1951	Surveyed
44-84142			9/12/1945	4121BU	8/19/1954	Reclaimed Kelly
44-84143		CIRCLE R	9/18/1945	9BG	8/19/1954	Reclaimed Mountain Home
44-84144		SQUARE A	9/19/1945	301BW	7/14/1954	Salvaged at Davis-Monthan
44-84145		TRIANGLE C	9/20/1945	509BW	10/5/1954	Reclaimed Hill
44-84146			9/21/1945	303BW	7/14/1954	Salvaged at Davis-Monthan

I. Master List

Serial #	Name	Identification	Delv	Assign	Off Inv	Circumstances
44-84147		TRIANGLE C	9/19/1945	509BW	11/24/1954	Reclaimed Tinker
44-84148			9/20/1945	611BEX	11/12/1953	Reclaimed Robins
44-84149		SQUARE T	9/30/1945	2BW	12/26/1956	Hit mountain in Alaska
44-84150						Unknown
44-84151		SQUARE I	9/26/1945	2BW	9/9/1954	Reclaimed Tinker
44-84152			9/25/1945	4141BU	10/12/1953	Reclaimed Pyote
44-84153						Unknown
44-84154				4105BU		Unknown
44-84155			9/30/1945	4105BU	5/10/1954	Reclaimed Davis-Monthan
44-84156		TRIANGLE O	9/30/1945	97BW	9/19/1954	Reclaimed Tinker
44-86242				611BASEX	7/15/1949	Class 01Z
44-86243		CIRCLE R		28BG		Unknown
44-86244			5/8/1945		7/14/1954	Reclaimed Davis-Monthan
44-86245			5/8/1945	611BASEX	10/12/1953	Reclaimed at Pyote
44-86246		CIRCLE R	5/9/1945	28BG	9/14/1953	Reclaimed Tinker
44-86247	M P I	SQUARE H	5/9/1945	98BG	11/18/1951	Crashed on takeoff from Yokota
44-86247	DRAGON BEHIND	TRI N 50 (2)		444BG	1/9/1952	Surveyed
44-86248			5/12/1945	XXIBC	10/12/1953	Reclaimed Pyote
44-86249			8/27/1945	XXIBC	10/12/1953	Reclaimed
44-86249		SQUARE I	5/11/1945	91SRC	7/19/1954	Reclaimed Tinker
44-86250		BLKSTRP	5/12/1945	19BG	12/4/1953	Reclaimed Robins
44-86251		TRIANGLE O	5/14/1945	97BW	6/27/1954	Reclaimed Dow
44-86252		BLK SQ M		19BG	10/17/1950	Surveyed
44-86252	MULE TRAIN	CIRCLE E		22BG		Unknown
44-86253					3/2/1947	Missing-circumstances unknown
44-86254	STATESIDE REJECT [9]	BLKSTRPRED		19BG	4/24/1951	Surveyed-Cr, burned 11/50 on Okinawa
44-86254	STAR DUSTER [9]	BLKSTRPBLU		19BG	4/24/1951	Surveyed-Cr, burned 11/50 on Okinawa
44-86255			5/14/1945	303BW	7/14/1954	Salvaged at Davis-Monthan
44-86256		CIRCLE E	5/16/1945	22BG	10/17/1950	Surveyed
44-86257		CIRCLE R		28BG	10/17/1950	Surveyed
44-86258		CIRCLE E YLW		22BG	6/14/1950	Surveyed
44-86259			5/17/1945	7RCUSQ	7/14/1954	Salvaged at Davis-Monthan
44-86260		CIRCLE R	5/19/1945	28BG	10/12/1953	Reclaimed Tinker
44-86261	BANANA BOAT	CIRC E BLUE	5/17/1945	22BG	10/12/1953	Reclaimed
44-86261	BANANA BOAT	TRI S	5/17/1945	40BG	9/14/1953	Reclaimed Tinker
44-86262			5/17/1945	4141BU	10/12/1953	Reclaimed Pyote
44-86263		TRIANGLE O	5/17/1945	97BG	3/30/1954	Reclaimed Tinker
44-86264			5/18/1945	4141BU	10/12/1953	Reclaimed Pyote
44-86265			5/23/1945	4141BU	6/23/1954	Reclaimed Pyote
44-86266			5/17/1945	4141BU	6/23/1954	Reclaimed Pyote
44-86267			5/18/1945	54WRS	7/5/1954	Reclaimed Robins
44-86268		SQUARE Y		307BG	5/14/1951	Surveyed, battle damage 4/7/51
44-86269			5/22/1945	55WRS	3/1/1954	Reclaimed McClellan
44-86270			5/23/1945	2750ABW	2/1/1955	Reclaimed Wright-Patterson
44-86271		SQUARE H		98BG	12/29/1952	Surveyed
44-86272	LONESOME POLECAT	SQUARE H	5/22/1945	98BG	7/14/1954	Reclaimed
44-86273	LIL DARLIN	SQUARE H		98BG	10/14/1951	Cr into bowling alley, laundary Yokota
44-86273		SQUARE H		98BG	10/21/1951	Surveyed
44-86274				XXIBC	8/29/1945	Crashed Toyama on POW drop
44-86275		SQUARE P	5/24/1945	306BW	9/27/1954	Reclaimed McClellan
44-86276					8/18/1945	Surveyed
44-86277		SQUARE A	5/25/1945	301BW	7/14/1954	Salvaged at Davis-Monthan
44-86278			5/25/1945	247BU	10/12/1953	Reclaimed
44-86279					4/1/1946	Surveyed
44-86280		CIRCLE E YLW	CIRCLE E	22BG	8/5/1950	Surveyed
44-86281		SQUARE A	5/25/1945	301BW	7/27/1954	Reclaimed Mountain Home
44-86282				4925TEST	11/21/1952	Crashed
44-86283			5/29/1945	4141BU	10/12/1953	Reclaimed Pyote
44-86284	DOWN'S CLOWNS	CIRCLE W	5/31/1945	92BG	7/14/1954	Reclaimed McClellan
44-86285			5/26/1945	4925SPG	8/16/1954	Reclaimed Laurel
44-86286	SIT N' GIT	SQUARE A	5/29/1945	301BW	6/17/1953	Reclaimed Davis-Monthan
44-86287		SQUARE A	5/29/1945	301BW	7/14/1954	Salvaged at Davis-Monthan
44-86288		SQUARE H		98BG	7/27/1945	Surveyed
44-86289		SQUARE T	5/30/1945	2BG	9/14/1954	Reclaimed Hill
44-86290	BABY SAN	CIRCLE R	5/31/1945	9BW	9/9/1954	Salvaged at Davis-Monthan

Serial #	Name	Identification	Delv	Assign	Off Inv	Circumstances
44-86290	PHIPPENS PIPPENS	SQUARE H	5/31/1945	98BG	7/14/1954	Salvaged at Davis-Monthan
44-86291		CIRCLE ARROW	3/28/1945	509CG	7/21/1953	Reclaimed Hill
44-86292	ENOLA GAY	CIRC R 82		509CG		Unknown
44-86293				4141BU		
44-86294		CIRCLE R	6/1/1945	28BG	10/12/1953	Reclaimed Tinker
44-86295		SQUARE Y	6/4/1945	307BG	8/16/1954	Reclaimed Davis-Monthan
44-86295		SQUARE H	6/4/1945	98BG	8/16/1954	Reclaimed Davis-Monthan
44-86296			6/4/1945	4141BU	10/12/1953	Reclaimed Pyote
44-86297		CIRCLE R	6/5/1945	28BG	10/12/1953	Reclaimed
44-86298		CIRCLE R	6/5/1945	28BG	9/14/1953	Reclaimed Tinker
44-86299		SQUARE P	6/6/1945	307BW	7/14/1954	Reclaimed McClellan
44-86300		CIRCLE R	6/8/1945	28BG	10/12/1953	Reclaimed Tinker
44-86301					6/8/1948	Surveyed
44-86302		CIRCLE R		28BG	10/17/1950	Surveyed
44-86303	EARLY BIRD	L TRI	6/8/1945	6BG	9/25/1954	Reclaimed
44-86303		YLW BANDS	6/8/1945	2ARS	9/25/1954	Reclaimed McClellan
44-86304		CIRCLE R	6/8/1945	28BG	10/12/1953	Reclaimed Tinker
44-86305					6/14/1950	Surveyed
44-86306			6/8/1945	4141BU	10/12/1953	Reclaimed Pyote
44-86307				4105BU	8/7/1947	Salvaged at Eglin
44-86308		YLW BANDS	6/11/1945	2ARS	5/5/1954	Salvaged Prestwick
44-86309		TRIANGLE O	6/13/1945	97BW	2/9/1955	Reclaimed Dow
44-86310			6/13/1945	4105BU	11/12/1953	Reclaimed Robins
44-86311			6/11/1945	4141BU	8/19/1954	Reclaimed Kelly
44-86312			6/12/1945	4141BU	10/12/1953	Reclaimed Pyote
44-86313		BLK SQ M		19BG	7/15/1946	Salvaged
44-86314			6/18/1945	310BW	3/30/1954	Reclaimed Davis-Monthan
44-86315					7/9/1945	Surveyed at Pueblo
44-86316	SPACE MISTRESS	BLKSTRPGRN	6/12/1945	19BG	10/12/1954	Reclaimed McClellan
44-86316	SPACE MISTRESS	SQUARE H	6/12/1945	98BG	10/12/1954	Reclaimed McClellan
44-86317		TRIANGLE O	6/14/1945	97BG	8/8/1954	Reclaimed Tinker
44-86318		SQUARE Y	6/13/1945	307BG		Reclaimed McClellan
44-86319			6/18/1945	XXIBC	8/22/1954	Reclaimed Mountain Home
44-86320			6/19/1945	4141BU	9/27/1945	Reclaimed
44-86320		BLK SQ M		19BG	11/23/1949	Salvaged
44-86321			6/19/1945	44BW	9/27/1945	Reclaimed
44-86322			6/20/1945	4141BU	10/12/1953	Reclaimed Pyote
44-86323	FOUR-A-BREAST	BLKSTRPGRN	6/19/1945	19BG	7/14/1954	Salvaged at Davis-Monthan
44-86324			6/18/1945	73BW	10/12/1953	Reclaimed Pyote
44-86325		SQUARE A	6/21/1945	301BW	7/15/1954	Reclaimed Mountain Home
44-86326		BLK SQ M		19BG	7/10/1945	Surveyed
44-86327		SQUARE H		98BG	7/24/1951	Surveyed 6/1/51 wing on fire fm Kwakson
44-86328		BLKSTRPGRN		19BG	10/4/1950	Surveyed, major accident 9/15/50
44-86329				XXIBC	7/15/1945	Crashed on takeoff
44-86330	(BULL BUTTING BOMB)	SQUARE H		98BG	12/18/1950	Surveyed
44-86330	APE SHIP	SQUARE H		98BG	12/18/1950	Surveyed
44-86330	APE SHIP	BLKSTRRED		19BG	12/18/1950	Surveyed
44-86331		BLKSTRP	7/9/1945	19BG		Salvaged at Davis-Monthan
44-86332			6/23/1945	611BSEX	8/8/1954	Reclaimed Tinker
44-86333		SQUARE A		301BG	9/26/1949	Crashed Talihina, Oklahoma
44-86335	MISS TAMPA X	SQUARE H	6/25/1945	98BG	7/14/1954	Salvaged at Davis-Monthan
44-86335	T.D.Y. WIDOW	SQUARE H	6/25/1945	98BG	7/14/1954	Salvaged at Davis-Monthan
44-86335	T.D.Y. WIDOW (WIFE)	BLKSTRP	6/25/1945	19BG	7/14/1954	Salvaged at Davis-Monthan
44-86336		CIRCLE R	6/26/1945	98BW	7/29/1954	Reclaimed Mountain Home
44-86337		TRIANGLE C	6/27/1945	509BW	7/14/1954	Salvaged at Davis-Monthan
44-86338		SQUARE H		98BG	4/7/1948	Surveyed
44-86339	BIG BLOW	SQUARE H	6/28/1945	98BG	10/24/1954	Reclaimed Kadena
44-86339		SQUARE Y	6/28/1945	307BG	10/24/1954	Reclaimed Kadena
44-86340	WOLF PACK	SQUARE H	6/26/1945	98BG	11/16/1954	Reclaimed Davis-Monthan
44-86341		SQUARE A	6/26/1945	301BW	8/19/1954	Reclaimed Kelly
44-86342			6/24/1945	XXIBC	10/12/1953	Reclaimed Pyote
44-86343		SQUARE Y		307BG	9/13/1952	Flak, exploded Suho
44-86343		CIRC X 56		9BG	9/14/1952	Salvaged at Kadena
44-86343		CIRC X 56		505BG	9/14/1952	Surveyed at Kadena
44-86343		CIRC X		9BG		Unknown

I. Master List

Serial #	Name	Identification	Delv	Assign	Off Inv	Circumstances
44-86344		TRI U 19		462BG	8/2/1945	Abandoned, fire in bomb bay
44-86345			6/29/1945	2750ABG	7/22/1954	Reclaimed Wright-Patterson
44-86346	LUKE THE SPOOK	CIRC W 94	6/15/1945	509CG	9/1/1954	Reclaimed Yokota
44-86346		SQUARE H		98BG		Unknown
44-86347	LAGGIN' DRAGON	BLK SQ O 95	6/15/1945	509CG	9/14/1954	Reclaimed Yokota
44-86348		SQUARE P	7/2/1945		9/27/1954	Reclaimed McClellan
44-86349				4141BU	10/17/1950	Surveyed
44-86349	JOHN'S OTHER WIFE	BLKSTRP		19BG		S/N not in Korea per ARC
44-86349	JOHN'S OTHER WIFE	CIRCLE E		22BG		S/N not in Korea per ARC
44-86350		SQUARE Y	7/3/1945	307BG	6/3/1954	Reclaimed Tinker
44-86351		CIRCLE E	7/3/1945	22BG	6/17/1953	Reclaimed Davis-Monthan
44-86352			7/5/1945	4141BU	10/12/1953	Reclaimed Pyote
44-86353		CIRCLE E		22BG	4/4/1949	Salvaged
44-86354			7/4/1954	4141BU	10/12/1953	Reclaimed Pyote
44-86355		YLW BANDS	7/4/1954			Reclaimed
44-86355		CIRCLE E YLW	7/4/1954	22BG		Reclaimed
44-86356				19BG	8/25/1948	Missing
44-86356		TRIANGLE O		97BG		Unknown
44-86356		SPEEDWING BLU		307BG		Unknown
44-86357		SQUARE Y		307BG	8/25/1951	Flak at Sunchow, abandoned, exploded
44-86357		SPEEDWING BLU		307BG	8/25/1951	Surveyed
44-86358		TRIANGLE S	7/7/1945	2BG	7/14/1954	Salvaged at Davis-Monthan
44-86359		BLKSTRPRED	7/6/1945	19BG	7/14/1954	Reclaimed McClellan
44-86360		SQUARE Y	7/7/1945	307BG	9/13/1954	Reclaimed Davis-Monthan
44-86360		SQUARE H	7/7/1945	98BG	9/13/1954	Reclaimed Davis-Monthan
44-86361	DICKERT'S DEMONS	SQUARE H	7/9/1945	98BG	7/14/1954	Salvaged at Davis-Monthan
44-86361	LONELY LADY	SQUARE H	7/9/1945	98BG	7/14/1954	Salvaged at Davis-Monthan
44-86361	LONESOME POLECAT	SQUARE H	7/9/1945	98BG	7/14/1954	Salvaged at Davis-Monthan
44-86362		BLK SQ P	7/9/1945	39BG	5/23/1945	MIA Tokyo
44-86363		TRIANGLE C	7/9/1945	509BW	2/3/1953	Reclaimed Bergstrom
44-86364			7/9/1945			Unknown
44-86365			7/11/1945	4141BU	10/12/1953	Reclaimed Pyote
44-86366		CIRCLE E YLW	7/10/1945	22BG	9/27/1954	Reclaimed McClellan
44-86367			7/11/1945	4141BU	6/23/1954	Reclaimed Pyote
44-86368			7/12/1945	4141BU	10/12/1953	Reclaimed Pyote
44-86369			7/12/1945	4141BU	10/12/1953	Reclaimed Pyote
44-86370	FUJIGMO	BLKSTRPRED		19BG	4/12/1951	Ditched, fighter damage
44-86371		SQUARE H		98BG	5/7/1951	Lost wing to flak at Pyongyang
44-86372			7/13/1945	4141BU	10/12/1953	Reclaimed Pyote
44-86373			7/17/1945	40BW	7/14/1954	Salvaged at Davis-Monthan
44-86374			7/20/1945	40BW	8/19/1954	Reclaimed
44-86375			7/16/1945	40BW	7/14/1954	Salvaged at Davis-Monthan
44-86376	MISS N.C.	BLKSTRPGRN	7/16/1945	19BG	7/14/1954	Salvaged at Davis-Monthan
44-86376	MISS N.C.	BLKSTRP	7/16/1945	19BG	7/14/1954	Salvaged at Davis-Monthan
44-86377			7/17/1945	40BW	7/14/1954	Salvaged at Davis-Monthan
44-86378			7/17/1945	40BW	7/14/1954	Salvaged at Davis-Monthan
44-86379			7/18/1945	513WRS	3/16/1954	Reclaimed Davis-Monthan
44-86380					12/19/1945	Salvaged at Walker
44-86381			7/19/1945	40BW	7/14/1954	Salvaged at Davis-Monthan
44-86382		TRIANGLE C	7/26/1945	509BW	3/15/1954	Reclaimed Hill
44-86383					8/12/1948	Surveyed
44-86384	FANNY-THE ATOM & I		7/27/1945	313BW	7/1/1954	Reclaimed Great Falls
44-86385		CIRCLE M	7/19/1945	93BW	7/14/1954	Reclaimed Davis-Monthan
44-86386		SQUARE P	7/20/1945	306BW	7/14/1954	Reclaimed McClellan
44-86387		BLKSTRP	7/21/1945	19BG	7/14/1954	Salvaged at Davis-Monthan
44-86387		CIRCLE W	7/21/1945	92BG	7/14/1954	Salvaged at Davis-Monthan
44-86387		SQUARE Y	7/21/1945	307BG	7/14/1954	Salvaged at Davis-Monthan
44-86388		SQUARE P	7/20/1945	306BW	7/14/1954	Salvaged at Davis-Monthan
44-86389		SQUARE A	7/24/1945	301BW	7/18/1954	Lost at Mt Home
44-86390	SNAKE BIT	SQUARE H		98BG	7/13/1952	Surveyed
44-86390	TROUBLE BREWER			SAC		unknown
44-86391		TRIANGLE O	7/23/1945	97BG	9/9/1954	Salvaged at Davis-Monthan
44-86392	WRIGHTS DELIGHT'S	SQUARE H		98BG	11/19/1952	Crashed 5 miles north of C'ho-do Island
44-86393		CIRCLE R	7/23/1945		6/23/1954	Reclaimed
44-86394		TRIANGLE O	7/24/1945	97BG	12/14/1953	Last reported at Randolph

Serial #	Name	Identification	Delv	Assign	Off Inv	Circumstances
44-86395		SQUARE Y	7/27/1945	307BG	8/1/1956	Dropped from inventory as surplus
44-86396		TRIANGLE O	7/26/1945	97BG	7/14/1954	Salvaged at Davis-Monthan
44-86397			7/25/1945	308RCN	4/18/1954	Last reported at McClellan
44-86398			7/25/1945	3203MSU	2/4/1955	Last reported at Olmsted
44-86399			7/26/1945		12/31/1954	Last reported at McClellan
44-86400	BIG GASS BIRD	SQUARE H		98BG	3/31/1952	Flak damage, hit hill, burned
44-86400	CHOTTO MATTE	SQUARE H		98BG	3/31/1952	Flak damage, hit hill, burned
44-86401		TRIANGLE O	7/30/1945	97BG	9/19/1954	Last reported Langley
44-86402			7/31/1945		7/19/1954	Last reported Lockheed Burbank
44-86403			8/3/1945		7/14/1954	Salvaged at Davis-Monthan
44-86404			8/3/1945		7/14/1954	Salvaged at Davis-Monthan
44-86405			8/2/1945		7/14/1954	Salvaged at Davis-Monthan
44-86406			8/1/1945	4141BU	7/14/1954	Salvaged at Davis-Monthan
44-86407			8/1/1945	4141BU	7/14/1954	Salvaged at Davis-Monthan
44-86408	HAGARTY'S HAG		8/6/1945		11/19/1953	Class 26 Dugway Proving Grounds
44-86409			8/2/1945		7/14/1954	Salvaged at Davis-Monthan
44-86410			8/6/1945	554BU	8/19/1954	Reclaimed Kelly
44-86411			8/4/1945		7/14/1954	Salvaged at Davis-Monthan
44-86412			8/4/1945		7/14/1954	Salvaged at Davis-Monthan
44-86413			8/8/1945		7/14/1954	Salvaged at Davis-Monthan
44-86414	OVERNITE BAG	BLKSTRPRED	8/8/1945	19BG	7/14/1954	Salvaged at Davis-Monthan
44-86414	OVERNITE BAG	CIRCLE E	8/8/1945	22BG	7/14/1954	Salvaged at Davis-Monthan
44-86415	SQUEEZE PLAY	SQUARE H		98BG	9/19/1951	Crashed in Sea of Japan
44-86416			8/8/1945		7/14/1954	Salvaged at Davis-Monthan
44-86417			8/10/1945		7/14/1954	Salvaged at Davis-Monthan
44-86418		RAINBOW STRP	8/9/1945		9/1/1954	Last reported at Yokota
44-86419			8/10/1945		7/14/1954	Salvaged at Davis-Monthan
44-86420					10/3/1952	Salvaged
44-86421		SQUARE P	8/13/1945	306BG	9/27/1954	Last reported at McClellan
44-86422		SQUARE Y	8/14/1945	307BG	7/14/1954	Salvaged at Davis-Monthan
44-86422		BLKSTRP	8/14/1945	19BG	7/14/1954	Salvaged at Davis-Monthan
44-86423			8/13/1945	4141BU	8/19/1954	Last reported at Kelly
44-86424		SQUARE Y	8/14/1945	307BG	9/27/1954	Last reported at McClellan
44-86425			8/21/1945		7/14/1954	Salvaged at Davis-Monthan
44-86426		SQUARE I	8/21/1945	92SRC	8/31/1953	Assigned Storage Sqdn at Davis-Monthan
44-86427		TRIANGLE O	8/22/1945	97BG	1/11/1955	Last reported at Dow
44-86428		TRIANGLE O	8/22/1945	97BG	9/5/1954	Last reported at Dow
44-86429		SQUARE H		98BG	5/2/1948	Salvaged
44-86430			8/27/1945		9/1/1954	Last reported at Yokota
44-86430		TRI S		40BS		Unknown
44-86431		CIRCLE W		92BG	7/23/1947	Surveyed
44-86432			8/30/1945		5/27/1954	Lastr reported at Elmendorf
44-86433	PEACE MAKER	BLKSTRPBLU	8/21/1945	19BG	8/8/1954	Reclaimed Tinker
44-86433	PEACE MAKER	CIRCLE W	8/21/1945	92BG	8/8/1954	Reclaimed Tinker
44-86433	PEACE MAKER	SQUARE H	8/21/1945	98BG	8/8/1954	Reclaimed Tinker
44-86434		CIRCLE M	8/21/1945	93BG	8/16/1954	Assigned Storage Sqdn at Davis-Monthan
44-86435		CIRCLE M	8/22/1945	93BG	9/14/1954	Last reported at Hamilton
44-86436	MAIS OUI	SQUARE H	8/24/1945	98BG	9/27/1954	Reclaimed McClellan
44-86437		TRIANGLE C	8/24/1945	509BG	12/17/1953	Last reported at Tinker
44-86438	FLYING PARTS	CIRCLE W	8/24/1945	92BG	2/10/1955	Last reported at McClellan
44-86439		TRIANGLE O	8/24/1945	97BG	5/10/1954	Reclaimed Davis-Monthan
44-86440		TRIANGLE C	8/27/1945	509BG	6/22/1952	Last reported at Randolph
44-86441		CIRCLE E	8/28/1945	22BG	7/14/1954	Reclaimed McClellan
44-86442		SQUARE Y	8/27/1945	307BG	6/17/1954	Class 26 Sheppard
44-86443		TRIANGLE C	8/30/1945	509BG	1/27/1955	Last reported Iceland
44-86444			8/29/1945		11/29/1954	Last reported at Randolph
44-86445		TRIANGLE O	8/29/1945	97BG	6/3/1954	Last reported at Randolph
44-86446	LADY IN DIS DRESS	SQUARE H	8/31/1945	98BG	7/8/1954	Assigned Storage Sqdn at Davis-Monthan
44-86446	LOS ANGELES CALLING	BLKSTRPRED	8/31/1945	19BG	7/8/1954	Assigned Storage Sqdn at Davis-Monthan
44-86446	LOS ANGELES CALLING	SQUARE H	8/31/1945	98BG	7/8/1954	Assigned Storage Sqdn at Davis-Monthan
44-86446	RANKLESS WRECK	SQUARE H	8/31/1945	98BG	7/8/1954	Assigned Storage Sqdn at Davis-Monthan
44-86446	RAPID RABBIT	SQUARE H	8/31/1945	98BG	7/8/1954	Assigned Storage Sqdn at Davis-Monthan
44-86447			9/10/1945		2/4/1953	Last reported at Tinker
44-86448						
44-86449		TRIANGLE O	8/31/1945	97BG	7/14/1954	Salvaged at Davis-Monthan

I. Master List

Serial #	Name	Identification	Delv	Assign	Off Inv	Circumstances
44-86450			8/31/1945		9/1/1954	Last reported at Yokota
44-86451		TRIANGLE O	8/31/1945	97BG	5/5/1954	Reclaimed Bergstrom
44-86452		SQUARE Y	8/29/1945	307BG	7/14/1954	Reclaimed Davis-Monthan
44-86453			8/31/1945		5/10/1954	Reclaimed Davis-Monthan
44-86454			9/10/1945	4105BU	5/10/1954	Reclaimed Davis-Monthan
44-86455			9/11/1945	4141BU	10/12/1953	Reclaimed Pyote
44-86456			9/4/1945	4141BU	10/12/1953	Reclaimed Pyote
44-86457			9/11/1945	4105BU	5/10/1954	Reclaimed Davis-Monthan
44-86458			9/12/1945	4141BU	10/12/1953	Reclaimed Pyote
44-86459			9/14/1945	4105BU	5/10/1954	Reclaimed Davis-Monthan
44-86460			9/12/1945	4105BU	5/10/1954	Reclaimed Davis-Monthan
44-86461			9/18/1945	4105BU	5/10/1954	Reclaimed Davis-Monthan
44-86462			9/12/1945	4105BU	5/10/1954	Reclaimed Davis-Monthan
44-86463			9/18/1945	4105BU	5/10/1954	Reclaimed Davis-Monthan
44-86464			9/20/1945	4105BU	5/10/1954	Reclaimed Davis-Monthan
44-86465			9/19/1945	4105BU	5/10/1954	Reclaimed Davis-Monthan
44-86466			9/20/1945	4141BU	10/12/1953	Reclaimed Pyote
44-86467			9/20/1945	4141BU	10/12/1953	Reclaimed Pyote
44-86468			9/21/1945	4141BU	10/12/1953	Reclaimed Pyote
44-86469			9/25/1945	4141BU	10/12/1953	Reclaimed Pyote
44-86470			9/26/1945	4141BU	8/19/1954	Reclaimed Kelly
44-86471			9/25/1945	4141BU	9/22/1953	Reclaimed Pyote
44-86472		SQUARE H		98BG	5/20/1947	Salvaged
44-86473					4/9/1948	Surveyed
44-86535		TRI S		40BG		Unknown
44-86370	LUCKY DOG	BLKSTRPRED		19BG	4/12/1951	Surveyed
44-87341	THE DREAMER	SQUARE H	4/30/1945	98BG	8/19/1954	Reclaimed
44-87584					7/15/1952	Surveyed
44-87585			5/15/1945	4105BU	5/5/1954	Reclaimed Davis-Monthan
44-87586			5/15/1945	4105BU	5/10/1954	Reclaimed Davis-Monthan
44-87587					7/15/1952	Surveyed
44-87588			5/19/1945	4141BU	6/23/1954	Reclaimed Pyote
44-87589			5/21/1945	4105BU	5/5/1954	Reclaimed Davis-Monthan
44-87590			5/21/1945	4141BU	6/23/1954	Reclaimed Pyote
44-87591		BLKSTRP	5/18/1945	19BG	7/14/1954	Salvaged at Davis-Monthan
44-87592			5/22/1945	324BU	8/27/1945	Surveyed Chatham
44-87593			5/18/1945		1/10/1954	Last reported at Tinker
44-87594			5/21/1945		12/6/1954	Last reported at Randolph
44-87595			5/26/1945		8/25/1953	Last reported Barksdale
44-87596		BLKSTRP	5/25/1945	19BG	7/14/1954	Salvaged at Davis-Monthan
44-87597		BLKSTRP	5/23/1945	19BG	7/14/1954	Salvaged at Davis-Monthan
44-87598		CIRCLE T	5/22/1945	44BG	9/24/1953	Last reported at Tinker
44-87599			5/23/1945		12/21/1955	Lost Eglin, unknown reason
44-87600					6/25/1952	Surveyed
44-87601	CHARLIE'S WAGON	SQUARE A	5/25/1945	301BG	11/15/1954	Last reported at Yokota
44-87601	TOUCH-N-GO		5/25/1945		11/15/1954	Last reported at Yokota
44-87602			5/24/1945		8/8/1954	Reclaimed Tinker
44-87603			5/23/1945		6/25/1952	Last reported at Randolph
44-87604			5/24/1945		8/8/1954	Reclaimed Davis-Monthan
44-87605		SQUARE A	5/25/1945	301BG	5/18/1953	Last reported at Tinker
44-87606		SQUARE A	5/24/1945	301BG	12/16/1953	Last reported Hunter
44-87607			5/25/1945		3/31/1954	Last reported Davis-Monthan
44-87608			5/26/1945		11/24/1953	Last reported at Tinker
44-87609			5/26/1945		1/2/1953	Last reported Travis
44-87610			5/26/1945		8/16/1954	Assigned Storage Sqdn at Davis-Monthan
44-87611			5/28/1945		8/16/1954	Assigned Storage Sqdn at Davis-Monthan
44-87612			5/28/1945		11/29/1953	Reclaimed Tinker
44-87613			5/29/1945		7/14/1954	Reclaimed McClellan
44-87614		CIRCLE Z	5/28/1945	90RCN	1/3/1953	Reclaimed Tinker
44-87615			5/28/1945	4136BU	5/10/1949	Reclaimed Tinker
44-87616			5/29/1945	4141BU	8/3/1953	Reclaimed Tinker
44-87617				19BG	5/9/1951	Surveyed
44-87617		CIRCLE E		92BG	5/9/1951	Surveyed
44-87617					5/9/1951	Surveyed
44-87618	NO SWEAT	BLKSTRPGRN		19BG	5/14/1951	Salvaged, collided with F-51 4/12/51

The B-29 Superfortress

Serial #	Name	Identification	Delv	Assign	Off Inv	Circumstances
44-87619			5/30/1945			Last reported Randolph 6/1/54
44-87620			5/29/1945	247BU	3/25/1953	Reclaimed McClellan
44-87620		CIRCLE W	5/29/1945	92BG	3/28/1948	Salvaged at Spokane
44-87621		CIRCLE W	6/30/1945	98BG	7/27/1948	Salvaged at Spokane
44-87622		SQUARE I	5/31/1945		7/14/1954	Salvaged at Davis-Monthan
44-87623			5/31/1945		9/8/1954	Reclaimed Davis-Monthan
44-87624			5/31/1945		4/29/1954	Salvaged at Smoky Hill
44-87625		SQUARE A	5/31/1945	301BW	9/8/1954	Reclaimed Davis-Monthan
44-87626		CIRCLE Z	6/1/1945	90RCN	12/1/1953	Reclaimed Forbes
44-87627			6/1/1945			Last reported Robins 4/54
44-87628			6/1/1945			Last reported Great Falls 2/54
44-87629			6/1/1945		8/4/1954	Reclaimed Robins
44-87630	UP N' COMIN'	TRI N 9	6/4/1945	444BG	8/5/1954	Lost Great Falls
44-87631			6/4/1945		2/15/1954	Reclaimed Randolph
44-87632		TRI S	6/4/1945	40BG	6/3/1954	Reclaimed Molesworth
44-87633		TRI S 05 (2)	6/4/1945	40BG	8/8/1954	Reclaimed Tinker
44-87634			6/6/1945		11/18/1948	Surveyed at Harmon
44-87635		TRI S 48	6/6/1945	40BG	9/8/1954	Reclaimed Davis-Monthan
44-87636		TRI U	6/5/1945	462BG	3/22/1945	Reclaimed Wheelus
44-87637			6/5/1945		5/18/1946	Sent to Aberdeen Proving Grounds
44-87638		CIRC X	6/5/1945	9BG	4/20/1947	Crashed
44-87639		BLK SQ M 44 (2)	6/5/1945	19BG	11/8/1950	Surveyed at Hill
44-87640	IRON GEORGE	TRI U		462BG	11/1/1951	Surveyed at Randolph
44-87641		CIRC X 31 (2)	6/6/1945	9BG	11/3/1953	Reclaimed Pyote
44-87642	CITY OF ATTLEBORO	BLK SQ P 44 (2)	6/7/1945	39BG	8/28/1946	Transferred to Army
44-87642	SLIM II	BLK SQ P 44 (2)	6/7/1945	39BG	8/28/1946	Transferred to Army
44-87643		TRI U	6/6/1945	462BG	8/8/1954	Reclaimed Tinker
44-87644		TRI I	6/12/1945	468BG		Last reported Kindley 12/31/54
44-87645		CIRC X	6/7/1945	9BG	8/8/1954	Reclaimed Tinker
44-87646	THE KICK-A-POO-JOY III	BLK SQ P 47	6/7/1945	39BG		Last reported Tinker 3/26/55
44-87647		TRI I 35	6/7/1945	468BG	8/16/1954	Reclaimed Griffis
44-87648		CIRC X	6/8/1945	9BG	6/8/1948	Surveyed at Harmon
44-87649		V SQ	6/9/1945	499BG	9/27/1954	Reclaimed McClellan
44-87649		SQUARE H	6/9/1945	98BG	9/27/1954	Reclaimed McClellan
44-87650	ANN GARRY V	CIRC R 37	6/8/1945	6BG	10/2/1948	Surveyed at Harmon
44-87651		V	6/8/1945	499BG	8/6/1950	Salvaged at Fairfield
44-87652		V 11 (3)	6/8/1945	499BG	7/20/1945	Shot down at Hitachi
44-87653	CITY OF RICHMOND III	BLK SQ M 48	6/8/1945	19BG	3/25/1954	Reclaimed Hill
44-87653	WANGO BANGO	BLK SQ M 48	6/8/1945	19BG	3/25/1954	Reclaimed Hill
44-87654		CIRC E	6/9/1945	504BG	8/8/1954	Reclaimed Tinker
44-87655		BLK SQ M	6/9/1945	19BG	11/3/1953	Reclaimed Pyote
44-87656		TRI I	6/12/1945	468BG	8/8/1954	Reclaimed Tinker
44-87657		V ?	6/9/1945	499BG		Survived war
44-87657		BLKSTRPGRN	6/9/1945	19BG	9/27/1954	Class 32 Wright-Patterson
44-87657	COMMAND DECISION	BLKSTRPGRN	6/9/1945	19BG		Unknown
44-87658	MISS LACE	TRI U	6/11/1945	462BG		Unknown
44-87659	MISS SANDY	TRI I 24	6/11/1945	468BG		Last reported Whellus 3/22/54
44-87660	CELESTIAL QUEEN	TRI S 49	6/11/1945	40BG		Last reported Barksdale 1/11/54
44-87661	AMERICAN BEAUTY III	TRI I 6	6/12/1945	468BG	7/14/1954	Reclaimed McClellan
44-87661	KOZA KID	BLKSTRPGRN	6/12/1945	19BG	7/14/1954	Reclaimed McClellan
44-87661	KOZA KID	BLKSTRPBLU	6/12/1945	19BG	7/14/1954	Reclaimed McClellan
44-87661	NIGHT-MARE!	BLKSTRPGRN	6/12/1945	19BG	7/14/1954	Reclaimed McClellan
44-87661	NIGHT-MARE!	BLKSTRPBLU	6/12/1945	19BG	7/14/1954	Reclaimed McClellan
44-87661	UGLY	BLKSTRPGRN	6/12/1945	19BG	7/14/1954	Reclaimed McClellan
44-87661	UGLY	BLKSTRPBLU	6/12/1945	19BG	7/14/1954	Reclaimed McClellan
44-87661	AMERICAN BEAUTY	BLKSTRP		19BG		
44-87662	SKY BLUES	TRI N	6/14/1945	444BG	5/13/1946	Transferred to Army
44-87662		BLK SQ O	6/14/1945	29BG	5/13/1946	Transferred to Army
44-87663		BLK SQ O	6/12/1945	29BG	5/13/1946	Transferred to Army
44-87664	THUNDER BIRD	BLK SQ O	6/15/1945	29BG	8/8/1945	Hit by flak, exploded, abandoned
44-87665		A	6/19/1945	497BG	9/9/1954	Reclaimed Robins
44-87666		TRI I 56	6/14/1945	468BG	3/25/1954	Reclaimed Hill
44-87667	BAD MEDICINE #2	CIRC W 19 (1)	6/15/1945	505BG	11/3/1953	Reclaimed Pyote
44-87668	PAPA TOM'S CABIN	TRI I 4		468BG		Unknown
44-87669		V ?	6/13/1945	499BG	8/8/1954	Reclaimed Tinker

I. Master List

Serial #	Name	Identification	Delv	Assign	Off Inv	Circumstances
44-87670			6/15/1945		5/26/1954	Reclaimed Randolph
44-87671		TRI I 12	6/13/1945	468BG	7/28/1948	Surveyed at Tinker
44-87672		CIRC R	6/13/1945	6BG	11/18/1953	Reclaimed Tinker
44-87673			6/12/1945		4/20/1954	Reclaimed Hill
44-87674	LADY CHOUTEAU	TRI I 22	6/15/1945	468BG	7/14/1954	Salvaged at Davis-Monthan
44-87675			6/14/1945		8/2/1954	Reclaimed Tinker
44-87676			6/18/1945			Last reported at Wheelus
44-87677			6/15/1945	326BU		Last reported Eglin 3/3/55
44-87678			6/19/1945	4117BU		Last reported Robins 1/4/55
44-87679			6/13/1945		8/8/1954	Reclaimed Tinker
44-87680		CIRCLE K	6/18/1945	43BG	7/14/1954	Salvaged at Davis-Monthan
44-87681			6/14/1945		8/8/1954	Reclaimed Tinker
44-87682			6/20/1945		7/14/1954	Reclaimed McClellan
44-87683			6/18/1945		8/16/1954	Reclaimed Molesworth
44-87684				8532BU	5/31/1952	Surveyed
44-87684				8532BU	5/31/1952	Surveyed
44-87685			6/18/1945	8532BU	7/13/1945	Surveyed at Randolph
44-87686			6/20/1945	4141BU	11/12/1953	Reclaimed Pyote
44-87687			6/20/1945	4141BU	11/12/1953	Reclaimed Pyote
44-87688			6/21/1945	4141BU	6/23/1954	Reclaimed Pyote
44-87689			6/19/1945	4141BU	11/12/1953	Reclaimed Pyote
44-87690			6/20/1945	4141BU	6/23/1954	Reclaimed Pyote
44-87691			6/21/1945	4141BU	6/23/1954	Reclaimed Pyote
44-87692			6/20/1945	2753AST	4/20/1954	Reclaimed Kelly
44-87693			6/25/1945	3705BU	8/25/1950	Surveyed McClellan
44-87694			6/21/1945	2753AST	4/20/1954	Reclaimed Kelly
44-87695			6/21/1945	4105BU	5/5/1954	Reclaimed Davis-Monthan
44-87696			6/22/1945	4141BU	6/23/1954	Reclaimed Pyote
44-87697			6/22/1945	4105BU	5/5/1954	Reclaimed Davis-Monthan
44-87698			6/22/1945			Unknown
44-87699				4141BU	3/25/1953	Reclaimed Pyote
44-87700			6/23/1945	4105BU	5/5/1954	Reclaimed Davis-Monthan
44-87701			6/23/1945	4141BU	3/25/1954	Reclaimed Kelly
44-87702			6/23/1945	4105BU	5/5/1954	Reclaimed Davis-Monthan
44-87703			6/23/1945	4105BU	5/5/1954	Reclaimed Davis-Monthan
44-87704		TRIANGLE C		509BG	12/12/1949	Surveyed
44-87705			6/25/1945	4141BU	6/23/1954	Reclaimed Pyote
44-87706			6/25/1945	4141BU	6/23/1954	Reclaimed Pyote
44-87707			6/27/1945	4141BU	11/12/1953	Reclaimed Pyote
44-87708			6/25/1945	4105BU	5/5/1954	Reclaimed Davis-Monthan
44-87709			6/26/1945	4141BU	11/12/1953	Reclaimed Pyote
44-87710			6/26/1945	4105BU	5/5/1954	Reclaimed Davis-Monthan
44-87711			6/27/1945	4141BU	3/25/1953	Reclaimed Pyote
44-87712			6/26/1945	4141BU	3/25/1953	Reclaimed Pyote
44-87713			6/26/1945	4141BU	6/23/1954	Reclaimed Pyote
44-87714			7/2/1945	4105BU	5/5/1954	Reclaimed Davis-Monthan
44-87715			6/27/1945	4105BU	5/10/1954	Reclaimed Davis-Monthan
44-87716			6/28/1945	4141BU	6/23/1954	Reclaimed Pyote
44-87717			6/27/1945	4127BU	5/7/1954	Reclaimed McClellan
44-87718			6/28/1945	4196BU	8/25/1950	Reclaimed McClellan
44-87719			6/28/1945	4136BU	9/11/1950	Reclaimed Tinker
44-87720			6/28/1945	4127BU	5/7/1954	Reclaimed McClellan
44-87721			6/28/1945	4105BU	5/5/1954	Reclaimed Davis-Monthan
44-87722			6/29/1945	4141BU	11/12/1953	Reclaimed Pyote
44-87723			6/29/1945	4141BU	6/23/1954	Reclaimed Pyote
44-87724			7/4/1945	4135BU	3/25/1954	Reclaimed Hill
44-87725		TRIANGLE C	6/29/1945	509BG	10/4/1954	Reclaimed Tinker
44-87726			6/29/1945		8/25/1950	Surveyed
44-87727			6/29/1945	4141BU	6/23/1954	Reclaimed Pyote
44-87728			7/2/1945	4127BU	6/25/1950	Surveyed McClellan
44-87729			7/3/1945	4105BU	5/5/1954	Reclaimed Davis-Monthan
44-87730			7/2/1945	4105BU	5/10/1954	Reclaimed Davis-Monthan
44-87731			7/3/1945	4105BU	5/5/1954	Reclaimed Davis-Monthan
44-87732			7/3/1945	3705BU	3/29/1953	Reclaimed Tinker
44-87733			7/2/1945	4127BU	12/6/1950	Reclaimed McClellan

THE B-29 SUPERFORTRESS

Serial #	Name	Identification	Delv	Assign	Off Inv	Circumstances
44-87734	DOUBLE WHAMMY	BLKSTRPRED		19BG	1/22/1952	#3 engine out, missing from Kadena
44-87734	ZZZUNNAMED	TRI U		462BG	1/22/1952	MIA Korea
44-87734	ZZZUNNAMED	CIRC R 15		6BG		Reclaimed
44-87735			7/9/1945		12/15/1949	Salvaged at Walker
44-87736					7/18/1954	Lost at Mt Home
44-87737	BATTLIN' BONNIE II	CIRC X 15		9BG	6/2/1952	Surveyed
44-87738			7/4/1945		8/8/1954	Reclaimed Tinker
44-87739		SQUARE H	7/5/1945	98BG	4/20/1954	Reclaimed Hill
44-87740			7/5/1945			Last reported Hickam 2/11/55
44-87741	PRIVATE LOVE WITCH	TRIANGLE C	7/5/1945	509BG	12/17/1953	Crash landed Andersen
44-87742		CIRCLE K	7/5/1945	43BG		Last reported Yokota 5/27/54
44-87743		TRIANGLE C	7/9/1945	509BG	7/18/1954	Lost at Mt Home
44-87744		TRIANGLE S		2BG	8/6/1951	Surveyed
44-87745		TRIANGLE O	7/9/1945	97BG	7/22/1954	Lost at Mt Home
44-87746		CIRCLE R	7/9/1945	9BG	8/9/1954	Reclaimed Laurel
44-87747			7/9/1945			Last reported at Yokoya 1/10/55
44-87748		TRIANGLE C	7/11/1945	509BG	8/8/1954	Reclaimed Tinker
44-87749			7/9/1945		10/4/1954	Reclaimed Tinker
44-87750	HURRICANE HATTIE		7/9/1945		4/26/1950	Surveyed Kindley
44-87751		TRIANGLE C	7/9/1945	509BG	8/8/1954	Reclaimed Tinker
44-87752		TRIANGLE O	7/10/1945	97BG	9/13/1954	Reclaimed Davis-Monthan
44-87753			7/9/1945		8/8/1954	Reclaimed Tinker
44-87754		TRIANGLE C	7/10/1945	509BG	7/29/1954	Lost at Mt Home
44-87755		TRIANGLE C	7/9/1945	509BG		Unknown
44-87756					4/5/1952	Crashed, prop reversal in pattern
44-87757		TRIANGLE C	7/9/1945	509BG	8/8/1954	Reclaimed Tinker
44-87758		CIRCLE K	7/9/1945	43BG	8/2/1954	Reclaimed Tinker
44-87759		TRIANGLE O	7/12/1945	97BG	5/12/1950	Salvaged at Barksdale
44-87760	NIP-PON-ESE	SQUARE Y	7/10/1945	307BG	8/8/1954	Reclaimed Tinker
44-87760	NIP-PON-ESE	CIRCLE W	7/10/1945	92BG	8/8/1954	Reclaimed Tinker
44-87761		TRIANGLE C	7/10/1945	509BG	8/16/1954	Assigned Storage Sqdn at Davis-Monthan
44-87761		CIRCLE R	7/10/1945	9BG	8/16/1954	Assigned Storage Sqdn at Davis-Monthan
44-87762			7/11/1945		9/13/1954	Reclaimed Tinker
44-87763		TRIANGLE C	7/11/1945	509BG	7/27/1954	Mountain Home
44-87764		TRIANGLE C	7/12/1945	509BG	8/8/1954	Reclaimed Tinker
44-87765		CIRCLE R	7/11/1945	9BG	7/18/1954	Lost at Mt Home
44-87766		TRIANGLE O	7/10/1945	97BG	5/14/1947	Transferred to Navy
44-87767		TRIANGLE C	7/13/1945	509BG	7/24/1954	Mountain Home
44-87768		YLW BANDS	7/12/1945		7/14/1954	Salvaged at Davis-Monthan
44-87769		TRIANGLE C	7/12/1945	509BG	11/29/1949	Salvaged at Walker
44-87770			7/13/1945	4187BU	12/21/1949	Salvaged at Marham
44-87771		TRIANGLE C	7/17/1945	509BG	8/16/1954	Assigned Storage Sqdn at Davis-Monthan
44-87772		TRIANGLE O	7/12/1945	97BG	1/14/1949	Surveyed Furstenfeldbuch
44-87773		TRIANGLE C	7/12/1945	509BG	7/18/1954	Lost at Mt Home
44-87774					8/1/1950	Surveyed
44-87775			7/17/1945			Last reported at Birmingham
44-87776		CIRCLE K	7/18/1945	43BG	7/14/1954	Salvaged at Davis-Monthan
44-87777		CIRCLE K	7/16/1945	43BG	7/14/1954	Salvaged at Davis-Monthan
44-87778		TRIANGLE C	7/18/1945	509BG	8/8/1954	Reclaimed Tinker
44-87779	LEGAL EAGLE II		7/16/1945			Unknown
44-87780		SQUARE A	7/16/1945	301BW	8/8/1954	Reclaimed Tinker
44-87781		TRIANGLE C	7/17/1945	509BG	8/16/1954	Assigned Storage Sqdn at Davis-Monthan
44-87782					3/13/1950	Surveyed
44-87783		TRIANGLE C	7/18/1945	509BG		Unknown
45-21693		CIRCLE K	7/14/1945	43BG	7/14/1954	Salvaged at Davis-Monthan
45-21694			7/19/1945		8/8/1954	Reclaimed Tinker
45-21695		TRIANGLE C	7/19/1945	509BG	8/2/1954	Reclaimed Tinker
45-21696		TRIANGLE C	7/20/1945	509BG	8/16/1954	Assigned Storage Sqdn at Davis-Monthan
45-21697		CIRCLE K		43BG	6/1/1950	Salvaged
45-21698			7/21/1945		1/31/1955	Last reported Wright-Patterson
45-21699		CIRCLE K	7/19/1945	43BG	8/8/1954	Reclaimed Tinker
45-21700		CIRCLE K	7/20/1945	43BG	8/2/1954	Reclaimed Tinker
45-21701		TRIANGLE C	7/19/1945	509BG	7/14/1954	Salvaged at Davis-Monthan
45-21702		CIRCLE K	7/21/1945	43BG	3/31/1954	Last reported at Forbes
45-21703		CIRCLE K	7/23/1945	43BG	7/14/1954	Salvaged at Davis-Monthan

I. Master List

Serial #	Name	Identification	Delv	Assign	Off Inv	Circumstances
45-21704		TRIANGLE C	7/21/1945	509BG	8/8/1954	Reclaimed Tinker
45-21705		TRIANGLE O		97BG	4/19/1949	Salvaged
45-21706				3510BPT	2/2/1950	Salvaged
45-21707		SQUARE Y		307BG	8/22/1949	Salvaged
45-21708			7/23/1945	611BU	9/13/1954	Last reported at Dow
45-21709			7/23/1945		8/16/1954	Assigned Storage Sqdn at Davis-Monthan
45-21710		SQUARE Y	7/24/1945	307BG	2/1/1955	Last reported at Davis-Monthan
45-21710		SQUARE Y	7/24/1945	307BG	2/1/1955	Last reported at Davis-Monthan
45-21711		CIRCLE T	7/21/1945	44BG	9/24/1953	Last reported at Tinker
45-21712			7/23/1945	4135BU	3/25/1954	Reclaimed Hill
45-21713		CIRCLE K	7/25/1945	43BG	7/14/1954	Salvaged at Davis-Monthan
45-21714		CIRCLE Z	7/24/1945	90BG	5/31/1954	Last reported at Tinker
45-21715				308WG	6/6/1952	Salvaged
45-21716	SHEEZA GOER	CIRCLE K	7/24/1945	43BG	7/14/1954	Salvaged at Davis-Monthan
45-21716	SKY OCTANE	CIRCLE K	7/24/1945	43BG	7/14/1954	Salvaged at Davis-Monthan
45-21716	SHEEZA GOER!	BLKSTRPBLU		19BG		Unknown
45-21717	WACHIN JO		7/25/1945	308WRCN	12/31/1954	Last reported at Eielsen
45-21718			7/26/1945		8/8/1954	Reclaimed Tinker
45-21719			7/26/1945		8/8/1954	Reclaimed Tinker
45-21720			7/25/1945		8/22/1954	Last reported at Randolph
45-21721	BURK'S JERKS	SQUARE H		98BG	2/7/1952	Crashed 5 miles north of Yokota
45-21721	SWEET JUDY II	SQUARE H		98BG	2/7/1952	Crashed 5 miles north of Yokota
45-21721	TAIL WIND	SQUARE H		98BG	3/19/1951	Surveyed
45-21722		SQUARE Y	7/27/1945	307BG	8/16/1954	Last reported at Davis-Monthan
45-21723		TRIANGLE C	7/26/1945	509BG	8/8/1954	Reclaimed Tinker
45-21724			7/27/1945	4135BU	4/20/1954	Last reported at Hill
45-21725		SQUARE Y		307BG	4/17/1951	Surveyed, April crash
45-21725		BLKSTRPRED		19BG	4/17/1951	Surveyed, crash landed Naha
45-21726		CIRCLE R		9BG	6/30/1946	Surveyed
45-21727		TRIANGLE C	7/27/1945	509BG	5/11/1953	Last reported at Tinker
45-21728			7/27/1945	4105BU	5/5/1954	Reclaimed Davis-Monthan
45-21729		CIRCLE A	7/30/1945	106BG	8/8/1954	Reclaimed Tinker
45-21730		TRIANGLE C	7/31/1945	509BG	7/18/1954	Reported lost at Mountain home
45-21731		CIRCLE K	8/1/1945	43BG	7/8/1954	Assigned Storage Sqdn at Davis-Monthan
45-21732			7/31/1945	4135BU	3/25/1954	Reclaimed Hill
45-21733		CIRCLE R	7/30/1945	509BG	8/16/1954	Assigned Storage Sqdn at Davis-Monthan
45-21734	CROSS OVER THE BRIDGE	RAINBOW STRP	7/31/1945		2/8/1955	Last reported at Andrews
45-21735		CIRCLE E	7/31/1945	22BG	5/18/1953	Last reported at Tinker
45-21736		TRIANGLE C		509BG	8/15/1950	Salvaged
45-21737		CIRCLE M	8/1/1945	93BW	8/8/1954	Reclaimed Tinker
45-21738		TRIANGLE C	8/1/1945	509BG	7/14/1954	Salvaged at Davis-Monthan
45-21739	UNIFICATION		8/3/1945		4/14/1954	Last reported at Randolph
45-21740			8/3/1945		8/18/1953	Last reported at Eielsen
45-21741		CIRCLE K	8/3/1945	43BG	8/8/1954	Reclaimed Tinker
45-21742		SQUARE I	8/3/1945	91SRW	7/14/1954	Salvaged at Davis-Monthan
45-21743		BLKSTRP	8/9/1945	19BG	7/13/1954	Last reported at Eglin
45-21744			8/9/1945	620EXGP	8/8/1954	Reclaimed Tinker
45-21745	LUCIFER	BLKSTRPBLU		19BG	2/14/1952	Surveyed
45-21746	RAZ N' HELL	BLKSTRPGRN	8/10/1945	19BG	8/16/1954	Last reported at Griffiss
45-21747		TRIANGLE C	8/9/1945	509BG	8/8/1954	Reclaimed Tinker
45-21748		TRIANGLE C	8/9/1945	509BG	6/23/1954	Class 26 At Chanute
45-21749		BLKSTRPRED		19BG	3/19/1951	#2 engine out, missing NW of Okinawa
45-21750		TRIANGLE O	8/14/1945	97BG	12/7/1954	Last reported at Tinker
45-21751		TRIANGLE C	8/17/1945	509BG	9/14/1954	Last reported at Griffiss
45-21752			8/13/1945	3208GOT	8/19/1954	Last reported at Edwards
45-21753						
45-21754		TRIANGLE C	8/13/1945	509BG	3/25/1954	Reclaimed Kelly
45-21755			8/14/1945		9/14/1954	Last reported at Griffiss
45-21756			8/17/1945	4105BU	3/25/1954	Reclaimed Kelly
45-21757			8/20/1945		6/23/1954	Class 26 At Chanute
45-21758		TRIANGLE O	8/9/1945	97BG	11/18/1952	Last Reported at Lowry
45-21759		TRIANGLE S	8/4/1945	28BG	9/14/1953	Last reported at Tinker
45-21760	AL-ASK-ER	CIRCLE M	8/3/1945	93BG	8/8/1954	Reclaimed Tinker
45-21761		CIRCLE K		43BG	8/18/1952	Salvaged
45-21762		CIRCLE X	8/3/1945	5SRCG	10/4/1954	Reclaimed Tinker

Serial #	Name	Identification	Delv	Assign	Off Inv	Circumstances
45-21763	THE KEE BIRD				4/14/1947	Crash landed Greenland
45-21764		CIRCLE K	8/4/1945	43BG	7/14/1954	Salvaged at Davis-Monthan
45-21765		TRIANGLE C	8/6/1945	509BG	8/8/1954	Reclaimed Tinker
45-21766		SQUARE I	8/9/1945	91RCN	10/21/1954	Last reported at Yokota
45-21767			8/6/1945		12/31/1954	Last reported at Eielsen
45-21768				91RCN		Sold civilian, crashed
45-21769		TRIANGLE C	8/6/1945	509BG	8/8/1954	Reclaimed Tinker
45-21770		CIRCLE M	8/10/1945	93BG	5/19/1953	Last reported at Tinker
45-21771		CIRCLE M		93BW	2/8/1951	Surveyed
45-21772		CIRCLE M	8/10/1945	93BG	8/8/1954	Reclaimed Tinker
45-21773			8/10/1945		7/14/1954	Salvaged at Davis-Monthan
45-21774			8/9/1945		12/31/1954	Last reported at McClellan
45-21775		CIRCLE K		43BG	12/31/1947	Surveyed
45-21776		CIRCLE M	8/14/1945	93BG	1/3/1953	Last reported at Randolph
45-21777		SQUARE I	8/14/1945		8/26/1954	Last reported at Yokota
45-21778		CIRCLE K	8/13/1945	43BG	8/2/1954	Reclaimed Tinker
45-21779		SQUARE A	8/13/1945	301BG	5/18/1953	Last reported at Tinker
45-21780		CIRCLE M	8/17/1945	93BG	1/20/1955	Last reported at Randolph
45-21781		CIRCLE W	8/20/1945	93BG	8/8/1954	Reclaimed Tinker
45-21782		TRIANGLE C	8/20/1945	509BG	7/27/1954	Last reported at Mt Home
45-21783		CIRCLE R	8/17/1945	9BG	8/16/1954	Assigned Storage Sqdn at Davis-Monthan
45-21784					5/31/1946	Salvaged
45-21785		TRIANGLE C	8/20/1945	509BG	8/8/1954	Reclaimed Tinker
45-21786		SQUARE I	8/20/1945	91SRC	9/8/1954	Reclaimed Davis-Monthan
45-21787	FERTILE MYRTLE	BUPER84029			5/5/1947	Transferred to Navy
45-21788		TRIANGLE C	8/20/1945	509BG	8/8/1954	Reclaimed Tinker
45-21789		BUPERS84028			5/6/1947	Transferred to Navy
45-21790					11/13/1950	Salvaged
45-21791		BUPERS84030			5/14/1947	Transferred to Navy
45-21792		TRIANGLE C	8/21/1945	509BG	7/14/1954	Salvaged at Davis-Monthan
45-21793	COSMIC RAY RESEARCH		8/21/1945	317ERD	7/18/1954	Last reported at Eglin
45-21793	OLE MISS VI		8/21/1945	317ERD	7/18/1954	Last reported at Eglin
45-21794			8/27/1945		11/22/1953	Last reported at Biggs
45-21795				4105BU	3/27/1947	Surveyed
45-21796			8/21/1945	4105BU	5/10/1954	Reclaimed Davis-Monthan
45-21797			8/22/1945	4105BU	5/10/1954	Reclaimed Davis-Monthan
45-21798			8/22/1945	4105BU	5/10/1954	Reclaimed Davis-Monthan
45-21799			8/22/1945	4105BU	5/10/1954	Reclaimed Davis-Monthan
45-21800	(STORK W/BABY)		8/23/1945		8/19/1954	Last reported at Edwards
45-21801				609BU	5/16/1951	Surveyed
45-21802			8/23/1945	4105BU	5/10/1954	Reclaimed Davis-Monthan
45-21803			8/23/1945	611BEX	8/8/1954	Reclaimed Tinker
45-21804			8/23/1945	611BEX	11/4/1953	Reclaimed Kelly
45-21805			8/23/1945	4105BU	5/10/1954	Reclaimed Davis-Monthan
45-21806			8/23/1945	4105BU	8/8/1954	Reclaimed Davis-Monthan
45-21807			8/24/1945	4105BU	8/2/1954	Reclaimed Tinker
45-21808			8/24/1945		5/12/1954	Last reported Boeing Seattle
45-21809			8/24/1945	4105BU	8/2/1954	Reclaimed Tinker
45-21810			8/24/1945	4105BU	10/4/1954	Reclaimed Tinker
45-21811			8/24/1945	4105BU	10/4/1954	Reclaimed Tinker
45-21812		TRIANGLE C		509BG	3/13/1950	Salvaged
45-21813	HEAVENLY BODY		8/27/1945		12/3/1955	Crash landing Yokota
45-21814		SQUARE Y		307BG	11/10/1950	shot down 7 miles SW Kusong
45-21815		TRIANGLE C	8/27/1945	509BG	10/31/1954	Assigned Storage Sqdn at Davis-Monthan
45-21816			8/28/1945		9/27/1954	Last reported at Tinker
45-21817		CIRCLE R	8/27/1945	9BG	6/2/1954	Last reported at Andersen
45-21818			8/28/1945	3078BU	11/19/1953	Last reported at Patrick
45-21819			8/28/1945		9/29/1954	Last reported at Eielsen
45-21820		TRIANGLE C	8/28/1945	509BG	7/27/1954	Last reported at Mt Home
45-21821		CIRCLE E	9/11/1945	22BG	8/8/1954	Reclaimed Tinker
45-21822	HEAVENLY LADEN	SQUARE H		98BG	1/24/1952	Engine fire, crashed in Tokyo
45-21823			8/29/1945		3/25/1954	Last reported at Andersen
45-21824	THE THING		8/31/1945		4/28/1954	Last reported at Andersen
45-21825		CIRCLE R		9BG	11/22/1948	Salvaged
45-21826			8/30/1945		2/9/1955	Last reported at Elmendorf

I. Master List

Serial #	Name	Identification	Delv	Assign	Off Inv	Circumstances
45-21827		TRIANGLE C	8/31/1945	509BG	7/14/1954	Salvaged at Davis-Monthan
45-21828			8/31/1945	559SR	8/8/1954	Reclaimed Tinker
45-21829		TRIANGLE C		509BG	2/10/1948	Salvaged
45-21830		TRIANGLE C	8/30/1945	509BG	7/18/1954	Reported lost at Mountain home
45-21831			8/31/1945		9/1/1954	Last reported at Yokota
45-21832		TRIANGLE C	8/31/1945	509BG	7/22/1954	Last reported at Mt Home
45-21833	HANGER QUEEN		9/7/1945		9/1/1954	Last reported at Yokota
45-21834		SQUARE H	8/31/1945	98BG	10/31/1954	Last reported at Davis-Monthan
45-21835					1/20/1949	Salvaged
45-21835					1/20/1949	Salvaged
45-21836					12/14/1949	Surveyed
45-21837					5/11/1948	Surveyed
45-21838			9/5/1945		9/22/1954	Last reported at Tinker
45-21839	TYPHOON GOON	TRIANGLE C		509BG	11/18/1949	Salvaged
45-21840					3/12/1951	Missing
45-21841			9/6/1945		4/4/1954	Last reported at Hickam
45-21842		TRIANGLE C	9/12/1945	9BG	8/16/1954	Assigned Storage Sqdn at Davis-Monthan
45-21843		CIRCLE Z		90RCN	2/25/1946	Salvaged
45-21844		TRIANGLE S		28BG	2/13/1946	Salvaged
45-21845		TRIANGLE C	9/12/1945	509BG	10/31/1954	Last reported at Davis-Monthan
45-21846		SQUARE I	9/12/1945	91SRC	7/14/1954	Salvaged at Davis-Monthan
45-21846		TRIANGLE S		28BG		Unknown
45-21847				2759EXPG	7/27/1948	Surveyed
45-21848		TRIANGLE C		509BG	10/3/1947	Surveyed
45-21849		TRIANGLE C	9/14/1945	509BG	7/25/1954	Last reported at Mt Home
45-21850			9/14/1945	2759BG	9/26/1954	Last reported at Laurel
45-21851		TRIANGLE C	9/14/1945	509BG	7/18/1954	Last reported at Mt Home
45-21852			9/14/1945	3203MSU	12/2/1954	Last reported at Wright Field
45-21853				55RCN	2/28/1947	Surveyed
45-21854					5/19/1950	Surveyed
45-21855				3902ABG	6/10/1948	Surveyed
45-21856		SQUARE I	9/14/1945	91SRC	7/14/1954	Salvaged at Davis-Monthan
45-21857					5/1/1950	Surveyed
45-21858		CIRCLE R	9/17/1945	9BG	2/6/1955	Last reported at Offutt
45-21859			9/17/1945		8/8/1954	Reclaimed Tinker
45-21860			9/17/1945		1/1/1955	Last reported at Laurel
45-21861		TRIANGLE C	9/18/1945	509BG	8/8/1954	Reclaimed Tinker
45-21862		TRIANGLE C	9/25/1945	509BG	8/8/1954	Reclaimed Tinker
45-21863			9/18/1945		8/16/1954	Last reported at Laurel
45-21864		TRIANGLE C	9/19/1945	509BG	10/31/1954	Last reported at Davis-Monthan
45-21865		TRIANGLE C	9/19/1945	509BG	8/8/1954	Reclaimed Tinker
45-21866				3171EREG	10/26/1948	Surveyed
45-21867					1/5/1948	Salvaged
45-21868		SQUARE H	9/20/1945	98BG	8/6/1954	Reclaimed Tinker
45-21869			9/24/1945	3171ERD	8/16/1954	Last reported at Griffiss
45-21870			9/20/1945		8/8/1954	Reclaimed Tinker
45-21871			9/20/1945		8/6/1953	Last reported at Niagara Falls
45-21872			10/3/1945		11/12/1953	Last reported at Eielsen

Notes

[1]. 42-24826. Unconfirmed single source references to this aircraft show two different names and assignments: "General Confusion" — 504th Bomb Group; "In the Mood" — 6th Bomb Group. A third, more reliable, reference is the Missing Air Crew Summary Report (MACR). It states that 42-24826 was missing on 26 May 1945 while assigned to the 505th Bomb Group.

[2]. 42-24855. MACR lists 42-24855 as ditching, out of fuel, on 13 July 1945 while assigned to the 468th Bomb Group. The 58th Bomb Wing Air Sea Rescue Report of 19 July 1945 also reports 42-24855 missing in the previous week.

The confusion occurs with the reported assignment of 42-24855 to the 497th Bomb Group as Jumbo II, A 50 (the second). There is little doubt that 42-24855 was assigned to the 497th Bomb Group. It is listed in the 27 January 1945 73rd Bomb Wing Consolidated Statistical Report, as well as in a 497th Bomb Group aircraft compilation list by a reliable source, as completing 32 missions.

The question, then, is when 42-24855 was transferred from the 497th Bomb Group to the 468th Bomb Group. 42-24855 does not appear on a list of the 58th Bomb Wing aircraft as of 23 May 1945, so any transfer had to occur after that date.

To further complicate the issue, the Aircraft Record Card for 42–24855, microfilm reel AC-1, frame 743, indicates that 42–24855 survived the war and was reclaimed (sent to scrapper) 10 May 1954. In light of the double hard data sources indicating the plane's loss on 13 July 1945, the author tends to look upon the ARC data as an erroneous input.

[3]. 42–63364. There are two conflicting reports as to the assignment of this plane. The first is the Aircraft Record Card, which does not show 42–63364 going overseas and being reclaimed at Pyote TX 12/21/49. The second is of dubious reliability and is not backed up by any other source. The author recommends discounting the assignment to the 504th Bomb Group.

[4]. 42–63445. There is no confirmation that 42–63445 was T 56 of the 498th Bomb Group. The author recommends that the researcher disregard this listing.

[5]. 42–65309. Group records show the aircraft went overseas as part of 19th Bomb Group 12 February 1945 and flew solely as part of the 19th between 9 March 1945 and 10 August 1945. Aircraft Record Card lists 42–65309 as salvaged 31 May 1945.

A reliable source indicates 42–65309 was assigned to the 497th Bomb Group (after end of hostilities?), but no details were available.

[6]. 42–93912. References unclear as to whether 42–93912 was assigned to the 29th Bomb Group or the 330th Bomb Group when it crashed.

[7]. 44–61776. Aircraft Record Card shows the aircraft converted to WB-29 in early 1950. If this is true, the plane would have been unable to fly bombardment mission with the 98th Bomb Group in 1952.

[8]. 44–69857. Identification of plane as "Merry Fortune" is from Japanese records. There is no USAAF confirmation. Plane is probably "Mary Fortune," 44–69867.

[9]. 44–86254. The citations showing 44–86254 named both "Stateside Reject" and "Star Duster" are erroneous. "Stateside Reject" is actually 44–62152, while "Star Duster" is 44–69818.

[10]. 42–93937. Aircraft Record Card says 42–93937 did not go overseas, and the MACR Summary Report does not show 42–93937 lost on 4/12/45. There is serious doubt regarding the assignment to the 330th Bomb Group and the loss while so assigned.

II

NAMES LIST

This list is an alphabetical cross-reference to the Master List, providing access to serial numbers through plane names. (Planes whose names begin with numerals, such as "110th NCB," are arranged in numerical order at the beginning of the list.)

Research for this book uncovered 357 named planes with group assignments but unknown serial numbers, and 134 named planes with no group assignments or serial numbers.

These planes are grouped in separate sections at the end of this list.

Research also turned up 22 aircrafts bearing nose art without a verifiable corresponding name. These, too, are grouped in their own sections at the end of the names list. A brief description of each plane's art—e.g., "Mule team," "Nude on right elbow"—is provided along with known serial numbers and group numbers.

110th NCB	44-69982	Al-Ask-Er	45-21760	Baby Gail	42-93917
112 Seabees	42-94010	Amarillo's Flying Solenoid	41-36959	Baby San	44-86290
121 Sea-Bees	42-24815	American Beauty	42-24686	Baby's Buggy	42-93964
1885 Aviation Engineers	44-69982	American Beauty	42-24703	Baby's Buggy	44-61618
1919th Company	42-24882	American Beauty III	44-87661	Bachelor Quarters	42-24507
20th Century Limited	42-24792	American Maid	42-24593	Bad Brew	42-24594
20th Century Limited	44-61797	Ancient Mariner	44-70113	Bad Brew Too	42-65292
20th Century Sweetheart	42-65251	The Ancient Mariner	42-65296	Bad Medicine	42-24850
20th Century Unlimited	42-6281	Andy Gump	42-24528	Bad Medicine #2	44-87667
27th N.C.B. Special	42-24823	Andy's Dandy's	42-65208	Bad Penny	42-65274
29 USN Construction Reg	42-24780	The Angellic Pig	44-61991	Bad Penny	44-69675
293	42-24574	Ann Dee	42-65249	Bainbridge Belle	42-63525
358th Air Service Group	42-93911	Ann Garry III	42-6351?	Bait me?	44-69802
38 Sea Bees	42-63499	Ann Garry V	44-87650	Ball of Fire	42-65344
421st Emblem	42-24814	Annie	42-6224	Banana Boat	42-63551
504BG Insignia	42-94002	The Antagonizer	44-69890	Banana Boat	44-69802
67 Sea Bee	42-24809	Antoinette	42-24751	Banana Boat	44-86261
The 8 Ball	44-70070	Antoinette II	44-70135	Barbara Ann	42-24652
890th Aviation Engr Co	44-69739	Ape Ship	44-86330	Barbara Ann	44-61658
92 Naval Const Batt	42-24784	Ape Ship II	44-69894	Barbara Ann	44-61685
The Able Fox	42-24466	Aphrodite	42-24821	The Barronness	42-24675
Abroad with 11 Yanks	42-24698	Arkansas Traveler	44-69892	Bataan Avenger	44-69753
Ace in the Hole	44-61872	Arkansas Traveller	44-65331	Battlin' Beauty	42-24457
Ace of the Base	44-61551	Arson Inc	44-70129	Battlin' Beauty	42-63457
Adams Eve	42-24600	Assid Test	42-24786	Battlin' Betty	42-24606
The Agitator	42-6399	Assid Test II	42-65336	Battlin' Betty	44-69847
The Agitator II	42-24899	Atomic Tom	44-69682	Battlin' Betty II	42-24760
Ah Sooooooo	44-61817	B. A. Bird	42-93896	Battlin' Betty III	44-69772
Airborn	42-65268	Baby Gail	42-24917	Battlin Bitch II	42-65367

Name	Serial	Name	Serial	Name	Serial
Battlin' Bonnie	42-24907	Booze Hound	44-69746	City of Charleston	44-69791
Battlin' Bonnie II	44-87737	Broken Heart	44-83777	City of Chattanooga	42-94062
Battlin Bulldozer	42-93908	Brooklyn Bessie	42-93854	City of Chicago	44-70003
Battling Betty III	44-69772	B-Sweet II	42-24522	City of Cincinnati	42-93989
Beats Me	42-93943	B-Sweet III	42-63498	City of Clayton	42-94024
Beats Me Too	42-93954	B-Sweet IV	44-70094	City of Cleveland	44-69768
Beaubomber II	42-63442	Bub	44-61815	City of Cleveland	44-70077
Bedroom Eyes	42-24610	Buckin' Bronco	44-70136	City of Clifton	42-65308
Beetle Bomb	44-69800	Bugger	42-63610	City of College Park	42-93996
Beetle Bomber	44-69800	Bug's (Ball) Buster	44-61638	City of Columbus	44-69741
Behrens Brood	42-93995	Burk's Jerks	45-21721	City of Cooperstown	44-70004
Bella Bortion	42-63355	Busty Babe Bomber	44-69795	City of Council Bluffs	42-93971
Belle of Baltimore	42-6332	Butch	44-69730	City of Covington	44-69875
The Belle of Bikini	44-61822	Butterfly Baby	44-61854	City of Dallas	42-24883
Belle of Martinez	42-63601	Caboose	44-69870	City of Denver	42-93913
Belle Ringer	42-63464	The Caboose	44-69870	City of Detroit	42-93923
Belle Ruth	42-24680	Cait Paornat	42-93829	City of Duluth Mn	42-93957
Bengal Lancer	42-24487	Cajun Queen	44-69982	City of Eagle Rock	44-69914
Bengal Lancer	42-6348	Calamity Jane	42-24589	City of El Paso	42-93974
Ben's Raiders	44-69725	Calamity Sue	42-6368	City of Eugene	44-69910
Beter 'N' Nutin	42-24538	Camel Caravan	42-6333	City of Evanston	42-94071
Betty Bee	42-65335	The Cannuck	42-24668	City of Farmington	42-94059
Betty Marian	44-69910	Capt Clay	42-63514	City of Flatbush	44-69682
Big Blow	44-86339	Carla Lani-Battle Baby	42-65213	City of Flint	44-61676
Big Boots	42-24865	Case Ace	42-6270	City of Fort Gibson	44-69762
Big Boots II	44-61550	Cat Girl	44-62053	City of Fort Worth	42-93982
Big Chief	42-6382	Celestial Princess	42-24590	City of Gainsville	42-94071
Big Fat Mama	42-93901	Celestial Queen	44-87660	City of Galveston	44-69785
Big Gass Bird	42-93896	Censored	44-69810	City of Glendale	42-93912
Big Gass Bird	44-86400	Censored Lady	44-69810	City of Grass Valley	44-69795
Big Joe	42-24885	The Challenger	42-6284	City of Griffin, Ga	44-70080
Big Mike	42-63619	The Challenger	42-63731	City of Grundy Center	44-69891
Big Poison	42-6353	Character Carriage	42-93917	City of Hartford	42-93928
Big Poison - 2nd Dose	42-65270	Charlie's Wagon	44-69746	City of Hatch	44-70010
Big Schmoo	44-69963	Charlie's Wagon	44-87601	City of Hershey	42-94037
Big Shmoo	44-62063	Chat'nooga Choo Choo	42-24471	City of Highland Falls	42-63539
The Big Stick	42-24661	Cheechako	44-61981	City of Indianapolis	44-69814
Big Stink	44-27354	Cherie	44-69763	City of Jackson	42-65369
The Big Time Operator	42-24791	Cherry-Horizontal Cat	42-63564	City of Jacksonville	42-93978
The Big Wheel	42-65283	Chicago Sal	44-61562	City of Jamestown NY	42-94047
Big! Ain't it	42-65273	Chief Mac's 10 Little Indians	44-62186	City of Jersey City	42-94016
Big! Ain't it?	44-70128	Chief Spokane the Red Eraser	44-61925	City of Jewett City	42-65361
Bigham	42-93974	The Chosen Few	44-86392	City of Kankakee	42-94029
Black Bart's Revenge	44-69706	Chotto Matte	44-86400	City of Knoxville	44-69995
Black Cat	42-24802	City of Aberdeen	42-93961	City of La Grange (Il)	44-69908
Black Hills Baby	44-84054	City of Akron	42-65363	City of Landsford Pa	42-93967
Black Jack	42-6292	City of Albuquerque	44-69796	City of Laredo	44-61795
Black Jack Too	42-63451	City of Allendale	44-69981	City of Las Vegas	42-65365
Black Magic	42-24672	City of Arcadia	42-93925	City of Lincoln	44-70103
Black Magic	42-6276	City of Asheville	42-93989	City of Lindsay	42-94023
Black Magic II	42-24718	City of Athens	42-94044	City of Los Angeles	42-65302
Black Sheep	42-65369	City of Atlanta	44-69788	City of Lynn	44-61664
Black Sheep	44-69681	City of Attleboro	44-87642	City of Martinez	42-63749
Blackjack	42-65367	City of Aurora	44-69870	City of Maywood	44-69901
Blind Date	42-24759	City of Austin	44-69681	City of Medford	44-69801
Blitz Buggy	42-24909	City of Bakersfield	44-69680	City of Memphis	42-93917
Blizzard Wizard	44-62216	City of Bedford	42-93935	City of Miami	42-65367
Blue Bonnet Belle	42-6307	City of Bel Air	42-63517	City of Miami Beach	42-93837
Bluetailfly	42-65272	City of Berkeley	44-70101	City of Miami Beach (II)	42-65370
Bockscar	44-27297	City of Birmingham	44-69790	City of Miles City	44-69962
Boeing Wichita 1000	44-69890	City of Boston	44-69685	City of Milwaukee	42-24917
Bombin' Buggy	42-6250	City of Boulder	44-69763	City of Montgomery	44-69779
Bombin' Buggy II	42-24541	City of Buffalo	44-69856	City of Oakland	44-69872
Bonnie Lee	42-6322	City of Burbank	44-69766	City of Ogden	42-65307
Bonnie Lee	44-84053	City of Burlington	42-63529	City of Oklahoma City	42-24917
Boomerang	44-61818	City of Burlington	42-65304	City of Omaha	42-65371
The Boomerang	42-63640	City of Cedar Rapids	42-93908	City of Orlando	44-69689

II. Names List

Name	Serial
City of Osceola	42-93995
City of Pacific Palisades	44-70010
City of Patterson	44-69774
City of Pittsburg	42-24895
City of Pittsfield	44-69773
City of Portsmouth Va	42-93943
City of Portsmouth Va	42-93954
City of Providence	42-63567
City of Quaker City	44-70016
City of Red Bank	44-69696
City of Redfeild	44-70079
City of Reno	44-69786
City of Richmond	42-65303
City of Richmond	42-65309
City of Richmond Ca	42-93996
City of Richmond Ca	44-69911
City of Richmond III	44-87653
City of Roanoke	44-69817
City of Rochester	42-94040
City of Rock Island	42-93964
City of Roswell	42-65364
City of Sam Bee Fla	44-84019
City of San Antonio	42-63565
City of San Francisco	44-69800
City of San Jose	42-63517
City of Santa Barbara	44-61532
City of Santa Fe	42-65365
City of Santa Fe	42-93975
City of Santa Monica	42-65303
City of Scotland Neck	44-69769
City of Spanish Forks	44-69997
City of Spokane	44-69910
City of Spokane	44-70018
City of Springfield	44-69797
City of Springfield Il	42-93943
City of St Petersburg	42-93980
City of St Petersburg	42-94032
City of Terra Haute	42-94052
City of Trenton	44-69680
City of Tulsa	44-69815
City of Tyler (Texas)	44-69884
City of University Park	44-69678
City of Vincennes	44-69986
City of Virginia Beach	42-65366
City of Virginia Beach	42-93975
City of West Palm Beach	42-93970
City of Williamsport	42-93980
City of Wilmington	42-65365
City of Youngstown	44-61524
Colleen	42-93955
Command Decision	44-84657
Confederate Soldier	44-61524
Connecticut Yankee	42-24783
Constant Nymph	42-63487
Constant Nymph	44-61693
Convincer	44-61521
Coral Queen	42-24615
Coral Queen	42-63499
Coral Queen	42-93875
Cosmic Ray Research	45-21793
Country Gentleman	42-24793
Cox's Army	42-63544
The Craig Comet	42-63445
Cramer's Crapper	44-83897
Cream of the Crop	44-61657
Cross Over the Bridge	45-21734
Cue Ball	44-70010
The Cultured Vulture	42-24901
Daddy Come Home	44-61823
Dangerous Lady	42-24823
Danny Mite	44-69771
Danny Mite	44-69777
Daring Donna III	42-24820
Dark Eyes	42-63555
Dark Slide	44-27326
Dauntless Dotty	42-24592
Dave's Dream	44-27354
The Deacon's Delight	42-24818
Deacon's Disciples	42-6215
Deacon's Disciples II	42-24492
Deal Me In	44-69805
Deaner Boy	42-24815
Dearly Beloved	44-70069
Dee-Fence Buster	44-62274
Defiant Lassie	44-69741
Delilah	44-70109
Destination Known	44-61822
Destiny's Tot	42-65284
Destiny's Tots	42-65293
Deuces Wild	42-6222
Deuces Wild	44-69809
Devilish Snooks	42-63527
Devil's Darlin'	42-24629
Devil's Delight	42-24652
Diajobo	44-61815
Dickert's Demons	44-86361
Dina Might	42-65280
Dina Might	44-69936
Dinah Might	42-65286
Ding Hao	42-6329
Ding Hao	42-6358
Ding How	42-6225
Ding How	42-6313
Dixie Darlin'	42-63413
Do It Again	42-65229
Doc	44-69972
Doc Said All I Needed —	42-65266
Doc's Deadly Dose	42-24780
Dode	44-83893
Don't Worry Abouta Thing	42-93957
Dopey	44-70016
Doris Ann	42-24677
Dottie's Dilemma	42-24796
Double Exposure	42-24877
Double Exposure	42-93855
Double Trouble	44-69901
Double Whammy	44-87734
Down's Clowns	44-86284
Draggin' Lady	42-24694
Draggin' Lady	42-63505
Dragon Behind	44-86247
Dragon Lady	42-24778
Dragon Lady	42-63525
Dragon Lady	42-65277
Dragon Lady	42-93892
Dragon Lady	44-61835
Dragon Lady	44-69663
The Dragon Lady	42-63425
Dream Girl	42-24673
Dream Girl	42-63480
The Dreamer	44-27341
Drunkard — Stagger Inn	44-61565
Duchess	42-63411
The Duchess Almost Ready	42-93880
Duffy's Tavern	44-62216
Duke of Albuquerque	44-69829
Dumbo	42-6257
Dyna-Mite II	44-69909
Eager Beaver III	42-24750
Early Bird	42-63556
Early Bird	44-86303
Earthquake McGoon	42-24866
Earthquake McGoon	42-65228
Eddie Allen	42-24579
Eddy Allen II	44-70151
Eight Ball	42-93966
Eight Ball	44-62237
Eight Ball Charlie	42-65328
El Pajaro De La Guerra	42-24874
Eleanor	42-65337
Ellie Barbara & Her Orphans	42-63605
Elsie	44-87341
Empire Express	42-63549
Enola Gay	44-86292
Ernie Pyle	44-69995
The Ernie Pyle	44-70118
Esso Express	42-6242
Every Man a Tiger	44-61830
Excalibur	42-6316
Experiment Perilous	44-69768
Faithful Faye	42-63533
Fancy Detail	42-24696
Fanny-The Atom & I	44-86384
Fast Company	42-24691
Fast Company	42-63495
Fay	42-65210
Feather Merchant	42-6308
Feather Merchants	42-94040
Fertile Myrtle	45-21787
Fever from the South	42-63497
Fickle Finger	42-63426
Fifi	44-62070
Fifty-Second Seabees	44-69792
Filthy Fay	42-65210
Filthy Fay II	42-93999
Filthy Fay III	44-69852
Fire Ball	44-62281
Fire Belle	44-61653
Fire Bug	42-63566
Fire Bug	44-69944
Flag Ship 500	44-61669
Flak Alley Sally	42-24878
Flak Magnet	44-61634
Flak Maid	44-70129
Fleet Admiral Nimitz	42-63650
Flyin' Home	44-24909
Flying Fool	44-70083
The Flying Guinea Pig	41-18335
Flying Jackass	42-24580
Flying Parts	44-86438
Flying Solenoid	41-36959
Flying Stud	42-6320
Flying Stud II	42-24464
Forbidden Fruit	42-24607
Forever Amber	44-69839
Forever Amber	44-70079
Forever Ambling	44-62231
Four Aces & Her Majesty	42-93975
Four-A-Breast	44-86323
Four of a Kind	44-69999
French's Kabazie Wagon	44-70011

Name	Serial	Name	Serial	Name	Serial
Friendly Undertaker	44-62213	Heavenly Body	45-21813	In the Mood	42-24826
Frisco Nanny	42-93889	Heavenly Flower	42-94025	Inchcliffe Castle	44-61543
The Fry'in Pan	44-69812	Heavenly Laden	44-61822	Indian Maid	42-24806
Fubar	42-63378	Heavenly Laden	45-21823	Indian Maid	42-24809
Fujigmo	44-62166	Hellbird Insignia	44-61639	Indiana	42-63546
Fujigmo	44-86370	Helles Belles	42-93878	Indiana II	42-94010
Fu-Kemal	42-6352	Hello Natural III	42-24640	Infant of Prague	42-63539
Fu-Kemal-Tu	42-24720	Hellon Wings	42-93857	Inspiration	42-63440
Full House	44-27298	Hell's Bell	42-63438	Irish Lassie	42-65246
Gallopin' Goose	42-6390	Hell's Belle	42-24680	Irish Lullaby	42-24830
The Gamecock	42-65266	Hell's Belle	42-24878	The Iron Bird	44-62295
Gas Gobbler	44-62314	Hell's Belle	42-63486	Iron George	44-27340
Gay Times	44-61815	Hell's Belle	42-65365	Iron George	44-87640
The Gear Box	42-24704	Her Majesty	42-63375	Iron Shillalah	42-63519
Geisha Gertie	42-24763	Her Majesty	42-93975	Island Girl	42-63614
General Andrews	44-69888	Herd of Bald Goats	44-70005	Island Queen	42-93982
General Confusion	42-24826	Here to Stay	44-69786	It Shouldn't Happen	44-69746
General HH Arnold Special	42-6365	Here's Hopin	44-70102	It's Hawg Wild	44-61748
Genie	42-63455	Here's Hopn	44-70127	Jabbitt III	44-27303
Genie II	44-61812	Hero Heater	44-61702	Jackpot	42-24797
Georgia Ann	42-24766	Hi Stepper	42-65275	Jack's Hack	44-61566
Georgia Ann	42-93875	High and Mighty	42-24730	Jake's Jalopy	44-69985
Georgia Peach	42-63356	High, eh doc?	42-24823	The Janice E	42-93947
Gertrude C	42-6334	Himalaya Hussy	42-6319	Janie	42-93943
The Ghastly Goose	42-63541	Ho Hum	44-70123	Janke's Jinx	44-61626
Global Glamor	44-70139	Hobo Queen	41-36963	Je Reviens	44-70101
Globe Girdle Myrtle	42-63386	Hog Wild	44-70136	Jo	42-65337
Globe Girtle Murtle	42-63386	Holly Hawk	42-2447X	Jo/Indiana	42-63546
God's Will	42-24831	Hollywood Commando	42-24724	Joe's Junk	42-24883
Goin' Jessie	42-24856	Holy Joe	42-63489	Johnny Rebel	44-61674
Going My Way	44-61638	Hombicrismus	42-6317	John's Other Wife	44-86349
Gone with the Wind	42-6331	Homer's Roamers	42-24794	Joker's Wild	42-24626
Gonna Mak'er	42-65231	Homing Bird	42-24824	Jokers Wild II	42-24897
Good Deal	42-24852	Homing Device	44-69751	Jolly Roger	42-24872
Good Humpin'	42-6268	Homing De-Vice	42-24785	Jolly Roger	42-63415
Gravel Gertie	42-63500	Homogenized Ethyl	44-69710	Jolly Roger	42-63444
Gravel Gertie	42-65221	Hon. Spy report	42-24876	Joltin' Josie	42-24614
The Great Artiste	44-27353	Honey	42-24669	Jook Girl	42-65255
Great Speckled Bird	42-24532	Honey Bucket Honshos	44-61929	Joy-Ous Venture	44-61821
Green Dragon	42-6248	Honey Bucket Honshos II	44-61742	The Judy Ann	42-94025
Grider Gal	42-24884	Honeywell Honey	42-24738	Jug Haid II	42-24650
Grumpy	44-69894	Hongchow	44-27308	Jug Haid III	44-69676
Guardians of Peace	44-62224	Honorable Cock Wagon	44-69739	The Juke Box	42-63353
Gunga Din	42-24504	Honorable TNT Wagon	42-63484	Jumbo II	42-24855
The Gusher	42-6356	Honshu Hawk	42-63444	Jumbo, King of the Show	42-63418
The Gypsy	44-61948	Honshu Hurricane	42-63481	The Jumping Stud	42-63414
Hagarty's Hag	44-86408	Hoodlum House II	42-24475	Kagu Tsuchi	42-24737
Haley's Comet	42-24616	Hore-Zontal Dream	42-24732	Kamode Head	44-62000
Hammer of Thor	42-94002	Hot Pants	42-63485	Kamra-Kaze	44-61583
Ham's Eggs	42-24670	Hot T' Trot	44-69727	Katie	42-6298
Hanger Queen	45-21833	Hot to Go	42-65352	Katie Ann	42-65300
Happy Savage	44-69696	Hot to Go	42-62183	Kayo Kid	44-69987
Hap's Characters	42-63424	Hot to Go (II)	44-69817	The Kee Bird	45-21763
Hap's Hope	42-6240	Houston Honey	42-63475	Kickapoo II	42-6232
Harry Miller	42-24740	Hoxie's Hoax	44-61923	Kickapoo Lou	42-24678
Hasta Luego	42-24647	Hull's Angel	42-63362	The Kick-A-Poo-Joy II	42-65361
Haulin' Ass	44-61600	Hump Happy Jr	42-6310	The Kick-A-Poo-Joy III	44-87640
Haulin' Ass	44-62103	Hump Happy Mammy	42-6241	King Size	42-6347
Haulinas	42-24461	Humpin Honey	42-6299	Kitten	42-63380
Hawg Wild	44-61748	Humpin' Honey	42-24463	Klondike Kutey	42-24612
Hearts Desire II	44-69656	Humpin' Honey	42-6444	Koehane's Kulprits	44-69774
The Heat's On	42-24605	Hump's Honey	42-24648	Koza Kid	44-87661
The Heat's On	42-65332	Hun-Da-Gee	44-61546	Kristy Ann	42-93886
Heavenly	44-69696	Hurricane Hattie	44-87750	Kritzer Blitzer	42-63542
Heavenly Body	42-6281	Hurricane Hunters	44-69987	Kro's Kids	42-24788
Heavenly Body	42-63510	Ichiban	44-61810	La Boheme	44-61851
Heavenly Body	44-69997	Iltywybad	44-84088	Lady Be Good	42-65227

II. Names List

Name	Serial
Lady Boomerang	42-6223
Lady Chouteau	44-87674
Lady Eve	42-65211
Lady Eve II	42-24663
Lady Francis	44-61670
Lady Hamilton	42-6274
Lady Hamilton II	42-24542
Lady in Dis Dress	44-86446
Lady in Waiting	44-84068
Lady Jane	42-65363
Lady Jayne	44-69874
Lady Marge	42-24485
Lady Marge II	42-63399
Lady Mary Anna	42-24625
Laggin' Dragon	44-86347
Laggin Wagon	42-65390
Lake Success Express	44-69980
Lancer	44-69844
Lassie	42-63356
Lassie Come Home	42-24609
Lassie II	42-24769
Lassie Too	42-63460
Lassy Too	42-93894
Last Resort	42-63394
Lazy Baby	42-63498
Leading Lady	42-24766
Legal Eagle	44-84118
Legal Eagle II	44-87779
The Lemon Drop Kid	44-70042
The Lemon	42-63462
Lethal Lady	42-6370
Life of Riley	42-65241
Lightning Lady	44-69801
Li'l Abner	44-69657
Lil Darlin	44-86273
Lil' Iodine	42-24875
Lil Iodine II	44-69748
Li'l Lassie	42-24693
Li'l Yutz	42-24892
Limber Dugan	42-6230
Limber Dugan II	42-65315
Limber Richard	44-70072
Little Bully	44-70077
Little Butch	42-94014
Little Clambert	42-24582
Little Evil	42-94025
Little Fellow	44-61697
Little Fellow	44-61782
Little Gem	42-24596
Little George Jr.	44-69791
Little Jeff	44-69855
Little Jo	42-24611
Little Jo	44-69932
Little Mike	42-63422
Little Miss	42-63496
Little Organ Annie	42-24893
Little Princess	42-24794
Live Wire	42-24853
Loaded Dice	42-63688
Loaded Lady	44-69944
Loaded Leven	44-69803
Lonely Lady	44-86361
Lonesome Polecat	42-94059
Lonesome Polecat	44-62152
Lonesome Polecat	44-86272
Lonesome Polecat	44-86361
Lonesum Poll Cat	44-62152
Long Distance	42-24544
Long John Silver	42-63502
Long Winded	42-63509
Look Homeward, Angel	44-69736
Lord's Prayer	44-69914
Los Angeles Calling	44-86446
Low & Lonely	44-69908
Lubricating Lady	44-61751
Lucifer	45-21745
Lucky 13	44-70149
Lucky Dog	44-86370
Lucky Eleven	42-24714
Lucky Eleven	44-69803
Lucky Irish	42-24622
Lucky Irish	42-63432
Lucky Irish II	42-63492
Lucky Lady	42-24584
Lucky Lady	42-24863
Lucky Lady	42-65272
Lucky Lady	44-62304
Lucky Leven	42-93951
Lucky Leven	42-93956
Lucky 'Leven	42-24695
Lucky Lynn	42-24591
Lucky Seven	42-6407
Lucky Seven	44-61734
Lucky Strike	42-63552
Lucky Strike	42-94029
Lucky Strike	44-62070
Lucky Strike II	44-62010
Lucky Strikes	42-94030
Luella Jean	44-61795
Luke the Spook	44-86346
M P I	44-86247
Mac's Effort	44-70073
Maiden USA	42-63548
Maiden's Prayer	44-61678
Mais Oui	44-86436
Male Call	42-63537
Malfunction Junction	42-6241
Mammy Yokum	42-63536
Man O' War II	42-63511
Maniuwa	42-63678
Man-O-War	42-6346
Many Happy Returns	44-69986
Margie's Mad Greek III	44-61843
Marianna Belle	44-69883
Marianna Belle	44-70015
Marianna Ram	44-69732
Marietta Belle	42-63396
Marietta Misfit	42-63363
Mark of Zorro	42-93888
Mary Ann	42-24494
Mary Ann	42-24550
Mary Ann	42-63550
Mary Anna	42-65253
Mary Anna II	44-69964
Mary Anne	42-24693
The Mary K	42-24525
Mary Kathleen	44-69814
Mary Lou	44-61960
The Marylin Gay	44-70113
Mary's Lil Lambs	42-6343
Mason's Honey	44-61721
Maximum Effort #3	44-69893
Maximum Load	42-63563
Maya's Dragon	42-94022
McNamara's Band	42-94016
Memphis Maid	44-70120
Merry Fortune	44-69857
Mighty Fine	42-93944
Mighty Fine II	44-61655
Million Dollar Baby	42-24660
Million Dollar Baby	42-6397
Million Dollar Baby	42-65247
Million Dollar Baby II	42-63532
Mis-Chief-Mak-Er	42-24896
Miss America '62	42-65281
Miss Behavin	44-69805
Miss Behavin'	42-24655
Miss Behavin'	44-69685
Miss Behavin' II	42-63523
Miss Carriage	42-63364
Miss Donna Lee	42-24915
Miss Hap	42-24774
Miss Hart of America	44-69730
Miss Judy	44-61555
Miss Lace	42-6354
Miss Lace	42-63554
Miss Lace	44-62304
Miss Lace	44-87658
Miss Lead	42-24734
Miss Margaret	42-63427
Miss Minette	42-6272
Miss Minooki	44-27332
Miss Mi-Nookie	44-69764
Miss N.C.	42-63376
Miss N.C.	44-86376
Miss North Carolina	44-61940
Miss Ogden	42-65307
Miss Peggy	44-69977
Miss Rosemary	42-24848
Miss Sandy	44-87659
Miss Shorty	42-65272
Miss Spokane	44-27332
Miss Su-Su	42-24812
Miss Take	42-93978
Miss Tampa X	44-86335
Miss Tittymouse	42-24649
Miss You	44-70100
Mission Inn	44-61669
Mission to Albuquerque	42-24849
Mission to Albuquerque II	42-93909
Missouri Belle	42-63557
Missouri Queen	42-6359
Monsoon	42-6294
Monsoon Goon	42-93828
Monsoon Goon II	42-24891
Monsoon II	42-24846
Monsoon Minnie	42-6295
Monstro	44-84111
Moon's Moonbeam	44-61815
Moonshine Raiders	44-61815
The Moose Is Loose	42-24851
Motley Crew	42-93912
Mrs Tittymouse	42-65212
Mule Train	44-86252
Musn Touch	42-24657
Mus'n Touch It	44-61548
My Buddy	42-65279
My Dragon	44-61830
My Gal	42-93980
My Gal II	42-94032
My Gal Sal	44-69660

Name	Serial	Name	Serial	Name	Serial
My Naked -	42-63725	Omaha One More Time	42-24851	Polar Queen	44-62157
Myakinas	44-62108	One Weakness	42-65365	Polar Queen	44-62163
Mya's Dragon	42-94042	The One You Love	44-69727	Ponderous Peg	42-63431
Myasas Draggin	42-63472	Oregon Express	44-69969	Porcupine	44-61558
Myasis Dragon	44-61830	O'Reilly's Daughter	42-6264	Postville Express	42-6279
Mysterious Mistress	42-6312	O'Reilly's Daughter II	44-61703	Power Play	42-24442
Naughty Nancy	42-63486	Ouija Bird	42-24508	Praying Mantis	42-6286
Naughty Nancy	42-63496	Our Baby	42-24597	Pretty Baby	42-63396
Necessary Evil	44-27291	Our Baby	42-94024	Pretty Baby	42-65349
Never Been Tried	44-83966	Our Gal	42-24484	Pride of the Yankees	42-24676
Never Hoppen	44-61562	Our Gal	44-61932	Pride of Tucson	42-65370
Never Hoppen	44-62196	Our L'Lass	44-61951	Princess Eileen	42-6323
New Glory	42-24756	Ours	42-6285	Princess Eileen II	42-24462
Next Objective	44-27299	The Outlaw	42-24685	Princess Eileen III	42-63559
Night Mare	42-6311	The Outlaw	42-65306	Princess Eileen IV	42-65327
Night-mare!	44-87661	The Outlaw	44-69668	Princess Pat	44-69680
Nina Ross	42-24689	Over Exposed	42-24877	Princess Pat II	44-70103
Nip Clipper	42-63512	Over Exposed	44-61951	Princess Pat III	44-83974
Nip Finale	44-69970	Over Exposed	44-61999	Princess Patsy	42-6344
Nip Finale II	44-70106	Overexposed	44-61813	Princess Pokey	42-63517
Nip Nemesis	44-69733	Overnite Bag	44-86414	Private Love Witch	44-87741
Nip On Ese Nipper	42-93917	Pacific Playboys	42-24764	Punch Bowl Queen	44-83974
Nip On Nees Nipper	42-24917	Pacific Princess	44-61813	Punchin' Judy	42-65219
Nip On Nees	44-62261	Pacific Princess (?)	44-61845	Purple Heartless	44-69760
Nipp On Ese	42-24917	Pacific Queen	42-63429	Purple Shaft	42-24802
Nippon Nipper	42-6289	Pacific Union	42-24595	Purple Shaft	42-65361
Nippon Nipper II	42-24503	Pacusan Dreamboat	44-84061	Quaker City	44-70016
Nippon Nipper III	42-24729	Paddy Daddy	44-62034	Quan Yin Cha Ara	42-93853
Nip-Pon-Ese	44-61617	Padre and His Angels	42-24839	Quantrell's Raiders	44-61990
Nip-Pon-Ese	44-87760	Pampered Lady	42-6306	Queen Bee	42-24840
Nipp-On-Nees	44-61617	Panchito, The Fighting	44-69724	Queen Cathy	42-94053
Nipponese Baby	44-62099	Papa Tom's Cabin	44-87668	Queen of the Air	42-6380
No Balls Atoll	42-93925	Pappy's Pullman	42-24882	Queen of the Neches	44-84065
No Name!	44-83897	Pappy's Pullman II	42-63481	Queenie	42-93831
No Papa	42-6325	Parker's Van	44-70003	Ragged But Right	42-6252
No Sweat	44-87618	Party Girl	42-6389	Ragged But Right	42-63593
No Sweat (II)	44-70134	Passion Wagon	42-24771	Raiden Maiden	42-6265
Noah's Ark	44-27334	Passion Wagon	42-63524	Raiden Maiden	44-61966
Oily Boid	42-24912	Passion Wagon	42-94043	Raiden Maiden II	42-65276
Ol' Smoker	42-93837	Patches	42-6338	Raiden Maiden III	42-63522
Ol' 297	42-6297	Patches	42-24624	Rainbow's End	42-93974
Old 574	42-63574	Patches	42-24822	Ramblin' Roscoe	42-24664
Old 900	42-24900	Patches	44-70085	Ramblin' Roscoe II	42-94049
Old-Bitch-U-Ary Bess	42-6273	Patience Reward	44-70131	Ramblin' Wreck	42-24471
Old Acquaintance	42-63457	Patricia Lynn	42-93901	Ramp Queen	42-63513
Old Battler	42-6251	Peace Maker	44-86433	Ramp Tramp	42-6256
The Old Campaigner	42-6272	Peace on Earth	42-63412	Ramp Tramp II	42-24904
Old Cracker Keg	42-6276	Peace on Earth	44-61790	The Ramp Tramp	44-70069
Old Double Deuce	42-94022	The Peacemaker	42-24916	Ramp Vamp	44-70069
Old Double Deuce	44-62022	Peachy	42-63508	Rankless Wreck	42-63420
Old Eighty One	44-69981	Peachy	44-62022	Rankless Wreck	44-86446
Old Firing Butt	42-6344	Pee Wee	42-24762	Rapid Rabbit	44-61776
The Old Girl	44-69889	Pepper	42-6217	Rapid Rabbit	44-86446
Old Ironsides	42-24436	Percussion Steamboat	44-84076	Rattle 'n' Roll	44-61803
Old Ironsides	42-63436	Petrol Packin' Mama	42-6219	Raz N' Hell	45-21746
Old P-7	44-69810	Phippens Pippens	44-86290	Ready Bettie	44-69817
Old Rusty	44-69693	Phony Express	42-24801	Ready Betty	42-24879
Old Soldier's Home	44-69766	Piece of Meanness	42-93975	Ready Teddy	42-63561
Old Wild Goose	44-69771	Piledriver	44-69796	Ready Willin', Wanton	44-62106
Old-Bitch-U-Ary Bess	42-6273	Pink Lady	42-63657	Real Rough	42-24714
Ole Boomerang	44-69790	Pioneer	42-6208	Reamatroid	44-69672
Ole Forty and Eight	42-94044	Pioneer II	42-24737	Rebel's Roost	42-93996
Ole Forty Eight	44-69794	Pioneer III	42-63534	Red Hot Rider	42-65338
Ole Gas Eater	42-24798	Pluto	42-94062	The Red	44-84122
Ole Miss IV	45-21793	Pocohantas Proud Pigeon	42-24601	Reddy Teddy	42-6408
Ole Smoker	44-69857	Poison Ivy	42-24585	Reddy Teddy II	42-63561
Ole Smoker II	42-65370	Pokahuntas	42-63517	Regina Coeli	44-83906

II. Names List

Name	Serial
Reluctant Drag'on	44-62253
Renegade	44-61552
Rip Van Winkle	42-24868
Road Apple	42-63600
Robert J Wilson	42-24714
Rock Happy	44-62053
The Rocket	42-24742
Rodger the Lodger	42-6243
Rosalia Rocket	42-24656
Round Robin	42-63489
Round Robin	42-63577
Round Trip Ticket	42-6262
Round Trip Ticket	42-65365
Rover Boys Express	42-24769
Rush Order	42-63393
Rushin' Rotashun	42-63417
Ruthless	42-94027
Sac's Appeal	44-61872
Sad Sac	42-61676
Sad Tomato	42-65285
Salem Witch	44-61533
Sally Delle	44-61623
Salome, Where She Danced	44-83893
Salt Censored Resistor	42-65307
Salty Dog	44-83897
Salvo Sally	42-24699
San Antonio Rose	42-24587
Sandman	42-24852
Sassy Lassy	42-24867
Satan's Angel	42-65202
Satan's Lady	42-24779
Satan's Lady	42-24806
Satan's Mate	44-62315
Satan's Sister	42-24665
Satan's Sister	42-63453
Sentimental Journey	44-61564
Sentimental Journey	44-61634
Sentimental Journey	44-70016
Sentimental Journey	44-70147
September Song	44-69746
Shack Rabbit	44-83934
Shady Lady	42-24619
Shady Lady	42-65357
Shaft Absorber	44-84068
Shag'n Home	42-93859
Shanghai Lil	42-6277
Shanghai Lil Rides Again	42-24723
Sharon Sue	42-63435
Sharon Sue	44-61692
Shasta	44-61549
She Wolf	42-93957
Sheer Madness	44-61948
Sheeza Goer	45-21716
Shillelagh	42-93935
Shinpainai	44-6221X
Shirley Dee	42-63443
Shoot You're Faded	42-63407
Shrewd Manuever	44-83937
The Shrike	42-24456
The Shrimper	42-93885
Shutterbug	42-93864
Shutterbug	44-61817
Sic 'em!	44-61809
Silver Streak	44-69756
Silver Thunder	44-61684
Sir Trofrepus	42-6237
Sir Trofrepus	42-6263
Sister Sue	42-6342
Sit 'n' Git	44-86287
Sit 'n' Git	44-61816
Sitting Pretty	42-24814
Sitting Pretty	42-24881
Sky Blues	44-87662
Sky Chief	42-24472
Sky Chief	44-70002
Sky Octane	45-21716
Sky Queen	44-69698
Sky Scrapper	42-63503
Skyscrapper	42-24599
Sky Scrapper	42-65364
Skyscrapper II	42-63463
Skyscrapper III	44-69745
Slave Girl	44-27307
Sleepy Time Gal	42-24620
Sleepy Time Gal	42-63469
The Slic Chic	44-69889
Slick Dick	42-24700
Slick's Chicks	42-24784
Slick's Chicks	42-24906
Slim	42-93979
Slim II	44-87642
Slow Freight	42-63360
Slow Freight IV	44-61834
Small Fry	42-24760
Smilin' Jack	42-24888
Smokey Stover	42-6269
Snafu Per Fort	42-63435
Snafu, Them Shif'les	42-24873
Snafuper Bomber	42-6275
Snake Bit	44-86390
Snatch Blach	42-65302
Snooky	42-24825
Snooky's Brats	42-93877
Snoopin' Kid	42-93865
Snoopy Droopy	44-69746
Snuffy	42-24873
Snugglebunny	44-69667
So Tired, 7 to 7	44-61727
Some Pumpkins	44-27296
Something to Fight For	44-69800
Sound and Fury	44-69678
South Sea Sinner	42-93874
Southern Belle	42-24717
Southern Bell	44-61677
Southern Belle	42-63478
Southern Comfort	44-61749
Southern Drawl	44-69769
Space Mistress	44-86316
The Spearhead	44-69975
Special Delivery	42-24628
Spirit of FDR	42-93846
The Spirit of FDR	44-69846
Spirit of Freeport	44-62060
Spirit of Lincoln	41-36954
Spittin Kittin	42-63380
Squeeze Play	44-87614
SS Annabelle	42-93980
Star Dust	42-94052
Star Duster	42-24782
Star Duster	42-63492
Star Duster	42-93858
Star Duster	44-86254
Star Duster	44-69818
Starduster	42-6305
The Starduster	42-94067
Stateside Reject	44-62152
Stateside Reject	44-86254
Steveadorable	42-65243
Sting Shift	44-61699
Sting Shift	44-69742
Stockett's Rocket	42-6261
Stork Club Boys	42-24864
Stormy Petrel	44-61660
Straight Flush	44-27301
Strange Cargo	44-27300
Strap Hanger	44-70111
Stripped For Action	42-63466
Su Su Baby	42-24721
Suchosi? Ni!	44-69812
Suella J	44-61577
Sukebe Girl	44-62089
Super Mouse	42-24524
Super Wabbit	42-65220
Super Wabbit	42-65222
Superstitious Aloysious	42-65233
Supine Sue	42-24653
Sure Thing	42-63557
Sweater Out	42-24513
Sweat'r Out	42-63471
Sweet Judy II	45-21721
Sweet Sixteen	44-61666
Sweet Sue	44-70001
Sweet Sue	44-70112
Sweet Thing	44-70108
Sweet'n Lola	44-61578
T.D.Y. Widow	44-86335
Tabooma	42-6342
Tabooma II	42-63374
Tabooma III	42-65233
Tabooma IV	44-69906
Tagalong	44-61651
Tail Wind	42-24761
Tail Wind	45-21721
Take It Off	42-93939
Talie ho!	44-70143
Tall in the Saddle	44-69686
Tamerlane	42-93888
Tanaka Termite	42-24749
Tarfu	42-6266
Target for Tonight	44-70007
Teaser	42-63526
Ten Under Parr	44-69741
Terrible Terry	42-63425
Texas Doll	42-24627
The Germ	42-93971
The One You Love	42-24851
There'll Always Be a Christmas	42-93939
The Thing	45-21824
This Is It	42-6321
This Is It!	44-61646
Three Feathers	42-24671
Three Feathers	42-6324
Throbbing Monster	42-94047
Thumper	42-24623
Thumper	42-63536
Thunder Bird	42-63454
Thunder Bird	44-87664
Thunderbird	42-63570
Thunderhead	42-24641
Thunderin' Loretta	42-24913
Tian Long (Sky Dragon)	42-65362

Tiger Lil	42-94000	Uncle Tom's Cabin II	42-24763	Weirite	42-93856
Timely Reminder	44-69716	Unconditional Surrender	44-70127	Wempy's Blitzburger	42-6290
Times-A-Wastin'	42-93894	Undecided	42-24580	Were Wolf	42-63423
Tinny Anne	44-69811	Undecided	44-83934	Wet Bulb Willy	42-93967
Tiny Tim	44-69691	Under-Exposed	42-93849	Wham Bam	42-24469
T-N-Teeny	42-65278	Unification	45-21739	What Happened	42-24829
T-N-Teeny II	44-69920	The Un-Invited II	42-24719	Wheel 'n' Deal (Lt Side)	42-24604
To Each His Own	44-62207	The Uninvited	42-6409	Where Next?	44-83992
Tojo's Nightmare	44-70124	The Uninvited	44-69754	Where's Kilroy	42-63565
Tokyo Local	42-24687	United Notions	44-27326	White Mistress/Huntress	42-24776
Tokyo Rose	42-93852	Unpredictable	44-84114	Whites Cargo	44-69872
Tokyo Twister	42-24682	Untouchable	42-63506	Who's Next	44-70019
Tokyo-Ko	42-24859	Up An' Atom	44-27304	Wichita Witch	42-24442
Tommy Hawk	42-24755	Up 'n' Comin'	44-87630	Wichita Witch	42-24752
Top of the Mark	44-69763	Upper Berth	42-24765	The Wichita Witch	42-24654
Top Secret	44-27302	Urgin' Virgin	42-6423	Wild Goose	44-27264
Torchy	42-24646	Urgin' Virgin II	42-93884	The Wild Goose	44-69668
Torrid Toby	42-93830	USS Comfort's Revenge	44-61556	Wild Hair	42-24505
Totin' to Tokyo	42-6454	USS Pintado	44-61756	Wild Westy's Wabbits	42-94114
Totin' to Tokyo II	42-63530	Valiant Lady	42-93870	The Wilful Witch	42-94037
Touch-N-Go	44-87601	Valkyrie Queen	42-63356	William Allen White	44-70078
Town Pump	44-27282	Vanishing Rae	44-69706	William Allen White	44-70121
Townswick's Terrors	44-62066	Vicious Roomer	44-62186	Willie Mae	42-24663
Tremlin Gremlins	44-62188	Victory Girl	42-24731	Willy's Ice Wagon	44-84110
Triflin' Gal	44-83919	Victory Jean	44-83946	Windy City	42-6253
Trigger Mortis	42-93911	Virginia Dare	44-83946	Windy City II	42-24486
Trigger Mortis II	44-69744	Virginia Tech	44-61529	Wing Ding	42-63458
Trojan Spirit	42-63360	Visiting Firemen	42-24834	Winged Victory II	42-63538
Trojan Spirit	42-63446	Vivacious Lady	42-93996	Winnie	42-6267
Trouble Brewer	44-86390	Vivacious Lady	44-69911	Winnie II	44-69963
Tumbling Tumbleweeds	42-93959	Wabash Cannonball	42-24743	Wolf Pack	44-86340
Twentieth Century Limited	44-61797	Wachin Jo	45-21717	The Wolf Pack	4294063
Twenty-Ninth USNCB	42-24781	Waddy's Wagon	42-24598	Wright's Delight's	44-62102
Two Passes and a Crap	44-69773	Waltzing Matilda	44-69774	Wright's Delight's	44-86392
Typhoon Goon	45-21838	The Wanderer	44-62224	Wugged Wascal	42-24658
Typhoon Goon II	44-69770	Wango Bango	44-87653	Yankee Made	42-24651
Typhoon Mcgoon	42-6303	War Weary	42-24633	Yellow Rose of Texas	42-6241
Ugly	44-87661	Warm Front	44-62128	Yokohama Yoyo	42-24621
Umbiago III, Dat's My Boy	42-63447	Warsaw Pigeon	44-69849	Yonkee Doll-Ah	42-65371
Umbriago	42-63545	We Dood It	44-69860	Yonkee Dollar	44-61643
Umbriago II	42-94041	Wee Miss America	42-93856	Yucatan Kids	44-61587
Uncle Tom's Cabin	42-24642	Wedding Belle	44-69779	Zero Avenger	42-63569

NAMES WITH KNOWN GROUP ASSIGNMENTS BUT UNKNOWN SERIAL NUMBERS

?*	330BG	Agonizer	39BG	Arbon's Angels	500BG
1000th B-29	39BG	Agony Wagon	497BG	B/Gen Powers	330BG
107 SeaBees	497BG	All Shook	19BG	The Bald Eagle	39BG
135th Sea Bees	505BG	All Shook	98BG	Ball O' Fire	505BG
13th Sea Bees	497BG	Alley Oop	92BG	Battlin' Betty II	29BG
18TH C.B.	505BG	Always Ready	504BG	The Beachcomber	505BG
20th Century Fox	501BG	Angel in Disguise	40BG	The Beeg Az Burd	16BG
35 'er Bust	19BG	Ann Garry IV	6BG	Betsy	9BG
9th NCB	504BG	Antagonizer	39BG	Big Ass Bird	9BG
A Good Deal	504BG	Any Time	98BG	Big Dick	330BG
Adams Eve	504BG	Apache	19BG	Big Dick	9BG

*This was the actual name of a plane: simply a question mark.

II. Names List

Name	Code	Name	Code	Name	Code
Big Hutch	502BG	Dirty Gertie	6BG	Island Queen	19BG
Boomerang	462BG	Dis-Ass-Ter	92BG	Jabbitt	505BG
Buccaneer	6BG	Dixie Babe	19BG	Jabbitt II	505BG
The Buffalo	19BG	Doc Jones	39BG	Janie	505BG
Burning Desire	39BG	Dolly Ann	500BG	Jeeter Bug	500BG
Bus-O-Bugs	40BG	Doris Ann	331BG	JITA**	19BG
Bust 'n the Blue	98BG	Dottie	462BG	Journey For Margaret	444BG
Cami Ningyo	98BG	Dottie's Baby	501BG	Jungle Folly	39BG
Cannibal	331BG	Double Or Nothin'	19BG	Jus' One Mo' Time	331BG
Can't Say No	19BG	Double Pleasure	39BG	Kansas Farmer	6BG
Censored	19BG	Eagar & Ready	39BG	Katie Ann II	19BG
Censored	444BG	Eight Ball II	39BG	K-Dink III	39BG
Charley's Haunt	9BG	Eman-On	40BG	Kens Men	43BG
Chi Gal	19BG	Ernie Pyle's Milk Wagon	6BG	Knipp's vs Nips	330BG
Chicago Cubs	16BG	Faded	40BG	La Cherie	6BG
City of Annapolis	29BG	Fancy Detail	498BG	The Laden Maiden	331BG
City of Athens	39BG	Fancy Nancy	330BG	Ladies Delight	6BG
City of Atlanta	29BG	Fickle Finger of Fate	444BG	Lady	6BG
City of Battle Creek	19BG	Fifty-Second SeaBees	39BG	Lady Annabelle	6BG
City of Boise	19BG	Flak Maid	6BG	Lady Beth	6BG
City of Brooklyn	19BG	Flak Shack	91RCN	Lady Jean	6BG
City of Charlottesville	19BG	The Fledgling	462BG	Lady Luck	462BG
City of Cooperstown	39BG	Fliffy Fuz II	501BG	Lady Luck	6BG
City of El Paso	39BG	Fluffy Fuz IV	501BG	Lane Girl	98BG
City of Gainsville	330BG	Flying Patches	504BG	Large Charge	444BG
City of Gainsville	505BG	For The Luvva Mike	331BG	Lassie Too	504BG
City of Houston	39BG	Freshly Maid	98BG	Lazy Daisy Mae	98BG
City of Hutchinson	505BG	Good Old Gus	98BG	Lazy Jane	468BG
City of Ketchikan	39BG	Gracious Oasis	2BG	Liberty Belle	501BG
City of Laramie	330BG	Grand Old Man	444BG	Liberty Belle	92BG
City of Lexington	19BG	Grand Slam	6BG	Liberty Belle	98BG
City of Long Beach	19BG	Green Hornet	500BG	Liberty Bell II	331BG
City of Lynchburg	330BG	Green Job	40BG	Li'l Herbert	98BG
City of Maywood ILL	19BG	Grimwood's Gremlins	331BG	Little David	29BG
City of Milwaukee II	19BG	Grym Gryphon	19BG	Loco Lobo	502BG
City of Minneapolis	29BG	The Gypsy	19BG	Lonesome Polecat II	98BG
City of Minneapolis	9BG	Had A Call	19BG	Louise	91RCN GP
City of Morehead	39BG	Hades Ex Alto	16BG	Lovely Lady	91RCN GP
City of Nashville	19BG	Headin' Home	502BG	Lucky Irish	497BG
City of Newbern	39BG	Hearts Desire	98BG	Lucky Lady	498BG
City of Niagara Falls	330BG	Hearts Desire II	98BG	Lucky Legs	499BG
City of Ottumwa	19BG	Heats On II	498BG	Ma Bel	498BG
City of Palm Beach	19BG	Heavenly Body	39BG	Mad Dog	444BG
City of Paterson	29BG	Hell n Gone	502BG	Madelyn II	468BG
City of Pensacola	19BG	Hells Bells	499BG	Mag Drop Myrtle	468BG
City of Philadelphia	330BG	Hell's Bells	331BG	Many Happy Returns	39BG
City of Pittsburg	19BG	Here's Lucky	501BG	Many Happy Returns	468BG
City of Rising Star	501BG	Hey Doc II	501BG	Margo	468BG
City of Seattle	19BG	Hey Doc II	98BG	Mary Ann	504BG
City of St Louis	29BG	High Tailin'	501BG	McNamara's Band	498BG
City of Sweetwater	29BG	Holton's Bar'l*	92BG	Me Worry?	331BG
City of Toledo	39BG	Holy Joe	501BG	Mighty Thor	499BG
City of Topeka	504BG	Honey Bucket Honshos II	504BG	Mingtoy	444BG
City of Virginia City	39BG	Honorable Sad Saki	505BG	Miss Behavin	98BG
City of Wichita	19BG	Honshu Hurricane	499BG	Miss Fortune	19BG
Clobbered Turkey	6BG	Hopeful Devil	504BG	Miss Fortune II	497BG
Co. B 92nd SeaBees	6BG	Horrible Monster	16BG	Miss Houston	497BG
Cock Sure	462BG	Hose Nose	98BG	Miss Jackie The Rebel	19BG
Cock Sure	6BG	Hot Box	19BG	Miss Lace	40BG
Come On A My House	98BG	The Hound of Heaven	40BG	Miss Leading Lady	499BG
Command Performance	499BG	Houston Flyer	9BG	Miss Manuki	499BG
Connecticut Yankee II	6BG	Idiot's Delite	29BG	Miss Megook	19BG
Contrary Mary	92BG	Indian Maid II	505BG	Miss Pacific	499BG
Cy	499BG	Island Girl	502BG	Miss Patches	499BG
Dead Jug	19BG	Island Princess	505BG	Miss Yankee Doodle	499BG
Deliver The Gods	9BG				
Dina Might	22BG	*Tail gunner's position		**Acronym for Jab In The Ass	

Miss-Leading Lady	499BG	Raunchy Rambler	500BG	Super Roc	500BG
The Missouri Belle	499BG	Reddy Teddy	497BG	Super Wabbitt II	499BG
Misty Christi	497BG	Reeeaal Rough	444BG	Super Wabbitt III	499BG
Moldy Fig	501BG	Regiment 30 – Sea Bees	504BG	Sweater Girl Beverley	505BG
Mona	505BG	The Renegades	39BG	Sweet Chariot	501BG
Moonshine Minny	40BG	Reserved	98BG	Sweet Jenny Lee	9BG
My Achin' Back	9BG	Ripple Springs	500BG	Sweet Marilyn	444BG
My Assam Dragon III	19BG	The Roc	499BG	Sweet Patootie	505BG
The Natural	9BG	Rose Marie	502BG	Sweet Sue	462BG
Night Mare	9BG	Rough Roman	19BG	Sweet Sue	498BG
Night Mission	9BG	Round Robin	509BG	Sweetwater	29BG
Night Mission II	9BG	Round Robin Rosie	444BG	T.D.Y. Widow II	98BG
Night Roamer	501BG	Round Trip	497BG	Take Off Or Bust	468BG
Oh Brother	502BG	The Rude Nude	501BG	Tale Of Misfortune	497BG
Old Charra	444BG	Rusty Dusty	499BG	Tale Of Misfortune II	497BG
Old Double Deuce	91RCN	SAC Mate	22BG	Tale Of The Texas Dozen	497BG
Old Eighty-One	39BG	Sand Bag	500BG	Ten Knights In A Barroom	501BG
Old Faithful	6BG	Scrapper	444BG	Ten Under Parr	497BG
Old Man Mose	500BG	Sea Biscuit	497BG	That's It	19BG
Old Man Mose	462BS	Sentimental Journey	444BG	Three Feathers III	500BG
Old Overcast	39BG	Shack Happy Pappy	9BG	Throttle Jockey	509CG
Old Sixty-Three	500BG	She Hasta	504BG	Tinny- Ann	6BG
The Old Tokyo Hitchhiker	504BG	The Shmoo	19BG	Tokyo Hitchhiker	504BG
Ole Forty Eight II	39BG	Silver Lady	505BG	Tokyo Trolley II	6BG
Ole Miss IV	500BG	The Silver Streak	499BG	Topeka	444BG
Omaha Two More Times	504BG	Skippy	444BG	Topsy Turvey	468BG
Our Baby	505BG	Sky Blue	505BG	Trouble Maker	40BG
Our Baby	98BG	Slick Chick	98BG	U. S. —(?)	444BG
Our Gal Sal	505BG	Snude's Dudes	39BG	Umbriago II	98BG
Pacific Pirate	499BG	Society Leader	98BG	The Uninvited	29BG
Peek Of Perfection	98BG	Soft Touch	19BG	The Uninvited	39BG
Persuade-Her	19BG	Speagle Eagle	6BG	The Uninvited	502BG
Poison Lil	499BG	The Spearhead	468BG	University of Oklahoma	498BG
Police Action	504BG	Spirit Of Minneapolis	9BG	USS Pintado	39BG
Pom Pom	331BG	Stagger Inn — Beer & Ale	40BG	Wambli	468BG
PT Special	462BG	The Stinker	501BG	War Admiral	462BG
Punky	504BG	Stork Club Boys II	6BG	Weirite	40BG
Rabbit Punch	500BG	The Strained Crane	501BG	Where Next?	91RG
Rainbow's End	39BG	Strato Wolfe	39BG	Winchester 73 Red Raider	98BG
Ramblin' Reck	504BG	Strato Wolfe II	39BG		

Unnamed Nose Art with Known Serial Number or Group Assignment

Description of Nose Art	Serial #	Group	Description of Nose Art	Serial #	Group
American chasing a Japanese soldier		468BG	Japanese nude with fan		19BG
Bull butting a bomb — later named "Ape Ship"	44-86330	98BG	Majorette	42-24625	498BG
Dragon and a lady	44-62253	98BG	Mule team		497BG
Eight ball	44-62237	98BG	Nude on right elbow		6BG
Girl flying		468BG	Nude running		497BG
Girl on arrowhead		497BG	Polynesian nude		497BG
Girl on world		504BG	Stork carrying baby	45-21800	
Sirl sitting on bomb		16BG	Varga girl	42-63481	504BG
Reclining girl with legs in air		499BG	Wasp with camera		497BG
Hitch-hiker to Tokyo		504BG	Winged nude with drink	44-69739	504BG
Indian maid sitting		497BG			

NAMES WITHOUT KNOWN SERIAL NUMBERS OR GROUP ASSIGNMENTS

17 AAA Bn
1ST Sep Engr Batt USMC
31st Air Refueling Squadron
358th Air Rescue
4th Marine Division
50th NCB
504th BG Insignia: Hammer of Thor
9th SeaBees
A Run on Suger
A Token to Tokyo
Abie's Babies
Already Reddy
Amiable
Ann's Raiders
Anticipation
Armwood's Gremlins
Aviation Engineers
Beat Me Daddy
Beats Me Daddy
Betsy
Bett n Babe
Bette D — w (?)
The Big — Bird
Black Bunny
Black Sheep
Boeing Boner
The Boomerang
Bopper II
Bouncing Betty
Brigade 5
Bushwacker
Captain Chuck
Chicago Queen
City of Muncie
City of Plainville
City of Queens
Cross Country
Dirty Nelly
Doris & DD
Dotty's Rival
Droopy
Dyin' Duck
Easy Maid

Els-Natcho
Expect'n
Fancy Detail
The Fat Cat
Fee Nix
Fire Plug
Firemen
Fluffy Fuz
Fluffy Fuz III
Fluffy Fuz V
Fluffy Fuz VI
Forever Amber II
Game Cock Charlie
General Ike
Gen'L Ike
Golden State Express
Hardstand Honcho
Heavenly Body
Heavenly Lady
Incendiary Blond
Jughound Jalopy
Kamikazi Miss
Kansas City Kitty
Les's Best
Li'l (picture of Eight Ball)
Little Giant
Little Wheels
Lovely Leta
Madoline
Maximum Effort #3
Meeting House
Miasis Dragon
Milk Run Special
Miss Daria Hope
Mission Accomplished
Mission Queen
Mizpah
Moon Happy
Neutral Spirits
Night Prowler
Nipponese Nuisance
Noah Borshuns
Norma Ray

Ogoshi Ni!
Ol' Matusalum
On Call
One Meat Ball
O-O-Oklahoma
Out House Mouse
Papa San's Parasites
Pilot's Dream
Pot Limit
Punch 'n' Judy
Queen of the Strip
Ramp Vamp
Rebel Queen
Rising Star
Ruff 'n' Ready
The Scalded Dog
Scuse Please
Shady Lady II
Short Time Only
Sizzling Susie
Snoopy
Snow White
Sophisticated Lady
Stormy Weather
Surprise Package
Susie
Sweet Eloise
Take It Easy
Tax Exempt
Tiny Fanny
Toddlin' Turtle
Tokyo Trolley
Tu Yung Tu
United Nations Justice
Virgin Squaw
Virginia Dare
Who Knows
Winnie Mae
Wolf Gal
Wolf Pack
The Worry Bird

III

Block Numbers and Construction Numbers

A *block number* consisted of one to three digits followed by a two-letter code. The digits represented aircraft built to the same engineering configuration. The two letters identified the factory at which the planes were built: BA Bell-Atlanta; BN Boeing Renton; BO Boeing Seattle; BW Boeing Wichita; MO Martin Omaha.

The block number combined with the basic model designation made up the aircraft model number, e.g., B-29-20-BW, B-29A-65-BN, B-29B-30-BA.

A *construction number* was simply a tracking number used during construction of some of the B-29s. It is not clear just who assigned these numbers. The construction numbers for serials 42-6205 through 42-6454 are sequential — 3339 through 3588. However, there were three different manufacturers producing planes in that sequence. Irregularities exist; for example, Martin Omaha was assigned construction numbers 3363 — 3366 and 3377 out of a build of 249 aircraft.

Bell Atlanta did not use construction numbers after 3371 for serial 42-6237, nor did Martin Omaha after 3377, serial 42-6243. Boeing Renton and Boeing Wichita, on the other hand, used them for all their B-29 builds.

Serial Numbers	Manufactuer	Model	Block Number	Construction Number
41-002	Boeing Seattle	XB-29-BO	BO	2482
41-003	Boeing Seattle	XB-29-BO	BO	2481
41-18335	Boeing Seattle	XB-29-BO	BO	2884
41-36954	Boeing Wichita	YB-29-BW	BW	3325
41-36955	Boeing Wichita	YB-29-BW	BW	3326
41-36956	Boeing Wichita	YB-29-BW	BW	3327
41-36957	Boeing Wichita	YB-29-BW	BW	3328
41-36958	Boeing Wichita	YB-29-BW	BW	3329
41-36959	Boeing Wichita	YB-29-BW	BW	3330
41-36960	Boeing Wichita	YB-29-BW	BW	3331
41-36961	Boeing Wichita	YB-29-BW	BW	3332
41-36962	Boeing Wichita	YB-29-BW	BW	3333
41-36963	Boeing Wichita	YB-29-BW	BW	3334
41-36964	Boeing Wichita	YB-29-BW	BW	3335
41-36965	Boeing Wichita	YB-29-BW	BW	3336
41-36966	Boeing Wichita	YB-29-BW	BW	3337
41-36967	Boeing Wichita	YB-29-BW	BW	3338
42-6205 –42-6221	Boeing Wichita	B-29-1-BW	1-BW	3339-3355
42-6222	Bell Atlanta	B-29-1-BA	1-BA	3356
42-6223	Boeing Wichita	B-29-1-BW	1-BW	3357
42-6224	Bell Atlanta	B-29-1-BA	1-BA	3358

III. Block Numbers and Construction Numbers

Serial Numbers	Manufactuer	Model	Block Number	Construction Number
42-6225–42-6228	Boeing Wichita	B-29-1-BW	1-BW	3359–3362
42-6229–42-6232	Martin Omaha	B-29-1-MO	1-MO	3363–3366
42-6233	Bell Atlanta	B-29-1-BA	1-BA	3367
42-6234	Boeing Wichita	B-29-1-BW	1-BW	3368
42-6235	Bell Atlanta	B-29-1-BA	1-BA	3369
42-6236	Boeing Wichita	B-29-1-BW	1-BW	3370
42-6237	Bell Atlanta	B-29-1-BA	1-BA	3371
42-6238–42-6242	Boeing Wichita	B-29-1-BW	1-BW	3372–3376
42-6243	Martin Omaha	B-29-1-MO	1-MO	3377
42-6244–42-6254	Boeing Wichita	B-29-1-BW	1-BW	3378–3388
42-6255–42-6304	Boeing Wichita	B-29-5-BW	5-BW	3389–3438
42-6305–42-6354	Boeing Wichita	B-29-10-BW	10-BW	3439–3488
42-6355–42-6404	Boeing Wichita	B-29-15-BW	15-BW	3489–3538
42-6405–42-6454	Boeing Wichita	B-29-20-BW	20-BW	3539–3588
42-14502–42-14701	North American	Contract Cancelled		
42-24420–42-24469	Boeing Wichita	B-29-25-BW	25-BW	4081–4130
42-24470–42-24519	Boeing Wichita	B-29-30-BW	30-BW	4131–4180
42-24520–42-24569	Boeing Wichita	B-29-35-BW	35-BW	4181–4230
42-24521–42-24669	Boeing Wichita	B-29-40-BW	40-BW	4231–4330
42-24670–42-24769	Boeing Wichita	B-29-45-BW	45-BW	4331–4430
42-24770–42-24869	Boeing Wichita	B-29-50-BW	50-BW	4431–4530
42-24870–42-24919	Boeing Wichita	B-29-55-BW	55-BW	4531–4580
42-63352–42-63365	Bell Atlanta	B-29-1-BA	1-BA	
42-63366–42-63381	Bell Atlanta	B-29-5-BA	5-BA	
42-62382–42-63401	Bell Atlanta	B-29-10-BA	10-BA	
42-63402–42-63451	Bell Atlanta	B-29-15-BA	15-BA	
42-63452–42-63501	Bell Atlanta	B-29-20-BA	20-BA	
42-63502–42-63551	Bell Atlanta	B-29-25-BA	25-BA	
42-63552–42-63580	Bell Atlanta	B-29-30-BA	30-BA	
42-63581–42-63621	Bell Atlanta	B-29B-30-BA	30-BA	
42-63622–42-63691	Bell Atlanta	B-29B-35-BA	35-BA	
42-62692–42-63736	Bell Atlanta	B-29B-40-BA	40-BA	
42-63737	Bell Atlanta	B-29-30-BA	40-BA	
42-63738–42-63743	Bell Atlanta	B-29B-40-BA	40-BA	
42-63744	Bell Atlanta	B-29-40-BA	40-BA	
42-63745–42-63749	Bell Atlanta	B-29B-40-BA	40-BA	
42-63750	Bell Atlanta	B-29-40-BA	40-BA	
42-63751	Bell Atlanta	B-29B-40-BA	40-BA	
42-65202–42-65204	Martin Omaha	B-29-1-MO	1-MO	
42-65205–42-65211	Martin Omaha	B-29-5-MO	5-MO	
42-65212–42-65219	Martin Omaha	B-29-10-MO	10-MO	
42-65220–42-65235	Martin Omaha	B-29-15-MO	15-MO	
42-65236–42-65263	Martin Omaha	B-29-20-MO	20-MO	
42-65264–42-65313	Martin Omaha	B-29-25-MO	25-MO	
42-65314	Martin Omaha	Contract Cancelled		
42-65315–42-65383	Martin Omaha	B-29-30-MO	30-MO	
42-65384–42-65401	Martin Omaha	B-29-35-MO	35-MO	
42-93824–42-93843	Boeing Renton	B-29A-1-BN	1-BN	7231–7250
42-93844–42-93873	Boeing Renton	B-29A-5-BN	5-BN	7251–7280
42-93874–42-93923	Boeing Renton	B-29A-10-BN	10-BN	7281–7330
42-93924–42-93973	Boeing Renton	B-29A-15-BN	15-BN	7331–7380
42-93974–42-94023	Boeing Renton	B-29A-20-BN	20-BN	7381–7430
42-94024–42-94073	Boeing Renton	B-29A-25-BN	25-BN	7431–7480
42-94074–42-94123	Boeing Renton	B-29A-30-BN	30-BN	7481–7530
44-27259–44-27325	Martin Omaha	B-29A-35-MO	35-MO	
44-27326–44-27358	Martin Omaha	B-29A-40-MO	40-MO	
44-61510–44-61609	Boeing Renton	B-29A-35-BN	35-BN	10987–11086
44-61610–44-61709	Boeing Renton	B-29A-40-BN	40-BN	11-87–11186
44-61710–44-61809	Boeing Renton	B-29A-45-BN	45-BN	11187–11286
44-61810–44-61909	Boeing Renton	B-29A-50-BN	50-BN	11287–11386
44-61910–44-62009	Boeing Renton	B-29A-55-BN	55-BN	11387–11486
44-62010–44-62109	Boeing Renton	B-29A-60-BN	60-BN	11487–11586
44-62110–44-62209	Boeing Renton	B-29A-65-BN	65-BN	11587–11686
44-62210–44-42309	Boeing Renton	B-29A-70-BN	70-BN	11687–11786
44-62310–44-62328	Boeing Renton	B-29A-75-BN	75-BN	11787–11805

The B-29 Superfortress

Serial Numbers	Manufactuer	Model	Block Number	Construction Number
44-62329–44-62909			Contract Cancelled	11806–12386
44-69655–44-69704	Boeing Wichita	B-29-55-BW	55-BW	10487–10536
44-69705–44-69804	Boeing Wichita	B-29-60-BW	60-BW	10537–10636
44-69805–44-69904	Boeing Wichita	B-29-65-BW	65-BW	10637–10736
44-69905–44-70004	Boeing Wichita	B-29-70-BW	70-BW	10737–10836
44-70005–44-70104	Boeing Wichita	B-29-75-BW	75-BW	10837–10936
44-70105–44-70154	Boeing Wichita	B-29-80-BW	80-BW	10937–10986
44-75027–44-76026			Contract Cancelled	12587–13586
44-83890–44-83893	Bell Atlanta	B-29B-40-BA	40-BA	
44-83894	Bell Atlanta	B-29-40-BA	40-BA	
44-83895	Bell Atlanta	B-29B-40-BA	40-BA	
44-83896–44-83899	Bell Atlanta	B-29B-45-BA	40-BA	
44-83900	Bell Atlanta	B-29-45-BA	45-BA	
44-83901–44-83903	Bell Atlanta	B-29B-45-BA	45-BA	
44-83904	Bell Atlanta	B-29-45-BA	45-BA	
44-83905–44-83907	Bell Atlanta	B-29B-45-BA	45-BA	
44-83908	Bell Atlanta	B-29-45-BA	45-BA	
44-83909–44-83910	Bell Atlanta	B-29B-45-BA	45-BA	
44-83911	Bell Atlanta	B-29-45-BA	45-BA	
44-83912–44-83913	Bell Atlanta	B-29B-45-BA	45-BA	
44-83914	Bell Atlanta	B-29-45-BA	45-BA	
44-83915–44-83916	Bell Atlanta	B-29B-45-BA	45-BA	
44-83917	Bell Atlanta	B-29-45-BA	45-BA	
44-83918–44-83919	Bell Atlanta	B-29B-45-BA	45-BA	
44-83920	Bell Atlanta	B-29-45-BA	45-BA	
44-83921–44-83922	Bell Atlanta	B-29B-45-BA	45-BA	
44-93923	Bell Atlanta	B-29-45-BA	45-BA	
44-83924–44-83925	Bell Atlanta	B-29B-45-BA odd #	45-BA	
44-83926–44-83962	Bell Atlanta	B-29-45-BA even #	45-BA	
	Bell Atlanta	B-29B-50-BW odd#	50-BA	
44-83963–44-84028	Bell Atlanta	B-29-50-BA even #	50-BA	
	Bell Atlanta	B-29B-55-BW odd#	55-BA	
44-84029–44-84056	Bell Atlanta	B-29-55-BA even #	55-BA	
	Bell Atlanta	B-29B-60-BA odd #	60-BA	
44-84057–44-84103	Bell Atlanta	B-29-60-BA even #	60-BA	
	Bell Atlanta	B-29B-65-BA odd #	65-BA	
44-84104–44-84149	Bell Atlanta	B-29-65-BA even #	65-BA	
	Bell Atlanta	B-29B-65-BA	65-BA	
44-84140		Contract Cancelled		
44-84150		Contract Cancelled		
44-84153		Contract Cancelled		
44-84154		Contract Cancelled		
44-84151	Bell Atlanta	B-29B-65-BA	65-BA	
44-84152	Bell Atlanta	B-29-65-BA	65-BA	
44-84155	Bell Atlanta	B-29B-65-BA	65-BA	
44-84156	Bell Atlanta	B-29-65-BA	65-BA	
44-86242–44-86276	Martin Omaha	B-29-40-MO	40-MO	
44-86277–44-86315	Martin Omaha	B-29-45-MO	45-MO	
44-86316–44-86370	Martin Omaha	B-29-50-MO	50-MO	
44-86371–44-86425	Martin Omaha	B-29-55-MO	55-MO	
44-86426–44-86473	Martin Omaha	B-29-60-MO	60-MO	
44-87584–44-87633	Boeing Wichita	B-29-80-BW	80-BW	12387–12436
44-87634–44-87683	Boeing Wichita	B-29-85-BW	85-BW	12437–12486
44-87684–44-87733	Boeing Wichita	B-29-86-BW	86-BW	12487–12536
44-87734–44-87783	Boeing Wichita	B-29-90-BW	90-BW	12537–12586
45-21693–45-21742	Boeing Wichita	B-29-90-BW	90-BW	13587–13636
45-21743–45-21757	Boeing Wichita	B-29-97-BW	97-BW	13637–13651
45-21758–45-21792	Boeing Wichita	B-29-95-BW	95-BW	13652–13686
45-21793–45-21812	Boeing Wichita	B-29-96-BW	96-BW	13687–13706
45-21813–45-21842	Boeing Wichita	B-29-95-BW	95-BW	13707–13736
45-21843–45-21842	Boeing Wichita	B-29-100-BW	100-BW	13737–13766
45-21873–45-22392		Contract Cancelled	13767–14286	
45-40911–45-41735		Contract Cancelled	14287–15111	
45-42480–45-43079		Contract Cancelled	15112–15711	

IV

Design Specifications and Performance Parameters

Based on the Boeing Model 345, the B-29 Superfortress is a four-engine mid-wing monoplane. The circular fuselage consists, from nose to tail, of the forward pressurized compartment, the forward bomb bay, the center wing tank, the rear bomb bay, the rear unpressurized compartment and the tail gunner's compartment.

The forward pressurized compartment contained positions for the bombardier, aircraft commander, pilot, flight engineer, navigator and radio operator. In the rear pressurized compartment were the assigned positions of the central fire control gunner, the left and right gunners and the radar operator. The tail gunner worked in isolation in a separate pressurized compartment under the rudder.

Dimensions

Wingspan: 141 feet 3 inches
Length: 99 feet
Height of Tail: 28 feet 7 inches
Wing Area: 1736 square feet
Flap Area: Approximately 360 square feet

Power Plant

Wright R-3350 18 cylinder radial engine
3350 cubic inches displacement
Two General Electric B-11 turbo superchargers
2200 Horsepower takeoff power at 2700 RPM

Performance

Maximum Speed: 365 MPH at 25,000 feet; 220 MPH cruise speed

Range: 1900 mile radius with 10,000 pound bomb load; 1550 mile radius with 20,000 pound load.

Service Ceiling: 31,850 feet

Weight:
　Empty: 72,000 pounds
　Maximum normal takeoff: 133,500 pounds
　Maximum overload takeoff: 140,000 pounds

Fuel Capacity: With maximum four 640 gallon bomb bay tanks 9,100 gallons

Armament: Four .50 caliber remotely aimed machine guns in forward dorsal turret; two remotely aimed .50 caliber machine guns each in aft dorsal, forward ventral and aft ventral turrets; two .50 caliber machine guns in the tail turret. Early models also had a 20MM cannon mounted between the tail machine guns. Due to sighting problems and trajectory differences the cannon was removed on later production models.

Total Built: 3,960

V

B-29 Variants

Specialized tasks required modifications to the standard production version of the B-29. This section lists twenty-seven such variations, distinct enough from the production model to require unique model designations.

Five of these variants — the F-13, KB-29M, KB-29P, SB-29 and WB-29 — were produced in significant numbers and are individually addressed in the following pages.

B-29C-BO	Upgrade of the B-29-BN with later model R-3350 engines. 5000 plane projected buy was cancelled.		KB-29M	Ninety-two B-29s converted to single point trailing hose refueling system.
B-29D-BN	Redesigned, re-engined B-29, became the B-50.		KB-29P	116 B-29s converted to 'Flying Boom' aerial refuelers.
B-29E-??-BW	One standard B-29 converted in 1946 to test fire control system.		QB-29	Controller of radio-controlled targets.
B-29F-BW	Six B-29-BWs temporarily modified for Alaska Cold Weather Trials.		SB-29	Sixteen B-29s usually cited as converted in 1944–45 to rescue mission with lifeboat carried external to the bomb bays. Further research shows that five (5) additional planes were converted. See information on the SB-29 in the following pages.
B-29L	Proposed designator for B-29s capable of accepting in-flight fueling.			
B-29-MR	Seventy-four B-29s converted to receive fueling from KB-29M.		TB-29	Trainer conversion. Usages included combat crew training, proficiency flying, target towing, radar calibration and evaluation and non-flying use as ground training aid.
CB-29K-BW	Single B-29 converted to freighter in 1949.			
DB-29	Conversion to Drone Directors.			
EB-29B-BA	S/N 44-84111. Converted to launch platform for the XF-85-MC parasite fighter.		WB-29	Weather reconnaissance conversions. Armament was removed, radio position moved to rear pressurized compartment; radar moved to the forward pressurized compartment, sharing a work station with the navigator on the left side of forward pressurized compartment. Airborne particulate collector moved to former position of CFC gunner.
ETB-29A-BN	S/N 44-62093. Experimental wingtip-to-wingtip hookups with two EF-84B-REs. Destroyed with EF-84Bs in crash, April 1953.			
F-13A	118 B-29s converted to reconnaissance role.			
FB-29J	Two YB-29Js converted for photo reconnaissance, became the RB-29J.		XB-29-BO	Prototypes — three built: 41-002, 41-003 and 41-18335.
GB-29	Conversion to launch platform for "X" Series vehicles.		XB-29G-BA	B-29 44-84083 converted to turbojet engine test bed.
KB-29K	Interim designation for the KB-29M tanker project.		XB-29H-BN	B-29A-BN used for special armament tests.

V. B-29 Variants 105

XB-39-BW	YB-29-BW s/n 41-36954 re-engined with four V-3420-11 in-line engines.		YB-29J	Six B-29s, including 44-84061.
YB-29-BW	Service Test Aircraft. Fourteen built, s/n's 41-36954 through 41-36967.		YKT-29T	KB-29M converted to a three point flight refueling station; hoses trailed from rear fuselage and wing tips. Prototype for KB-50D.

F-13 Photo Reconnaissance Configuration

The F-13 came to life in March of 1944 when the Air Staff in Washington directed Air Materiel Command to do an engineering study to determine the practicality of mounting a suite of cameras in the B-29. As a result, Boeing, in association with Fairchild, started work on the design of a photo reconnaissance version of the B-29.

The design called for an array of six off-the-shelf cameras to be mounted in the rear pressurized compartment aft of the central fire control station in three separate installations. The first held a single vertical Fairchild K-18 camera for generalized photorecon work. The K-18 mount was also capable of holding the K-17, K-19 or K-22 cameras.

The second installation held two K-22 cameras for photo interpretation work. A strip of ground three miles wide was photographed with these narrow-angle cameras.

The third installation consisted of three K-17B cameras, the middle unit mounted vertically, the two outer units aimed outward, producing a set of three exposures covering a thirty mile strip.

The crew of the F-13 consisted of aircraft commander and pilot, flight engineer, navigator, radio operator, photo navigator, a cameraman, and in some cases, a radar operator. The photo navigator occupied the bombardier's position and was responsible for the selection of photo targets.

When the F-13 reached the combat theater it was used extensively for mission planning, gathering photographs of potential targets and post-strike damage assessment photography.

An F-13A named *Tokyo Rose* of the 3rd Photo Reconnaissance Squadron, attached to the 73rd Bomb Wing on Saipan, in November of 1944 became the first American plane over Tokyo since Doolittle's B-25 raid of 1942. The airplane was also the first land-based American aircraft over Tokyo in World War II. The Doolittle raiders had taken off from the carrier *Hornet*.

In late 1945 the F-13A was designated FB-29. In 1948 it was redesignated RB-29.

42-6412	F-13-20-BW	Boeing Wichita	42-93852	F-13A-5-BN	Boeing Renton	42-93992	F-13A-20-BN	Boeing Renton
42-24566	F-13A-35-BW	Boeing Wichita	42-93853	F-13A-5-BN	Boeing Renton	42-93993	F-13A-20-BN	Boeing Renton
42-24567	F-13A-35-BW	Boeing Wichita	42-93854	F-13A-5-BN	Boeing Renton	42-94000	F-13A-20-BN	Boeing Renton
42-24583	F-13A-40-BW	Boeing Wichita	42-93855	F-13A-5-BN	Boeing Renton	42-94022	F-13A-20-BN	Boeing Renton
42-24585	F-13A-40-BW	Boeing Wichita	42-93856	F-13A-5-BN	Boeing Renton	42-94054	F-13A-25-BN	Boeing Renton
42-24586	F-13A-40-BW	Boeing Wichita	42-93863	F-13A-5-BN	Boeing Renton	42-94074	F-13A-30-BN	Boeing Renton
42-24588	F-13A-40-BW	Boeing Wichita	42-93864	F-13A-5-BN	Boeing Renton	42-94080	F-13A-30-BN	Boeing Renton
42-24589	F-13A-40-BW	Boeing Wichita	42-93865	F-13A-5-BN	Boeing Renton	42-94081	F-13A-30-BN	Boeing Renton
42-24621	F-13A-40-BW	Boeing Wichita	42-93866	F-13A-5-BN	Boeing Renton	42-94113	F-13A-30-BN	Boeing Renton
42-24803	F-13A-50-BW	Boeing Wichita	42-93867	F-13A-5-BN	Boeing Renton	42-94114	F-13A-30-BN	Boeing Renton
42-24805	F-13A-50-BW	Boeing Wichita	42-93868	F-13A-5-BN	Boeing Renton	44-27326	F-13A-40-MO	Martin Omaha
42-24810	F-13A-50-BW	Boeing Wichita	42-93869	F-13A-5-BN	Boeing Renton	44-61528	F-13A-35-BN	Boeing Renton
42-24811	F-13A-50-BW	Boeing Wichita	42-93870	F-13A-5-BN	Boeing Renton	44-61531	F-13A-35-BN	Boeing Renton
42-24813	F-13A-50-BW	Boeing Wichita	42-93871	F-13A-5-BN	Boeing Renton	44-61533	F-13A-35-BN	Boeing Renton
42-24816	F-13A-50-BW	Boeing Wichita	42-93872	F-13A-5-BN	Boeing Renton	44-61577	F-13A-35-BN	Boeing Renton
42-24817	F-13A-50-BW	Boeing Wichita	42-93874	F-13A-5-BN	Boeing Renton	44-61578	F-13A-35-BN	Boeing Renton
42-24819	F-13A-50-BW	Boeing Wichita	42-93879	F-13A-10-BN	Boeing Renton	44-61583	F-13A-35-BN	Boeing Renton
42-24821	F-13A-50-BW	Boeing Wichita	42-93880	F-13A-10-BN	Boeing Renton	44-61659	F-13A-40-BN	Boeing Renton
42-24829	F-13A-1-BN	Boeing Renton	42-93903	F-13A-10-BN	Boeing Renton	44-61684	F-13A-40-BN	Boeing Renton
42-24833	F-13A-50-BW	Boeing Wichita	42-93912	F-13A-10-BN	Boeing Renton	44-61727	F-13A-45-BN	Boeing Renton
42-24860	F-13A-50-BW	Boeing Wichita	42-93914	F-13A-10-BN	Boeing Renton	44-61734	F-13A-45-BN	Boeing Renton
42-24869	F-13A-50-BW	Boeing Wichita	42-93919	F-13A-10-BN	Boeing Renton	44-61810	F-13A-50-BN	Boeing Renton
42-24870	F-13A-55-BW	Boeing Wichita	42-93926	F-13A-15-BN	Boeing Renton	44-61813	F-13A-50-BN	Boeing Renton
42-24871	F-13A-55-BW	Boeing Wichita	42-93933	F-13A-15-BN	Boeing Renton	44-61815	F-13A-50-BN	Boeing Renton
42-24877	F-13A-55-BW	Boeing Wichita	42-93965	F-13A-15-BN	Boeing Renton	44-61817	F-13A-50-BN	Boeing Renton
42-93849	F-13A-5-BN	Boeing Renton	42-93967	F-13A-15-BN	Boeing Renton	44-61818	F-13A-50-BN	Boeing Renton
42-93850	F-13A-5-BN	Boeing Renton	42-93968	F-13A-15-BN	Boeing Renton	44-61819	F-13A-50-BN	Boeing Renton
42-93851	F-13A-5-BN	Boeing Renton	42-93987	F-13A-20-BN	Boeing Renton	44-61822	F-13A-50-BN	Boeing Renton

44-61832	F-13A-50-BN	Boeing Renton	44-61947	F-13A-55-BN	Boeing Renton	44-62296	F-13A-70-BN	Boeing Renton
44-61843	F-13A-50-BN	Boeing Renton	44-61948	F-13A-55-BN	Boeing Renton	45-21742	F-13A-90-BW	Boeing Wichita
44-61847	F-13A-50-BN	Boeing Renton	44-61951	F-13A-55-BN	Boeing Renton	45-21761	F-13A-95-BW	Boeing Wichita
44-61854	F-13A-50-BN	Boeing Renton	44-61960	F-13A-55-BN	Boeing Renton	45-21762	F-13A-95-BW	Boeing Wichita
44-61855	F-13A-50-BN	Boeing Renton	44-61961	F-13A-55-BN	Boeing Renton	45-21763	F-13A-95-BW	Boeing Wichita
44-61857	F-13A-50-BN	Boeing Renton	44-61981	F-13A-55-BN	Boeing Renton	45-21766	F-13A-95-BW	Boeing Wichita
44-61860	F-13A-50-BN	Boeing Renton	44-61986	F-13A-55-BN	Boeing Renton	45-21768	F-13A-95-BW	Boeing Wichita
44-61862	F-13A-50-BN	Boeing Renton	44-61989	F-13A-55-BN	Boeing Renton	45-21773	F-13A-95-BW	Boeing Wichita
44-61866	F-13A-50-BN	Boeing Renton	44-61991	F-13A-55-BN	Boeing Renton	45-21775	F-13A-95-BW	Boeing Wichita
44-61924	F-13A-55-BN	Boeing Renton	44-61999	F-13A-55-BN	Boeing Renton	45-21777	F-13A-95-BW	Boeing Wichita
44-61929	F-13A-55-BN	Boeing Renton	44-62000	F-13A-55-BN	Boeing Renton	45-21790	F-13A-95-BW	Boeing Wichita
44-61930	F-13A-55-BN	Boeing Renton	44-62282	F-13A-70-BN	Boeing Renton	45-21812	F-13-95-BW	Boeing Wichita
44-61931	F-13A-55-BN	Boeing Renton	44-62284	F-13A-70-BN	Boeing Renton	45-21846	F-13-95-BW	Boeing Wichita
44-61933	F-13A-55-BN	Boeing Renton	44-62285	F-13A-70-BN	Boeing Renton	45-21847	F-13-95-BW	Boeing Wichita
44-61934	F-13A-55-BN	Boeing Renton	44-62287	F-13A-70-BN	Boeing Renton	45-21848	F-13-95-BW	Boeing Wichita
44-61939	F-13A-55-BN	Boeing Renton	44-62288	F-13A-70-BN	Boeing Renton	45-21856	F-13-95-BW	Boeing Wichita
44-61945	F-13A-55-BN	Boeing Renton	44-62289	F-13A-70-BN	Boeing Renton	45-21859	F-13-95-BW	Boeing Wichita
44-61946	F-13A-55-BN	Boeing Renton	44-62290	F-13A-70-BN	Boeing Renton			

KB-29M Hose and Reel Tanker Configuration

A few years after the end of World War II, the Strategic Air Command was seeking a way to upgrade its operational envelope. The solution to the problem lay in the aerial refueling of its bomber fleet.

Boeing reopened a plant in Wichita, Kansas, and immediately started modifying B-29s to a tanker configuration using a hose system developed by Flight Refueling, Limited, in Great Britain. Eventually ninety-two airplanes were converted to this configuration and designated KB-29M. A 2300 gallon jettisonable fuel tank was installled in each bomb bay, resulting in a fuel capacity of over 12,000 gallons.

On the other end of the hose, seventy-two B-29s were modified into receivers for the hose refueling system and redesignated B-29MR. The advantages of the increased range were, however, offset by a loss in bomb-carrying capacity, since the modifications reduced the effective bomb load by half. The system required the permanent installation of an aft bomb bay tank as a receiving vessel for the incoming fuel. Fuel in the aft bomb bay tank was then transferred to operational tanks using the plane's internal fuel transfer system.

While workable, the system was slow and complicated, and the non-rigid hose presented some aerodynamic problems.

44-27268	KB-29M-35-MO	44-69841	KB-29M-65-BW	44-87601	KB-29M-80-BW	45-21700	KB-29M-90-BW
44-27280	KB-29M-35-MO	44-69853	KB-29M-65-BW	44-87610	KB-29M-80-BW	45-21701	KB-29M-90-BW
44-27282	KB-29M-35-MO	44-69860	KB-29M-65-BW	44-87611	KB-29M-80-BW	45-21702	KB-29M-90-BW
44-27325	KB-29M-35-MO	44-69875	KB-29M-65-BW	44-87622	KB-29M-80-BW	45-21703	KB-29M-90-BW
44-27329	KB-29M-40-MO	44-69951	KB-29M-70-BW	44-87680	KB-29M-85-BW	45-21704	KB-29M-90-BW
44-27330	KB-29M-40-MO	44-69953	KB-29M-70-BW	44-87725	KB-29M-86-BW	45-21705	KB-29M-90-BW
44-27333	KB-29M-40-MO	44-69957	KB-29M-70-BW	44-87742	KB-29M-90-BW	45-21706	KB-29M-90-BW
44-27338	KB-29M-40-MO	44-69958	KB-29M-70-BW	44-87747	KB-29M-90-BW	45-21713	KB-29M-90-BW
44-27340	KB-29M-40-MO	44-69960	KB-29M-70-BW	44-87758	KB-29M-90-BW	45-21716	KB-29M-90-BW
44-69681	KB-29M-55-BW	44-69962	KB-29M-70-BW	44-87770	KB-29M-90-BW	45-21731	KB-29M-90-BW
44-69685	KB-29M-55-BW	44-69981	KB-29M-70-BW	44-87776	KB-29M-90-BW	45-21734	KB-29M-90-BW
44-69699	KB-29M-55-BW	44-70019	KB-29M-75-BW	44-87777	KB-29M-90-BW	45-21738	KB-29M-90-BW
44-69704	KB-29M-55-BW	44-70024	KB-29M-75-BW	44-87778	KB-29M-86-BW	45-21741	KB-29M-90-BW
44-69709	KB-29M-60-BW	44-70044	KB-29M-75-BW	44-87779	KB-29M-90-BW	45-21764	KB-29M-90-BW
44-69710	KB-29M-60-BW	44-70047	KB-29M-75-BW	44-87780	KB-29M-90-BW	45-21765	KB-29M-90-BW
44-69729	KB-29M-60-BW	44-70081	KB-29M-75-BW	44-87781	KB-29M-90-BW	45-21769	KB-29M-90-BW
44-69731	KB-29M-60-BW	44-70144	KB-29M-80-BW	44-87782	KB-29M-90-BW	45-21778	KB-29M-90-BW
44-69732	KB-29M-60-BW	44-84144	KB-29M-75-BA	44-87783	KB-29M-90-BW	45-21785	KB-29M-90-BW
44-69798	KB-29M-60-BW	44-86270	KB-29M-40-MO	45-21693	KB-29M-90-BW	45-21788	KB-29M-90-BW
44-69806	KB-29M-65-BW	44-86277	KB-29M-45-MO	45-21695	KB-29M-90-BW	45-21792	KB-29M-90-BW
44-69807	KB-29M-65-BW	44-86389	KB-29M-55-MO	45-21696	KB-29M-90-BW	45-21864	KB-29M-90-BW
44-69809	KB-29M-65-BW	44-86418	KB-29M-55-MO	45-21697	KB-29M-90-BW	45-21865	KB-29M-90-BW
44-69815	KB-29M-65-BW	44-86420	KB-29M-55-MO	45-21699	KB-29M-90-BW		

KB-29P Boom Tanker Configuration

The KB-29M hose and drogue refueling method worked, but it was very slow and time-consuming. Searching for a better way, Boeing engineers came up with the concept of the flying boom.

A telescoping boom, mounted on the lower aft fuselage, was made "flyable" by means of a pair of rudimentary elevons controlled by the boom operator in the rear gunner's position. The receiving aircraft moved up on the tanker from behind and below. When he arrived in the proper position, guided by indicator lights on the belly of the tanker, the boom operator "flew" the boom into the receiver receptacle and the tanker began pumping fuel. When the receiver had taken on his required load, he dropped off the boom. The boom operator purged the refueling system with gaseous nitrogen to clear the system of gas fumes, retracted the boom to the maximum and stowed it in a horizontal position for flight.

With the use of form-fitting bomb bay tanks, the KB-29P carried just under 12,000 gallons. Assuming a stripped plane gross weight of 70,000 pounds, the 12,000 gallons would put the takeoff weight in the vicinity of 142,000 pounds, well above the specified 137,500 pound maximum takeoff weight. Takeoff weights over 140,000 pounds were, however, quite common in World War II and Korea.

The normal crew of a KB-29P consisted of aircraft commander, pilot, flight engineer, navigator, radio operator, two scanners, a radar operator and the boom operator.

The initial contract for conversion of planes to KB-29P was let in November of 1949. Sometime prior to April 1950, the first tests of the new system was made with the successful transfer of fuel to an F-84.

Serial#	Model	Serial#	Model	Serial#	Model	Serial#	Model
42-65389	KB-29P-30-M0	44-69822	KB-29P-65-BW	44-84071	KB-29P-60-BA	44-86363	KB-29P-50-MO
42-93921	KB-29P-10-BN	44-69823	KB-29P-65-BW	44-84079	KB-29P-60-BA	44-86384	KB-29P-55-MO
44-27346	KB-29P-40-MO	44-69826	KB-29P-65-BW	44-84107	KB-29P-65-BA	44-86427	KB-29P-60-MO
44-27348	KB-29P-40-MO	44-69837	KB-29P-65-BW	44-84118	KB-29P-65-BA	44-86428	KB-29P-60-MO
44-27353	KB-29P-40-MO	44-69858	KB-29P-65-BW	44-84123	KB-29P-65-BA	45-21827	KB-29P-95-BW
44-61651	KB-29P-40-BN	44-69878	KB-29P-65-BW	44-84134	KB-29P-65-BA	44-69700	KB-29P-55-BW
44-69672	KB-29P-55-BW	44-69904	KB-29P-65-BW	44-84139	KB-29P-65-BA	44-69828	KB-29P-65-BW
44-69674	KB-29P-55-BW	44-69915	KB-29P-70-BW	44-84145	KB-29P-65-BA	44-69846	KB-29P-65-BW
44-69687	KB-29P-55-BW	44-70019	KB-29P-75-BW	44-84147	KB-29P-65-BA	44-69847	KB-29P-65-BW
44-69693	KB-29P-55-BW	44-70026	KB-29P-75-BW	44-84149	KB-29P-65-BA	44-69874	KB-29P-65-BW
44-69701	KB-29P-55-BW	44-83905	KB-29P-45-BA	44-84151	KB-29P-65-BA	44-69914	KB-29P-70-BW
44-69702	KB-29P-55-BW	44-83906	KB-29P-45-BA	44-84156	KB-29P-65-BA	44-83937	KB-29P-45-BA
44-69709	KB-29P-60-BW	44-83943	KB-29P-45-BA	44-86244	KB-29P-40-MO	44-83950	KB-29P-45-BA
44-69716	KB-29P-60-BW	44-83971	KB-29P-50-BA	44-86249	KB-29P-40-MO	44-83985	KB-29P-50-BA
44-69797	KB-29P-60-BW	44-84025	KB-29P-55-BA	44-86251	KB-29P-40-MO	44-84029	KB-29P-55-BA
44-69819	KB-29P-65-BW	44-84047	KB-29P-55-BA	44-86309	KB-29P-45-MO		
44-69821	KB-29P-65-BW	44-84055	KB-29P-55-BA	44-86350	KB-29P-50-MO		

SB-29 Rescue Configuration

While most sources cite sixteen as the number of B-29s converted to the SB-29 rescue configuration, some controversy does surround that number. Alwyn Lloyd, for example, in his *B-29 Superfortress in Detail and Scale, Part 2, Derivatives* lists the following yearly inventory of B-29s assigned to the Air Rescue Service:

1947–49	3	1953	16
1950	7	1954	17
1951	9	1955	14
1952	19		

Yet three lines later, following the sentence "The following airplanes were converted into SB-29's," he lists sixteen serial numbers.

In an attempt to clarify the situation, I made an SB-29 data dump from my B-29 data base. By this means I identified twenty-three serial numbers assigned SB-29 model designations. A check of my sources showed that for 21 of the 23 serial numbers listed, the SB-29 designation was derived from entries on the Aircraft Record Cards. Four photographs in my collection showed specific B-29s in Air Rescue markings.

When the twenty-three SB-29s from my data base are

combined with Lloyd's sixteen serial numbers, the lists only partially match (see details following). The combined lists produce twenty-five serial numbers with a legitimate claim to SB-29 configuration.

Regardless of the number of conversions, the conversion itself produced three changes not found in production B-29s. These were (1) the moving of the AN/APQ-13 radome from between the bomb bays to the location normally occupied by the lower forward turret; (2) the mounting of the 30-foot, droppable Edo A-3 lifeboat exterior to the bomb bays; and (3) the move of the radio operator's position from the right-rear corner of the forward compartment to the rear pressurized compartment.

The remaining armament, the upper forward turret, two aft turrets and the tail guns were in the aircraft and in operating condition.

Published List of Conversions	Author's List of Conversions	Source	Listed on A/C Record Cards
44-61671 SB-29A-40-BN*			44-61671 Not Active 1951 m/f
44-69957 SB-29-70-BW	44-69957 SB-29-70-BW	ARC	44-69957 Converted Tinker 7/52
44-69971 SB-29-70-BW	44-69971 SB-29-70-BW	ARC & Photo	44-69971 Converted Tinker 10/52
44-70119 SB-29-80-BW	44-70119 SB-29-80-BW	Photo	44-70119 Converted Tinker 7/51
44-84030 SB-29-55-BA		ARC	44-84030 Converted Tinker 8/50
44-84078 SB-29-60-BA	44-84078 SB-29-60-BA	ARC & Photo	44-84078 Converted Tinker 1/50
44-84084 SB-29-60-BA	44-84084 SB-29-60-BA	ARC	44-84084 Converted Tinker 1/50
44-84086 SB-29-60-BA	44-84086 SB-29-60-BA	ARC	44-84086 Converted Tinker 3/50
44-84088 SB-29-60-BA	44-84088 SB-29-60-BA	ARC & Photo	44-84088 Converted Tinker 2/50
44-84098 SB-29-60-BA†			44-84088 Not converted per ARC
44-84112 SB-29-65-BA	44-84112 SB-29-65-BA	ARC	44-84112 Converted Tinker 3/50
44-86303 BS-29-45-MO	44-86303 BS-29-45-MO	ARC	44-86303 Converted Tinker 2/50
44-86308 SB-29-45-MO	44-86308 SB-29-45-MO	ARC	44-86308 Converted Tinker 5/50
44-87644 SB-29-85-BW		ARC	44-87644 Converted Tinker 4/52
44-87665 SB-29-85-BW		ARC	44-87665 Converted Tinker 10/52
44-87761 SB-29-90-BW†			44-87761 Not converted per ARC
	44-27308 SB-29-35-MO	ARC & Photo	44-47308 Converted Tinker 6/50
	44-27312 SB-29-35-MO	ARC	44-27312 Not converted per ARC
	44-62190 SB-29A-65-BN	ARC	44-62190 Converted Tinker 9/52
	44-62194 SB-29A-65-BN	ARC	44-62194 Converted Tinker 9/52
	44-62210 SB-29A-65-BN		44-62210 Converted Tinker 8/52
	44-70089 SB-29-75-BW	ARC	44-70089 Converted Tinker 12/51
	44-70101 SB-29-75-BW	ARC	44-70101 Converted Tinker 12/51
	44-70117 SB-29-80-BW	ARC	44-70117 Converted Tinker 12/52
	44-70131 SB-29-80-BW	ARC	44-70131 Converted Tinker 12/52
	44-84054 SB-29-55-BA		44-84034 Converted Tinker 1/50
	44-84096 SB-29-60-BA	ARC	44-84096 Converted Tinker 1/50
	44-86259 SB-29-40-MO	ARC	44-86259 Converted Tinker 1/50
	44-86355 SB-29-50-MO	ARC	44-86355 Converted Yokota 11/50
	23 Conversions identified by R. Mann		*25 conversions from combined A. Lloyd List and Mann List confirmed by Record Cards*

*44-61671 Not listed on 'Active in 1951' ARC microfilm

†44-84098, 44-87761 were not converted to SB-29 per ARC

13 of 16 planes in above list confirmed by ARC

WB-29 WEATHER RECONNAISSANCE CONFIGURATION

Because the Air Force wanted to retain a bombardment capability in all its air crews, on November 26, 1946, they assigned the weather reconnaissance squadrons a secondary mission of bombardment.

The weather squadrons of the time were flying B/RB-29s fully equipped, with one exception, for the bombardment role. Because the recon units were sometimes asked to fly nuclear sampling missions, airborne particulate filter assemblies were installed at the location of the lower aft turret.

This requirement persisted until mid-1948. On July 5th of that year the 514th Weather Recon on Guam joined with the 19th Bomb Group on a practice mission to "bomb" Tokyo.

Ten days later, on July 15th, the Air Weather Service was relieved of its bombardment responsibility.

Although relieved of the bombardment requirement, the weather squadrons were still flying B/RB-29 aircraft and would continue to do so until 1950, when airplanes specifically configured for weather reconnaissance started to reach the squadrons from the stateside depots.

The new airplanes had all armament and armament-related equipment removed. The radar position was moved to the forward cabin, sharing a table along the left side of the fuselage with the navigator. The radio position, in turn, was moved from the right-rear corner of the forward compartment to the rear pressurized compartment.

A dropsonde chamber was installed in the floor of the rear pressurized compartment, and the nuclear particulate filtering assembly, known to the crews as the "bug-catcher" or "crackerbox," was mounted on the top of the fuselage at the point where the CFC gunner's blister had been on the armed ships. The movement of these two pieces of equipment to the interior of the rear compartment meant that the dropsondes could be deployed and the filter pads changed without depressurizing the aircraft.

While each weather track had precise points where the filters were to be changed, the general rule was that new filters would be installed at every change in course or altitude, or every two hours. On the B/RB-29s, whenever the track was flown at altitude, a filter change required depressurization of the plane with the entire crew going on oxygen and the dropsonde operator climbing out into the rear unpressurized compartment with a clip-on walk-around oxygen bottle. Once he had the new filters in place and the airman returned to the rear cabin, repressurization was started. The tedious process would begin again in another hour and a half.

Specialized weather equipment installed for use by the weather observer included a MIL-313 psychrometer utilizing wet and dry bulb thermometers to measure water vapor in the air; an ID/AMQ-2 aerograph for determining relative humidity; an SCR-718 radio altimeter; an AN/APN radar altimeter; and a number of AN/AMT radiosonde (dropsonde) units. These units were dropped from the plane and contained instruments that recorded, and then transmitted back to the plane, temperature, humidity and pressure data.

The crew of the WB-29s consisted of the weather observer, two pilots, flight engineer, navigator and radar observer in the forward compartment. In the rear compartment, the dropsonde operator also functioned as the left scanner, while the crew chief occupied the right blister position. Also in the rear compartment was the radio operator and the radiosonde operator, who transcibed the weather data being transmitted by the dropsonde.

Known WB-29 conversions are listed by serial number below. In addition to the serial number, columns list the model, factory where built and air weather service assignments.

The location of the bases cited, and the units assigned to them, are as follows:

Andersen AFB. Located at the northeast end of Guam. Formerly North Guam AFB, North Field. 514th Reconnaissance Squadron (VLR) Weather redesignated 54th Strategic Reconnaissance Squadron (Medium) Weather.

Eielson AFB. Located 26 miles southeast of Fairbanks, Alaska. 375th Reconnaissance Squadron (VLR) Weather.

Elmendorf AFB. Located just outside Anchorage, Alaska.

Fairfield AFB. Full name was Fairfield-Suisun AFB, located at Fairfield, California. Name later changed to Travis AFB. 374th Reconnaissance Squadron (VLR) Weather; Flight "B" 375th Reconnaissance Squadron (VLR) Weather, Detached Service.

Hickam AFB. Shares runways with Honolulu International Airport. 57th Strategic Reconnaissance Squadron (Medium) Weather.

Kindley AFB. Located on Bermuda. 3753rd Reconnaissance Squadron (VLR) Weather, redesignated 53rd Strategic Reconnaissance Squadron (Medium) Weather.

Kwajalein. Located in the Caroline Islands between Guam and Hawaii. Detachment of the 57th Strategic Reconnaissance Squadron (Medium) Weather.

Ladd AFB. Adjacent to Fairbanks, Alaska. Flight "A" 375th Reconnaissance Squadron (VLR) Weather, Detached Service.

McClellan AFB. Located just northeast of Sacramento. Air Materiel Depot. 55th Strategic Reconnaissance Squadron (Medium) Weather.

Misawa Air Base. Located at the northern end of Honshu, Japan. 512th Reconnaissance Squadron (VLR) Weather redesignated 56th Strategic Reconnaissance Squadron (Medium) Weather.

Morrison. Morrison Field, West Palm Beach, Florida. 54th Reconnaissance Squadron (VLR) Weather.

North Field. Located on northeast end of Guam. Later became North Guam AFB, Andersen AFB. 54th Reconnaissance Squadron (VLR) Weather redesignated 514th Reconnaissance Squadron (VLR) Weather.

Tinker AFB. Located in area of Oklahoma City. Air Materiel Depot. 513th Reconnaissance Squadron (VLR) Weather.

Yokota Air Base. Yokota is located outside Tachikawa, 30-35 miles west of Tokyo. 512th Reconnaissance Squadron (VLR) Weather redesignated 56th Strategic Reconnaissance Squadron (Medium) Weather.

Serial #	Model	Factory	Assignments
42-63459	WB-29-20-BA	Bell Atlanta	No Details Available
42-65281	WB-29-55-MO	Martin Omaha	No Details Available
42-93951	WB-29A-15-BN	Boeing Renton	Kindley, Yokota
42-93975	WB-29A-20-BN	Boeing Renton	Andersen, McClellan, Kindley
42-93980	WB-29A-20-BN	Boeing Renton	Andersen, Kindley
42-94040	WB-29A-25-BN	Boeing Renton	Kindley (three tours)
42-94047	WB-29A-25-BN	Boeing Renton	Kindley, Yokota
44-27269	WB-29-35-MO	Martin Omaha	Hickam (two tours)
44-27271	WB-29-35-MO	Martin Omaha	McClellan Hickam
44-27321	WB-29-35-MO	Martin Omaha	Ladd, Eielsen (two tours)
44-27324	WB-29-35-MO	Martin Omaha	Eielsen, McClellan (two tours)
44-27335	WB-29-40-MO	Martin Omaha	Fairfield, Hickam (two tours)
44-27337	WB-29-40-MO	Martin Omaha	McClellan, Tinker
44-27343	WB-29-40-MO	Martin Omaha	Fairfield, McClellan, Tinker, Hickam (two tours)
44-27344	WB-29-40-MO	Martin Omaha	Fairfield, McClellan, Hickam
44-61540	WB-29A-35-BN	Boeing Renton	No Details Available
44-61556	WB-29A-35-BN	Boeing Renton	Andersen (three tours)
44-61600	WB-29A-35-BN	Boeing Renton	Yokota, Andersen (two tours), England
44-61640	WB-29A-40-BN	Boeing Renton	Andersen, lost on weather mission 2/27/52
44-61697	WB-29A-40-BN	Boeing Renton	Morrison, North Field, Fairfield
44-61710	WB-29A-45-BN	Boeing Renton	Kindley (two tours)
44-61728	WB-29A-45-BN	Boeing Renton	No Details Available
44-61734	WB-29A-45-BN	Boeing Renton	Fairfield, Andersen, Kindley
44-61738	WB-29A-45-BN	Boeing Renton	Fairfield (two tours), Tinker
44-61776	WB-29A-45-BN	Boeing Renton	Kindley
44-61788	WB-29A-45-BN	Boeing Renton	No Details Available
44-61792	WB-29A-45-BN	Boeing Renton	No Details Available
44-61800	WB-29A-45-BN	Boeing Renton	Fairfield
44-61808	WB-29A-50-BN	Boeing Renton	Morrison
44-61882	WB-29A-50-BN	Boeing Renton	Morrison, Fairfield
44-61885	WB-29A-50-BN	Boeing Renton	Yokota (two tours), Misawa
44-61888	WB-29A-50-BN	Boeing Renton	Andersen (three tours)
44-61891	WB-29A-50-BN	Boeing Renton	Morrison, North Field, Kindley, Fairfield
44-61893	WB-29A-50-BN	Boeing Renton	Yokota (two tours), Misawa
44-61901	WB-29A-50-BN	Boeing Renton	Yokota, Andersen
44-61909	WB-29A-50-BN	Boeing Renton	Andersen (two tours)
44-61964	WB-29A-55-BN	Boeing Renton	Andersen (two tours), McClellan, Andersen
44-61974	WB-29A-55-BN	Boeing Renton	Kindley (two tours)
44-62034	WB-29A-60-BN	Boeing Renton	Yokota, Misawa
44-62061	WB-29A-60-BN	Boeing Renton	Yokota (two tours), Misawa
44-62077	WB-29A-60-BN	Boeing Renton	Morrison, Fairfield, Tinker, McClellan (3 tours), Hickam
44-62079	WB-29A-60-BN	Boeing Renton	Kindley, Andersen
44-62080	WB-29A-60-BN	Boeing Renton	Kindley (two tours), McClellan, Andersen
44-62089	WB-29A-60-BN	Boeing Renton	Yokota, Misawa, Kindley
44-62090	WB-29A-60-BN	Boeing Renton	Yokota, Andersen, Kindley
44-62094	WB-29A-60-BN	Boeing Renton	Andersen, (three tours)
44-62125	WB-29A-65-BN	Boeing Renton	Tinker, Yokota, Eielsen
44-62126	WB-29A-65-BN	Boeing Renton	No Details Available
44-62128	WB-29A-65-BN	Boeing Renton	No Details Available
44-62151	WB-29A-65-BN	Boeing Renton	Ladd, Eiellsen (two tours)
44-62155	WB-29A-65-BN	Boeing Renton	No Details Available
44-62156	WB-29A-65-BN	Boeing Renton	Ladd, Eielsen, Hickam
44-62157	WB-29A-65-BN	Boeing Renton	No Details Available
44-62163	WB-29A-65-BN	Boeing Renton	Eielsen (two tours), Yokota, Hickam
44-62195	WB-29A-65-BN	Boeing Renton	Ladd, Eielsen, Hickam
44-62197	WB-29A-65-BN	Boeing Renton	Fairfield, Eielsen (two tours)
44-62200	WB-29A-65-BN	Boeing Renton	No Details Available
44-62201	WB-29A-65-BN	Boeing Renton	No Details Available
44-62202	WB-29A-65-BN	Boeing Renton	Fairfield, Hickam (two tours)
44-62214	WB-29A-70-BN	Boeing Renton	Fairfield, Ladd, Eielson (two tours)
44-62216	WB-29A-70-BN	Boeing Renton	No Details Available
44-62219	WB-29A-70-BN	Boeing Renton	Elmendorf, Eielson (two tours) Yokota
44-62220	WB-29A-70-BN	Boeing Renton	Kindley, Andersen, Kwajalein, Hickam (two tours)
44-62225	WB-29A-70-BN	Boeing Renton	Kindley, Eielesen
44-62229	WB-29A-70-BN	Boeing Renton	Kindley, Andersen, Alaska (three tours)

V. B-29 Variants

Serial #	Model	Factory	Assignments
44-62232	WB-29A-70-BN	Boeing Renton	Fairfield, Elmendorf, Eielson (three tours)
44-62233	WB-29A-70-BN	Boeing Renton	No Details Available
44-62273	WB-29A-70-BN	Boeing Renton	Yokota, Misawa
44-62277	WB-29A-70-BN	Boeing Renton	Yokota, Misawa, Kindley
44-62297	WB-29A-70-BN	Boeing Renton	Fairfield
44-62306	WB-29A-70-BN	Boeing Renton	No Details Available
44-69676	WB-29-55-BW	Boeing Wichita	Kindley
44-69769	WB-29-60-BN	Boeing Renton	Kindley (three tours)
44-69770	WB-29-60-BN	Boeing Renton	Andersen, lost in typhoon 10/27/52
44-69791	WB-29-60-BN	Boeing Renton	Kindley
44-69987	WB-29-60-BN	Boeing Renton	Kindley
44-70153	WB-29-70-BW	Boeing Wichita	No Details Available
44-86267	WB-29-40-MO	Martin Omaha	Andersen (three tours)
44-86269	WB-29-40-MO	Martin Omaha	Fairfield, Hickam, McClellan (two tours)
44-86379	WB-29-55-MO	Martin Omaha	Fairfield, McClellan, Yokota
44-86399	WB-29-55-MO	Martin Omaha	Kindley (two tours), Andersen, Hickam (three tours) Yokota, McClellan
44-86450	WB-29-60-MO	Martin Omaha	Andersen (two tours), Yokota (three tours)
44-87740	WB-29-90-BW	Boeing Wichita	No Details Available
44-87744	WB-29-90-BW	Boeing Wichita	No Details Available
44-87747	WB-29-90-BW	Boeing Wichita	Fasirfield, Andersen, Kindley (two tours), Yokota
44-87750	WB-29-90-BW	Boeing Wichita	No Details Available
44-87756	WB-29-90-BW	Boeing Wichita	Fairfield, Andersen, McClellan. Crashed 4/5/52
44-87762	WB-29-90-BW	Boeing Wichita	Kindley, Andersen, Eielson (three tours)
45-21717	WB-29-90-BW	Boeing Wichita	Tinker, Eielsen (two tours)
45-21740	WB-29-90-BW	Boeing Wichita	Fairfield, Andersen, Eielsen, Hickam
45-21767	WB-29-95-BW	Boeing Wichita	Fairfield, McClellan, Eielson (two tours), Hickam
45-21774	WB-29-95-BW	Boeing Wichita	McClellan (two tours), Hickam
45-21813	WB-29-95BW	Boeing Wichita	Yokota, Misawa
45-21816	WB-29-95BW	Boeing Wichita	Kindley, Tinker, Hickam, Eielson
45-21817	WB-29-95BW	Boeing Wichita	Andersen (three tours)
45-21819	WB-29-95BW	Boeing Wichita	Tinker, Hickam (two tours), Eielson
45-21823	WB-29-95BW	Boeing Wichita	Fairfield, McClellan, Yokota, Andersen
45-21824	WB-29-95BW	Boeing Wichita	Andersen (two tours) Yokota
45-21826	WB-29-95BW	Boeing Wichita	Fairfield, McClellan, Andersen
45-21831	WB-29-95BW	Boeing Wichita	Fairfield, Andersen (two tours) Kindley, Yokota (two tours)
45-21833	WB-29-95BW	Boeing Wichita	Kindley, Andersen (two tours), Yokota
45-21838	WB-29-95BW	Boeing Wichita	Andersen, McClellan (two tours), Hickam (two tours)
45-21841	WB-29-95BW	Boeing Wichita	Andersen, Kindley (two tours), Hickam (two tours)
45-21872	WB-29-100-BW	Boeing Wichita	Kindley, Andersen, McClellan, Hickam, Alaska

VI

THE XX BOMBER COMMAND

The XX Bomber Command, consisting of the four B-29 groups — the 40th, 444th, 462nd and the 468th — of the 58th Bomb Wing, conducted forty-nine missions in the China-Burma-India Theater before flying to Tinian in May 1945 and becoming a part of the XXI Bomber Command.

The XX BC was assigned to the Twentieth Air Force, command of which resided in Washington, D.C., in the person of General H.H. "Hap" Arnold, the Army Air Force chief of staff. This was a deliberate move on Arnold's part to keep operational control out of the hands of General Claire Chennault's Fourteenth Air Force. Arnold was afraid that Chennault would waste the B-29s long range capability and assign them targets that were within the range of his B-24s.

Chennault was not the only one who wanted a piece of the B-29 pie. In March General Brereton wanted control of all B-29s sent to India; General Harmon wanted some in the South Pacific; in April the Antisubmarine Command tried to get some. General Kenny flew to Washington in April, wanting B-29s in Australia. MacArthur wanted all the B-29s in his Southwest Pacific Theater. Lord Mountbatten, Supreme Allied Commander, Southeast Asia, wanted some under his thumb, and last but not least, Chiang Kai-Shek wanted the B-29s under his command. Finally, the Twentieth Air Force was activated April 4, 1944, with General Arnold in command.

Four bases in southern Bengal, built by the British to handle B-24 Liberators, were earmarked for the B-29s. These fields, at Kharagapur, Chakulia, Piardoba and Dudhkundi, west of Calcutta, were brought up to B-29 standards using native hand labor.

Forward bases were required in China for operations against the Japanese Home Islands. The first problem that arose was the price tag. Generalissimo Chiang Kai-shek set the cost at more than $2 billion Chinese dollars. At the exchange rate set by the Chinese government, this translated to U.S. $200 million.

While base construction proceeded in India and China, serious airplane problems were arising back in the States. Service tests in the field were still being conducted while the factories at Wichita and Atlanta were turning out production models. Too often, the production models did not receive the modifications required to fix problems found in the service testing, and many remained grounded for mechanical reasons. This, in turn, had a major impact on crew training.

General Arnold decided that the XX Bomber Command would need 175 airplanes. He went to Wichita, found the 175th plane, and told the Wichita executives that he wanted that plane on March 1. He got it on February 28. It didn't help much; the first departures for India, scheduled for March 9, didn't get off the ground for mechanical reasons.

One solution was to fly a crew to a modification center, assign them an airplane and have them ferry the plane to whichever other modification center had fixes for them. It took weeks of long hours, and a lot of repairs accomplished outdoors in atrocious weather, but eventually the planes were ready. Departure started in the end of March with a flight plan of Kansas–Presque Isle–Gander–Newfoundland–Marrakech–French Morroco–Cairo–Karachi–Calcutta.

The heat of India almost did in the B-29s again. Engine fires became almost commonplace, and field modifications, including cowl flap modifications and a crossover oil feed line between the intake and exhaust cylinder rocker boxes, were rushed into implementation. Engines were not allowed to idle in long takeoff lines but were shut down and started only to move the plane.

These problems were compounded by overweight takeoffs when the planes were loaded with supplies for the bases in China. The Himalayan Mountains had to be crossed; there was no effective way around them. "The Hump," as the overmountain route was known, soon acquired a second name,

VI. The XX Bomber Command

"The Aluminum Trail," for the crashed planes all over the mountains. Conventional wisdom assumed that five B-29 trips over the Hump were required to deliver enough gas and bombs to prepare a single B-29 for a mission against Japan — and those missions were coming.

Despite the problems, the XX Bomber Command's first mission was launched on June 5, 1944, when all four groups sent planes to attack the Makasan Rail Yards in Bangkok, Thailand. It was not a textbook demonstration. Of the ninety-eight planes airborne, fourteen aborted before reaching the target and seven more did not bomb the primary. Overall accuracy of bombs dropped on the target was poor to moderate. Five planes were lost to various causes, twelve landed at the wrong B-29 bases and thirty more landed at other fields.

There was intense political pressure from Washington to mount a mission against Japan. General Arnold insisted on at least 70 planes being involved, and on June 15, sixty-eight were airborne for a night attack on the Imperial Iron and Steel Works at Yawata, on the island of Kyushu. Again, not very effective. Only fifty-one planes bombed the primary target and only a single bomb hit the target area.

But they learned, and in the days that followed the ratio of planes up to planes bombing slowly increased; losses, with the exception of the daylight return to Yawata on August 20, when fifteen planes did not return, slowly decreased. The last raid against the Japanese Home Islands was on January 6 against Omura. After that date XX Bomber Command targets were almost exclusively in the Southeast Asia region.

MISSION LIST

The data presented in the mission list need very little explanation. The columns of data are defined as follows:

MSN. This is the XX Bomber Command mission number. (Note: Because missions were sometimes postponed, they were not always flown in numbered order. The list below organizes the missions chronologically, so some numbers will appear out of order.)

Date. Date of mission.

Primary Target. Name of the target.

Groups. Groups participating in the attack.

No. Up. Refers to the number of aircraft airborne and attacking. A notation of "in above" signifies that the number of planes attacking that particular target is included in the group total cited just above.

Bombed. Number of aircraft attacking the primary target.

Lost. Number of aircraft lost on mission. Includes aircraft surveyed or salvaged.

MSN	Date	Primary Target	Groups	No. Up	Bombed	Lost
1	6/5	Makasan Rail Yards, Bangkok, Thailand	40/444/462/468	98	77	5
2	June 15/16	Imperial Iron & Steel Works, Yawata, Japan	40/444/462/468	68	47	7
3	7/7	Sasebo Shipyard, Nagasaki, Omura & Yawata	444/468	18	14	0
4	7/29	Showa Metal Factory, Anshan, Manchuria	40/444/462/468	96	75	3
5	August 10/11	Baraban Oil Refineries/Moesi River, Palembang	444/468	54	39	2
6	August 10/11	Nagasaki	444/468	29	24	1
7	8/20	Yawata Metal Factory, Yawata, Japan	40/444/462/468	88	71	14
8	9/8	Showa Metal Factory, Anshan, Manchuria	40/444/462/468	88	72	3
9	9/26	Showa Metal Factory, Anshan, Manchuria	40/444/462/468	109	83	0
10	10/14	Takao NAS, Okayama, Formosa	40/444/462/468	131	106	2
11	10/16	Okayama, Formosa	444/462	49	38	0
11	10/16	Heito, Formosa	468	24	24	
12	10/17	Einansho Airfield, Tainan, Formosa	40	30	10	
13	10/25	Omura	40/444/462/468	75	58	0
14	11/3	Malagon Railroad Yards, Rangoon	40/444/462/468	50	44	1
15	11/5	Singapore	40/444/462/468	74	53	3
16	11/11	Omura Aircraft Factory	40/444/462/468	93	29	2
17	11/21	Omura Aircraft Factory	40/444/462/468	96	63	2
18	11/27	Bangkok	40/444/462/468	60	55	5
19	12/7	Manchuria Airplance Co., Mukden	40/444/462/468	73	40	9
20	12/14	Rama VI Railroad Bridge, Bangkok	40/444/462/468	45	33	1
21	12/18	Hankow, China	40/444/462/468	94	85	6
22	12/19	Omura	40/444/462/468	36	17	5
23	12/21	Manchuria Airplane Company, Mukden	40/444/462/468	55	40	0
24	1/2	Rama VI Railroad Bridge, Bangkok	40/444/462/468	49	44	2
25	1/6	Omura	40/444/462/468	48	29	2
26	1/9	Formosa	40/444/462/468	46	40	0
27	1/11	Mission Totals — Two Primaries	40/444/462/468	43	27	0
27	1/11	Floating Dry Dock, Singapore	40/444/462/468	In above	16	0
27	1/11	King's Dry Dock, Singapore	40/444/462/468	In above	11	0
28	1/14	Kagi Airbase, Formosa	40/444/462/468	82	54	0

29	1/17	Shinchiku, Formosa	40/444/462/468	90	78	2
30	January 25/26	Mine-laying Operations, Indo-China Area	462	26	25	0
31	January 25/26	Mine-laying Operations, Singapore Area	444/462	45	41	1
32	1/27	Saigon Navy Yard and Arsenal	40	25	22	0
33	2/1	Singapore Floating Drydock & Drydock Gates	40/444/462/468	104	78	0
34	2/7	Saigon Navy Yard	444/462	66	44	0
34	2/7	Phnom Penn	444/462		17	0
35	2/7	Rama VI Railroad Bridge, Bangkok	40/468	65	59	2
36	2/11	Rangoon	40/444/462/468	60	56	1
37	2/19	Kuala Lumpur, Malaya	444/468	58	48	0
38	2/24	Empire Dock, Singapore	40/444/462/468	117	105	0
40	February 27/28	Johore Strait, Singapore Mining Mission	444	12	10	0
41	3/2	Singapore Naval Base	40/444/462/468	62	48	1
39	March 4/5	Mining Yangtze River Near Shanghai	468	30	24	0
43	3/10	RR Yards, Kuala Lumpur, Malaya	468	29	23	2
42	3/12	Samboe Island Oil Storage, Singapore	40	15	11	0
42	3/12	Bukum Island Oil Storage, Singapore	444	30	21	0
42	3/12	Sebarok Island Oil Storage, Singapore	462	15	11	0
44	3/17	Rangoon	40/444/462/468	78	70	0
45	3/22	Rangoon	40/468	39	39	0
45	3/22	Mingaladon	444/462	30	28	0
46	March 28/29	Mine Lower Yangtze River	468	10	10	0
47	March 28/29	Mine Saigon Area and Cam Ranh Bay	468	18	16	0
48	March 28/29	Mine Johore Channel and Rhio Straits, Singapore	444	33	32	0
49	March 29/30	Bukum Island Oil Storage, Singapore	40/462	26	24	0

MISSION DETAILS

MISSION NUMBER 1

5 June 1944 • Target: Bangkok, Thailand

	40BG	444BG	462BG	468BG	Total
Planes Up	Unknown	Unknown	Unknown	Unknown	98
Bombed Primary	Unknown	9	Unknown	Unknown	77
Bombed Other	Unknown	Unknown	Unknown	Unknown	Unknown
Did Not Bomb	Unknown	Unknown	Unknown	Unknown	20
Ground Abort	Unknown	Unknown	Unknown	Unknown	Unknown
Lost	42-6282 42-6304 42-6318	42-6361	42-6336	None	5

MISSION NUMBER 2

15/16 June 1944 • Target: Imperial Iron & Steel Works, Yawata, Kyushu, Japan

	40BG	444BG	462BG	468BG	Total
Planes Up	Unknown	Unknown	Unknown	18	68
Bombed Primary	Unknown	11	Unknown	Unknown	47
Bombed Other	Unknown	Unknown	Unknown	Unknown	Unknown
Did Not Bomb	Unknown	Unknown	Unknown	Unknown	11
Ground Abort	Unknown	Unknown	Unknown	Unknown	Unknown
Lost	42-6261	42-6220 42-6293	None	42-6229 42-6230 42-6231 42-93826	7

MISSION NUMBER 3

7/8 July 1944 • Target: Sasebo, Omura, Tobata and Yawata, Kyushu, Japan

	40BG	444BG	462BG	468BG	Total
Planes Up		Unknown		Unknown	18
Bombed Primary		5		9	14
Bombed Other		Unknown		Unknown	Unknown
Did Not Bomb		Unknown		Unknown	4
Ground Abort		Unknown		Unknown	Unknown
Lost		None		None	0

MISSION NUMBER 4

29 July 1944 • Target: Showa Steel Works, Anshan, Manchuria

	40BG	444BG	462BG	468BG	Total
Planes Up	Unknown	Unknown	Unknown	Unknown	96
Bombed Primary	Unknown	15	Unknown	Unknown	75
Bombed Other	Unknown	Unknown	Unknown	Unknown	Unknown
Did Not Bomb	Unknown	Unknown	Unknown	Unknown	21
Ground Abort	Unknown	Unknown	Unknown	Unknown	Unknown
Lost	42-6351	None	42-6526	42-6274	3

MISSION NUMBER 5

10/11 August 1944 • Target: Barabon Oil Refinery & Mining of Moesi River, Palembang, Borneo

	40BG	444BG	462BG	468BG	Total
Planes Up		Unknown		Unknown	54
Bombed Primary		9		30	39
Bombed Other		Unknown		Unknown	Unknown
Did Not Bomb		Unknown		Unknown	21
Ground Abort		Unknown		Unknown	Unknown
Lost		42-24420		42-6243	2

MISSION NUMBER 6

10/11 August 1944 • Target: Nagasaki, Kyushu, Japan, Diversion for Palembang Mission, Palembang, Borneo

	40BG	444BG	462BG	468BG	Total
Planes Up		Unknown		Unknown	29
Bombed Primary		7		17	24
Bombed Other		Unknown		Unknown	Unknown
Did Not Bomb		Unknown		Unknown	Unknown
Ground Abort		Unknown		Unknown	Unknown
Lost		None		1	1

MISSION NUMBER 7

20 August 1944 • Target: Imperial Iron & Steel Works, Yawata, Kyushu, Japan

	40BG	444BG	462BG	468BG	Total
Planes Up	Unknown	Unknown	Unknown	24	88
Bombed Primary	Unknown	14	Unknown	Unknown	71

(cont'd)

	40BG	444BG	462BG	468BG	Total
Bombed Other	Unknown	Unknown	Unknown	Unknown	Unknown
Did Not Bomb	Unknown	Unknown	Unknown	Unknown	15
Ground Abort	Unknown	Unknown	Unknown	Unknown	Unknown
Lost	42-6301	42-6286	42-6305	42-6264	14
	42-6308	42-6320	42-6392	42-6334	
	42-6425	42-6330	42-24474	42-6368	
	42-93829			42-6408	

MISSION NUMBER 8

8 September 1944 • Target: Showa Metal Factory, Anshan, Manchuria

	40BG	444BG	462BG	468BG	Total
Planes Up	Unknown	Unknown	Unknown	Unknown	88
Bombed Primary	Unknown	Unknown	Unknown	Unknown	72
Bombed Other	Unknown	Unknown	Unknown	Unknown	Unknown
Did Not Bomb	Unknown	Unknown	Unknown	Unknown	15
Ground Abort	Unknown	Unknown	Unknown	Unknown	Unknown
Lost	None	42-6212	42-6360	None	3
		42-6234			

MISSION NUMBER 9

26 September 1944 • Target: Showa Metal Factory, Anshan, Manchuria

	40BG	444BG	462BG	468BG	Total
Planes Up	Unknown	Unknown	Unknown	Unknown	109
Bombed Primary	Unknown	26	Unknown	Unknown	83
Bombed Other	Unknown	Unknown	Unknown	Unknown	Unknown
Did Not Bomb	Unknown	Unknown	Unknown	Unknown	24
Ground Abort	Unknown	Unknown	Unknown	Unknown	Unknown
Lost	None	None	None	None	0

MISSION NUMBER 10

14 October 1944 • Target: Takao NAS, Okayama, Formosa

	40BG	444BG	462BG	468BG	Total
Bombed Primary	42-6237	42-6215	42-6209	42-6217	106
	42-6269	42-6225	42-6270	42-6265	
	42-6275	42-6251	42-6278	42-6362	
	42-6276	42-6262	42-6285	42-6370	
	42-6281	42-6267	42-6299	42-6389	
	42-6290	42-6292	42-6311	42-6390	
	42-6294	42-6300	42-6312	42-6397	
	42-6295	42-6307	42-6316	42-6409	
	42-6298	42-6324	42-6329	42-6411	
	42-6303	42-6340	42-6338	42-24429	
	42-6313	42-6341	42-6346	42-24442	
	42-6319	42-6352	42-6347	42-24469	
	42-6322	42-6353	42-6382	42-24471	
	42-6331	42-6399	42-6444	42-24486	
	42-6342	42-6423	42-24461	42-24494	
	42-6344	42-24462	42-24463	42-24504	
	42-6348	42-24472	42-24475	42-24525	
	42-6418	42-24485	42-24479	42-24576	
	42-24452	42-24506	42-24484	42-63353	
	42-24466	42-24538	42-24505	42-63354	

(cont'd)

VI. The XX Bomber Command

	40BG	444BG	462BG	468BG	Total
		42-24503	42-24580	42-24506	42-63395
		42-24508	42-63360	42-24531	42-65208
		42-24541	42-63375	42-24581	42-93828
		42-24582	42-63378	42-63362	
		42-24587	42-63403	42-63393	
		42-24589	42-65202	42-93827	
		42-63363	42-65204	42-93830	
		42-63396			
		42-93831			
Bombed Other		42-6276	42-24510	42-63386	42-6365	12
		42-24457	42-24524		42-6407
		42-24522			42-6454
					42-24487
					42-63355
					42-63356
Did Not Bomb		42-6297	42-6343	42-6273	42-6272	12
		42-24579	42-24464		42-6279
			42-24494		42-6284
			42-24507		42-6358
					42-6411
Ground Abort			42-6321	42-24456	42-24542	3
Lost		42-24513	42-6280			2

MISSION NUMBER 11

16 October 1944 • Target: Okayama and Heiti, Formosa

	40BG	444BG	462BG	468BG	Total
Bombed Primary		Okayama	Okayama	Okayama	Okayama
		42-6225	42-6209	42-6265	31
		42-6251	42-6213	42-6411	
		42-6267	42-6270	42-24442	Heito
		42-6340	42-6311	42-24504	25
		42-24518	42-6329	42-63395	
		42-24580	42-6338		
		42-63403	42-6346	Heito	
		42-65202	42-6347	42-6279	
			42-6444	42-6362	
		Heito	42-24456	42-6389	
		42-6251	42-24461	42-6409	
		42-6252	42-24463	42-24429	
		42-6292	42-24475	42-24469	
		42-6324	42-24479	42-24484	
		42-6353	42-24484	42-24486	
		42-6399	42-24505	42-24525	
		42-24462	42-24506	42-24542	
		42-24464	42-63393	42-24546	
		42-24485		42-63354	
		42-24538		42-63355	
		42-63360		42-65208	
Bombed Other		42-6262	42-6285	42-6390	11
		42-6300	42-6382		
		42-6307	42-24581		
		42-24472	42-63362		
		42-65204	42-63386		
Did Not Bomb		42-24492		42-6217	3
				42-63353	
Ground Abort					0
Lost					0

MISSION NUMBER 12

17 October 1944 • Target: Einansho Airfield, Tainan, Formosa

	40BG	444BG	462BG	468BG	Total
Bombed Primary	42-6237				10

(cont'd)

Bombed Other	42-6269	14
	42-6276	
	42-6294	
	42-6298	
	42-6331	
	42-6344	
	42-24579	
	42-24589	
	42-63396	
	42-6275	
	42-6290	
	42-6295	
	42-6303	
	42-6313	
	42-6319	
	42-6322	
	42-6418	
	42-24452	
	42-24466	
	42-24522	
	42-24582	
	42-24587	
	42-63363	
Did Not Bomb	42-6281	5
	42-6297	
	42-24503	
	42-24541	
	42-63831	
Lost	42-6342	5

MISSION NUMBER 13

25 October 1944 • Target: Omura, Kyushu, Japan,

	40BG	444BG	462BG	468BG	Total
Bombed Primary	42-6237	42-6215	42-6278	42-6284	56-0
	42-6276	42-6225	42-6316	42-6362	
	42-6281	42-6251	42-6329	42-6365	
	42-6306	42-6267	42-6347	42-6397	
	42-6313	42-6300	42-6382	42-6407	
	42-6418	42-6307	42-24463	42-6454	
	42-24457	42-6324	42-24479	42-24486	
	42-24508	42-6340	42-24484	42-24487	
	42-24522	42-6352	42-24505	42-24546	
	42-63407	42-6353	42-24506	42-63353	
	42-93831	42-6423	42-63393	42-63354	
		42-24462	42-93827	42-63355	
		42-24472	42-93830	42-63395	
		42-24485		42-63424	
		42-24518			
		42-24538			
		42-63378			
		42-63492			
Bombed Other	42-24579	42-24492	42-6213	42-6272	12
		42-24507	42-6338		
		42-63411	42-24456		
		42-63419	42-24462		
		42-63422			
		42-65202			
Did Not Bomb	42-6331		42-6311	42-6358	6
	42-24589		42-63386	42-6370	
Lost	None	None	None	None	0

VI. The XX Bomber Command

MISSION NUMBER 14

3 November 1944 • Target: Malagon Railroad Yards, Rangoon, Burma

	40BG	444BG	462BG	468BG	Total
Bombed Primary	42-24452	42-24485	42-6213	42-6279	56
	42-24503	42-24492	42-6329	42-24429	
	42-24508	42-24507	42-6444	42-24442	
	42-24541	42-24510	42-24461	42-24469	
	42-24574	42-24524	42-24463	42-24486	
	42-24579	42-24580	42-24475	42-24494	
	42-24582	42-24584	42-24479	42-24525	
	42-63394	42-63375	42-24505	42-24542	
	42-63396	42-63411	42-24506	42-24546	
	42-63407	42-63419	42-24531	42-63353	
			42-24581	42-63415	
			42-63393	42-65208	
Bombed Other	42-63407	42-65202			2
Did Not Bomb	42-24457	42-24515			4
	42-24466				
	42-24522				
Lost	None	42-24518	None	None	1

MISSION NUMBER 15

5 November 1944 • Target: Singapore, Malaya

	40BG	444BG	462BG	468BG	Total
Bombed Primary	42-6237	42-6300	42-6213	42-6265	53
	42-6269	42-6352	42-6270	42-6279	
	42-6276	42-6353	42-6278	42-6362	
	42-6290	42-24462	42-6338	42-6397	
	42-6294	42-24472	42-6382	42-6454	
	42-6298	42-24485	42-24456	42-24486	
	42-6313	42-24492	42-24475	42-24487	
	42-6322	42-24507	42-24479	42-24494	
	42-6331	42-24538	42-24506	42-24542	
	42-24503	42-24584	42-24531	42-63354	
	42-24574	42-63411	42-24581	42-63355	
	42-24587	42-63419	42-63393	42-63417	
	42-24589	42-65204		42-63424	
	42-93831			42-65208	
Bombed Other	42-24508	42-6267	42-6329	42-6284	11
	42-24582	42-24524	42-93830	42-6365	
	42-63394	42-24580		42-24429	
Did Not Bomb		42-24464	42-6285	42-6208	10
		42-24510	42-6299	42-6407	
		42-65202	42-24461		
			42-24463		
			42-63386		
Lost	None	None	None	None	0

MISSION NUMBER 16

11 November 1944 • Target: Omura Aircraft Factory, Omura, Kyushu, Japan

	40BG	444BG	462BG	468BG	Total
Bombed Primary	42-6276	42-6225	42-6270	42-6217	30
	42-6344	42-6262	42-6312	42-6279	

(cont'd)

	40BG	444BG	462BG	468BG	Total
		42-6300	42-24463	42-6358	
		42-6321	42-24506	42-6365	
		42-6353	42-65213	42-6390	
		42-6399		42-24469	
		42-24462		42-24486	
		42-24485		42-24494	
		42-24510		42-24525	
		42-63422		42-63353	
				42-63354	
				42-63424	
				42-65208	
Bombed Other	42-6237	42-6251	42-6311	42-6272	50
	42-6290	42-6324	42-24456	42-6265	
	42-6292	42-6352	42-24475	42-6272	
	42-6294	42-6390	42-93830	42-6284	
	42-6298	42-24472		42-6362	
	42-6303	42-24492		42-6397	
	42-6306	42-24505		42-6407	
	42-6313	42-24507		42-6409	
	42-6322	42-63378		42-24429	
	42-6331	42-63411		42-24487	
	42-24457	42-65204		42-24525	
	42-24466			42-24546	
	42-24503			42-63395	
	42-24508			42-63415	
	42-24522				
	42-24574				
	42-24582				
	42-24589				
	42-63394				
	42-63396				
	42-63407				
Did Not Bomb	42-6297	42-6267	42-6231	42-6208	12
	63-6319	42-6340	42-24461	42-24542	
	42-93831	42-24538	42-24505		
			42-24581		
Lost	42-6237	42-6300			2

MISSION NUMBER 17

21 November 1944 • **Target: Omura Aircraft Factory, Omura, Kyushu, Japan**

	40BG	444BG	462BG	468BG	Total
Bombed Primary	42-6275	42-6262	42-6338	42-6217	63-2
	42-6276	42-6267	42-6346	42-6284	
	42-6290	42-6292	42-6382	42-6358	
	42-6294	42-6321	42-24461	42-6409	
	42-6297	42-6341	42-24475	42-24529	
	42-6306	42-6343	42-24479	42-63353	
	42-6319	42-6353	42-24484		
	42-6322	42-6399	42-63393		
	42-24452	42-24462	42-93827		
	42-24457	42-24464	42-93870		
	42-24466	42-24472			
	42-24503	42-24485			
	42-24522	42-24492			
	42-24541	42-24507			
	42-24574	42-24510			
	42-24579	42-24524			
	42-24582	42-24538			
	42-63396	42024580			
	42-63404	42-24584			
	42-63420	42-63360			
		42-63375			

(cont'd)

VI. The XX Bomber Command 121

		42-63378			
		42-63411			
		42-63451			
		42-65202			
		42-65204			
		42-65226			
Bombed Other	42-6269	42-6423	42-6209	42-6208	28
	42-6303		42-6346	42-6272	
	42-6313		42-6382	42-6279	
	42-6331		42-24461	42-6407	
	42063407		42-24475	42-6454	
	42-63396		42-24479	42-24494	
	42-63407		42-24484	42-24542	
			42-63393	42-24546	
			42-93827	42-63354	
			42-93980	42-63355	
				42-63559	
				42-93828	
Lost	None	42-6321	None	42-6358	2

MISSION NUMBER 18

27 November 1944 • Target: Bangkok, Thailand

	40BG	444BG	462BG	468BG	Total
Bombed Primary	42-6269	42-6292	42-6311	42-6284	55
	42-6276	42-24462	42-6329	42-6390	
	42-6295	42-24464	42-6346	42-6411	
	42-6297	42-24472	42-6382	42-24456	
	42-6298	42-24492	42-24456	42-24469	
	42-6313	42-24507	42-24461	42-24471	
	42-6322	42-24524	42-24463	42-24486	
	42-6331	42-24538	42-24475	42-24487	
	42-24452	42-24723	42-24484	42-24525	
	42-24522	42-63375	42-24505	42-24542	
	42-24587	42-63411	42-24506	42-63356	
	42-24729	42-63422	42-24728	42-63395	
	42-93831	42-63451	42-63393	42-65208	
		42-65226	42-65213	42-93828	
Bombed Other	42-6319			42-6217	3
	42-24457				
Did Not Bomb		42-63378	42-6359		2
Lost	None	None	None	None	0

MISSION NUMBER 19

7 December 1944 • Target: Manchuria Airplane Company Bombed Primary, Mukden, Manchuria

	40BG	444BG	462BG	468BG	Total
	42-6269	42-6292	42-6213	42-6208	91
	42-6276	42-6225	42-6273	42-6217	
	42-6297	42-6341	42-6299	42-6265	
	42-6313	42-6343	42-6311	42-6279	
	42-6322	42-6353	42-6312	42-6389	
	42-6348	42-6399	42-6316	42-6390	
	42-24457	42-6423	42-6329	42-6397	
	42-24466	42-24462	42-6346	42-6411	
	42-24508	42-24472	42-6359	42-6454	
	42-24522	42-24485	42-24456	42-24429	
	42-24541	42-24492	42-24463	42-24442	
	42-24579	42-24507	42-24484	42-24469	

(cont'd)

	40BG	444BG	462BG	468BG	Total
	42-24582	42-24538	42-24505	42-24471	
	42-24587	42-24580	42-24506	42-24486	
	42-24729	42-24584	42-24581	42-24487	
	42-24738	42-24724	42-24728	42-24494	
	42-63363	42-63378	42-63362	42-24525	
	42-63394	42-63411	42-63393	42-24542	
	42-63396	42-65202	42-63457	42-24546	
	42-63404	42-65226	42-93827	42-63353	
	42-63407	42-65262	42-93830	42-63354	
	42-65225			42-63355	
	42-93831			42-63395	
				42-63417	
				42-63424	
				42-65208	
Bombed Other	42-6294	42-6324		42-6272	9
	42-6331	42-6352		42-6284	
	42-63420			42-6409	
				42-63356	
Did Not Bomb		42-24464	42-6209	42-63415	7
		42-24524	42-6270		
		42-63360			
		42-63451			
Lost	42-63363	42-65262	42-6299		4
			42-6359		

MISSION NUMBER 20

14 December 1944 • Target: Rama VI Railroad Bridge, Bangkok, Thailand

	40BG	444BG	462BG	468BG	Total
Bombed Primary		42-6352	42-6338	42-6272	33
		42-24462	42-24463	42-6409	
		42-24485	42-24475	42-6411	
		42-24507	42-24479	42-24469	
		42-24724	42-24505	42-24471	
		42-24732	42-24581	42-24525	
		42-63378	42-24711	42-24542	
		42-63422	42-24728	42-63353	
		42-63451	42-63386	42-63354	
		42-65202	42-63393	42-63443	
		42-65226	42-63457		
			42-63472		
Bombed Other				42-63415	2
				42-65227	
Did Not Bomb		42-6340			2
Lost		None	None	None	0

Target: Rangoon, Burma

	40BG	444BG	462BG	468BG	Total
Bombed Primary	42-24457				12
	42-24508				
	42-24574				
	42-24587				
	42-24589				
	42-24685				
	42-24726				
	42-24729				
	42-63407				
	42-65225				
	42-93831				
	42-93859				
Bombed Other	42-93859				1
Lost	42-24457				5

(cont'd)

VI. The XX Bomber Command   123

```
42-24508
42-24574
42-24726
42-93859
```

MISSION NUMBER 21

18 December 1944 • Target: Hankow, China

	40BG	444BG	462BG	468BG	Total
Bombed Primary	42-6269	42-6292	42-6273	42-6265	85
	42-6297	42-6324	42-6285	42-6272	
	42-6319	42-6340	42-6287	42-6284	
	42-6348	42-6341	42-6311	42-6397	
	42-24466	42-6352	42-6312	42-6407	
	42-24541	42-6399	42-6316	42-6409	
	42-24579	42-6423	42-24456	42-6454	
	42-24620	42-24462	42-24463	42-24442	
	42-24729	42-24464	42-24475	42-24469	
	42-24738	42-24472	42-24479	42-24486	
	42-24752	42-24485	42-24505	42-24487	
	42-63404	42-24492	42-24506	42-24525	
	42-63420	42-24524	42-24581	42-24542	
	42-65233	42-24538	42-24711	42-24691	
	42-93859	42-24584	42-63362	42-24704	
		42-24724	42-63386	42-24715	
		42-24731	42-63393	42-24737	
		42-24732	42-63448	42-63354	
		42-63375	42-63457	42-63355	
		42-63378	42-63472	42-63417	
		42-63422	42-65232	42-63445	
		42-63451	42-93830	42-63460	
		42-65202		42-65208	
		42-93857		42-65227	
Bombed Other			42-24461	42-6411	7
			42-63362	42-24494	
				42-24546	
				42-63354	
				42-63424	
Did Not Bomb	42-6294	42-63376			2
Lost	42-24466	None	None	None	1

MISSION NUMBER 22

19 December 1944 • Target: Omura, Kyushu, Japan

	40BG	444BG	462BG	468BG	Total
Bombed Primary	42-24466	42-24464		42-24678	18
	42-24541	42-24524		42-24703	
	42-24579	42-24538		42-24704	
	42-24729	42-24584		42-24719	
	42-24738			42-63445	
	42-24739				
	42-63404				
	42-63420				
	42-93859				
Bombed Other	42-24582	42-24462	42-63393	42-6411	21
	42-24752	42-24732	42-63448	42-24494	
	42-65223	42-63451	42-63452	42-24546	
		42-65202	42-63454	42-63354	
		42-65226	42-63457	42-63424	

(cont'd)

			42-65228	42-63472	
				42-65232	
Did Not Bomb				42-24461	2
				42-24484	
Lost			42-24466	42-63452	2

MISSION NUMBER 23

21 December 1944 • **Target: Manchuria Airplane Company, Mukden, Manchuria**

	40BG	444BG	462BG	468BG	Total
Bombed Primary	42-24541	42-24485	42-24456	42-24456	40
	42-24579	42-24580	42-24479	42-24469	
	42-24620	42-24584	42-24505	42-24486	
	42-24729	42-24724	42-24506	42-24691	
	42-24738	42-24730	42-24728	42-24703	
	42-24739	42-24731	42-63393	42-24715	
	42-24752	42-24732	42-63454	42-24737	
	42-63394	42-63422	42-63457	42-63417	
	42-63404	42-65226	42-63473	42-63460	
	42-65233	42-93857	42-65232	42-63464	
Bombed Other	42-24718	42-24538	42-24461	42-24487	8
		42-65228	42-24463	42-24704	
				42-65208	
Did Not Bomb	42-24589				1
Ground Aborts	42-24582		42-24581		4
			42-24590		
			42-24711		
Lost	None	None	42-24505	42-24715	2

MISSION NUMBER 24

2 January 1945 • **Target: Rama VI Railroad Bridge, Bangkok, Thailand**

	40BG	444BG	462BG	468BG	Total
Bombed Primary	42-24589	42-24485	42-24463	42-24429	44
	42-24620	42-24492	42-24506	42-24442	
	42-24739	42-24507	42-24800	42-24469	
	42-24752	42-24538	42-24801	42-24486	
	42-24798	42-24580	42-63450	42-24487	
	42-63396	42-24724	42-63454	42-24494	
	42-63462	42-24730	42-63457	42-24542	
	42-63498	42-24731	42-63503	42-24691	
Participating Aircraft	42-65233	42-63376	42-65252	42-24703	
	42-93859	42-63451	42-65254	42-24704	
		42-65202		42-24714	
		42-65226		42-65208	
Bombed Other	42-24579		42-24590		2
Did Not Bomb	42-24582		42-63472		3
	42-24738				
Lost	None	None	None	None	0

MISSION NUMBER 25

6 January 1945 • **Target: Omura, Kyushu, Japan**

	40BG	444BG	462BG	468BG	Total
Bombed Primary	42-24620	42-24462	42-24800	42-24442	29

(cont'd)

VI. The XX Bomber Command

	40BG	444BG	462BG	468BG	Total
	42-24685	42-24464	42-63448	42-24469	
	42-24738	42-24492	42-63457	42-24494	
	42-24739	42-24507	42-65232	42-24525	
	42-24798	42-24524	42-65254	42-63456	
	42-63396	42-24538		42-63500	
	42-63498	42-24584			
	42-63505	42-24724			
		42-63411			
		42-63496			
Bombed Other	42-24522	42-24485	42-24484	42-24542	15
	42-24589	42-24730	42-24506	42-24691	
	42-24718		42-24590	42-24734	
			42-24728	42-24703	
			42-24786		
			42-63503		
Did Not Bomb	42-62462		42-63450	42-24714	3
Lost	None	None	None	None	0

MISSION NUMBER 26

9 January 1945 • Target: Formosa

	40BG	444BG	462BG	468BG	Total
Bombed Primary	42-24541	42-24492	42-24461	42-24471	39
	42-24589	42-24524	42-24590	42-24494	
	42-24620	42-24538	42-24786	42-24704	
	42-24718	42-24584	42-24800	42-24714	
	42-24738	42-24723	42-63448	42-24734	
	42-24752	42-24724	42-63457	42-63424	
	42-24798	42-24730	42-63459	42-63456	
	42-63396	42-63411	42-63473	42-63464	
	42-63462	42-63496	42-63503	42-63500	
	42-63498	42-65228			
	42-63505				
Bombed Other	42-24740	42-24507	42-24456	42-24525	6
		42-63376		42-24719	
Did Not Bomb				42-24691	1
Lost		42-65226		42-24704	2

MISSION NUMBER 27

11 January 1945 • Target: Floating Drydock, Singapore, Malaya

	40BG	444BG	462BG	468BG	Total
Bombed Primary	42-24579	42-24462	42-24475	42-24546	16
	42-24582	42-24723	42-24581	42-24678	
		42-24732	42-24800	42-24734	
		42-63496	42-24801		
			42-63472		
			42-63480		
			42-63540		
Bombed Primary	42-24587	42-24580	42-63502	42-24704	11
	42-24757	42-63375			
	42-63374	42-65226			
	42-63394				
	42-63407				
	42-63455				

Target: King's Drydock, Singapore, Malaya

	40BG	444BG	462BG	468BG	Total
Bombed Other	42-24503	42-63378	42-24463	42-24469	16

(cont'd)

	40BG	444BG	462BG	468BG	Total
	42-24757	42-63446	42-24838	42-24486	
	42-65233		42-63454	42-24546	
			42-65230	42-24691	
				42-24714	
				42-63445	
				42-63464	
Did Not Bomb		42-63451		42-24471	5
		42-65202		42-24487	
				42-63417	

MISSION NUMBER 28

14 January 1945 • Target: Kagi Air Base, Formosa

	40BG	444BG	462BG	468BG	Total
Bombed Primary	42-24541	42-24462	42-24484	42-24429	54
	42-24579	42-24485	42-24590	42-24542	
	42-24740	42-24492	42-24711	42-24703	
	42-24798	42-24538	42-24786	42-24714	
	42-63374	42-24580	42-24800	42-24719	
	42-63404	42-24720	42-24801	42-24734	
	42-65274	42-24724	42-24838	42-24494	
	42-93859	42-24732	42-63448	42-63417	
		42-63375	42-63450	42-63424	
		42-63376	42-63459	42-63445	
		42-63378	42-63473	42-63456	
		42-63411	42-63476	42-63534	
		42-63422	42-63540		
		42-63446	42-93873		
		42-63496			
		42-63533			
		42-65202			
		42-65228			
		42-65268			
		42-65277			

Target: Heito Air Base, Formosa

	40BG	444BG	462BG	468BG	Total
Bombed Primary	42-24739	42-63386	42-63386		16
	42-24522	42-63474	42-63474		
	42-24587	42-65232	42-65232		
	42-24620				
	42-24729				
	42-24846				
	42-63396				
	42-63420				
	42-63498				
	42-63505				
Bombed Other	42-24738	42-24464	42-24463	42-2447110	
	42-24795	42-24524	42-24728	42-24525	
		42-24723		42-65275	
Did Not Bomb		42-24736		42-24486	5
		42-93857		42-63415	
		42-93884			
Lost	None	None	None	None	0

MISSION NUMBER 29

17 January 1945 • Target: Shinchiku, Formosa

	40BG	444BG	462BG	468BG	Total
Bombed Primary	42-24522	42-24472	42-24463	42-24469	78
	42-24541	42-24485	42-24484	42-24486	

(cont'd)

VI. The XX Bomber Command

	40BG	444BG	462BG	468BG	Total
	42-24579	42-24492	42-24506	42-24525	
	42-24587	42-24507	42-24590	42-24542	
	42-24620	42-24524	42-24711	42-24691	
	42-24729	42-24538	42-24728	42-24714	
	42-24739	42-24580	42-24786	42-24719	
	42-24740	42-24720	42-24800	42-24734	
	42-24795	42-24723	42-24801	42-63417	
	42-24798	42-24731	42-24838	42-63424	
	42-24846	42-24732	42-63393	42-63445	
	42-63374	42-63376	42-63521	42-63456	
	42-63396	42-63378	42-63454	42-63464	
	42-63404	42-63411	42-63457	42-63500	
	42-63455	42-63422	42-63459	42-63534	
	42-65233	42-63446	42-63473	42-65208	
	42-65274	42-63496	42-63474	42-65272	
	42-93859	42-65202	42-63476	42-65275	
		42-65228	42-63480		
		42-65268	42-63503		
		42-65270	42-65232		
Bombed Other	42-24757	42-65277	42-24461	42-65276	9
			42-24475		
			42-63502		
			42-63531		
			42-65230		
			42-93873		
Did Not Bomb	42-63505			42-65227	3
	42-63407				
Lost	None	None	None	None	0

MISSION NUMBER 30

25–26 January 1945 • Target: Mine-laying, Saigon Area, French Indo-China

	40BG	444BG	462BG	468BG	Total
Mined Primary			42-24484		18
			42-24506		
			42-24590		
			42-24711		
			42-24728		
			42-63393		
			42-63450		
			42-63454		
			42-63473		
			42-63476		
			42-63480		
			42-63488		
			42-63502		
			42-63503		
			42-63521		
			42-63531		
			42-65220		
			42-65232		

Target: Cam Ranh Bay Area, French Indo-China

	40BG	444BG	462BG	468BG	Total
Mined Primary			42-24463		6
			42-24786		
			42-24800		
			42-24838		
			42-63459		
			42-63540		
Mined Other			42-24801		2
			42-93873		
Lost			None		0

MISSION NUMBER 31

25–26 January 1945 • Target: Mine-laying, Singapore Malaya

	40BG	444BG	462BG	468BG	Total
Mined Primary		42-24462		42-24429	39
		42-24464		42-24469	
		42-24720		42-24471	
		42-24730		42-24486	
		42-24492		42-24525	
		42-24524		42-24691	
		42-24861		42-24698	
		42-63-376		42-24703	
		42-63378		42-24714	
		42-63411		42-24719	
		42-63422		42-24879	
		42-63496		42-24892	
		42-63533		42-24895	
		42-65270		42-63415	
		42-93884		42-63424	
				42-63456	
				42-63500	
				42-63529	
				42-63530	
				42-63534	
				42-65208	
				42-65272	
				42-65275	
				42-65276	
Mined - Other		42-24485		42-63417	4
		42-24736		42-63532	
Lost		None		None	0

MISSION NUMBER 32

27 January 1945 • Target: Saigon Navy Yard, French Indo-China

	40BG	444BG	462BG	468BG	Total
Bombed Primary		42-24503			21
		42-24522			
		42-24541			
		42-24587			
		42-24620			
		42-24738			
		42-24740			
		42-24757			
		42-24795			
		42-24804			
		42-24846			
		42-63374			
		42-63407			
		42-63420			
		42-63455			
		42-63498			
		42-63505			
		42-63542			
		42-65223			
		42-65269			
		42-65274			
Mined - Other		42-24485		42-63417	4
		42-24736		42-63532	
Lost		None		None	0

MISSION NUMBER 33

1 February 1945 • Target: Floating Dry Dock, Singapore, Malaya

	40BG	444BG	462BG	468BG	Total
	42-24503	42-24464	42-24463	42-24429	78
	42-24541	42-24485	42-24475	42-24546	
	42-24579	42-24492	42-24484	42-24542	
	42-24589	42-24507	42-24506	42-24714	
	42-24620	42-24524	42-24590	42-24719	
	42-24685	42-24538	42-24711	42-24895	
	42-24718	42-24584	42-24786	42-63415	
	42-24738	42-24720	42-24801	42-63445	
	42-24740	42-24732	42-24904	42-63464	
	42-24795	42-24736	42-63503	42-63529	
	42-24804	42-63376	42-63450	42-63530	
	42-24888	42-63378	42-63454	42-65275	
	42-63396	42-63422	42-63473		
	42-63404	42-63451	42-63474		
	42-63407	42-63496	42-63476		
	42-63498	42-63537	42-63480		
	42-63505	42-63559	42-63488		
	42-63527	42-65228	42-63521		
	42-63538	42-65268	42-63540		
	42-65271	42-65273	42-63459		
	42-65274	42-93884	42-63531		
			42-63560		
			42-65230		
			42-65299		
Bombed Other	42-24508	42-24472	42-24800	42-24487	21
	42-24522	42-24730	42-24838	42-24734	
	42-24729	42-24861	42-63502	42-63417	
	42-63455	42-24873		42-63456	
	42-65233	42-63375		42-63460	
		42-93857		42-63532	
				42-65208	
Did Not Bomb	42-24587	42-24580	42-65232		3
Lost	42-24589	42-24736			2

MISSION NUMBER 34

7 February 1945 • Target: Saigon Navy Yard

40BG	444BG	462BG	468BG	Total
	42-24472	42-24461		44
	42-24485	42-24463		
	42-24492	42-24475		
	42-24507	42-24590		
	42-24524	42-24711		
	42-24538	42-24728		
	42-24580	42-24800		
	42-24584	42-63448		
	42-24731	42-63476		
	42-24732	42-63480		
	42-24862	42-63521		
	42-24897	42-65299		
	42-24899	42-93873		
	42-63375			
	42-63378			
	42-63411			
	42-63422			
	42-63446			

(cont'd)

THE B-29 SUPERFORTRESS

```
42-63451
42-63496
42-63533
42-63537
42-63559
42-65202
42-65228
42-65268
42-65270
42-65273
42-65277
42-93873
42-93884
```

Target: Phnom Penh

Bombed Alternate		42-24456	19
		42-24479	
		42-24801	
		42-24484	
		42-24838	
		42-24904	
		42-63386	
		42-63454	
		42-63472	
		42-63450	
		42-63459	
		42-63473	
		42-63502	
		42-63503	
		42-63531	
		42-63540	
		42-63560	
		42-65230	
		42-65232	
Bombed Other	42-63376	42-24786	2
Lost	None	None	0

MISSION NUMBER 35

7 February 1945 • Target: Rama VI Railroad Bridge, Bangkok, Thailand

	40BG	444BG	462BG	468BG	Total
Bombed Primary	42-24541			42-24429	56
	42-24579			42-24442	
	42-24587			42-24469	
	42-24620			42-24486	
	42-24685			42-24487	
	42-24718			42-24525	
	42-24508			42-24542	
	42-24738			42-24546	
	42-24740			42-24678	
	42-24752			42-24691	
	42-24757			42-24703	
	42-24888			42-24714	
	42-24795			42-24719	
	42-24798			42-24734	
	42-24846			42-24879	
	42-24894			42-24892	
	42-24908			42-63424	
	42-63374			42-63456	
	42-63404			42-63460	
	42-63420			42-63464	

(cont'd)

VI. The XX Bomber Command

	42-63462		42-63529		
	42-63498		42-63534		
	42-63505		42-63536		
	42-63527		42-65208		
	42-63538		42-65272		
	42-63542		42-65276		
	42-65233		42-93877		
	42-65269		44-69663		
Bombed Other			42-24895	2	
			42-63445		
Did Not Bomb	42-63455		42-24471	4	
			42-63532		
			44-69663		
Lost	None		None	0	

MISSION NUMBER 36

11 February 1945 • Target: Rangoon, Burma

	40BG	*444BG*	*462BG*	*468BG*	*Total*
Bombed Primary	42-24541	42-24464	42-24463	42-24487	56
	42-24579	42-24524	42-24590	42-24678	
	42-24587	42-24861	42-24711	42-24734	
	42-24740	42-24492	42-24728	42-24858	
	42-24795	42-24743	42-24838	42-24879	
	42-24804	42-24724	42-24898	42-63445	
	42-24846	42-24897	42-24904	42-63460	
	42-24888	42-63375	42-63386	42-63464	
	42-63407	42-63378	42-63450	42-63500	
	42-63420	42-63451	42-63450	42-63529	
	42-63505	42-63496	42-63476	42-63532	
	42-63542	42-63533	42-63502	42-65276	
	42-65269	42-63559	42-63503	44-69660	
	42-65271	42-65270	42-63540		
		42-65277			
Did Not Bomb	42-24757		42-63560	42-63417	3
Ground Abort				42-63530	1
Lost	None	None	None	None	0
Bombed Other					
Did Not Bomb					
Lost					

MISSION NUMBER 37

19 February 1945 • Target: Kuala Lumpur, Malaya

	40BG	*444BG*	*462BG*	*468BG*	*Total*
Bombed Primary		42-24472		42-24486	48
		42-24492		42-24487	
		42-24507		42-24542	
		42-24538		42-24678	
		42-24720		42-24691	
		42-24723		42-24525	
		42-24724		42-24714	
		42-24732		42-24719	
		42-24861		42-24734	
		42-24873		42-24858	
		42-24899		42-24892	
		42-63375		42-24893	
		42-63376		42-24895	
		42-63411		42-24909	

(cont'd)

		42-63451		42-63415	
		42-63533		42-63424	
		42-63537		42-63445	
		42-63559		42-63456	
		42-65268		42-63500	
		42-65270		42-63529	
		42-65277		42-63534	
		42-93884		42-63536	
				42-65227	
				42-65276	
				42-65315	
				44-69663	
Bombed Other		42-24524		42-63417	8
		42-24580		42-63532	
		42-24730		42-65272	
		42-63422			
		42-65273			
Did Not Bomb		42-24891			5
		42-24897			

MISSION NUMBER 38

24 February 1945 • Target: Empire Dock, Singapore, Malaya

	40BG	444BG	462BG	468BG	Total
Bombed Primary	42-24541	42-24462	42-24479	42-24471	105
	42-24620	42-24464	42-24711	42-24486	
	42-24684	42-24472	42-24728	42-24487	
	42-24718	42-24485	42-24786	42-24525	
	42-24729	42-24492	42-24800	42-24542	
	42-24738	42-24524	42-24801	42-24678	
	42-24739	42-24538	42-24838	42-24703	
	42-24752	42-24584	42-24898	42-24734	
	42-24795	42-24723	42-63386	42-24858	
	42-24886	42-24724	42-63448	42-24879	
	42-24888	42-24730	42-63450	42-24892	
	42-24894	42-24731	42-63457	42-24893	
	42-24908	42-24732	42-63459	42-24909	
	42-24915	42-24861	42-63472	42-63445	
	42-24918	42-24873	42-63474	42-63456	
	42-63396	42-24891	42-63476	42-63464	
	42-63462	42-24897	42-63480	42-63529	
	42-63498	42-24899	42-63488	42-63532	
	42-63505	42-63422	42-63503	42-63534	
	42-63527	42-63446	42-63540	42-63536	
	42-63528	42-63451	42-63560	42-65227	
	42-63535	42-63496	42-65299	42-65276	
	42-63542	42-63537	44-69661	42-65279	
	42-63555	42-63557		42-65315	
	42-65269	42-63559		44-69660	
	42-65271	42-63268		44-69665	
	42-93859	42-65270			
		42-65273			
		42-65277			
Bombed Other	44-69668	42-24580	42-63502	42-63424	6
				44-69663	
				42-63460	
Did Not Bomb	42-63420		42-24896	42-63460	6
	42-63455		42-63531		
			42-65232		
Lost	None	None	42-24479	None	1

MISSION NUMBER 39

4–5 March 1945 • Target: Yangtze River Mining, China

	40BG	444BG	462BG	468BG	Total
Mined Primary				42-24714	12
			42-24719	42-24895	
				42-63415	
				42-63417	
				42-63500	
				42-63529	
				42-63530	
				42-63536	
				42-65272	
				44-69660	
				44-69663	

MISSION NUMBER 40

27–28 February 1945 • Target: Johore Strait Mining, Singapore, Malaya

	40BG	444BG	462BG	468BG	Total
Mined Primary		42-24507			11
		42-24524			
		42-24538			
		42-24731			
		42-24861			
		42-24873			
		42-63451			
		42-63533			
		42-63557			
		42-65273			
		42-93857			
Did Not Bomb		42-65277			

MISSION NUMBER 41

2 March 1945 • Target: Singapore Naval Base, Singapore, Malaya

	40BG	444BG	462BG	468BG	Total
Bombed Primary	42-24541	42-24462	42-24786	42-24486	50
	42-24718	42-24464	42-24800	42-24887	
	42-24740	42-24472	42-24801	42-24678	
	42-24752	42-24485	42-63457	42-24691	
	42-24795	42-24524	42-63473	42-24703	
	42-24846	42-24538	42-63540	42-24879	
	42-63455	42-24584	42-65299	42-24892	
	42-63462	42-24731	44-69694	42-24893	
	42-63542	42-24732		42-63424	
	42-63555	42-24861		42-63445	
	42-65233	42-24891		42-63534	
		42-63451		42-64276	
		42-63496		42-65279	
		42-63557		42-65315	
		42-65228		44-69665	
		42-65327			
Bombed Other	42-63420	42-24897		42-63460	4
		42-65273			

(cont'd)

		42-63447	42-24728	42-24469	10
Did Not Bomb		42-65337	42-24919	42-24734	
			42-63474	42-63464	
			42-63502	42-65227	
Lost	None	None	None	42-24469	2
				42-24678	

MISSION NUMBER 42

12 March 1945 • Target: Oil Storage Areas, Singapore, Malaya

	40BG	444BG	462BG	468BG	Total
Bombed Primary	Samboe Is. Oil Storage	Bukum Is. Oil Storage	Sebarok Is. Oil Storage		44
	42-24685	42-24462	42-24711		
	42-24729	42-24492	42-24786		
	42-24752	42-24507	42-24801		
	42-24913	42-24524	42-24838		
	42-24914	42-24580	42-24896		
	42-63396	42-24584	42-63450		
	42-63455	42-24720	42-63454		
	42-63498	42-24723	42-63472		
	42-63505	42-24730	42-63476		
	42-63580	42-24899	42-65299		
	44-69668	42-63446	42-63531		
		42-63496	44-69730		
		42-63533			
		42-63537			
		42-63557			
		42-63559			
		42-65228			
		42-65270			
		42-65277			
		42-65337			
		42-93884			
Bombed Other		42-24538			1
Did Not Bomb	42-24738		42-63459		4
			42-63503		
			42-65336		
Ground Aborts	42-24620	42-24472			11
	42-24739	42-24731			
	42-63527	42-24861			
		42-24897			
		42-63411			
		42-63424			
		42-65202			
		42-93857			
Lost	None	None	None		0

MISSION NUMBER 43

10 March 1945 • Target: Kuala Lumpur Railroad Yards, Kuala Lumpur, Malaya

	40BG	444BG	462BG	468BG	Total
Bombed Primary				42-24486	23
				42-24525	
				42-24703	
				42-24714	
				42-24719	
				42-24734	
				42-24858	

(cont'd)

VI. The XX Bomber Command

	40BG	444BG	462BG	468BG	Total
				42-24879	
				42-24892	
				42-24893	
				42-24895	
				42-24909	
				42-63415	
				42-63456	
				42-63460	
				42-63500	
				42-63529	
				42-63530	
				42-65315	
				42-93877	
				44-69660	
				44-69663	
				44-69665	
				44-69701	
Bombed Other				42-24542	3
				42-65227	
				42-65272	
Did Not Bomb				42-63424	2
				42-65279	
Lost				None	0

MISSION NUMBER 44

17 March 1945 • Target: Rangoon, Burma

	40BG	444BG	462BG	468BG	Total
Bombed Primary	42-24738	42-24485	42-24786	42-24471	70
	42-24740	42-24584	42-24801	42-24542	
	42-24752	42-24720	42-24838	42-24714	
	42-24757	42-24724	42-24919	42-24858	
	42-24795	42-24731	42-63454	42-24879	
	42-24846	42-24732	42-63459	42-24893	
	42-24908	42-24740	42-63473	42-24909	
	42-24914	42-24861	42-63474	42-63445	
	42-63527	42-24899	42-63480	42-63456	
	42-63542	42-63446	42-63531	42-63464	
	42-63580	42-63533	42-65299	42-63500	
	42-65233	42-63537	42-65329	42-63529	
	42-65328	42-63559	44-69661	42-63530	
		42-65228	44-69734	42-63534	
		42-65268		42-65227	
		42-65277		42-65272	
		42-65270		42-65276	
		42-65273		42-93877	
		42-65337		44-69660	
		42-93857		44-69663	
		42-93884		44-69665	
				44-69701	
Bombed Other		42-65202		42-63424	2
Did Not Bomb	42-24739	42-24873	42-65252	42-24719	5
	44-69659				
Ground Abort		42-24507			1
Lost	None	None	None	None	0

MISSION NUMBER 45

22 March 1945 • Target: Rangoon and Mingaladon, Burma

	40BG	444BG	462BG	468BG	Total
Bombed Primary	Rangoon	Mingaladon	Mingaladon	Rangoon	76
	42-24620	42-24464	42-24711	42-24486	

(cont'd)

	42-24685	42-24485	42-24728	42-24525	
	42-24729	42-24507	42-24786	42-24703	Rangoon
	42-24738	42-24861	42-24800	42-24734	39
	42-24739	42-24873	42-24801	42-24892	
	42-24740	42-24899	42-24838	42-63415	Mingaladon
	42-24752	42-63533	42-24896	42-63530	37
	42-24757	42-63548	42-24919	42-63534	
	42-24795	42-63557	42-63450	42-63536	
	42-24846	42-65202	42-63457	42-63315	
	42-24886	42-65228	42-63473	42-63529	
	42-24914	42-65268	42-63474	42-65277	
	42-24915	42-65273	42-63488	42-65279	
	42-63396		42-63502	44-69665	
	42-63455		42-63503	44-69701	
	42-63462		42-63531		
	42-63498		42-63540		
	42-63580		42-65232		
	42-65233		42-65299		
	42-65269		42-65329		
	42-65328		42-93873		
	42-93859		44-69694		
	44-69659		44-69730		
	44-69668		44-69734		
Did Not Bomb		42-24897			2
		42-65337			
Lost	None	None	None	None	0

MISSION NUMBER 46

28–29 March 1945 • Target: Mine-laying, Lower Yangtze River, China

	40BG	444BG	462BG	468BG	Total
Bombed Primary				42-24714	10
				42-24719	
				42-24858	
				42-24894	
				42-24909	
				42-63456	
				42-63529	
				42-63536	
				42-65276	
				42-65279	
Lost				None	0

MISSION NUMBER 47

28–29 March 1945 • Target: Mine-laying Saigon and Cam Ranh Bay, French Indo-China

	40BG	444BG	462BG	468BG	Total
Mined Primary Area				42-24471	18
				42-24486	
				42-24542	
				42-24703	
				42-24879	
				42-24892	
				42-24893	
				42-63415	
				42-63460	
				42-63445	
				42-63530	
				42-63534	

(cont'd)

VI. The XX Bomber Command

	42-65227	
	42-65272	
	42-65315	
	42-93877	
	44-69660	
	44-69663	
Lost	None	0

MISSION NUMBER 48

28–29 March 1945 • Target: Mine-laying Singapore Area

	40BG	444BG	462BG	468BG	Total
Mined Primary Area		42-24462			32
		42-24464			
		42-24485			
		42-24507			
		42-24524			
		42-24538			
		42-24580			
		42-24584			
		42-24720			
		42-24724			
		42-24730			
		42-24731			
		42-24732			
		42-24861			
		42-24873			
		42-24897			
		42-24899			
		42-63422			
		42-63446			
		42-63496			
		42-63533			
		42-63537			
		42-63559			
		42-43577			
		42-65228			
		42-65268			
		42-65270			
		42-65273			
		42-65277			
		42-65337			
		42-93857			
		42-93884			
Did Not Bomb		42-63548			1
Lost		None			0

MISSION NUMBER 49

29–30 March 1945 • Target: Bukum Island Oil Storage, Singapore, Malaya

	40BG	444BG	462BG	468BG	Total
Bombed Primary	42-24579		42-24711		23
	42-24685		42-24786		
	42-24738		42-24838		
	42-24757		42-24896		
	42-24846		42-24909		
	42-24919		42-63472		
	42-63420		42-63480		
	42-63527		42-63521		

(cont'd)

	42-65233	42-65232	
	42-65269	44-69661	
	44-69659	44-69730	
	44-69668		
Bombed Other		42-63488	2
		44-69734	
Ground Abort	42-65274		1
Lost	None	None	0

LOSSES

42-6209	42-6249	42-6289	42-6323	42-6362	42-24461	42-63352
42-6211	42-6253	42-6290	42-6326	42-6365	42-24466	42-63357
42-6212	42-6255	42-6291	42-6327	42-6368	42-24469	42-63363
42-6215	42-6256	42-6292	42-6328	42-6369	42-24474	42-63394
42-6220	42-6261	42-6293	42-6330	42-6370	42-24479	42-63395
42-6222	42-6262	42-6296	42-6331	42-24589	42-24482	42-63419
42-6223	42-6263	42-6299	42-6332	42-24678	42-24494	42-63452
42-6226	42-6264	42-6300	42-6334	42-6383	42-24504	42-63458
42-6228	42-6268	42-6301	42-6335	42-6389	42-24505	42-65203
42-6229	42-6271	42-6302	42-6336	42-6390	42-24506	42-65204
42-6230	42-6274	42-6304	42-6342	42-6408	42-24510	42-65213
42-6231	42-6275	42-6305	42-6343	42-6418	42-24513	42-65226
42-6234	42-6276	42-6307	42-6345	42-6425	42-24513	42-65254
42-6235	42-6277	42-6308	42-6350	42-6444	42-24574	42-93825
42-6237	42-6278	42-6309	42-6351	42-6452	42-24582	42-93826
42-6238	42-6280	42-6214	42-6356	42-24420	42-24704	42-93829
42-6240	42-6281	42-6318	42-6358	42-24446	42-24706
42-6243	42-6282	42-6320	42-6359	42-24452	42-24715
42-6246	42-6286	42-6321	42-6360	42-24457	42-24726
42-6247	42-6288	42-6322	42-6361	42-24458	42-24736

VII

THE XXI BOMBER COMMAND

Serious planning for the movement of the XXI Bomber Command's B-29s to Saipan, Tinian and Guam began in April 1944. The construction and defense of the bases would be the Navy's responsibility, as would logistical support.

General H.H. "Hap" Arnold selected General Walter Frank to head the Army Air Forces planning delegation sent to Pearl Harbor to work with the Navy planners on the staff of Admiral Chester Nimitz, the Commander in Chief, Pacific (CINCPAC). Arnold told Frank to be diplomatic, not demanding. Frank took those instructions so much to heart that the relationship between the two services was a model of cooperation.

Before the B-29s could begin operating against Japan from the Marianas, the islands first had to be taken away from the Japanese. This began on 15 June 1944 when the Marines landed on Saipan.

A week later, Marines and the 77th Infantry Division's 305th Regimental Combat Team landed on Guam. By August 10 the island was secured.

Tinian was assaulted on July 24 and secured by August 1.

Hard on the heels of the combat troops came the SeaBees — the Naval Construction Battalions (N-C-B). The SeaBees did not meet their schedule but came very close. In a little over three months they had built a base and air field on Saipan capable of supporting 240 B-29s and their logistical "tail."

On Tinian, the SeaBees built the largest bomber base ever. The 6th Naval Construction Brigade built four 8500-foot runways for the 313th Bomb Wing, and all required infrastructure; then went to the west end of the island and at West Field laid down two 8500-foot runways for the 58th Bomb Wing.

On Guam the 5th NCB built North Field for the 314th Bomb Wing and Northwest Field for the 315th. On Iwo Jima the 9th NCB built a 9800-foot runway in increments, extending the runway as the defending Japanese were beaten back.

The man responsible for implementing the plans drawn up in response to the Frank report was General Millard Harmon. Harmon was in an unenviable position. He was dealing with four major headquarters: General Arnold in Washington, Admiral Nimitz in Pearl Harbor and Guam, the Army in Hawaii and the Twentieth Air Force.

And deal with them he did, despite major revisions to the original plan. Delays made the Saipan strips barely usable when *Joltin' Josie* landed on October 12, 1944. Josie was commanded by Brigadier General Haywood Hansell, whose presence brought the XXI Bomber Command to the Marianas.

Eight days later Brigadier General Emmett "Rosie" O'Donnell arrived to set up 73rd Bomb Wing Headquarters. Following him, in a stream reaching back to the States, were B-29s of Colonel Will Ganey's 498th Bomb Group; Colonel Stuart Wright's 497th; Colonel Sam Harris' 499th; and Colonel Dick King's 500th.

Gen. Hansell started the combat training against lightly defended Japanese targets. On October 18, nine planes each of the 497th and 498th Bomb Groups took off on the 73rd Bomb Wing's first mission. The submarine pens at Truk were the target, and the results were mediocre, with only nine planes bombing the primary target and few hitting it.

The same two groups returned to Truk on October 30 with even fewer bombs landing on the target. The third try, on November 2, was briefed as a radar bombing mission. Again the results were indifferent, with bombs scattered all over the general target area.

General Hansell stepped up the combat bombing training on November 5 and 8, sending the 497th and 498th to bomb airfields on Iwo Jima. On November 11, the 500th

Bomb Group made its initial combat mission to the Submarine Base on Truk.

In order to properly plan missions to Japan, up-to-date photos of the proposed targets were needed. So on November 1, two days after arriving on Saipan, Captain Ralph Steakley of the 3rd Photo Reconnaissance Squadron took off in his F-13A (photo reconnaissance-configured B-29) bound for Tokyo. Steakley cruised over Tokyo at 32,000 feet for 35 minutes taking picture after picture. A few fighters made it up to the camera plane's altitude but did not attack. Steakley's plane was the first land-based American plane over Tokyo since Jimmy Doolittle's 1942 raid. In honor of his mission, Steakley's plane was named *Tokyo Rose*.

The 3rd PRS continued to fly single plane missions to Japan, taking high altitude pre-strike target planning photos and poststrike damage assessment photos. The losses incurred by the 3rd in this type of activity were relatively light. Seldom were the Japanese able to get fighters up to 30,000 feet in time to intercept the photo ships.

Ready or not, all four groups of the 73rd Bomb Wing made their first trip to Japan on November 24 with 111 planes airborne. The target was the Nakajima Aircraft Engine Plant at Musashino in the arsenal sector of Tokyo. Because of bad weather, only 24 planes attacked the primary target; the majority dropped on the secondary target of the Tokyo Docks.

The group returned to Musashino on the 27th and closed out November with a raid on the Tokyo Industrial Area and Docks.

The pace picked up the next month with a return to Musashino on December 3. December 8 saw 82 planes attacking the Iwo Jima Air Fields. Then it was back to Japan, hitting Nagoya on December 18 and 22. Iwo Jima was revisited on December 24 and Tokyo the final target of the month on December 27.

In addition to the wing strikes, December saw the initiation of another type of single plane mission: the weather strike. These were missions to Japan to collect weather data and drop nuisance bombs on Japanese cities. Again, these were high altitude missions with surprisingly low losses.

The 3rd PRS lost two planes in December; 42-93866 ditched on December 7 and 42-93867 ditched two days later on December 9. Out of the 75 weather strike missions flown in December, there were only three losses. Plane 42-24628 was overdue and missing on December 18. Plane 42-24733 ditched on the same day, and 42-24686 on December 28. During the same time period the 73rd Wing lost 21 planes to all causes on the six multi-plane raids by the wing as a whole.

The pace continued to quicken for the 73rd in January. Musashino on January 9 cost six planes; Nagoya on January 14, another five. One weather strike, out of 83 flown, was lost when 42-24744 ditched on January 10.

But help was on the way. The pipeline was getting up to speed and new groups were arriving in the Marianas. The 313th Bomb Wing, operating off of Tinian, sent 44 planes to Pagan Island on January 16, another 33 to Truk on January 21, and 28 against the Iwo Jima Air fields on January 24. Thirty-thrre more returned to Iwo Jima on January 29.

The 313th's 504th and 505th groups joined in the attack on Kobe on February 4, while on February 9 their sister groups, the 6th and 9th, made their initial training mission against the Moen Island Air Field on Truk.

The 6th and 9th Bomb Groups joined in the attack on Tokyo on February 25, as did the 314th Bomb Wing's 19th Bomb Group, flying out of North Field, Guam.

On March 9th, the 314th's 29th Bomb Group became a part of the bomber force attacking Tokyo. A month later, on April 12, the 314th wing's remaining two groups, the 39th and 330th, joined in the attack on the Hodagaya Chemical Works in Koriyama.

With the addition of the 39th and 330th, the XXI Bomber Command now had three wings, twelve groups, thirty-six squadrons of 15 B-29s each at their disposal. In May, the 58th Bomb Wing completed its move from India to Tinian, adding four more groups to the XXI Bomber Command.

Late May saw the arrival of the first of the 315th Bomb Wing, whose planes were equipped with the new AN/APQ Eagle radar. The antenna for this radar was an 18-foot, wing-shaped unit mounted under the fuselage. The antenna swept a 60-degree arc along the flight path of the plane, and a higher frequency signal gave a much-improved radarscope picture.

The 315th was given a dedicated task of destroying the facilities of Japan's oil industry. Not counting five single plane missions in May, the 315th's initial mission was on June 26 against the Utsube oil refinery at Yokkaichi.

By war's end, the XXI Bomber Command had flown 331 misions against the Japanese Home Islands from the Marianas.

BOMBER COMMAND MISSION LIST

Research for this book revealed that the existing mission lists for the XXI Bomber Command — developed at the wing or mission number level — did not reflect the diversity of targets actually bombed under the umbrella of a single XXI Bomber Command mission number.

For example, mission number 308 as listed in the *20th Air Force Album* shows the target as the Nagaoka Urban Area. Yet when one examines the histories of the groups participating in that mission, one finds that Tanabe, Fukui, Toyama and Hamamatsu were also bombed.

VII. The XXI Bomber Command

In order to reflect this reality, this section offers a new, more detailed, mission list based on group data.

The data included in the mission list require very little explanation. There are nine columns of data:

MSN. This is the XXI Bomber Command mission number.

Date. Date of mission.

Target. Name of the target. If secondary targets or targets of opportunity were hit on that mission, those targets are listed below the first (primary) target and indented for clarity.

Location. Geographic location of the target.

Wing. Parent Bomb Wing of the group bombing. If more than one wing participated in the attack, the word "MISSION" appears in this column. In such cases, the numbers to the right of "MISSION" in the *No. Up, Bombed* and *Lost* columns are the totals for all wings/groups airborne, and numbers specific to a particular wing will follow on the next line. A notation of "in above" signifies that the number of planes attacking that particular target — almost always a target of opportunity — is included in the group total cited just above.

Grps. Unit(s) doing the bombing.

No. Up. Number of aircraft airborne and attacking.

Bombed. Number of aircraft attacking the primary target. Wing total should equal sum of groups bombing the primary target. However, in many cases, something happened to the numbers from the groups on their arrival at wing, and the totals reported by the wing did not always match the sum of the reportage by the groups.

Lost. Number of aircraft lost on mission. Includes aircraft surveyed or salvaged.

Following this composite mission list for the XXI Bomber Command are mission lists broken down by individual bomb group.

XXI Bomber Command Detailed Mission List

MSN	Date	Target	Location	Wings	Grps	No. Up	Bombed	Lost
1	Oct 28, 1944	Submarine Base	Truk	73BW		18	9	0
1	Oct 28, 1944	Submarine Base	Truk		497BG	9	7	0
1	Oct 28, 1944	Submarine Base	Truk		498BG	9	2	0
2	October 30	Submarine Base	Truk	73BW		18	17	0
2	October 30	Submarine Base	Truk		497BG	9	9	0
2	October 30	Submarine Base	Truk		498BG	9	8	0
3	November 2	Submarine Base	Truk	73BW		20	17	0
3	November 2	Submarine Base	Truk		497BG	11	9	0
3	November 2	Submarine Base	Truk		498BG	9	8	0
4	November 5	Airfields	Iwo Jima	73BW		36	25	0
4	November 5	Airfields	Iwo Jima		497BG	18	16	0
4	November 5	Airfields	Iwo Jima		498BG	18	9	0
5	November 8	Airfields	Iwo Jima	73BW		17	6	1
5	November 8	Airfields	Iwo Jima		497BG	7	6	0
5	November 8	Airfields	Iwo Jima		498BG	10	0	1
6	November 11	Submarine Base	Truk	73BW	500BG	9	8	0
7	November 24	Musashino — Nakajima AC Eng	Tokyo	73BW		111	24	2
7	November 24	Secondary Urban & Dock Area	Tokyo		WING	in above	59	0
7	November 24	Targets of Opportunity	Matsuzaki Village		WING	in above	5	0
7	November 24	Docks & Urban Area — Secondary	Tokyo		497BG	28	20	1
7	November 24	Docks & Urban Area — Secondary	Tokyo		498BG	29	24	0
7	November 24	Docks & Urban Area — Secondary	Tokyo		499BG	not avail	27	1
7	November 24	Docks & Urban Area — Secondary	Tokyo		500BG	not avail	16	0
8	November 27	Musashino — Nakajima AC Eng	Tokyo	73BW		81	0	1
8	November 27	Secondary Urban & Dock Area	Tokyo		WING	in above	49	0
8	November 27	Target of Opportunity	Various		WING	in above	13	0
8	November 27	Musashino — Bombed Secondary	Tokyo		497BG	26	14	0
8	November 27	Docks Area — Secondary	Tokyo		498BG	27	20	0
8	November 27	Docks Area — Secondary	Tokyo		499BG	in above	13	0
8	November 27	Docks Area — Secondary	Tokyo		500BG	in above	15	1
9	November 29/30	Industrial Area	Tokyo	73BW		28	24	1
9	November 29/30	Last Resort Targets	Tokyo		WING	in above	2	
9	November 29/30	Industrial Area	Tokyo		497BG	9	7	0
9	Nov 29/30	Light Industrial & Dock Area	Tokyo		498BG	10	8	0
9	Nov 29/30	Light Industrial & Dock Area	Tokyo		499BG	not avail	5	0
9	Nov 29/30	Light Industrial & Dock Area	Tokyo		500BG	not avail	6	1
10	December 3	Musashino — Nakajima AC Eng	Tokyo	73BW		86	60	6
10	December 3	Secondary Targets	Tokyo		WING	in above	15	0
10	December 3	Musashino — Nakajima AC Eng	Tokyo		497BG	24	11	1
10	December 3	Musashino — Nakajima AC Eng	Tokyo		498BG	24	20	3

MSN	Date	Target	Location	Wings	Grps	No. Up	Bombed	Lost
10	December 3	Musashino — Nakajima AC Eng	Tokyo		499BG	not avail	10	0
10	December 3	Musashino — Nakajima AC Eng	Tokyo		500BG	not avail	19	1
11	December 8	Airfields	Iwo Jima	73BW		82	61	0
11	December 8	Secondary Target	unknown		WING	in above	4	0
11	December 8	Airfields	Iwo Jima		497BG	16	12	0
11	December 8	Airfields	Iwo Jima		498BG	23	18	0
11	December 8	Airfields	Iwo Jima		499BG	not avail	17	0
11	December 8	Airfields	Iwo Jima		500BG	not avail	18	0
12	December 13	Mitsubishi Aircraft Engine Plant	Nagoya	73BW		90	71	5
12	December 13	Secondary Targets	Nagoya Region		WING	in above	9	0
12	December 13	Mitsubishi Aircraft Engine Plant	Nagoya		497BG	18	16	0
12	December 13	Mitsubishi Aircraft Engine Plant	Nagoya		498BG	25	22	2
12	December 13	Mitsubishi Aircraft Engine Plant	Nagoya		499BG	not avail	21	1
12	December 13	Mitsubishi Aircraft Engine Plant	Nagoya		500BG	not avail	21	1
13	December 18	Mitsubishi Aircraft Factory	Nagoya	73BW		89	63	4
13	December 18	Targets of Opportunity	Nagoya Area		WING	in above	10	0
13	December 18	Mitsubishi Aircraft Factory	Nagoya		497BG	18	14	1
13	December 18	Mitsubishi Aircraft Factory	Nagoya		498BG	29	24	1
13	December 18	Mitsubishi Aircraft Factory	Nagoya		499BG	not avail	21	1
13	December 18	Mitsubishi Aircraft Factory	Nagoya		500BG	not avail	14	1
14	December 22	Mitsubishi Aircraft Engine Plant	Nagoya	73BW		79	48	3
14	December 22	Targets of Opportunity	Nagoya Area		WING	in above	14	0
14	December 22	Mitsubishi Aircraft Engine Plant	Nagoya		497BG	14	7	1
14	December 22	Mitsubishi Aircraft Engine Plant	Nagoya		498BG	25	13	0
14	December 22	Secondary Target	Nagoya		498BG	in above	4	0
14	December 22	Mitsubishi Aircraft Engine Plant	Nagoya		499BG	not avail	32	1
14	December 22	Mitsubishi Aircraft Engine Plant	Nagoya		500BG	not avail	6	0
15	December 24	Airfields	Iwo Jima	73BW		29	23	0
15	December 24	Navigation mission	Iwo Jima		497BG	3	n/a	0
15	December 24	Airfields	Iwo Jima		498BG	2	2	0
15	December 24	Airfields	Iwo Jima		499BG	not avail	12	1
15	December 24	Airfields	Iwo Jima		500BG	not avail	9	0
16	December 27	Nakajima AC Engine Plant	Tokyo	73BW		72	39	3
16	December 27	Secondary target	Tokyo Area		WING	in above	6	0
16	December 27	Targets of Opportunity	Tokyo Area		WING	in above	7	0
16	December 27	Nakajima AC Engine Plant	Tokyo		497BG	21	14	0
16	December 27	Nakajima AC Engine Plant	Tokyo		498BG	23	15	3
16	December 27	Secondary Target	Tokyo		498BG	in above	1	0
16	December 27	Nakajima AC Engine Plant	Tokyo		499BG	not avail	17	0
16	December 27	Nakajima AC Engine Plant	Tokyo		500BG	not avail	5	0
17	Jan 3, 1945	Urban Area	Nagoya	73BW		97	57	5
17	Jan 3, 1945	Targets of Opportunity	Nagoya Area		WING	in above	21	0
17	Jan 3, 1945	Urban Area	Nagoya		497BG	21	17	1
17	Jan 3, 1945	Urban Area	Nagoya		498BG	24	13	1
17	Jan 3, 1945	Secondary Target	Unreported		498BG	in above	1	0
17	Jan 3, 1945	Urban Area	Nagoya		499BG	not avail	28	0
17	Jan 3, 1945	Urban Area	Nagoya		500BG	not avail	19	2
18	January 9	Musashino — Nakajima AC Engine Tokyo	Tokyo	73BW		72	18	6
18	January 9	Targets of Opportunity	Tokyo Area		WING	in above	34	0
18	January 9	Musashino — Nakajima AC Engine Tokyo	Tokyo		497BG	17	10	3
18	January 9	Musashino — Nakajima AC Engine Tokyo	Tokyo		498BG	18	2	0
18	January 9	Secondary Targets	Tokyo		498BG	in above	6	0
18	January 9	Musashino — Nakajima AC Engine Tokyo	Tokyo		499BG	not avail	6	2
18	January 9	Target of Opportunity	Tokyo Area		500BG	not avail	10	1
19	January 14	Mitsubishi AC Frame Plant	Nagoya	73BW		73	40	5
19	January 14	Secondary target	Nagoya		WING	in above	22	
19	January 14	Mitsubishi Aircraft Factory	Nagoya		497BG	21	17	2
19	January 14	Mitsubishi Aircraft Factory	Nagoya		498BG	22	9	1
19	January 14	Mitsubishi Aircraft Factory	Nagoya		499BG	not avail	28	1
19	January 14	Mitsubishi Aircraft Factory	Nagoya		500BG	not avail	8	0
	January 16	Pagan Island	Pagan Island	313BW	WING	44	32	0
20	January 19	Kawasaki Aircraft Factory	Akashi	73BW		80	62	0
20	January 19	Targets of Opportunity	Akashi Area		WING	in above	7	0
20	January 19	Kawasaki Aircraft Factory	Akashi		497BG	19	13	0
20	January 19	Kawasaki Aircraft Factory	Akashi		499BG	not avail	24	0

VII. The XXI Bomber Command

MSN	Date	Target	Location	Wings	Grps	No. Up	Bombed	Lost
20	January 19	Kawasaki Aircraft Factory	Akashi		499BG	not avail	24	0
20	January 19	Kawasaki Aircraft Factory	Akashi		500BG	not avail	16	0
21	January 21	Airfield	Truk	313BW	WING	33	30	0
22	January 23	Mitsubishi Aircraft Engine Plant	Nagoya			75	28	2
22	January 23	Secondary Target	Nagoya U.A.		WING	in above	27	0
22	January 23	Targets of Opportunity	Nagoya Area		WING	in above	5	0
22	January 23	Mitsubishi Aircraft Engine Plant	Nagoya		497BG	17	11	1
22	January 23	Secondary Target	Nagoya City		498BG	22	16	0
22	January 23	Mitsubishi Aircraft Engine Plant	Nagoya		499BG	not avail	20	1
22	January 23	Mitsubishi Aircraft Engine Plant	Nagoya		500BG	not avail	13	1
23	January 24	Airfields	Iwo Jima	313BW	WING	28	20	0
24	January 27	Musashi — Nakajima Aircraft Factory	Tokyo	73BW		76	0	9
24	January 27	Secondary Target	Tokyo U.A.		WING	in above	56	0
24	January 27	Targets of Opportunity			WING	in above	6	0
24	January 27	Musashi — Nakajima Aircraft Factory	Tokyo		497BG	19	13	5
24	January 27	Docks & Urban Area — Secondary	Tokyo		498BG	21	18	2
24	January 27	Musashi — Nakajima Aircraft Factory	Tokyo		499BG	not avail	21	1
24	January 27	Docks & Urban area — Secondary	Tokyo		500BG	not avail	10	0
25	January 29	Airfields	Iwo Jima	313BW	WING	33	28	0
26	February 4	Urban Area	Kobe	Mission		110	69	2
26	February 4	Urban Area	Kobe	73BW		74	37	2
26	February 4	Secondary Target			WING	in above	29	0
26	February 4	Target of Opportunity			WING	in above	1	0
26	February 4	Urban Area	Kobe		497BG	12	10	0
26	February 4	Urban Area	Kobe		498BG	23	19	1
26	February 4	Secondary Target	Kobe		498BG	in above	1	0
26	February 4	Urban Area	Kobe		499BG	not avail	22	0
26	February 4	Matsuzaka	Kobe		500BG	not avail	15	0
26	February 4	Urban Area	Kobe	313BW		not avail	32	1
26	February 4	Urban Area	Kobe		504BG	14	12	0
26	February 4	Urban Area	Kobe		505BG	not avail	20	1
28	February 9	Airfields	Truk	313BW		60	60	0
28	February 9	Moen Island Airfield	Truk		6BG	30	30	0
28	February 9	Moen Island Airfield	Truk		9BG	30	29	0
29	February 10	Nakajima Aircraft Assy Plant	Ota	MISSION		120	84	12
29	February 10	Nakajima Aircraft Assy Plant	Ota	73BW		85	55	3
29	February 10	Target of Opportunity	Ota		WING	in above	13	0
29	February 10	Nakajima Aircraft Assy Plant	Ota		497BG	11	9	0
29	February 10	Nakajima Aircraft Assy Plant	Ota		498BG	24	16	2
29	February 10	Secondary target	Ota		498BG	in above	6	0
29	February 10	Nakajima Aircraft Assy Plant	Ota		499BG	not avail	19	1
29	February 10	Nakajima Aircraft Assy Plant	Ota		500BG	not avail	18	1
29	February 10	Nakajima Aircraft Assy Plant	Ota	313BW		35	29	9
29	February 10	Nakajima Aircraft Assy Plant	Ota		504BG	14	11	1
29	February 10	Nakajima Aircraft Assy Plant	Ota		505BG	21	18	8
30	February 11	Sea Search		313BW	6BG	9	8	0
31	February 12	Antiaircraft	Iwo Jima	313BW	9BG	21	21	0
32	February 12	Sea Search		313BW	6BG	10	8	0
33	February 14	Sea Search		313BW	6BG	6	5	1
34	February 15	Mitsubishi Aircraft Engine Factory	Nagoya	MISSION		117	33	1
34	February 15	Secondary Target	Nagoya		WINGS	in above	68	0
34	February 15	Target of Opportunity	Nagoya		WINGS	in above	16	0
34	February 15	Mitsubishi Aircraft Engine Factory	Nagoya	73BW		91	27	0
34	February 15	Targets of Opportunity	Nagoya		WING	in above	51	0
34	February 15	Mitsubishi Aircraft Engine Factory	Nagoya		497BG	20	6	0
34	February 15	Target Last Resort	Nagoya City		497BG	in above	11	0
34	February 15	Mitsubishi Aircraft Engine Factory	Nagoya		498BG	24	10	0
34	February 15	Mitsubishi Aircraft Engine Factory	Nagoya		499BG	not avail	not avail	1
34	February 15	Mitsubishi Aircraft Engine Factory	Nagoya		500BG	not avail	7	0
34	February 15	Mitsubishi Aircraft Engine Factory	Nagoya	313BW		28	4	0
34	February 15	Mitsubishi Aircraft Engine Factory	Nagoya		504BG	13	0	1
34	February 15	Mitsubishi Aircraft Engine Factory	Nagoya		505BG	15	4	0
34	February 15	Last Resort Target	Hachijo Jima		505BG	in above	Unk	0
34	February 15	Last Resort Target	Hamamatsu		505BG	in above	Unk	0
34	February 15	Last Resort Target	Maisaka		505BG	in above	Unk	0

MSN	Date	Target	Location	Wings	Grps	No. Up	Bombed	Lost
34	February 15	Last Resort Target	Fukude		505BG	in above	Unk	0
35	February 17	Submarine Base	Truk	73BW	WING	9	8	0
36	February 18	Airfields	Truk	313BW		38	37	0
36	February 18	Moen Airfield #2	Truk		6BG	19	19	0
36	February 18	Moen Airfield #2	Truk		9BG	18	18	0
37	February 19	Nakajima Aircraft Engine Factory	Tokyo	MISSION		150	0	6
37	February 19	Secondary Target	Tokyo Urban		XXIBC	in above	131	
37	February 19	Last Resort Target	Tokyo		XXIBC	in above	7	
37	February 19	Target of Opportunity	Tokyo		XXIBC	in above	5	
37	February 19	Tokyo Urban Area	Tokyo	73BW		101	0	5
37	February 19	Secondary Target	Tokyo		WING	in above	81	0
37	February 19	Target of Opportunity	Tokyo		WING	in above	5	0
37	February 19	Nakajima Aircraft Engine Factory	Tokyo		497BG	23	20	1
37	February 19	Secondary Target	Tokyo Docks		498BG	33	27	0
37	February 19	Secondary Target	Tokyo Docks		499BG	not avail	22	1
37	February 19	Industrial Area	Tokyo		500BG	not avail	17	2
37	February 19	Urban Area	Tokyo	313BW		38	0	0
37	February 19	Urban Area	Tokyo		6BG		3	
37	February 19	Secondary Target	Tokyo Urban		504BG	18	0	0
37	February 19	Nakajima Aircraft Engine Factory	Tokyo		505BG	0	0	0
37	February 19	Port of Tokyo — Secondary	Tokyo		505BG	20	17	0
37	February 19	Last Resort Target	Kawasaki		505BG	in above	1	0
37	February 19	Last Resort Target	Toyohashi		505BG	in above	1	0
37	February 19	Last Resort Target	Shizuoka		505BG	in above	1	0
38	February 25	Urban Area	Tokyo	MISSION		229	172	3
38	February 25	Last Resort Target	Tokyo		XXIBC	in above	28	
38	February 25	Target of Opportunity	Tokyo		XXIBC	in above	1	
38	February 25	Urban Area	Tokyo	73BW		116	94	3
38	February 25	Targets of Opportunity	Tokyo		WING	in above	8	0
38	February 25	Urban Area	Tokyo		497BG	24	19	2
38	February 25	Urban area & Docks	Tokyo		498BG	37	27	0
38	February 25	Urban area & Docks	Tokyo		499BG	not avail	29	
38	February 25	Urban area & Docks	Tokyo		500BG	not avail	27	0
38	February 25	Urban Area	Tokyo	313BW		not avail	60	0
38	February 25	Urban Area	Tokyo		6BG	not avail	21	0
38	February 25	Urban Area	Tokyo		9BG	32	23	0
38	February 25	Targets of Opportunity	Tokyo Area		9BG	in above	5	0
38	February 25	Urban Area	Tokyo		504BG	14	8	0
38	February 25	Urban Area	Tokyo		505BG	15	8	0
38	February 25	Last Resort Target	Toyohashi		505BG	in above	2*	0
38	February 25	Last Resort Target	Hamamatsu		505BG	in above	2*	0
38	February 25	Last Resort Target	Numazu		505BG	in above	2*	0
38	February 25	Last Resort Target	Kega		505BG	in above	1*	0
38	February 25	NE Tokyo	NE Tokyo	314BW	19BG	11	10	0
38	February 25	Target of opportunity	Nagoya		19BG	in above	1	
39	March 4	Nakajima AC Engine Plant	Tokyo	MISSION		192	0	1
39	March 4	Secondary Target	Tokyo		WINGS	in above	177	0
39	March 4	Last Resort Target	Tokyo		WINGS	in above	17	0
39	March 4	Target of Opportunity	Tokyo		WINGS	in above	1	0
39	March 4	Nakajima AC Engine Plant	Tokyo	73BW		116	0	0
39	March 4	Secondary Target	Tokyo Docks		WING	in above	94	1
39	March 4	Last Resort Target	Unknown		WING	in above	12	0
39	March 4	Secondary Target	Tokyo Docks		497BG	34	32	1
39	March 4	Secondary Target	Tokyo Docks		498BG	37	30	0
39	March 4	Secondary Target	Tokyo Urban		499BG	not avail	15	0
39	March 4	Secondary Target	Tokyo Urban		500BG	not avail	19	0
39	March 4	Nakajima AC Engine Plant	Tokyo	313BW		76	0	0
39	March 4	Secondary Target	Tokyo Urban		WING	76	68	0
39	March 4	Secondary Target	Tokyo Urban		6BG	20	20	0
39	March 4	Secondary Target	Tokyo Urban		9BG	23	20	0
39	March 4	Target of Opportunity	Shizuoka		9BG	in above	1	0
39	March 4	Target of Opportunity	Maug		9BG	in above	1	0
39	March 4	Secondary Target	Tokyo Urban		504BG	11	10	0
39	March 4	Tokyo Port & Urban —(Second	Tokyo		505BG	21	19	0
39	March 4	Hamamatsu (Last Resort)	Hamamatsu		505BG	in above	2*	0

VII. The XXI Bomber Command

MSN	Date	Target	Location	Wings	Grps	No. Up	Bombed	Lost
39	March 4	Nakajima Aircraft Factory	Tokyo	314BW	19BG	9	not avail	
40	March 9/10	Urban Area	Tokyo	MISSION		325	279	14
40	March 9/10	Urban Area	Tokyo	73BW		165	137	1
40	March 9/10	Target of Opportunity	Tokyo		WING	in above	9	0
40	March 9/10	Urban Area	Tokyo		497BG	37	34	0
40	March 9/10	Urban Area	Tokyo		498BG	42	34	1
40	March 9/10	Secondary Target	Tokyo		498BG	in above	3	0
40	March 9/10	Urban Area	Tokyo		499BG	not avail	44	0
40	March 9/10	Urban Area	Tokyo		500BG	42	31	1
40	March 9/10	Urban Area	Tokyo	313BW		not avail	99	3
40	March 9/10	Urban Area	Tokyo		6BG	not avail	32	0
40	March 9/10	Urban Area	Tokyo		9BG	32	29	2
40	March 9/10	Target of Opportunity	Unknown		9BG	in above	1	0
40	March 9/10	Tokyo Urban Area	Tokyo		504BG	21	18	0
40	March 9/10	Tokyo Urban Area	Tokyo		505BG	not avail	20	1
40	March 9/10	Target of Last Resort	Kasumigoura AF		505BG	in above	1	0
40	March 9/10	Target of Last Resort	Haha Jima		505BG	in above	1	0
40	March 9/10	Target of Last Resort	Chichi Jima		505BG	in above	1	0
40	March 9/10	Target of Last Resort	Guguan		505BG	in above	1	0
40	March 9/10	Urban Area	Tokyo	314BW		54	49	8
40	March 9/10	Urban Area	Tokyo		19BG	28	27	3
40	March 9/10	Urban Area	Tokyo		29BG	26	22	5
41	March 11/12	Urban Area	Nagoya	MISSION		310	285	1
41	March 11/12	Urban Area	Nagoya	73BW		160	145	1
41	March 11/12	Target of Opportunity	Nagoya		WING	in above	3	0
41	March 11/12	Urban Area	Nagoya		497BG	34	34	0
41	March 11/12	Urban Area	Nagoya		498BG	39	38	0
41	March 11/12	Urban Area	Nagoya		499BG	19	40	1
41	March 11/12	Urban Area	Nagoya		500BG	39	36	
41	March 11/12	Urban Area	Nagoya	313BW		not avail	101	
41	March 11/12	Urban Area	Nagoya		6BG	not avail	32	
41	March 11/12	Urban Area	Nagoya		9BG	31	27	2
41	March 11/12	Urban Area	Nagoya		504BG	21	17	
41	March 11/12	Urban Area	Nagoya		505BG	25	25	1
41	March 11/12	Urban Area	Nagoya	314BW		42	39	0
41	March 11/12	Urban Area	Nagoya		19BG	21	20	0
41	March 11/12	Urban Area	Nagoya		29BG	21	19	0
42	March 13/14	Urban Area	Osaka	MISSION		298	274	2
42	March 13/14	Urban Area	Osaka	73BW		141	124	2
42	March 13/14	Target of Opportunity	Osaka		WING	in above	2	0
42	March 13/14	Urban Area	Osaka		497BG	37	35	0
42	March 13/14	Urban Area	Osaka		498BG	30	28	1
42	March 13/14	Urban Area	Osaka		499BG	not avail	not avail	1
42	March 13/14	Urban Area	Osaka		500BG	32	32	0
42	March 13/14	Urban Area	Osaka	313BW		115	109	0
42	March 13/14	Urban Area	Osaka		6BG	not avail	30	0
42	March 13/14	Urban Area	Osaka		9BG	33	33	0
42	March 13/14	Urban Area	Osaka		504BG	21	20	0
42	March 13/14	Urban Area	Osaka		505BG	not avail	26	0
42	March 13/14	Urban Area	Osaka	314BW		45	43	0
42	March 13/14	Urban Area	Osaka		19BG	22	21	0
42	March 13/14	Urban Area	Osaka		29BG	23	22	0
43	March 16/17	Urban Area	Kobe	MISSION		330	308	3
43	March 16/17	Urban Area	Kobe	73BW		154	142	1
43	March 16/17	Target of Opportunity	Kobe		WING	in above	2	0
43	March 16/17	Urban Area	Kobe		497BG	38	38	0
43	March 16/17	Urban Area	Kobe		498BG	41	39	0
43	March 16/17	Urban Area	Kobe		499BG	32	28	0
43	March 16/17	Urban Area	Kobe		500BG	39	37	1
43	March 16/17	Urban Area	Kobe	313BW		128	117	2
43	March 16/17	Urban Area	Kobe		6BG	34	33	0
43	March 16/17	Urban Area	Kobe		9BG	34	30	1
43	March 16/17	Urban Area	Kobe		504BG	24	23	1
43	March 16/17	Urban Area	Kobe		505BG	36	32	0
43	March 16/17	Urban Area	Kobe	314BW		52	47	0

MSN	Date	Target	Location	Wings	Grps	No. Up	Bombed	Lost
43	March 16/17	Urban Area	Kobe		19BG	26	24	0
43	March 16/17	Urban Area	Kobe		29BG	26	23	0
44	March 18/19	Urban Area	Nagoya	MISSION		310	290	2
44	March 18/19	Urban Area	Nagoya	73BW		142	128	1
44	March 18/19	Urban Area	Nagoya		497BG	39	37	0
44	March 18/19	Urban Area	Nagoya		498BG	38	34	0
44	March 18/19	Urban Area	Nagoya		499BG	31	28	0
44	March 18/19	Urban Area	Nagoya		500BG	32	29	0
44	March 18/19	Urban Area	Nagoya	313BW		121	114	1
44	March 18/19	Urban Area	Nagoya		6BG	34	32	0
44	March 18/19	Urban Area	Nagoya		9BG	31	30	0
44	March 18/19	Urban Area	Nagoya		504BG	23	21	0
44	March 18/19	Urban Area	Nagoya		505BG	33	31	1
44	March 18/19	Urban Area	Nagoya	314BW		49	48	0
44	March 18/19	Urban Area	Nagoya		19BG	26	25	0
44	March 18/19	Urban Area	Nagoya		29BG	23	23	0
45	March 24/25	Mitsubishi Aircraft Engine Plant	Nagoya	MISSION		251	223	5
45	March 24/25	Mitsubishi Aircraft Engine Plant	Nagoya	73BW		122	106	4
45	March 24/25	Target of Opportunity	Toyohashi		WING	in above	1	0
45	March 24/25	Target of Opportunity	Shingu		WING	in above	1	0
45	March 24/25	Mitsubishi Aircraft Engine Plant	Nagoya		497BG	30	27	1
45	March 24/25	Mitsubishi Aircraft Engine Plant	Nagoya		498BG	30	26	1
45	March 24/25	Mitsubishi Aircraft Engine Plant	Nagoya		499BG	not avail	28	1
45	March 24/25	Mitsubishi Aircraft Engine Plant	Nagoya		500BG	not avail	25	0
45	March 24/25	Mitsubishi Aircraft Engine Plant	Nagoya	313BW		79	71	2
45	March 24/25	Target of Opportunity	Unknown		WING	in above	1	0
45	March 24/25	Mitsubishi Aircraft Engine Plant	Nagoya		6BG	not avail	18	0
45	March 24/25	Mitsubishi Aircraft Engine Plant	Nagoya		9BG	22	19	1
45	March 24/25	Mitsubishi Aircraft Engine Plant	Nagoya		504BG	14	13	1
45	March 24/25	Mitsubishi Aircraft Engine Plant	Nagoya		505BG	not avail	21	0
45	March 24/25	Mitsubishi Aircraft Engine Plant	Nagoya	314BW		50	48	0
45	March 24/25	Mitsubishi Aircraft Engine Plant	Nagoya		19BG	22	22	0
45	March 24/25	Mitsubishi Aircraft Engine Plant	Nagoya		29BG	28	26	0
46	March 27	Kyushu Airfields	Kyushu	MISSION		161	146	0
46	March 27	Kyushu Airfields	Kyushu	73BW		117	112	0
46	March 27	Tachiari Airport	Kyushu		497BG	30	29	0
46	March 27	Tachiari Airport	Kyushu		498BG	30	29	0
46	March 27	Oita Airfield	Kyushu		499BG	not avail	not avail	0
46	March 27	Oita Airfield	Kyushu		500BG	39	30	0
46	March 27	Kanoya Airfield	Kyushu	314BW		44	0	0
46	March 27	Omura Airfield (Secondary)	Omura		WING	in above	40	0
46	March 27	Omura Airfield	Omura		19BG	22	21	0
46	March 27	Omura Airfield	Omura		29BG	22	22	0
46	March 27	Target of Opportunity	Kami Kushita		19BG	in above	1	0
46	March 27	Target of Opportunity	Chichi Jima		29BG	in above	1	0
46	March 27	Target of Opportunity	Uchinoura		29BG	in above	1	0
47	March 27/28	Mining Mission 1	Shimonoseki Strait	313BW		unknown	105	3
47	March 27/28	Mine Field Mike	Shimonoseki Strait		505BG	24	24	2
47	March 27/28	Mine Field Mike	Shimonoseki Strait		6BG	not avail	30	0
47	March 27/28	Mine Field Mike	Shimonoseki Strait		504BG	49	20	1
47	March 27/28	Mine Field Love	Shimonoseki Strait		9BG	31	31	0
48	March 30/31	Mitsubishi Aircraft Engine Plant	Nagoya	314BW	19BG	14	12	0
48	March 30/31	Target of Opportunity	Seto		19BG	in above	1	0
49	March 30/31	Mining Mission 2	Shimonoseki	313BW		95	87	2
49	March 30/31	Mine Field Item	Kure		6BG	23	23	0
49	March 30/31	Mine Field Item	Kure		9BG	8	in above	0
49	March 30/31	Mine Field Jig	Hiroshima Entry		9BG	11	10	1
49	March 30/31	Mine Field Love	East Shimonoseki		504BG	14	14	0
49	March 30/31	Mine Field Love	East Shimonoseki		505BG	29	in above	0
49	March 30/31	Mine Field Charlie	Hiroshima		9BG	1	1	0
49	March 30/31	Mine Field Roger	Sasebo Approach		504BG	3	3	0
49	March 30/31	Mine Field Roger	Sasebo Approach		9BG	4	4	0
49	March 30/31	Mine Field Roger	Sasebo Approach		6BG	2	2	0
50	March 31	Kyushu Airfields	Kyushu	73/314		149	137	1
	March 31	Oita Airfield	Kyushu		WING	34	32	0

VII. The XXI Bomber Command

MSN	Date	Target	Location	Wings	Grps	No. Up	Bombed	Lost
50	March 31	Tachiari Machine Works	Kyushu		WING	117	105	1
50	March 31	Tachiari Machine Works	Kyushu		497BG	31	29	0
50	March 31	Tachiari Machine Works	Kyushu		498BG	30	27	1
50	March 31	Tachiari Machine Works	Kyushu		499BG	not avail	not avail	0
50	March 31	Tachiari Machine Works	Kyushu		500BG	not avail	24	0
50	March 31	Omura Airfield	Kyushu	314BW		34	32	0
50	March 31	Omura Airfield	Kyushu		19BG	12	10	0
50	March 31	Omura Airfield	Kyushu		29BG	22	22	0
51	April 1/2	Nakajima Aircraft Engine Plant	Tokyo	73BW		124	115	7
51	April 1/2	Nakajima Aircraft Engine Plant	Tokyo		497BG	26	25	1
51	April 1/2	Nakajima Aircraft Engine Plant	Tokyo		498BG	31	29	3
51	April 1/2	Nakajima Aircraft Engine Plant	Tokyo		499BG	not avail	27	1
51	April 1/2	Nakajima Aircraft Engine Plant	Tokyo		500BG	not avail	34	1
52	April 1/2	Mining Mission 3	Kure	313BW	9BG	6	6	0
53	April 2/3	Mine Field Jig- Mission 4	Hiroshima	313BW	504BG	10	9	0
54	April 3/4	Mine Field Item — Mission 5	Hiroshima	313BW	505BG	9	9	0
55	April 3/4	Aircraft Engine Factory	Shizuoka	314W		49	48	0
55	April 3/4	Aircraft Engine Factory	Shizuoka		19BG	25	24	0
55	April 3/4	Aircraft Engine Factory	Shizuoka		29BG	24	24	0
56	April 3/4	Nakajima Aircraft Plant	Koizuma	313BW		78	43	0
56	April 3/4	Secondary Target	Unknown		WING	in above	18	0
56	April 3/4	Nakajima Aircraft Plant	Koizama		6BG	not avail	19	0
56	April 3/4	Nakajima Aircraft Plant	Koizama		9BG	28	22	0
56	April 3/4	Secondary Targets	Koizama		9BG	in above	3	0
56	April 3/4	Nakajima Aircraft Plant	Koizama		504BG	11	6	0
56	April 3/4	Nakajima Aircraft Plant	Koizama		505BG	11	4	0
56	April 3/4	Secondary Target	Tokyo Urban		505BG	in above	16	0
57	April 3/4	Tachikawa Aircraft Mfg Factory	Tachikawa	73BW		115	61	1
57	April 3/4	Secondary — Urban Area	Tachikawa		WING	in above	33	0
57	April 3/4	Last Resort Target	Tachikawa		WING	in above	1	0
57	April 3/4	Aircraft Frame Plant	Tachikawa		497BG	26	17	1
57	April 3/4	Secondary — Urban Area	Tachikawa		497BG	in above	7	0
57	April 3/4	Aircraft Frame Plant	Tachikawa		498BG	29	13	0
57	April 3/4	Secondary Target	Tachikawa		498BG	in above	14	0
57	April 3/4	Kawasaki Aircraft Factory	Tachikawa		499BG	not avail	14	0
57	April 3/4	Kawasaki Aircraft Factory	Tachikawa		500BG	not avail	30	1
58	April 7	Nakajima Aircraft Factory	Tokyo	73BW		111	101	6
58	April 7	Target of Opportunity	Tokyo		WING	in above	2	0
58	April 7	Nakajima Aircraft Factory	Tokyo		497BG	35	29	0
58	April 7	Nakajima Aircraft Factory	Tokyo		498BG	32	31	1
58	April 7	Nakajima Aircraft Factory	Tokyo		499BG	not avail	not avail	2
58	April 7	Nakajima Aircraft Factory	Tokyo		500BG	not avail	9	1
53	April 7	Mitsubishi Aircraft Engine Plant	Nagoya	313BW		132	98	0
53	April 7	Mitsubishi Aircraft Engine Plant	Nagoya		6BG	30	30	1Pirates Log
53	April 7	Mitsubishi Aircraft Engine Plant	Nagoya		9BG	39	31	0
53	April 7	Mitsubishi Aircraft Works	Nagoya		504BG	31	28	0
53	April 7	Mitsubishi Aircraft Works	Nagoya		505BG	32	21	0
53	April 7	Last Resort Target	Hamamatsu		505BG	in above	1	0
53	April 7	Last Resort Target	Ujiyamada		505BG	in above	1	0
53	April 7	Last Resort Target	Akazaki		505BG	in above	1	0
53	April 7	Last Resort Target	Kushimoto		505BG	in above	1	0
53	April 7	Last Resort Target	Owase		505BG	in above	1	0
59	April 7	Mitsubishi Aircraft Engine Plant	Nagoya	MISSION		194	153	2
59	April 7	Secondary Target	Nagoya		WINGS	in above	29	
59	April 7	Mitsubishi Aircraft Engine Plant	Nagoya	313BW		63	55	0
59	April 7	Mitsubishi Aircraft Engine Plant	Nagoya		6BG	not avail	30	0
59	April 7	Mitsubishi Aircraft Engine Plant	Nagoya		9BG	39	31	0
59	April 7	Secondary Target	Ujiyamada		WINGS	194	153	2
59	April 7	Secondary Target	Unknown		WINGS	in above	29	0
59	April 7	Mitsubishi Aircraft Works	Nagoya		504BG	not avail	not avail	
59	April 7	Mitsubishi Aircraft Works	Nagoya		505BG	32	21	0
59	April 7	Last Resort Target	Hamamatsu		505BG	in above	1	0
59	April 7	Last Resort Target	Ujiyamada		505BG	in above	1	0
59	April 7	Last Resort Target	Akazaki		505BG	in above	1	0
59	April 7	Last Resort Target	Kushimoto		505BG	in above	1	0

MSN	Date	Target	Location	Wings	Grps	No. Up	Bombed	Lost
59	April 7	Last Resort Target	Owase		505BG	in above	1	0
59	April 7	Mitsubishi Aircraft Engine Plant	Nagoya	314BW		63	57	2
59	April 7	Mitsubishi Aircraft Engine Plant	Nagoya		19BG	32	29	0
59	April 7	Target of Opportunity	Shingu		19BG	in above	1	0
59	April 7	Mitsubishi Aircraft Engine Plant	Nagoya		29BG	31	28	2
59	April 7	Target of Opportunity	Oitsu		29BG	in above	1	0
59	April 7	Target of Opportunity	Shingu		29BG	in above	1	0
59	April 7	Target of Opportunity	Tamaski A/F		29BG	in above	1	0
59	April 7	Target of Opportunity	Chichi Jima		29BG	in above	1	0
59	April 7	Target of Opportunity	Nakiri		29BG	in above	1	0
60	April 8	Kagoshima	Kyushu	73BW		33	29	0
60	April 8	Kanoya Airfield	Kyushu		500BG	in above	3	1
61	April 8	Kanoya East Airfield	Kyushu	313BW		21	6	0
61	April 8	Kagoshima	Kyushu		WING	in above	13	1
61	April 8	Kanoya East Airfield	Kyushu		6BG	10	10	0
61	April 8	Kagoshima	Kyushu		500BG	not avail	28	
62	April 9/10	Mining Mission 6	Shimonoseki	313BW		20	16	0
62	April 9/10	Shimonoseki Mining	Shimonoseki		6BG	10	10	0
62	April 9/10	Mine Field Mike	Shomonoseki		505BG	10	10	0
63	April 12	Nakajima Aircraft Factory	Tokyo	73BW		119	94	0
63	April 12	Secondary Target	Shizuoka		WING	in above	13	0
63	April 12	Target of Opportunity	Nii Shima A/F		WING	in above	1	0
63	April 12	Target of Opportunity	Miyeke Jima		Wing	in above	1	0
63	April 12	Nakajima Aircraft Factory	Tokyo		497BG	30	24	0
63	April 12	Nakajima Aircraft Factory	Tokyo		498BG	28	17	0
63	April 12	Nakajima Aircraft Factory	Tokyo		499BG	not avail	not avail	0
63	April 12	Nakajima Aircraft Factory	Tokyo		500BG	not avail	29	0
64	April 12	Hodagaya Chemical Works	Koriyama	313BW		75	66	0
64	April 12	Targets of Opportunity	Motomiya		WING	in above	7	0
64	April 12	Targets of Opportunity	Nii Shima A/F		WING	in above	1	0
64	April 12	Targets of Opportunity	Shimoda		WING	in above	1	0
64	April 12	Hodagaya Chemical Works	Koriyama		6BG	not avail	20	0
64	April 12	Hodagaya Chemical Works	Koriyama		9BG	23	20	0
64	April 12	Hodagaya Chemical Works	Koriyama		504BG	18	18	0
64	April 12	Hodagaya Chemical Works	Koriyama		505BG	not avail	not avail	0
65	April 12	Hodagaya Chemical Works	Koriyama	314BW		85	81	3
65	April 12	Target of Opportunity	Koriyama		WING	in above	5	
65	April 12	Hodagaya Chemical Works	Koriyama		19BG	21	22	0
65	April 12	Hodagaya Chemical Works	Koriyama		29BG	22	22	0
65	April 12	Hodagaya Chemical Works	Koriyama		39BG	22	20	0
65	April 12	Hodagaya Chemical Works	Koriyama		330BG	20	17	3
65	April 12	Target of Opportunity	Onehame		330BG	in above	1	0
66	April 12/13	Mining Mission 7	Shimonoseki	313BW	9BG	6	5	0
67	April 13/14	Tokyo Arsenal	Tokyo	MISSION		348	327	7
67	April 13/14	Tokyo Arsenal	Tokyo	73BW		122	115	5
67	April 13/14	Tokyo Arsenal	Tokyo		497BG	30	29	1
67	April 13/14	Tokyo Arsenal	Tokyo		498BG	32	30	0
67	April 13/14	Tokyo Arsenal	Tokyo		499BG	not avail	28	2
67	April 13/14	Tokyo Arsenal	Tokyo		500BG	not avail	28	0
67	April 13/14	Tokyo Arsenal	Tokyo	313BW		115	110	1
67	April 13/14	Tokyo Arsenal	Tokyo		6BG	29	29	0
67	April 13/14	Tokyo Arsenal	Tokyo		9BG	28	28	0
67	April 13/14	Tokyo Arsenal	Tokyo		504BG	26	24	0
67	April 13/14	Tokyo Arsenal	Tokyo		505BG	not avail	35	1
67	April 13/14	Tokyo Arsenal	Tokyo	314BW		113	102	3
67	April 13/14	Tokyo Arsenal	Tokyo		19BG	35	34	1
67	April 13/14	Tokyo Arsenal	Tokyo		29BG	38	36	1
67	April 13/14	Tokyo Arsenal	Tokyo		39BG	24	21	0
67	April 13/14	Tokyo Arsenal	Tokyo		330BG	16	11	1
67	April 13/14	T of O, Targets 213, 214	Tokyo		330BG	in above	1	0
67	April 13/14	T of O — Matsuda Airfield	Hachijo Jima		330BG	in above	2	0
68	April 15/16	Urban Area	Kawasaki	MISSION		219	194	12
68	April 15/16	Urban Area	Kawasaki	313BW		111	95	6
68	April 15/16	Urban Area	Kawasaki		6BG	24	24	0
68	April 15/16	Urban Area	Kawasaki		9BG	33	26	4

VII. The XXI Bomber Command

MSN	Date	Target	Location	Wings	Grps	No. Up	Bombed	Lost
68	April 15/16	Targets of Opportunity	Kawasaki		9BG	in above	4	0
68	April 15/16	Urban Area	Kawasaki		504BG	22	20	1
68	April 15/16	Urban Area	Kawasaki		505BG	32	25	1
68	April 15/16	Urban Area	Kawasaki	314BW		108	102	6
68	April 15/16	Urban Area	Kawasaki		19BG	33	33	3
68	April 15/16	Urban Area	Kawasaki		29BG	32	30	2
68	April 15/16	Urban Area	Kawasaki		39BG	23	23	1
68	April 15/16	Urban Area	Kawasaki		330BG	20	20	0
69	April 15/16	Tokyo Urban Area	Tokyo	73BW		120	109	1
69	April 15/16	Urban Area	Tokyo		497BG	28	23	0
69	April 15/16	Urban Area	Tokyo		498BG	33	30	0
69	April 15/16	Urban Area	Tokyo		499BG	not avail	26	1
69	April 15/16	Urban Area	Tokyo		500BG	34	30	1
70	April 17	Izumi Airfield	Kyushu	73BW		22	20	0
70	April 17	Target of Opportunity	Chichi Jima		WING	in above	1	0
70	April 17	Target of Opportunity	Pagan Island		WING	in above	1	0
70	April 17	Izumi Airfield	Kyushu		499BG	not avail	9	0
70	April 17	Izumi Airfield	Kyushu		500BG	not avail	11	0
71	April 17	Tachiari Airfield	Kyushu		WING	21	21	0
71	April 17	Tachiari Airfield	Kyushu		497BG	11	11	0
71	April 17	Tachiari Airfield	Kyushu		498BG	10	10	0
72	April 17	Kokubu Airfield	Kyushu	313BW		45	32	0
72	April 17	Kokubu Airfield	Kyushu		505BG	24	20	0
72	April 17	Target of Opportunity	Nittigahara A/F		505BG	in above	2	0
73	April 17	Kanoya East Airfield	Kyushu		6BG	21	10	0
73	April 17	Kanoya East Airfield	Kyushu		504BG	in above	in above	0
74	April 17	Nittagahara Airfield	Kyushu	314BW	19BG	10	6	0
75	April 17	Kanoya Airfield	Kyushu	314BW		34	30	0
75	April 17	Kanoya Airfield	Kyushu		29BG	12	9	0
75	April 17	Kanoya Airfield	Kyushu		330BG	11	11	0
75	April 17	Kanoya Airfield	Kyushu		39BG	11	10	0
76	April 18	Tachiari Airfield	Kyushu	73BW		22	20	2
76	April 18	Target of Opportunity	Miyazaki A/F		WING	in above	1	0
76	April 18	Tachiari Airfield	Kyushu		497BG	11	10	2
76	April 18	Tachiari Airfield	Kyushu		498BG	11	10	0
77	April 18	Kanoya East Airfield	Kyushu	313BW		20*	19*	0
77	April 18	Target of Opportunity	Kushira A/F		WING	in above	7	0
77	April 18	Kushira Airfield	Kyushu		6BG	not avail	10	0
77	April 18	Kanoya East Airfield	Kyushu		504BG	5	5	0
78	April 18	Kanoya Airfield	Kyushu	314BW		33	30	0
78	April 18	Kanoya Airfield	Kyushu		29BG	not avail	not avail	
78	April 18	Kanoya Airfield	Kyushu		39BG	11	7	0
78	April 18	Kanoya East with 313BW	Kyushu		39BG	in above	2	0
78	April 18	Kanoya Airfield	Kyushu		330BG	11	10	0
79	April 18	Izumi Airfield	Kyushu	73BW		23	21	0
79	April 18	Target of Opportunity	Kaimon Dake A/F		WING	in above	1	0
79	April 18	Target of Opportunity	Unknown		WING	in above	1	0
79	April 18	Izumi Airfield	Kyushu		499BG	not avail	10	0
79	April 18	Izumi Airfield	Kyushu		500BG	not avail	11	0
80	April 18	Kokubu Airfield	Kyushu	313BW	9BG	22	18	0
80	April 18	Target of Opportunity	Kanoya East A/F		9BG	in above	1	0
80	April 18	Target of Opportunity	Izumi A/F		9BG	in above	1	0
81	April 18	Nittagahara Airfield	Kyushu	314BW	19BG	11	11	0
	April 20	Izumi Airfield	Kyushu	313BW	504BG	6	5	0
82	April 21	Oita Airfield	Kyushu	73BW		30	17	0
82	April 21	Target of Opportunity	Kagoshima		WING	in above	11	0
82	April 21	Target of Opportunity	Tomitaki A/F		WING	in above	1	0
82	April 21	Oita Airfield	Kyushu		499BG	not avail	not avail	0
82	April 21	Oita Airfield	Kyushu		500BG	not avail	9	0
83	April 21	Kanoya East Airfield	Kyushu	313BW	WING	33	27	0
83	April 21	Target of Opportunity	Kanoya		WING	in above	2	0
83	April 21	Target of Opportunity	Kokubu A/F		WING	in above	1	0
83	April 21	Target of Opportunity	Yamakawa Harbor		WING	in above	1	0
83	April 21	Kanoya East Airfield	Kyushu		6BG	23	22	0
83	April 21	Kanoya Airfield	Kyushu		9BG	10	7	0

MSN	Date	Target	Location	Wings	Grps	No. Up	Bombed	Lost
83	April 21	Secondary Target	Kyushu		9BG	in above	1	0
83	April 21	Last Resort Target	Kyushu		9BG	in above	1	0
84	April 21	Kanoya Airfield	Kyushu	314BW		33	27	0
84	April 21	Target of Opportunity	Kaimon Dake A/F		WING	in above	1	0
84	April 21	Kanoya Airfield	Kyushu		29BG	11	10	0
84	April 21	Kanoya Airfield	Kyushu		39BG	11	9	0
84	April 21	Simon DakeA/F	Kyushu		39BG	in above	1	0
84	April 21	Kanoya Airfield	Kyushu		330BG	11	11	0
85	April 21	Usa Airfield	Kyushu	73BW		30	29	0
85	April 21	Usa Airfield	Kyushu		497BG	22	20	0
85	April 21	Usa Airfield	Kyushu		500BG	9	9	0
86	April 21	Kokubu Airfield	Kyushu	313BW		35	34	0
86	April 21	Kokubu Airfield	Kyushu		505BG	23	22	0
86	April 21	Kokubu Airfield	Kyushu		9BG	12	12	0
87	April 21	Kushira Airfield	Kyushu	314BW		31	27	0
87	April 21	Kushira Airfield	Kyushu		29BG	9	8	0
87	April 21	Kushira Airfield	Kyushu		39BG	11	9	0
87	April 21	Kushira Airfield	Kyushu		330BG	11	11	0
88	April 21	Tachiari Airfield	Kyushu	73BW		21	17	0
88	April 21	Tachiari Airfield	Kyushu		498BG	21	17	0
88	April 21	Target of Opportunity	Oyodo A/F		498BG	in above	1	0
89	April 21	Izumi Airfield	Kyushu	313BW	504BG	16	13	0
89	April 21	Target of Opportunity	Miyazake A/F		504BG	in above	2	0
89	April 21	Target of Opportunity	Kanoya East A/F		504BG	in above	1	0
90	April 21	Nittagahara Airfield	Kyushu	314BW	19BG	23	22	0
91	April 22	Izumi Airfield	Kyushu	73BW		22	19	0
91	April 22	Target of Opportunity	Shimizu		WING	in above	1	0
91	April22	Izumi Airfield	Kyushu		499BG	not avail	9	0
91	April22	Izumi Airfield	Kyushu		500BG	not avail	10	0
92	April22	Kushira Airfield	Kyushu	313BW		18	10	0
92	April22	Target of Opportunity	Kanoya A/F		505BG	in above	5	0
92	April22	Kushira Airfield	Kyushu		9BG	9	7	0
92	April22	Kushira Airfield	Kyushu		505BG	not avail	2	0
93	April22	Miyazaki Airfield	Kyushu	314BW		22	21	0
93	April 22	Miyazaki Airfield	Kyushu		19BG	11	11	
93	April 22	Miyazaki Airfield	Kyushu		29BG	11	11	
94	April22	Tomitaka Airfield	Kyushu	73BW		19	18	0
94	April22	Tomitaka Airfield	Kyushu		497BG	11	8	0
94	April22	Tomitaka Airfield	Kyushu		498BG	10	10	0
95	April 22	Kanoya Airfield	Kyushu	313BW		36	28	1
95	April22	Kanoya Airfield	Kyushu		9BG	4	3	0
95	April 22	Kanoya Airfield	Kyushu		6BG	21	16	
95	April22	Kanoya Airfield	Kyushu		505BG	5	5	
96	April 24	Hitachi Aircraft Factory	Tachikawa	MISSION		133	101	5
96	April 24	Hitachi Aircraft Factory	Tachikawa	73BW		45	34	1
96	April 24	Secondary-A/C Plant	Shizuoka		WING	in above	8	0
96	April 24	Hitachi Aircraft Factory	Tachikawa		497BG	11	11	0
96	April 24	Hitachi AircraftEngine Plant	Tachikawa		498BG	11	9	0
96	April 24	Hitachi Aircraft Factory	Tachikawa		499BG	not avail	no avail	1
96	April 24	Hitachi Aircraft Factory	Tachikawa		500BG	11	9	0
96	April 24	Hitachi Aircraft Factory	Tachikawa	313BW		45	29	2
96	April 24	Hitachi Aircraft Factory	Tachikawa		6BG	12	9	0
96	April 24	Hitachi Aircraft Factory	Tachikawa		9BG	12	1	0
96	April 24	Last Resort — Urban Area	Tachikawa		9BG	in above	9	0
96	April 24	Hitachi Aircraft Factory	Tachikawa		504BG	11	9	2
96	April 24	Hitachi Aircraft Factory	Tachikawa		505BG	10	10	0
96	April 24	Bombed Tof O with 313BW	Unknown	314BW	39BG	in above	3	0
96	April 24	Target of Opportunity	Unknown		39BG	in above	1	0
96	April 24	Hitachi Aircraft Factory	Tachikawa	314BG		42	38	2
96	April 24	Hitachi Aircraft Factory	Tachikawa		19BG	11	11	0
96	April 24	Hitachi Aircraft Factory	Tachikawa		29BG	11	11	0
96	April 24	Hitachi Aircraft Factory	Tachikawa		39BG	10	6	0
96	April 24	Hitachi Aircraft Factory	Tachikawa		330BG	10	10	1
96	April 24	Secondary Target	Sukumo A/F	313BW	504BG	16	14	0
96	April 24	Secondary Target	Kochi A/F		504BG	in above	in above	0

VII. The XXI Bomber Command

MSN	Date	Target	Location	Wings	Grps	No. Up	Bombed	Lost
97	April 26	Usa Airfield	Kyushu	73BW	497BG	22	18	0
97	April 26	Target of Opportunity	Tomitaki A/F		497BG	in above	1	0
97	April 26	Target of Opportunity	Hi-Saki A/F		497BG	in above	1	0
98	April 26	Oita Airfield	Kyushu	73BW		22	19	0
98	April 26	Oita Airfield	Kyushu		499BG	22	19	0
98	April 26	Target of Opportunity	Tomitaki A/F		499BG	in above	2	0
99	April 26	Saeki Airfield	Kyushu	73BW		23	19	0
99	April 26	Saeki Airfield	Kyushu		500BG	23	19	0
100	April 26	Tomitaka Airfield	Kyushu	73BW		22	21	0
100	April 26	Tomitaka Airfield	Kyushu		498BG	22	21	0
100	April 26	Target of Opportunity	Usa A/F		498BG	22	21	0
101	April 26	Matsuyama Airfield	Kyushu	313BW		37	18 plus	0
101	April 26	Targets of Opportunity	Kochi A/F		WING	in above	16	0
101	April 26	Targets of Opportunity	Susaki		WING	in above	1	0
101	April 26	Targets of Opportunity	Masaki		WING	in above	1	0
101	April 26	Targets of Opportunity	Susaki A/F		WING	in above	1	0
101	April 26	Targets of Opportunity	Sakumo		WING	in above	1	0
101	April 26	Matsuyama Airfield	Kyushu		6BG	not avail	18	0
101	April 26	Matsuyama Airfield	Kyushu		505BG	not avail	not avail	0
102	April 26	Nittagahara Airfield	Kyushu	313BW		18	18	0
102	April 26	Nittagahara Airfield	Kyushu		505BG	18	18	0
102	April 26	Targets of Opportunity	Kaimon Dake AF		505BG	in above	1	0
102	April 26	Targets of Opportunity	Miyazaki A/F		505BG	in above	1	0
103	April 26	Miyazaki Airfield	Kyushu	313BW	9BG	21	19	0
103	April 26	Secondary Target	Kokubu A/F		9BG	in above	1	0
104	April 26	Kanoya Airfield	Kyushu	314BW	19BG	22	19	0
104	April 26	Targets of Opportunity	Sakamoto		19BG	in above	1	0
104	April 26	Targets of Opportunity	Kaimon Dace		19BG	in above	1	0
105	April 26	Kushira Airfield	Kyushu	314BW	29BG	22	13	0
105	April 26	Target of Opportunity	Kokubu A/F		29BG	in above	5	0
105	April 26	Target of Opportunity	Chiran A/F		29BG	in above	1	0
105	April 26	Target of Opportunity	Kanoya A/F		29BG	in above	1	0
105	April 26	Target of Opportunity	Tanaga Shima		29BG	in above	2	0
106	April 26	Kokubu Airfield	Kyushu	314BW	39BG	22	17	0
106	April 26	Targets of Opportunity	Miyakonojo A/F		39BG	in above	2	0
106	April 26	Targets of Opportunity	Nagasaki Docks		39BG	in above	1	0
106	April 26	Targets of Opportunity	Shibuski A/F		39BG	in above	1	0
107	April 26	Miyakonojo Airfield	Kyushu	313BW		21	17	0
107	April 26	Miyakonojo Airfield	Kyushu		504BG	9	8	0
107	April 26	Miyakonojo Air Field	Kyushu	314BW	330BG	20	2	0
107	April 26	Secondary Target	Miyazaki AF		330BG	in above	15	0
107	April 26	Target of Opportunity	Yanikawa		330BG	in above	1	0
107	April 26	Target of Opportunity	Makurazaki		330BG	in above	1	0
108	April 27	Izumi Airfield	Kyushu	73BW		22	21	1
108	April 27	Izumi Airfield	Kyushu		499BG	11	10	
108	April 27	Izumi Airfield	Kyushu		500BG	11	11	
109	April 27	Miyazaki Airfield	Kyushu	73BW		23	21	0
109	April 27	Miyazaki Airfield	Kyushu		497BG	11	10	0
109	April 27	Miyazaki Airfield	Kyushu		498BG	11	11	0
110	April 27	Kokubu Airfield	Kyushu	313BW		22	17	0
110	April 27	Target of Opportunity	Kanoya A/F		WING	in above	1	0
110	April 27	Kokubu Airfield	Kyushu		9BG	10	9	0
110	April 27	Kokubu Airfield	Kyushu		505BG	12	8	0
111	April 27	Miyakonojo Airfield	Kyushu	313BW		18	14	0
111	April 27	Miyakonojo Airfield	Kyushu		6BG	not avail	6	0
111	April 27	Miyakonojo Airfield	Kyushu		504BG	not avail	8	0
112	April 27	Kanoya Airfield	Kyushu	314BW		21	21	0
112	April 27	Kanoya Airfield	Kyushu		19BG	11	10	0
112	April 27	Kanoya Airfield	Kyushu		330BG	10	10	0
113	April 27	Kushira Airfield	Kyushu	314BW		19	15	1
113	April 27	Target of Opportunity	Kanoya A/F		WING	in above	1	0
113	April 27	Target of Opportunity	Makurazaki A/F		WING	in above	1	0
113	April 27	Kushira Airfield	Kyushu		29BG	7	4	0
113	April 27	Kushira Airfield	Kyushu		39BG	12	11	1
114	April 28	Izumi Airfield	Kyushu	73BW	500BG	25	23	0

MSN	Date	Target	Location	Wings	Grps	No. Up	Bombed	Lost
115	April 28	Miyazaki Airfield	Kyushu	73BW	499BG	21	20	1
116	April 28	Kokubu Airfield	Kyushu	313BW	9BG	20	17	1
117	April 28	Miyakonojo Airfield	Kyushu	313BW	6BG	19	18	0
117	April 28	Target of Opportunity	Tanega Shima AF		6BG	in above	1	0
118	April 28	Kanoya Airfield	Kyushu	314BW		23	22	1
118	April 28	Kanoya Airfield	Kyushu		19BG	11	11	
118	April 28	Kanoya Airfield	Kyushu		330BG	12	10	0
119	April 28	Kushira Airfield	Kyushu		WING	23	22	2
119	April 28	Target of Opportunity	Kanoya A/F		WING	in above	1	0
119	April 28	Target of Opportunity	Naka Fukara		WING	in above	1	0
119	April 28	Kushira Airfield	Kyushu		29BG	11	10	
119	April 28	Kushira Airfield	Kyushu		39BG	12	11	
120	April 29	Miyazaki Airfield	Kyushu	73BW	497BG	22	19	2
121	April 29	Miyakonojo Airfield	Kyushu	73BW	498BG	23	22	0
122	April 29	Kokubu Airfield	Kyushu	313BW	505BG	22	22	0
123	April 29	Kanoya East Airfield	Kyushu	313BW	504BG	15	14	0
124	April 29	Kanoya Airfield	Kyushu	314BW		20	18	0
124	April 28	Kanoya Airfield	Kyushu		19BG	11	9	0
124	April 28	Kanoya	Kyushu		330BG	9	9	0
125	April 29	Kushira Airfield	Kyushu	314BW		20	16	0
125	April 29	Kushira Airfield	Kyushu		29BG	9	9	0
125	April 29	Kushira Airfield	Kyushu		39BG	11	7	0
126	April 30	Tachikawa Air Arsenal	Tachikawa	MISSION		106	69	0
126	April 30	Tachikawa Air Arsenal	Tachikawa	73BW		70	55	0
126	April 30	Target of Opportunity	Torishima		WING	in above	2	0
126	April 30	Tachikawa Air Arsenal	Hamamatsu		497BG	11	9	0
126	April 30	Tachikawa Air Arsenal	Hamamatsu		498BG	11	11	0
126	April 30	Tachikawa Air Arsenal	Hamamatsu		499BG	not avail	15	0
126	April 30	Tachikawa Air Arsenal	Hamamatsu		500BG	not avail	20	0
126	April 30	Tachikawa Air Depot	Tachikawa	313BW		40	14	0
126	April 30	Secondary Target	Hamamatsu		WING	in above	9	0
126	April 30	Target of Opportunity	Hamamatsu A/F		WING	in above	8	0
126	April 30	Target of Opportunity	Toyohashi		WING	in above	5	0
126	April 30	Target of Opportunity	Hachijo Jima		WING	in above	1	0
126	April 30	Tachikawa Air Arsenal	Hamamatsu		6BG	not avail	7	0
126	April 30	Tachikawa Air Arsenal	Hamamatsu		9BG	17	15	0
126	April 30	Tachikawa Air Arsenal	Hamamatsu		504BG	5	0	0
126	April 30	Secondary Target	Unknown		504BG	in above	5	0
126	April 30	Tachikawa Air Arsenal	Hamamatsu		505BG	not avail	7	0
127	April 30	Kanoya Airfield	Kyushu	314BW	19BG	11	10	0
128	April 30	Kanoya East Airfield	Kyushu	314BW	19BG	11	10	0
129	April 30	Kokubu Airfield	Kyushu	314BW	39BG	10	7	0
130	April 30	Oita Airfield	Kyushu	314BW	29BG	11	10	0
131	April 30	Tomitaka Airfield	Kyushu	314BW	330BG	10	10	0
132	April 30	Saeki Airfield	Kyushu	314BW	29BG	11	11	0
133	April 30	Tachiarai Airfield	Kyushu	314BW	29BG	11	9	0
133	April 30	Target of Opportunity	Kochi A/F		29BG	in above	1	0
134	April 30	Miyazaki Airfield	Kyushu	314BW	29BG	11	11	0
135	April 30	Miyakonojo Airfield	Kyushu	314BW	29BG	11	11	0
136	April 30	Kanoya Airfield	Kyushu	314BW	19BG	11	8	0
136	April 30	Target of Opportunity	Kokubu Airfield		19BG	in above	2	0
136	April 30	Target of Opportunity	Tanega Shima AF		19BG	in above	1	0
137	May 3	Kanoya East Airfield	Kyushu	314BW	19BG	11	11	1
138	May 3	Kokubu Airfield	Kyushu		19BG	11	9	1
138	May 3	Target of Opportunity	Tanega Shima AF		19BG	in above	1	0
139	May 3/4	Mining Mission 8	Shimonoseki	313BW		100	91	0
139	May 3/4	Mine Field Able	Kobe/Osaka		504BG	16	16	0
139	May 3/4	Mine Field Able	Kobe/Osaka		505BG	22	19	0
139	May 3/4	Mine Field Mike	Shimonoseki		505BG	4	4	0
139	May 3/4	Mine Field Mike	Shimonoseki		9BG	24	19	0
139	May 3/4	Secondary Target	Shimonoseki		9BG	in above	1	0
139	May 3/4	Mine Field Love	Moji Area		6BG	14	14	0
139	May 3/4	Mine Field Love	Suo Nada		6BG	20	19	0
140	May 3/4	Oita Airfield	Kyushu	314BW	39BG	22	11	1
141	May 4	Omura Airfield	Kyushu	314BW	330BG	10	10	0

VII. The XXI Bomber Command

MSN	Date	Target	Location	Wings	Grps	No. Up	Bombed	Lost
141	May 4	Target of Opportunity	Kushikino A/F		330BG	in above	1	0
142	May 4	Saeki Airfield	Kyushu	314BW	39BG	9	9	0
143	May 4	Matsuyama Naval Air Station	Kyushu	314BW	330BG	19	17	0
143	May 4	Target of Opportunity	Lomuta		330BG	in above	1	0
143	May 4	Target of Opportunity	Wakagama		330BG	in above	1	0
144	May 4	Oita Airfield	Kyushu	314BW	29BG	17	17	0
145	May 4	Tachiari A/F	Kyushu	314BW	29BG	11	10	2
146	May 5	Hiro Naval Aircraft Factory	Kure	MISSION		172	148	2
146	May 5	Hiro Naval Aircraft Factory	Kure	58BW		42	37	0
146	May 5	Hiro Naval Aircraft Factory	Kure		40BG	31	27	0
146	May 5	Target of Opportunity	Nageoka		40BG	in above	1	0
146	May 5	Hiro Naval Aircraft Factory	Kure		462BG	11	10	0
146	May 5	Hiro Naval Aircraft Factory	Kure	73BW		130	111	3
146	May 5	Target of Opportunity	Kure		WING	in above	3	
146	May 5	Hiro Naval Aircraft Factory	Kure		497BG	33	25	1
146	May 5	Hiro Naval Aircraft Factory	Kure		498BG	36	33	1
146	May 5	Hiro Naval Aircraft Factory	Kure		499BG	not avail	24	0
146	May 5	Hiro Naval Aircraft Factory	Kure		500BG	not avail	32	0
147	May 5	Kanoya Airfield	Kyushu	314BW	19BG	11	10	1
148	May 5	Chiran Airfield	Kyushu		19BG	11	8	0
149	May 5	Ibusuki Airfield	Kyushu		19BG	10	10	0
150	May 5	Mining Mission 9	See Below	313BW		99	90	0
150	May 5	Mine Field Able	Kobe-Osaka		505BG	12	in above	0
150	May 5/6	Mine Field Baker	Aki/Nada		6BG	5	in above	0
150	May 5/6	Mine Field Dog	Shoda Shima		9BG	28	in above	0
150	May 5/6	Mine Field Easy	Tokyo & Ise Bays		504BG	10	in above	0
150	May 5/6	Mine Field Fox	Bingo Nada		6BG	20	in above	0
150	May 5/6	Mine Field Jig/Item	Hiroshima/Kure		504BG	7	in above	0
150	May 5/6	Mine Field King	Tokuyama		505BG	8	in above	0
150	May 5/6	Mine Field Oboe	Tokyo		504BG	4	in above	0
150	May 5/6	Mine Field Tare	Nagoya		505BG	5	in above	0
151	May 5/6	Kanoya Airfield	Kyushu		6BG	10	10	0
152	May 7	Ibusuki Airfield	Kyushu		6BG	11	10	0
153	May 7	Oita Airfield	Kyushu		505BG	10	10	2
154	May 7	Usa Airfield	Kyushu		505BG	11	11	1
155	May 7	Kanoya Airfield	Kyushu	314BW	504BG	21	17	0
155	May 7	Target of Opportunity	Omura City		504BG	in above	1	0
156	May 8	Miyakonojo Airfield	Kyushu	313BW	504BG	12	12	0
157	May 8	Oita Airfield	Kyushu	313BW	9BG	12	11	0
158	May 8	Matsuyama West Airfield	Shikoku	313BW	9BG	22	16	0
159	May 8	Matsuyama West Airfield	Shikoku	313BW	9BG	23	15	0
159	May 8	Target of Opportunity	Kumamota A/F		9BG	in above	2	0
160	May 10	Usa Airfield	Kyushu	313BW	6BG	22	15	0
160	May 10	Target of Opportunity	Matsuyama A/F		6BG	in above	2	0
160	May 10	Target of Opportunity	Nakamura		6BG	in above	1	0
160	May 10	Target of Opportunity	Nobeoka		6BG	in above	1	0
160	May 10	Target of Opportunity	Susaki Docks		6BG	in above	1	0
	May 10	Kyushu Airfields	Unknown	313BW	9BG	23	16	0
	May 10	Target of Opportunity	Unknown		9BG	in above	2	0
161	May 10	Miyazaki Airfield	Kyushu	313BW	505BG	12	7	0
161	May 10	Target of Opportunity	Tanoga Naval		505BG	in above	1	0
162	May 10	Kanoya Airfield	Kyushu	313BW	505BG	12	4	0
162	May 10	Target of Opportunity	Kyushu		505BG	in above	6	0
163	May 10	Naval Fueling Station	Tokuyama	73BW		64	56	0
163	May 10	Target of Opportunity	Iwakuni		WING	in above	1	0
163	May 10	Target of Opportunity	Otake Refinery		WING	in above	1	0
163	May 10	Naval Fueling Station	Tokuyama		497BG	33	28	0
163	May 10	Naval Fueling Station	Tokuyama		498BG	34	29	0
164	May 10	Naval Coal Yard/Briquette Plant	Tokuyama	73BW		65	56	0
164	May 10	Target of Opportunity	Coastal Town		WING	in above	1	0
164	May 10	Target of Opportunity	Chichi Jima		WING	in above	1	0
164	May 10	Naval Coal Yard/Briquette Plant	Tokuyama		499BG	not avail	28	0
164	May 10	Naval Coal Yard/Briquette Plant	Tokuyama		500BG	not avail	28	0
165	May 10	Oil Refinery	Otake	314BW		132	112	1
165	May 10	Target of Opportunity	Tokugama		WING	in above	7	0

MSN	Date	Target	Location	Wings	Grps	No. Up	Bombed	Lost
165	May 10	Target of Opportunity	Oshima Oil		WING	in above	3	0
165	May 10	Target of Opportunity	Shimuzo		WING	in above	1	0
165	May 10	Target of Opportunity	Saeki Airfield		WING	in above	1	0
165	May 10	Target of Opportunity	Uwa Jima		WING	in above	1	0
165	May 10	Target of Opportunity	Bofu Airfield		WING	in above	1	0
165	May 10	Oil Refinery	Otake		19BG	not avail	33	0
165	May 10	Oil Refinery	Otake		29BG	not avail	31	0
165	May 10	Oil Refinery	Otake		39BG	33	29	0
165	May 10	Oil Refinery	Otake		330BG	33	33	0
166	May 10	Oshima Oil Storage	Oshima	58BW		88	80	0
166	May 10	Naval Oil Storage	Oshima		40BG	31	28	0
166	May 10	Target of Opportunity	Sukumo Seaplane		40BG	in above	1	0
166	May 10	Naval Oil Storage	Oshima		444BG	24	20	0
166	May 10	Target of Opportunity — Ship	32-57N, 132-58E		444BG	in above	1	0
166	May 10	Target of Opportunity	Otake Oil Refinery		444BG	in above	2	0
166	May 10	Naval Oil Storage	Oshima		462BG	33	32	0
167	May 10	Oita Airfield	Kyushu	313BW	504BG	20	17	0
167	May 10	Target of Opportunity	Miyachi		504BG	in above	1	0
168	May 11	Saeki Airfield	Kyushu	313BW	505BG	11	7	0
168	May 11	Target of Opportunity	shimo Kawaguchi		505BG	in above	1	0
169	May 11	Nittagahara Airfield	Kyushu	313BW	6BG	11	5	0
169	May 11	Targets of Opportunity	Hoso Shima NB		6BG	in above	6	0
169	May 11	Targets of Opportunity	Kochi A/F		6BG	in above	1	0
169	May 11	Targets of Opportunity	Kubokawa		6BG	in above	1	0
169	May 11	Targets of Opportunity	Suma		6BG	in above	1	0
169	May 11	Targets of Opportunity	Shibushi		6BG	in above	1	0
169	May 11	Targets of Opportunity	Tanega Shima		6BG	in above	1	0
170	May 11	Miyazaki Airfield	Kyushu	313BW	9BG	12	11	0
171	May 11	Miyakonojo Airfield	Kyushu	313BW	9BG	11	10	0
172	May 11	Kawanishi Aircraft Factory	Kobe	MISSION		102	92	1
172	May 11	Kawanishi Aircraft Factory	Kobe	58BW		38	35	0
172	May 11	Kawanishi Aircraft Factory	Kobe		444BG	13	12	0
172	May 11	Target of Opportunity	Sakimahara		444BG	in above	1	0
172	May 11	Kawanishi Aircraft Factory	Kobe		468BG	25	23	0
172	May 11	Kawanishi Aircraft Factory	Fukae	314BW		40	40	1
172	May 11	Kawanishi Aircraft Factory	Fukae		19BG	not avail	33	0
172	May 11	Kawanishi Aircraft Factory	Fukae		29BG	not avail	9	0
172	May 11	Kawanishi Aircraft Factory	Fukae		39BG	11	10	0
172	May 11	Kawanishi Aircraft Factory	Fukae		330BG	11	11	0
172	May 11	Kawanishi Aircraft Factory	Kobe	73BW		24	17	0
172	May 11	Kawanishi Aircraft Factory	Kobe		497BG	11	7	0
172	May 11	Kawanishi Aircraft Factory	Kobe		500BG	not avail	10	0
173	May 13/14	Mining Mission 10	See Below	313BW	9BG	12	12	0
173	May 13/14	Mine Field Mike	Shimonoseki		9BG	3	3	0
173	May 13/14	Mine Field Love	Shimonoseki		9BG	5	5	0
173	May 13/14	Mine Field Uncle	Niigata		9BG	4	4	0
174	May 14	Urban Area	Nagoya	MISSION		524	472	11
174	May 14	Urban Area	Nagoya	58BW		141	124	2
174	May 14	Target of Opportunity	Biwa Lake Area		WING	in above	1	
174	May 14	Target of Opportunity	Unknown		WING	in above	1	
174	May 14	Urban Area	Nagoya		40BG	35	32	
174	May 14	Target of Opportunity	Unknown		40BG	in above	1	
174	May 14	Urban Area	Nagoya		444BG	32	28	
174	May 14	Urban Area	Nagoya		462BG	36	33	
174	May 14	Urban Area	Nagoya		468BG	38	31	
174	May 14	Target of Opportunity	Neve (sp?) Lake		468BG	in above	1	
174	May 14	Urban Area	Nagoya	73BW		162	145	2
174	May 14	Target of Opportunity	Kushimoto		WING	in above	1	0
174	May 14	Urban Area	Nagoya		497BG	44	36	1
174	May 14	Urban Area	Nagoya		498BG	31	29	0
174	May 14	Urban Area	Nagoya		499BG	Not avail	40	1
174	May 14	Urban Area	Nagoya		500BG	not avail	41	
174	May 14	Urban Area	Nagoya	313BW		79	68	3
174	May 14	Target of Opportunity	Nagashima		WING	in above	2	0
174	May 14	Target of Opportunity	Nishii		WING	in above	1	0

VII. The XXI Bomber Command

MSN	Date	Target	Location	Wings	Grps	No. Up	Bombed	Lost
174	May 14	Target of Opportunity	Suita		WING	in above	1	0
174	May 14	Urban Area	Nagoya		6BG	32	25	0
174	May 14	Urban Area	Nagoya		504BG	17	13	0
174	May 14	Secondary Target	Nagoya		504BG	in above	2	0-
174	May 14	Urban Area	Nagoya		505BG	not avail	not avail	2
174	May 14	North Urban Area	Nagoya	314BW		144	135	4
174	May 14	Urban Area	Nagoya		19BG	not avail	39	
174	May 14	Urban Area	Nagoya		29BG	Not avail	38	
174	May 14	Urban Area	North Nagoya		39BG	33	32	1
174	May 14	Urban Area	North Nagoya		330BG	32	32	0
175	May 16/17	Mining Mission 11	Shimonoseki	313BW	WING	30	25	0
175	May 16/17	Secondary Target	Shimonoseki		WING	in above	2	0
175	May 16/17	Mine Field Mike	Shimonoseki		9BG	24	20	0
175	May 16/17	Mine Field Zebra	Maizuru		9BG	6	5	
176	May 16/17	South Urban Area	Nagoya	MISSION		516	439	3
176	May 16/17	Urban Area	Nagoya	58BW		140	122	7
176	May 16/17	Urban Area	Nagoya		40BG	32	29	0
176	May 16/17	Target of Opportunity	Matsuyaka		40BG	in above	1	0
176	May 16/17	Urban Area	Nagoya		444BG	32	26	0
176	May 16/17	Target of Opportunity	Koshimoto		444BG	in above	1	0
176	May 16/17	Urban Area	Nagoya		462BG	37	33	0
176	May 16/17	Urban Area	Nagoya		468BG	39	34	0
176	May 16/17	Target of Opportunity	Yokosuka		468BG	in above	1	0
176	May 16/17	Target of Opportunity	Osaka		468BG	in above	1	0
176	May 16/17	Target of Opportunity	Okura		468BG	in above	1	0
176	May 16/17	Urban Area	Nagoya	73BW		154	137	0
176	May 16/17	Target of Opportunity	Shirahama		WING	in above	1	0
176	May 16/17	South Nagoya Urban	Nagoya		497BG	38	35	0
176	May 16/17	Urban Area	Nagoya		498BG	40	38	0
176	May 16/17	Urban Area	Nagoya		499BG	not avail	25	
176	May 16/17	Urban Area	Nagoya		500BG	not avail	40	
176	May 16/17	Urban Area	Nagoya	313BW		93	79	5
176	May 16/17	Target of Opportunity	Shingu		WING	in above	1	0
176	May 16/17	Target of Opportunity	Kushimoto N.B.		WING	in above	1	0
176	May 16/17	Target of Opportunity	Kushimoto		WING	in above	1	0
176	May 16/17	Urban Area	Nagoya		6BG	33	33	0
176	May 16/17	Urban Area	Nagoya		504BG	not avail	not avail	
176	May 16/17	South Nagoya Urban	Nagoya	314BW		131	127	0
176	May 16/17	Target of Opportunity	Toyohashi		WING	in above	1	0
176	May 16/17	Target of Opportunity	Kushimoto		WING	in above	1	0
176	May 16/17	South Nagoya Urban	Nagoya		19BG	not avail	37	
176	May 16/17	South Nagoya Urban	Nagoya		29BG	not avail	30	
176	May 16/17	South Nagoya Urban	Nagoya		39BG`	32	28	0
176	May 16/17	South Nagoya Urban	Nagoya		330BG	32	32	1
177	May 18/19	Mining Mission 12	See Below	313BW	9BG	34	30	0
177	May 18/19	Mine Field Love	Shimonoseki		9BG	22	18	0
177	May 18/19	Mine Field Zebra	Tsuruga		9BG	12	12	0
178	May 19	Aircraft Factory	Tachikawa	MISSION		309	272	18
178	May 19	Aircraft Factory	Tachikawa	58BW		67	60	4
178	May 19	Aircraft Factory	Tachikawa		40BG	24	23	1
178	May 19	Target of Opportunity	Kofu		40BG	in above	1	
178	May 19	Aircraft Factory	Tachikawa		444BG	21	17	
178	May 19	Target of Opportunity	Matsugami		444BG	in above	1	
178	May 19	Target of Opportunity	Shimata		444BG	in above	1	
178	May 19	Aircraft Factory	Tachikawa		462BG	22	20	
178	May 19	Target of Opportunity	Tokyo		462BG	inabove	1	
178	May 19	Tachikawa Arsenal	Hamamatsu	73BW		94	83	0
178	May 19	Target of Opportunity	Coastal Area		WING	in above	1	0
178	May 19	Target of Opportunity	Toyohashi		WING	in above	1	0
178	May 19	Tachikawa Arsenal	Hamamatsu		497BG	22	22	0
178	May 19	Tachikawa Arsenal	Hamamatsu		498BG	23	20	0
178	May 19	Tachikawa Arsenal	Hamamatsu		499BG	23	20	0
178	May 19	Tachikawa Aircraft Compamnu	Hamamatsu		500BG	23	23	
178	May 19	Tachikawa Arsenal	Hamamatsu	313BW		64	53	7
178	May 19	Target of Opportunity	Shimada Airfield		WING	in above	1	0

MSN	Date	Target	Location	Wings	Grps	No. Up	Bombed	Lost
178	May 19	Target of Opportunity	Shizuoka		WING	in above	1	0
178	May 19	Target of Opportunity	Hachijo Jima		WING	in above	1	0
178	May 19	Tachikawa Arsenal	Hamamatsu		6BG	30	25	
178	May 19	City of Hamamatsu	Hamamatsu	314BW		88	82	5
178	May 19	Target of Opportunity	Shizuoka		WING	in above	1	0
178	May 19	Target of Opportunity	Tatesama		WING	in above	1	0
178	May 19	Target of Opportunity	Shimada Airfield		WING	in above	1	0
178	May 19	Target of Opportunity	Tokyo		WING	in above	1	0
178	May 19	City of Hamamatsu	Hamamatsu		19BG	not avail	22	0
178	May 19	City of Hamamatsu	Hamamatsu		29BG	not avail	22	
178	May 19	City of Hamamatsu	Hamamatsu		39BG	22	16	0
178	May 19	Target of Opportunity	Shizuiko Urban		39BG	in above	1	0
178	May 19	Target of Opportunity	Shimoda A/F		39BG	in above	1	0
178	May 19	City of Hamamatsu	Hamamatsu		330BG	22	22	
179	May 20/21	Mining-Mission 13	Shimonoseki	313BW	9BG	32	30	3
179	May 20/21	Mine field Mike	Shimonoseki		9BG	23	22	0
179	May 20/21	Mine Field Love	Shimonoseki		9BG	5	4	0
179	May 20/21	Mine Field Zebra	Maizura Bay		9BG	4	4	0
180	May 22/23	Mining Mission 14	Shimonoseki	313BW	9BG	32	30	1
180	May 22/23	Mine Field Mike	Shimonoseki		9BG	9	8	0
180	May 22/23	Mine Field Mike	Shimonoseki		9BG	9	8	0
180	May 22/23	Mine Field Love	Shimonoseki		9BG	14	14	0
181	May 23/24	Urban Area	Tokyo	MISSION		558	520	17
181	May 23/24	Urban Area	Tokyo	58BW		127	119	4
181	May 23/24	Urban Area	Tokyo		40BG	32	30	1
181	May 23/24	Target of Opportunity	Sagara		40BG	in above	1	0
181	May 23/24	Urban Area	Tokyo		444BG	34	32	0
181	May 23/24	Urban Area	Tokyo		462BG	30	27	1
181	May 23/24	Urban Area	Tokyo		468BG	32	29	1
181	May 23/24	Target of Opportunity	Shizuoka		468BG	in above	1	0
181	May 23/24	South Tokyo Urban	Tokyo	73BW		170	156	6
181	May 23/24	Target of Opportunity	Shizuoka		WING	in above	2	0
181	May 23/24	South Tokyo Urban	Tokyo		497BG	41	35	2
181	May 23/24	South Tokyo Urban	Tokyo		498BG	44	44	4
181	May 23/24	Urban Area	Tokyo		499BG	Not avail	37	0
181	May 23/24	Urban Area	South Tokyo		500BG	not avail	42	2
181	May 23/24	Urban Area	Tokyo	313BW		103	87	5
181	May 23/24	Urban Area	Tokyo		6BG	35	31	3
181	May 23/24	Urban Area	Tokyo		504BG	31	29	2
181	May 23/24	Urban Area	Tokyo		505BG	not avail	35	
181	May 23/24	Urban Area	Tokyo	314BW		160	148	2
181	May 23/24	Target of Opportunity	Hamamatsu		WING	in above	1	0
181	May 23/24	Urban Area	Tokyo		29BG	not avail	44	
181	May 23/24	South Tokyo Urban	Tokyo		WING	not avail	157	2
181	May 23/24	South Tokyo Urban	Tokyo		19BG	43	43	0
181	May 23/24	South Tokyo Urban	Tokyo		29BG	not avail	44	0
181	May 23/24	South Tokyo Urban	Tokyo		39BG	38	35	1
181	May 23/24	South Tokyo Urban	Tokyo		330BG	35	35	1
182	May 24/25	Mining Mission 15	Fushiki/Niigata	313BW	9BG	30	27	0
182	May 24/25	Secondary Targets	Fushiki/Niigata		9BG	in above	2	0
182	May 24/25	Mine Field Uncle	Niigata		9BG	11	9	0
182	May 24/25	Mine field Nan	Fushiki		9BG	10	9	0
182	May 24/25	Mine Field Mike	Shimonoseki		9BG	9	7	0
183	May 25/26	Urban Area	Tokyo	MISSION		498	464	26
183	May 25/26	Urban Area	Tokyo	58BW		133	119	11
183	May 25/26	Urban Area	Tokyo		40BG	33	30	2
183	May 25/26	Urban Area	Tokyo		444BG	34	33	5
183	May 25/26	Urban Area	Tokyo		462BG	31	30	1
183	May 25/26	Urban Area	Tokyo		468BG	35	26	1
183	May 25/26	Target of Opportunity	Hamamatsu		468BG	in above	1	0
183	May 25/26	Target of Opportunity	Tateyama Hoto		468BG	in above	1	0
183	May 25/26	Urban Area	Tokyo	73BW		144	132	4
183	May 25/26	Target of Opportunity	Hamamatsu		WING	in above	1	0
183	May 25/26	Target of Opportunity	Shizuoka		WING	in above	1	0
183	May 25/26	Target of Opportunity	Hachijo Jima		WING	in above	1	0

VII. The XXI Bomber Command

MSN	Date	Target	Location	Wings	Grps	No. Up	Bombed	Lost
183	May 25/26	Tokyo Palace Area	Tokyo		497BG	35	30	1
183	May 25/26	Urban Area	Tokyo		498BG	33	30	1
183	May 25/26	Urban Area	Tokyo		499BG	not avail	35	1
183	May 25/26	Urban Area	Tokyo		500BG	not avail	40	1
183	May 25/26	Urban Area	Tokyo	313BW		86	82	9
183	May 25/26	Urban Area	Tokyo		6BG	not avail	24	1
183	May 25/26	Urban Area	Tokyo		504BG	26	26	2
183	May 25/26	Urban Area	Tokyo		505BG	not avail	not avail	4
183	May 25/26	South Central Tokyo Urban Area	Tokyo	314BW		138	125	2
183	May 25/26	South Central Tokyo Urban Area	Tokyo		19BG	not avail	34	
183	May 25/26	South Central Tokyo Urban Area	Tokyo		29BG	not avail	40	
183	May 25/26	South Central Tokyo Urban Area	Tokyo		39BG	36	33	1
183	May 25/26	South Central Tokyo Urban Area	Tokyo		330BG	28	18	0
184	May 25/26	Mining Mission 16	See Below	313BW		30	29	0
184	May 25/26	Mine Field Mike	Shimonoseki		9BG	2	2	0
184	May 25/26	Mine Field Nan	Fushiki		9BG	6	6	0
184	May 25/26	Mine Field Love	Shimonoseki		9BG	7	7	0
184	May 25/26	Mine Field Charlie	Fukuoka Bay		9BG	15	14	0
185	May 27	Mining Mission 17	See Below	313BW	9BG	11	9	1
185	May 27	Mine Field Charlie	Fukuoka Bay		9BG	1	1	0
185	May 27	Mine Field Love	Shimonoseki		9BG	10	8	0
186	May 29	Urban Area	Yokohama	MISSION		510	454	7
186	May 29	Urban Area	Yokohama	58BW		132	122	2
186	May 29	Urban Area	Yokohama		40BG	31	30	
186	May 29	Urban Area	Yokohama		444BG	29	26	
186	May 29	Urban Area	Yokohama		462BG	34	30	
186	May 29	Target of Opportunity	Hachijo Jima		462BG	in above	1	
186	May 29	Urban Area	Yokohama		468BG	38	36	
186	May 29	Target of Opportunity	Hachijo Jima		468BG	in above	1	
186	May 29	Urban Area	Yokohama	73BW		152	136	0
186	May 29	Target of Opportunity	Aoga Shima		WING	in above	1	0
186	May 29	Target of Opportunity	Hamamatsu		WING	in above	1	0
186	May 29	Target of Opportunity	Kawasaki		WING	in above	1	0
186	May 29	Target of Opportunity	Chichi Jima		WING	in above	1	0
186	May 29	Urban Area	Yokohama		497BG	39	34	
186	May 29	Urban Area	Yokohama		498BG	41	34	
186	May 29	Urban Area	Yokohama		499BG	not avail	26	
186	May 29	Urban Area	Yokohama		500BG	not avail	42	
186	May 29	Urban Area	Yokohama	313BW		83	65	2
186	May 29	Target of Opportunity	Tokyo U.A.		WING	in above	10	
186	May 29	Urban Area	Yokohama		6BG	27	25	
186	May 29	Urban Area	Yokohama		504BG	25	22	2
186	May 29	Urban Area	Yokohama	314BG		146	131	3
186	May 29	Target of Opportunity	Shimoda A/F		WING	in above	2	
186	May 29	Target of Opportunity	Nii Shima		WING	in above	1	
186	May 29	Target of Opportunity	Hamamatsu		WING	in above	1	
186	May 29	Target of Opportunity	Sagara		WING	in above	1	
186	May 29	Target of Opportunity	Kisarazu Nav Sta		WING	in above	1	
186	May 29	Urban Area	Yokohama	314BW	29BG	not avail	37	
186	May 29	Urban Area	Yokohama	314BW	19BG	not avail	39	
186	May 29	Urban Area	Yokohama	314BW	39BG	31	25	2
186	May 29	Target of Opportunity	Sagara		39BG	in above	1	0
186	May 29	Target of Opportunity	Shimoda A/F		39BG	in above	1	0
186	May 29	Urban Area	Yokohama	314BW	330BG	38	34	0
Weather Strikes	Month of May	Weather Strikes	Home Islands	All		93	91	0
Weather Strikes	Total to Date	Weather Strikes	Home Isalnds	All		498	459	6
Photo Recon	Month of May	Photo Recon	Home Islands	All		71	44	0
Photo Recon	Total to Date	Photo Recon	Home Isalnds	All		303	209	3
187	June 1	Urban Area	Osaka	MISSION		509	487	10
187	June 1	Urban Area	Osaka		WING	119	107	4
187	June 1	Target of Opportunity	Katsuuri		WING	in above	1	
187	June 1	Target of Opportunity	Shingu		WING	in above	1	
187	June 1	Urban Area	Osaka		40BG	29	27	0
187	June 1	Urban Area	Osaka		444BG	27	22	3
187	June 1	Target of Opportunity	Matsuke		444BG	in above	1	0

MSN	Date	Target	Location	Wings	Grps	No. Up	Bombed	Lost
187	June 1	Urban Area	Osaka		462BG	28	26	0
187	June 1	Urban Area	Osaka		468BG	35	32	1
187	June 1	Target of Opportunity	Shingu		468BG	in above	1	0
187	June 1	Urban Area	Osaka	73BW		158	140	4
187	June 1	Target of Opportunity	Chichi Jima		WING	in above	2	0
187	June 1	Target of Opportunity	Muki Tomika		WING	in above	1	0
187	June 1	Target of Opportunity	Nara Kushimoto		WING	in above	1	0
187	June 1	Target of Opportunity	Tanabe		WING	in above	1	0
187	June 1	Urban Area	Osaka		497BG	34	32	2
187	June 1	Urban Area	Osaka		498BG	39	37	1
187	June 1	Urban Area	Osaka		499BG	not avail	36	1
187	June 1	Urban Area	Osaka		500BG	not avail	35	
187	June 1	Urban Area	Osaka	314BG		118	103	2
187	June 1	Target of Opportunity	Takashima		WING	in above	1	
187	June 1	Target of Opportunity	Kushimoto A/F		WING	in above	1	
187	June 1	Urban Area	Osaka		29BG	30	29	0
187	June 1	Urban Area	Osaka		19BG	36	32	0
187	June 1	Urban Area	Osaka		39BG	15	13	1
187	June 1	Target of Opportunity	Wakayama		39BG	in above	1	0
187	June 1	Urban Area	Osaka		330BG	37	34	0
187	June 1	Urban Area	Osaka	313BW		119	108	1
187	June 1	Target of Opportunity	Nagashima		WING	in above	1	
187	June 1	Target of Opportunity	Kochi		WING	in above	1	
187	June 1	Target of Opportunity	Shiona Misaki		WING	in above	1	
187	June 1	Urban Area	Osaka		6BG	27	27	1
187	June 1	Urban Area	Osaka		9BG	36	31	0
187	June 1	Target of Opportunity	Osaka		9BG	in above	1	0
187	June 1	Urban Area	Osaka		504BG	24	20	0
187	June 1	Urban Area	Osaka		505BG	31	33	0
188	June 5	Urban Area	Kobe	MISSION		524	494	12
188	June 5	Urban Area	Kobe	58BW		120	107	5
188	June 5	Urban Area	Kobe		40BG	29	24	0
188	June 5	Target of Opportunity	Shizuki		40BG	in above	1	0
188	June 5	Urban Area	Kobe		444BG	27	25	1
188	June 5	Urban Area	Kobe		462BG	31	26	4
188	June 5	Target of Opportunity	Wakayama		462BG	in above	1	0
188	June 5	Urban Area	Kobe		468BG	33	32	0
188	June 5	Target of Opportunity	Komatsushima		468BG	in above	1	0
188	June 5	Urban Area	Kobe	73BW		154	139	2
188	June 5	Urban Area	Kobe		497BG	36	31	1
188	June 5	Urban Area	Kobe		498BG	37	32	1
188	June 5	Urban Area	Kobe		499BG	not avail	35	0
188	June 5	Urban Area	Kobe		500BG	not avail	41	0
188	June 5	Urban Area	Kobe	313BW		124	113	2
188	June 5	Target of Opportunity	Shingu		WING	in above	1	
188	June 5	Target of Opportunity	Komatushima		WING	in above	1	
188	June 5	Urban Area	Kobe		6BG	29	28	
188	June 5	Urban Area	Kobe		9BG	35	32	0
188	June 5	Urban Area	Kobe		504BG	20	18	
188	June 5	Secondary Target	Kobe		504BG	in above	2	0
188	June 5	Urban Area	Kobe		505BG	not avail	not avail	
188	June 5	Urban Area	Kobe	314BW		129	120	2
188	June 5	Target of Opportunity	Futaminora A/F		WING	in above	1	
188	June 5	Target of Opportunity	Saki Nohana		WING	in above	1	
188	June 5	Target of Opportunity	Chichi Jima		WING	in above	1	
188	June 5	Urban Area	Kobe		29BG	38	35	1
188	June 5	Urban Area	Kobe		19BG	33	32	0
188	June 5	Urban Area	Kobe		39BG	27	24	0
188	June 5	Urban Area	Kobe		330BG	31	25	1
189	June 7	Urban Area	Osaka	MISSION		449	409	2
189	June 7	Urban Area	Osaka	58BW		115	107	0
189	June 7	Urban Area	Osaka		40BG	32	31	0
189	June 7	Urban Area	Osaka		444BG	26	24	0
189	June 7	Target of Opportunity	Shingo		444BG	in above	1	0
189	June 7	Urban Area	Osaka		462BG	26	25	0

VII. The XXI Bomber Command

MSN	Date	Target	Location	Wings	Grps	No. Up	Bombed	Lost
189	June 7	Target of Opportunity	Shirahama		462BG	in above	1	0
189	June 7	Urban Area	Osaka		468BG	31	27	0
189	June 7	Target of Opportunity	Chichi Jima		468BG	in above	11	0
189	June 7	Target of Opportunity	Tokushima		468BG	in above	1	0
189	June 7	Target of Opportunity	None'		468BG	in above	1	0
189	June 7	Urban Area	Osaka	73BW		133	120	0
189	June 7	Urban Area	Osaka		497BG	30	24	0
189	June 7	Urban Area	Osaka		498BG	33	30	0
189	June 7	Urban Area	Osaka		499BG	not avail	31	0
189	June 7	Urban Area	Osaka		500BG	not avail	35	0
189	June 7	Urban Area	Osaka	313BW		83	83	
189	June 7	Urban Area	Osaka		6BG	27	26	
189	June 7	Urban Area	Osaka		9BG	32	32	0
189	June 7	Urban Area	Osaka		504BG	23	19	1
189	June 7	Secondary Target	Osaka		504BG	in above	1	0
189	June 7	Urban Area	Osaka	314BW		119	107	1
189	June 7	Urban Area	Osaka		29BG	31	28	0
189	June 7	Urban Area	Osaka		19BG	32	31	1
189	June 7	Urban Area	Osaka		39BG	29	21	0
189	June 7	Urban Area	Osaka		330BG	27	27	0
190	June 7/8	Mining Mission 18	See Below	313BW		31	26	0
190	June 7/8	Mine Field Mike	Shimonoseki		505BG	10	9	0
190	June 7/8	Mine Field Love	Shimonoseki		505BG	10	7	0
190	June 7/8	Mine Field Charlie	Fukuoka Bay		505BG	5	4	0
190	June 7/8	Mine Field Mike	Shimonoseki		505BG	10	6	0
191	June 9	Kawanishi Aircraft Factory	Osaka	58BW		46	44	0
191	June 9	Kawanishi Aircraft Factory	Osaka		444BG	29	28	0
191	June 9	Target of Opportunity	Aichi Plant-Nagoya		444BG	in above	1	0
191	June 9	Kawanishi Aircraft Factory	Naruo		462BG	17	16	0
192	June 9	Kawasaki Aircraft Factory	Akashi	313BW		26	24	0
192	June 9	Kawasaki Aircraft Factory	Akashi		6BG	26	24	
192	June 9	Kawasaki Aircraft Factory	Akashi		504BG	20	19	0
193	June 9	Aichi Aircraft Engine Plant	Atsuta	313BW		44*	42*	
193	June 9	Aichi Aircraft Engine Plant	Atsuta		9BG	26	24	0
193	June 9	Target of Opportunity	Unknown		9BG	in above	1	0
193	June 9	Hitachi Aircraft Plant	Chiba	314BW	39BG	27	26	0
194	June 9/10	Mining Mission 19	Shimonoseki	313BW		28	26	0
194	June 9/10	Mine Field Mike	Shimonoseki		505BG	20	19	0
194	June 9/10	Mine Field Love	Shimonoseki		505BG	8	7	0
195	June 10	Kasumigahara Seaplane Base	Kasumiguara	58BW		62	55	0
195	June 10	Kasumigahara Seaplane Base	Kasumiguara		40BG	29	23	29
195	June 10	Target of Opportunity	Hachido Jima		40BG	in above	1	0
195	June 10	Target of Opportunity	Shinoda		40BG	in above	1	0
195	June 10	Kasumigahara Seaplane Base	Kasumiguara	314BW	330BG	32	25	0
195	June 10	Secondary Target	Gifu		330BG	in above	1	0
196	June 10	Tomioka Engine Plant	Tomioka	58BW	468BG	33	32	0
196	June 10	Target of Opportunity	Shimizu		468BG	in above	1	0
197	June 10	Hitachi Aircraft Engines	Kaigan	73BW		126	118	0
197	June 10	Hitachi Engine Works (2ndary)	Kaigan		497BG	30	29	0
197	June 10	Hitachi Engine Works (2ndary)	Kaigan		498BG	33	32	0
197	June 10	Musashi—Nakajima Aircraft Factory	Tokyo		499BG		24	0
197	June 10	Musashi—Nakajima Aircraft Factory	Tokyo		500BG		33	0
198	June 10	Hitachi—Aircraft Parts	Chiba	314BW	39BG	32	26	0
199	June 10	Ogikubu—Nakajima Aircraft Factory	Ogikubu	314BW		65	7	1
199	June 10	Kasumiguara Air Base	Kasumiguara		WING	in above	45	1
199	June 10	Ogikubu—Nakajima Aircraft Factory	Ogikubu		330BG	32	25	0
199	June 10	Kasumiguara Air Base	Kasumiguara		19BG	33	26	1
200	June 10	Tachikawa Air Depot-Parts Factory	Tachikawa		WING	34	29	0
200	June 10	Tachikawa Air Depot-Parts Factory	Tachikawa		29BG	34	32	0
201	June 11/12	Mining Mission 20	See Below	313BW		26	26	0
201	June 11/12	Mine Field Mike	Shimonoseki		505BG	15	15	0
201	June 11/12	Mine Field Zebra	Tsuruga		505BG	11	11	0
202	June 13/14	Mining Mission 21	See Below		WING	30	29	0
202	June 13/14	Mine Field Love	Shimonoseki		505BG	17	17	0
202	June 13/14	Mine Field Uncle	Niigata		505BG	12	12	0

MSN	Date	Target	Location	Wings	Grps	No. Up	Bombed	Lost
203	June 15	Osaka/Amagasaki Urban Area	Osaka/Amagasaki	MISSION		511	476	2
203	June 15	Osaka/Amagasaki Urban Area	Osaka/Amagasaki	58BW		138	119	1
203	June 15	Urban Area	Osaka		40BG	33	29	0
203	June 15	Target of Opportunity	Kimomoto		40BG	in above	1	0
203	June 15	Target of Opportunity	Kobe		40BG	in above	1	0
203	June 15	Urban Area	Osaka		444BG	35	32	0
203	June 15	Target of Opportunity	Kushimoto		444BG	in above	1	0
203	June 15	Urban Area	Amagasaki		462BG	30	25	0
203	June 15	Target of Opportunity	Shirahama		462BG	in above	1	0
203	June 15	Urban Area	Osaka		468BG	40	33	0
203	June 15	Target of Opportunity	Kushimoto		468BG	in above	1	0
203	June 15	Target of Opportunity	Katsaura		468BG	in above	1	0
203	June 15	Target of Opportunity	Tokushima		468BG	in above	1	0
203	June 15	Target of Opportunity	24:15N; 138:15E		468BG	in above	1	0
203	June 15	Osaka/Amagasaki Urban Area	Osaka/Amagasaki	73BW		145	132	0
203	June 15	Target of Opportunity	Wakayama		WING	in above	1	0
203	June 15	Osaka/Amagasaki Urban Area	Osaka/Amagasaki		497BG	40	36	0
203	June 15	Osaka/Amagasaki Urban Area	Osaka/Amagasaki		498BG	36	34	0
203	June 15	Osaka/Amagasaki Urban Area	Osaka/Amagasaki		499BG	35	30	0
203	June 15	Osaka/Amagasaki Urban Area	Osaka/Amagasaki		500BG	33	32	0
203	June 15	Osaka/Amagasaki Urban Area	Osaka/Amagasaki	314BW		135	108	1
203	June 15	Target of Opportunity	Wakayama		WING	in above	1	0
203	June 15	Target of Opportunity	Mugi		WING	in above	1	0
203	June 15	Target of Opportunity	Tokushima		WING	in above	1	0
203	June 15	Target of Opportunity	Shiono Misaki		WING	in above	1	0
203	June 15	Target of Opportunity	Chichi Jima		WING	in above	2	0
203	June 15	Target of Opportunity	Matsune		WING	in above	1	0
203	June 15	Target of Opportunity	Katsuura		WING	in above	1	0
203	June 15	Osaka/Amagasaki Urban Area	Osaka/Amagasaki		19BG	34	27	0
203	June 15	Osaka/Amagasaki Urban Area	Osaka/Amagasaki		29BG	37	34	1
203	June 15	Urban Area	Osaka/Amagasaki		39BG	31	23	0
203	June 15	Target of Opportunity	Uwano		39BG	in above	1	0
203	June 15	Target of Opportunity	Kushimoto		39BG	in above	2	0
203	June 15	Osaka/Amagasaki Urban Area	Osaka/Amagasaki		330BG	33	24	0
203	June 15	Osaka/Amagasaki Urban Area	Osaka/Amagasaki	313BW		94	85	0
203	June 15	Target of Opportunity	Ujiyamada		WING	in above	1	0
203	June 15	Target of Opportunity	Fukido		WING	in above	1	0
203	June 15	Target of Opportunity	Wakayama		WING	in above	1	0
203	June 15	Target of Opportunity	Tokushima		WING	in above	1	0
203	June 15	Target of Opportunity	Shingu		WING	in above	1	0
203	June 15	Target of Opportunity	Kochi		WING	in above	1	0
203	June 15	Urban Area	Osaka/Amagasaki		6BG	35	35	0
203	June 15	Urban Area	Osaka/Amagasaki		9BG	35	33	0
203	June 15	Target of Opportunity	Osaka/Amagasaki		9BG	in above	2	0
203	June 15	Urban Area	Osaka/Amagasaki		504BG	24	20	0
203	June 15	Secondary Target	Osaka/Amagasaki		504BG	in above	1	0
204	June 15/16	Mining Nission 22	See Below		WING	30	30	0
204	June 15/16	Mine Field Mike	Shimonoseki		505BG	13	13	0
204	June 15/16	Mine Field Nan	Fushiki		505BG	9	9	0
204	June 15/16	Mine Field Charlie	Fukuoka		505BG	8	8	0
205	June 17/18	Mining Mission 23	See Below	313BW		25	25	0
205	June 17/18	Mine Field Mike	Shimonoseki		505BG	18	18	0
205	June 17/18	Mine Field Able	Kobe/Osaka		505BG	7	7	0
206	June 17/18	Urban Area	Kagoshima	73BW	WING	137	131	
206	June 17/18	Urban Area	Kagoshima	313BW	WING	94	89	0
206	June 17/18	Urban Area	Kagoshima	314BW		120	117	1
206	June 17/18	Target of Opportunity	Yamakawa		WING	in above	1	0
206	June 17/18	Target of Opportunity	Iwakawa A/F		WING	in above	1	0
206	June 17/18	Urban Area	Kagoshima		19BG	32	32	
206	June 17/18	Urban Area	Kagoshima		29BG	33	32	
206	June 17/18	Urban Area	Kagoshima		39BG	27	26	
206	June 17/18	Urban Area	Kagoshima		330BG	28	28	0
207	June 17/18	Urban Area	Omuta	58BW		126	116	0
207	June 17/18	Urban Area	Omuta		40BG	32	30	0
207	June 17/18	Target of Opportunity	Nagazu		40BG	in above	1	0

VII. The XXI Bomber Command

MSN	Date	Target	Location	Wings	Grps	No. Up	Bombed	Lost
207	June 17/18	Urban Area	Omuta		444BG	35	32	0
207	June 17/18	Target of Opportunity	Nokeoka		444BG	in above	1	0
207	June 17/18	Urban Area	Omuta		462BG	22	20	0
207	June 17/18	Urban Area	Omuta		468BG	37	34	0
207	June 17/18	Target of Opportunity	Hatamara		468BG	in above	1	0
208	June 17/18	Urban Area	Hamamatsu	73BW		139	130	0
208	June 17/18	Urban Area	Hamamatsu		497BG	36	35	0
208	June 17/18	Urban Area	Hamamatsu		498BG	31	31	0
208	June 17/18	Urban Area	Hamamatsu		499BG	34	31	0
208	June 17/18	Urban Area	Hamamatsu		500BG	37	33	0
209	June 17/18	Urban Area	Yokkaichi	313BW		94	89	0
209	June 17/18	Urban Area	Yokkaichi		9BG	38	36	0
209	June 17/18	Urban Area	Yokkaichi		6BG	30	28	0
209	June 17/18	Urban Area	Yokkaichi		504BG	25	23	0
210	June 19/20	Urban Area	Toyohashi	MISSION		515	486	2
210	June 19/20	Urban Area	Toyohashi	58BW		141	136	0
210	June 19/20	Urban Area	Toyohashi		40BG	35	34	0
210	June 19/20	Urban Area	Toyohashi		444BG	35	34	0
210	June 19/20	Urban Area	Toyohashi		462BG	33	33	0
210	June 19/20	Urban Area	Toyohashi		468BG	38	35	0
211	June 19/20	Urban Area	Fukuoka	73/313		237	225	0
211	June 19/20	Urban Area	Fukuoka		WING	144	131	0
211	June 19/20	Target of Opportunity	Tomitaka		WING	in above	1	0
211	June 19/20	Target of Opportunity	Chichi Jima		WING	in above	1	0
211	June 19/20	Urban Area	Fukuoka		497BG	37	34	0
211	June 19/20	Urban Area	Fukuoka		498BG	34	29	0
211	June 19/20	Urban Area	Fukuoka		499BG	not avail	38	0
211	June 19/20	Urban Area	Fukuoka		500BG	not avail	32	0
211	June 19/20	Urban Area	Fukuoka	313BW		95	92	0
211	June 19/20	Target of Opportunity	Miyazaki		WING	in above	1	0
211	June 19/20	Urban Area	Fukuoka		6BG	29	29	0
211	June 19/20	Urban Area	Fukuoka		9BG	37	34	0
211	June 19/20	Target of Opportunity	Fukuoka		9BG	in above	1	0
211	June 19/20	Urban Area	Fukuoka		504BG	27	27	0
212	June 19/20	Urban Area	Shizuoka	314BW		137	123	2
212	June 19/20	Target of Opportunity	Hachijo Jima		WING	in above	1	0
212	June 19/20	Target of Opportunity	Chichi Jima		WING	in above	1	0
212	June 19/20	Urban Area	Shizuoka		19BG	35	33	1
212	June 19/20	Urban Area	Shizuoka		29BG	37	32	0
212	June 19/20	Urban Area	Shizuoka		39BG	32	28	1
212	June 19/20	Urban Area	Shizuoka		330BG	33	30	0
213	June 19/20	Mining Mission 24	See Below	313BW		28	28	0
213	June 19/20	Mine Field Mike	Shimonoseki		505BG	10	10	0
213	June 19/20	Mine Field Uncle	Niigata		505BG	8	8	0
213	June 19/20	Mine Field Zebra	Miazuru/Miyazu		505BG	10	10	0
214	June 21/22	Mining Mission 25	See Below	WING		30	27	0
214	June 21/22	Mine Field X-Ray	Senzaki		505BG	9	9	0
214	June 21/22	Mine Field Nan	Fushiki		505BG	10	10	0
214	June 21/22	Mime Field Able	Kobe/Osaka		505BG	6	6	0
215	June 22	Kure Arsenal	Kure	58BW		64	58	1
215	June 22	Kure Arsenal	Kure		462BG	31	27	1
215	June 22	Kure Arsenal	Uwajima		468BG	33	31	0
215	June 22	Target of Opportunity	Kure		468BG	in above	1	0
215	June 22	Target of Opportunity	Nobeoka		468BG	in above	1	0
215	June 22	Target of Opportunity	Sukumo N.B		468BG	in above	1	0
215	June 22	Target of Opportunity	Hoseshima		468BG	in above	1	0
215	June 22	Target of Opportunity	Oita A/F		468BG	in above	1	0
215	June 22	Target of Opportunity	Sagonoseki		468BG	in above	1	0
215	June 22	Target of Opportunity	Sadei		468BG	in above	1	0
215	June 22	Target of Opportunity	Ochimura		468BG	in above	1	0
215	June 22	Target of Opportunity	Shimazu RR Brdg		468BG	in above	1	0
215	June 22	Target of Opportunity	Meisho		468BG	in above	1	0
215	June 22	Target of Opportunity	Okina Shima		468BG	in above	1	0
215	June 22	Target of Opportunity	Tomitaki A/F		468BG	in above	1	0
215	June 22	Kure Arsenal	Kure	73BW		126	104	1

MSN	Date	Target	Location	Wings	Grps	No. Up	Bombed	Lost
215	June 22	Kure Arsenal	Kure		497BG	30	24	0
215	June 22	Kure Arsenal	Kure		498BG	30	27	1
215	June 22	Target of Opportunity	Kure		498BG	in above	1	0
215	June 22	Kure Arsenal	Kure		499BG	not avail	29	0
215	June 22	Kure Arsenal	Kure		500BG	not avail	26	0
216	June 22	Mitsubishi Aircraft Factory	Tamashima	314BW		124	114	2
216	June 22	Targets of Opportunity	Chichi Jima		WING	in above	2	0
216	June 22	Targets of Opportunity	Taname		WING	in above	1	0
216	June 22	Targets of Opportunity	Habeshima		WING	in above	1	0
216	June 22	Targets of Opportunity	Kita Shioya		WING	in above	10	
216	June 22	Targets of Opportunity	Gobo		WING	in above	2	0
216	June 22	Targets of Opportunity	Wakayama		WING	in above	1	0
216	June 22	Targets of Opportunity	Tanabe		WING	in above	1	0
216	June 22	Mitsubishi Aircraft Factory	Tamashima		19BG	33	29	0
216	June 22	Mitsubishi Aircraft Factory	Tamashima		29BG	28	28	0
216	June 22	Mitsubishi Aircraft Factory	Tamashima		39BG	29	27	0
216	June 22	Target of Opportunity	Tamashima		39BG	in above	1	0
216	June 22	Target of Opportunity	Kushimoto		39BG	in above	2	0
216	June 22	Mitsubishi Aircraft Factory	Gobo		330BG	34	30	1
216	June 22	Kagamigahara/Akashi	Kagamigahara	313BW		80	73	0
216	June 22	Akashi — Kawasaki Aircraft Factory	Akashi		6BG	not avail	29	
216	June 22	Utsube Oil refinery	Unknown		9BG	28	17	0
216	June 22	Secondary target	Himeji		9BG	in above	6	0
218	June 22	Mitsubishi Aircraft Factory	Kagamigahara	313BW		28	17	1
218	June 22	Mitsubishi Aircraft Factory	Kagamigahara		504BG	22	20	1
216	June 22	Kagamihara — Aircraft Factory	Akashi		WING	21	17	1
216	June 22	Akashi — Kawasaki Aircraft Factory	Akashi		WING	30	25	0
217	June 22	Himeji — Kawanishi Aircraft Factory	Himeji	58BW		58	52	0
217	June 22	Himeji — Kawanishi Aircraft Factory	Tokushima		40BG	28	24	0
217	June 22	Target of Opportunity	Wadashima		40BG	in above	1	0
217	June 22	Target of Opportunity	Tokoshima		40BG	in above	1	0
217	June 22	Himeji — Kawanishi Aircraft Factory	Tomashima		444BG	30	28	0
217	June 22	Target of Opportunity	Tomashima		444BG	in above	2	0
218	June 22	Kagashima Aircraft Factory	Akashi	313BW	9BG	37	32	0
218	June 22	Secondary Target	Unknown		9BG	in above	3	0
219	June 22	Kagoshima Aircraft Factory	Akashi	313BW	6BG	not avail	29	0
219	June 22	Target of Opportunity	Kagamigahara		6BG	in above	1	0
219	June 22	Target of Opportunity	Hamamatsu		6BG	in above	1	0
219	June 22	Target of Opportunity	Kawasaki AC		6BG	in above	1	0
219	June 22	Target of Opportunity	Hachionan		6BG	in above	1	0
220	June 22	Akashi Aircraft Factory	Akashi		Unk	30	25	0
220	June 22	Target of Opportunity	Kagamigahara	313BW		in above	1	
221	June 23/24	Mining Mission 26	Fukuoka		WING	27	26	1
221	June 23/24	Mine Field Charlie	Fukuoka		505BG	9	9	
221	June 23/24	Mine Field Yoke	Sakai		505BG	8	8	
221	June 23/24	Mine Field Uncle	Niigata		505BG	9	9	
222	June 25/26	Mining Mission 27	See Below		WING	27	26	0
222	June 25/26	Mine Field Mike	Shimonoseki		505BG	11	11	0
222	June 25/26	Mine Field Love	Shimonoseki		505BG	6	6	0
222	June 25/26	Mine Field Zebra	Tsuruga Bay		505BG	9	9	0
223	June 26	See Below	Various	MISSION		510	493	
223	June 26	Sumitomo Light Metal Industry	Osaka	58BW		71	64	0
223	June 26	Sumitomo Light Metal Industry	Osaka		462BG	35	34	0
223	June 26	Sumitomo Light Metal Industry	Osaka		468BG	36	30	0
223	June 26	Target of Opportunity	Kushimoto		468BG	in above	2	0
223	June 26	Target of Opportunity	Oshimo		468BG	in above	1	0
223	June 26	Target of Opportunity	Tokushima		468BG	in above	3	0
223	June 26	Target of Opportunity	Ujiyamada		468BG	in above	6	0
224	June 26	Osaka Arsenal	Oaska	73BW		122	109	1
224	June 26	Target of Opportunity	Tsuta		WING	in above	1	0
224	June 26	Target of Opportunity	Tsuruga Bay		WING	in above	1	0
224	June 26	Target of Opportunity	Kochi		WING	in above	1	0
224	June 26	Osaka Arsenal	Oaska		497BG	30	24	0
224	June 26	Osaka Arsenal	Oaska		498BG	31	28	0
224	June 26	Osaka Arsenal	Oaska		499BG		28	1

VII. The XXI Bomber Command

MSN	Date	Target	Location	Wings	Grps	No. Up	Bombed	Lost
224	June 26	Osaka Arsenal	Oaska		500BG		29	0
225	June 26	Kawasaki Aircraft Factory	Ahashi	58BW	40BG	38	31	1
225	June 26	Secondary Target	Tsu		40BG	in above	4	0
225	June 26	Kawasaki Aircraft Factory	Ahashi	58BW	444BG	41	27	1
225	June 26	Secondary Target	Tsu		444BG	in above	11	0
225	June 26	Target of Opportunity	Kochi A/F		444BG	in above	2	0
225	June 26	Target of Opportunity	Banzai		444BG	in above	1	0
225	June 26	Various	Nagoya Region	314BW		102	91	1
225	June 26	Kawasaki Aircraft Factory	Akashi		19BG	35	33	0
226	June 26	Chigura Factory, Nagoya Arsenal	Nagoya		29BG	35	31	0
227	June 26	Atsuta Factory NagoyaArsenal	Nagoya		39BG	32	16	1
227	June 26	Secondary target	City of Tsu		39BG	in above	9	0
227	June 26	Target of Opportunity	City of Tsu		39BG	in above	2	0
227	June 26	Target of Opportunity	Hamamatsu		39BG	in above	4	0
227	June 26	Target of Opportunity	Shiroki		39BG	in above	1	0
227	June 26	Target of Opportunity	Shingu		39BG	in above	2	0
228	June 26	Mitsubishi Aircraft Factory	Shiroko	58BW		79	60	1
228	June 26	Mitsubishi Aircraft Factory	Kagamigahara		40BG	38	33	1
228	June 26	Mitsubishi Aircraft Factory	Kagamigahara		444BG	41	27	0
228	June 26	Target of Opportunity	Komatsushima		444BG	in above	2	0
228	June 26	Target of Opportunity	Yokkaichi		444BG	in above	2	0
229	June 26	Aichi Aircraft Works	Eitoku	313BW		67	60	2
229	June 26	Aichi Aircraft Works	Eitoku		9BG	37	32	0
229	June 26	Aichi Aircraft Works	Eitoku		504BG	30	18	2
229	June 26	Secondary Target	Eitoku		504BG	in above	11	0
230	June 26	Sumitome Light metals	Nagoya	314BG	330BG	33	29	0
230	June 26	Target of Opportunity	Owase		330BG	in above	1	0
230	June 26	Target of Opportunity	Hachijo Jima		330BG	in above	1	0
230	June 26	Target of Opportunity	Hamamatsu A/F		330BG	in above	1	0
231	June 26	Kita-Shioya Aircraft Factory	Kagamihara	314BG	19BG	35	25	2
231	June 26	Target of Opportunity	Matsusaka		19BG	in above	2	0
231	June 26	Target of Opportunity	Toyohashi		19BG	in above	1	0
231	June 26	Target of Opportunity	Katada		19BG	in above	1	0
231	June 26	Target of Opportunity	Kita Shioya		19BG	in above	1	0
232	June 26/27	Utsube Oil Refinery	Yokkaichi	315BW		34	33	0
232	June 26/27	Target of Opportunity	Kagata		WING	in above	1	0
232	June 26/27	Utsube Oil Refinery	Yokkaichi		16BG	15	15	0
232	June 26/27	Utsube Oil Refinery	Yokkaichi		501BG	19	17	0
233	June 27/28	Mining Mission 28	See Below	313BW		30	29	0
233	June 27/28	Mine Field X-Ray	Hagi		505BG	14	14	0
233	June 27/28	Mine Field Able	Kobe/Osaka		505BG	8	8	0
233	June 27/28	Mine Field Uncle	Niigata		505BG	7	7	0
234	June 28/29	Okayama Urban Area	Okiyama	58BW		141	138	1
234	June 28/29	Okayama Urban Area	Okiyama		40BG	36	34	0
234	June 28/29	Okayama Urban Area	Okiyama		444BG	35	35	0
234	June 28/29	Okayama Urban Area	Okiyama		462BG	34	33	0
234	June 28/29	Okayama Urban Area	Okiyama		468BG	36	36	1
235	June 28/29	Sasebo Urban Area	Sasebo	73BW		145	141	0
235	June 28/29	Target of Opportunity	Isahaya		WING	in above	1	0
235	June 28/29	Target of Opportunity	Makurazaki		WING	in above	2	0
235	June 28/29	Sasebo Urban Area	Sasebo		497BG	42	41	0
235	June 28/29	Sasebo Urban Area	Sasebo		498BG	31	28	0
235	June 28/29	Sasebo Urban Area	Sasebo		499BG	not avail	38	0
235	June 28/29	Sasebo Urban Area	Sasebo		500BG	not avail	34	0
236	June 28/29	Urban Area	Moji	313BW		95	91	0
236	June 28/29	Target of Opportunity	Shimonoseki		WING	in above	1	0
236	June 28/29	Target of Opportunity	Uwajima		WING	in above	3	0
236	June 28/29	Urban Area	Moji		6BG	not avail	30	0
236	June 28/29	Urban Area	Moji		9BG	36	32	0
236	June 28/29	Target of Opportunity	Moji		9BG	in above	1	0
236	June 28/29	Urban Area	Moji		504BG	35	32	0
237	June 28/29	Urban Area	Nobeoka	314BW		122	117	0
237	June 28/29	Urban Area	Nobeoka		19G	30	30	0
237	June 28/29	Urban Area	Nobeoka		29BG	28	27	0
237	June 28/29	Urban Area	Nobeoka		39BG	29	28	0

MSN	Date	Target	Location	Wings	Grps	No. Up	Bombed	Lost
237	June 28/29	Urban Area	Nobeoka		330BG	35	32	0
238	June 29/30	Nippon Oil Refinery	Kadamatsu	315BW		36	32	0
238	June 29/30	Nippon Oil Refinery	Kadamatsu	315BW	16BG	18	16	0
238	June 29/30	Nippon Oil Refinery	Kadamatsu	315BW	501BG	18	16	0
239	June 29/30	Mining — Mission 29	See Below	313BW		29	25	0
239	June 29/30	Mine Field Mike	Yawata		505BG	12	12	0
239	June 29/30	Mine Field Zebra	Maizura Bay		505BG	4	4	0
239	June 29/30	Mine Field Uncle	Miyahoura		505BG	9	9	0
240	July 1/2	Kure Urban Area	Kure	58BW		160	152	0
240	July 1/2	Kure Urban Area	Kure		40BG	41	39	0
240	July 1/2	Kure Urban Area	Kure		444BG	39	38	0
240	July 1/2	Kure Urban Area	Kure		462BG	40	37	0
240	July 1/2	Kure Urban Area	Kure		468BG	40	38	0
241	July 1/2	Urban Area	Kumamoto	73BW		164	154	1
241	July 1/2	Urban Area	Kumamoto		497BG	40	38	0
241	July 1/2	Urban Area	Kumamoto		498BG	41	39	0
241	July 1/2	Urban Area	Kumamoto		499BG	not avail	36	0
241	July 1/2	Urban Area	Kumamoto		500BG	not avail	41	0
242	July 1/2	Secondary Target	Ube	313BG		112	100	0
242	July 1/2	Secondary Target	Ube		6BG	39	35	0
242	July 1/2	Secondary Target	Ube		9BG	38	30	0
242	July 1/2	Secondary Target	Ube		504BG	35	33	0
243	July 1/2	Urban Area	Shimonoseki City	314BW		142	128	1
243	July 1/2	Target of Opportunity	Unknown		WING	in above	3	0
243	July 1/2	Urban Area	Shimonoseki City		19BG	34	25	
243	July 1/2	Urban Area	Shimonoseki City		29BG	32	34	
243	July 1/2	Urban Area	Shimonoseki City		39BG	38	32	0
243	July 1/2	Target of Opportunity	Shimonoseki City		39BG	in above	1	0
243	July 1/2	Target of Opportunity	Kumamoto		39BG	in above	1	0
243	July 1/2	Urban Area	Shimonoseki		330BG	38	37	0
244	July 1/2	Mining Mission 30	Shimonoseki City	313BW		30	24	0
244	July 1/2	Mine Field Love	Shimonoseki City		505BG	9	2	0
244	July 1/2	Mine Field Nan	Fushiki, Nanao		505BG	21	22	0
245	July 2/3	Maruzan Oil Refirnery	Shimotsu	315BW		40	39	0
245	July 2/3	Target of Opportunity	Sakinohama		WING	in above	1	0
245	July 2/3	Maruzan Oil Refirnery	Shimotsu		16BG	19	19	0
245	July 2/3	Maruzan Oil Refirnery	Shimotsu		501BG	21	20	0
246	July 3/4	Mining Mission 31	See Below	313BW		31	28	0
246	July 3/4	Mine Field William	Funakoshi		505BG	6	10	0
246	July 3/4	Mine Field Zebra	Maizuru		505BG	4	3	0
246	July 3/4	Mine Field Mike	Shimonoseki		505BG	16	13	0
247	July 3/4	Urban Area	Takamatsu	58BW		128	116	3
247	July 3/4	Urban Area	Takamatsu		40BG	31	28	0
247	July 3/4	Urban Area	Takamatsu		444BG	32	28	0
247	July 3/4	Urban Area	Takamatsu		462BG	32	31	1
247	July 3/4	Target of Opportunity	Nobeoka		462BG	in above	1	0
247	July 3/4	Urban Area	Takamatsu		468BG	33	29	2
247	July 3/4	Target of Opportunity	Chichi Jima		468BG	in above	1	0
247	July 3/4	Target of Opportunity	Sukumo		468BG	in above	1	0
248	July 3/4	Urban Area	Kochi	73BW		129	125	1
248	July 3/4	Urban Area	Kochi		497BG	33	30	1
248	July 3/4	Urban Area	Kochi		498BG	33	31	0
248	July 3/4	Urban Area	Kochi		499BG		32	0
248	July 3/4	Urban Area	Kochi		500BG		32	0
249	July 3/4	Urban Area	Himeji	313BW		95	94	0
249	July 3/4	Urban Area	Himeji		6BG	35		
249	July 3/4	Urban Area	Himeji		9BG	36	36	0
249	July 3/4	Urban Area	Himeji		504BG	35	35	0
250	July 3/4	Urban Area	Tokushima	314BW		137	129	0
250	July 3/4	Target of Opportunity	Shingu		WING	in above	1	0
250	July 3/4	Target of Opportunity	Tenuna		WING	in above	1	0
250	July 3/4	Urban Area	Tokushima		19BG	31	29	0
250	July 3/4	Urban Area	Tokushima		29BG		35	0
250	July 3/4	Urban Area	Tokushima		39BG	34	32	0
250	July 3/4	Urban Area	Tokushima		330BG	35	35	0

VII. The XXI Bomber Command

MSN	Date	Target	Location	Wings	Grps	No. Up	Bombed	Lost
	July 5	Marcus Island	Tokushima	313BW	6BG	3	3	0
251	July 6/7	Chiba Urban Area	Chiba	58BW		129	124	0
251	July 6/7	Chiba Urban Area	Chiba		40BG	32	30	0
251	July 6/7	Target of Opportunity	Unknown		40BG	in above	1	0
251	July 6/7	Chiba Urban Area	Chiba		444BG	31	30	0
251	July 6/7	Chiba Urban Area	Chiba		462BG	33	33	0
251	July 6/7	Chiba Urban Area	Akashi		468BG	33	31	0
252	July 6/7	Urban Area	Akashi	73BW		131	123	0
252	July 6/7	Urban Area	Akashi		497BG	33	30	0
252	July 6/7	Urban Area	Akashi		498BG	33	30	0
252	July 6/7	Target of Opportunity	Akashi		498BG	in above	1	0
252	July 6/7	Urban Area	Akashi		499BG	not avail	not avail	
252	July 6/7	Urban Area	Shimizu		500BG	not avail	3	
253	July 6/7	Urban Area	Shimizu	313BW		136	133	1
253	July 6/7	Urban Area	Shimizu		6BG	not avail	36	
253	July 6/7	Urban Area	Shimizu		9BG	34	34	0
253	July 6/7	Urban Area	Shimizu		504BG	32	30	0
253	July 6/7	Urban Area	Kofu		505BG	not avail	33	0
254	July 6/7	Urban Area	Kofu	314BW		138	131	0
254	July 6/7	Urban Area	Kofu		19BG	not avail	30	
254	July 6/7	Urban Area	Kofu		29BG	not avail	35	0
254	July 6/7	Urban Area	Kofu		39BG	not avail	not avail	0
254	July 6/7	Urban Area	Kofu		330BG	33	32	0
255	July 6/7	Maruzen Oil Refinery	Shimotsu	315BW		59	58	0
255	July 6/7	Maruzen Oil Refinery	Shimotsu		501BG	28	28	0
255	July 6/7	Maruzen Oil Refinery	Shimotsu		16BG	31	30	0
256	July 9/10	Mining Mission 32	See Below	313BW		30	29	1
256	July 9/10	Mine Field Mike	Shimonoseki		6BG	14	not avail	not avail
256	July 9/10	Mine Field Uncle	Niigata/Nanao		6BG	9	not avail	not avail
256	July 9/10	Mine Field Nan	Fushiki		6BG	7	not avail	not avail
257	July 9/10	Urban Area	Sendai	58BW		131	123	1
257	July 9/10	Urban Area	Sendai		40BG	33	31	0
257	July 9/10	Urban Area	Sendai		444BG	33	32	0
257	July 9/10	Urban Area	Sendai		462BG	33	29	1
257	July 9/10	Urban Area	Sendai		468BG	33	32	0
258	July 9/10	Urban Area	Sakai	73BW		127	116	0
258	July 9/10	Urban Area	Sakai		497BG	33	29	0
258	July 9/10	Urban Area	Sakai		498BG	30	30	0
258	July 9/10	Urban Area	Sakai		499BG	not avail	26	0
258	July 9/10	Urban Area	Sakai		500BG	no avail	30	
259	July 9/10	Urban Area	Wakayama	313BW		109	108	0
259	July 9/10	Urban Area	Wakayama		9BG	36	36	0
259	July 9/10	Urban Area	Wakayama		504BG	37	37	0
259	July 9/10	Urban Area	Wakayama		505BG	not avail	35	0
260	July 9/10	Urban Area	Gifu	314BW		135	129	2
260	July 9/10	Urban Area	Gifu		19BG	32	not avail	
260	July 9/10	Urban Area	Gifu		29BG	not avail	34	
260	July 9/10	Urban Area	Gifu		39BG	34	33	1
260	July 9/10	Urban Area	Gifu		330BG	34	31	0
261	July 9/10	Utsube Oil refinery	Yokkaichi	315BW		64	61	0
261	July 9/10	Utsube Oil refinery	Yokkaichi		16BG	30	33	0
261	July 9/10	Utsube Oil refinery	Yokkaichi		501BG	34	28	0
262	July 11	Mining Mission 33	See below	313BW		30	27	0
262	July 11	Mine Field Mike	Shimonoseki		6BG	7	not avail	0
262	July 11	Mine Field Rashin	Korea		6BG	6	not avail	0
262	July 11	Mine Field Fusan	Korea		6BG	9	not avail	0
262	July 11	Mine Field Zebra	Maizura Bay		6BG	8	not avail	0
263	July 12	Urban Area	Utsonomiya	58BW		130	115	1
263	July 12	Urban Area	Utsonomiya		40BG	33	29	0
263	July 12	Urban Area	Utsonomiya		444BG	33	29	0
263	July 12	Urban Area	Utsonomiya		462BG	32	27	0
263	July 12	Urban Area	Utsonomiya		468BG	33	30	1
264	July 12/13	Urban Area	Ichinomiya	73BW		133	124	0
264	July 12/13	Urban Area	Ichinomiya		497BG	33	33	0
264	July 12/13	Urban Area	Ichinomiya		498BG	33	31	0

MSN	Date	Target	Location	Wings	Grps	No. Up	Bombed	Lost
264	July 12/13	Urban Area	Ichinomiya		499BG	not avail	30	0
264	July 12/13	Urban Area	Ichinomiya		500BG	not avail	31	0
265	July 12/13	Urban Area	Tsuruga	313BW		98	92	0
265	July 12/13	Urban Area	Tsuruga		9BG	34	30	0
265	July 12/13	Target of Opportunity	Yamada		9BG	in above	1	0
265	July 12/13	Target of Opportunity	Shingu		9BG	in above	1	0
265	July 12/13	Urban Area	Tsuruga		504BG	33	30	0
265	July 12/13	Urban Area	Tsuruga		505BG	not avail	not avail	
266	July 12/13	Urban Area	Uwajima	314BW		130	123	0
266	July 12/13	Urban Area	Uwajima		19BG	not avail	26	
266	July 12/13	Urban Area	Uwajima		29BG	not avail	32	0
266	July 12/13	Urban Area	Uwajima		39BG	33	30	0
266	July 12/13	Target of Opportunity	Sukumo		39BG	in above	1	0
266	July 12/13	Urban Area	Uwajima		330BG	33	33	
267	July 12/13	Petroleum Oil Center	Kawasaki	315BW		62	53	2
267	July 12/13	Petroleum Oil Center	Kawasaki		16BG	28	28	
267	July 12/13	Petroleum Oil Center	Kawasaki		501BG	34	25	
268	July 13	Mining Mission 34	See Below	313BW		32	27	0
268	July 13	Mine Field Mike	Shimonoseki		6BG	9	not avail	0
268	July 13	Mine Field Seishin	Korea		6BG	6	not avail	0
268	July 13	Mine Field Masan-Reisu	Korea		6BG	13	not avail	0
268	July 15/16	Mine Field Charlie	Fukuoka		6BG	3	not avail	0
269	July 15/16	Mining Mission 35	See Below		6BG	28	26	0
269	July 15/16	Mine Field Uncle/Nan	Niigata/Naoetsu		6BG	9	not avail	0
269	July 15/16	Mine Field Rashin	Rashin, Korea		6BG	8	not avail	0
269	July 15/16	Mine Field Genzen-Konan	Genzen, Korea		6BG	4	not avail	0
269	July 15/16	Mine Field Fusan	Fusan, E1079Korea		6BG	7	not avail	0
270	July 15/16	Kudamatsu Oil Refinery	Kudamatsu	315BW		69	59	0
270	July 15/16	Kudamatsu Oil Refinery	Kudamatsu		16BG	30	30	0
270	July 15/16	Kudamatsu Oil Refinery	Kudamatsu		501BG	39	27	0
271	July 15/16	Urban Area	Namazu	58BW		148	99	0
271	July 15/16	Urban Area	Namazu		40BG	40	26	0
271	July 15/16	Urban Area	Namazu		444BG	40	31	0
271	July 15/16	Urban Area	Namazu		462BG	26	21	0
271	July 15/16	Urban Area	Namazu		468BG	42	21	0
272	July 16/17	Urban Area	Oita	73BW		134	127	0
272	July 16/17	Urban Area	Oita		497BG	33	30	0
272	July 16/17	Urban Area	Oita		498BG	35	34	0
272	July 16/17	Urban Area	Oita		499BG	29	31	0
272	July 16/17	Urban Area	Oita		500BG	32	32	0
273	July 16/17	Urban Area	Kuwana	313BW		99	94	0
273	July 16/17	Urban Area	Kuwana		9BG	33	32	0
273	July 16/17	Target of Opportunity	Shingu		9BG	in above	1	0
273	July 16/17	Urban Area	Kuwana		504BG	34	31	0
273	July 16/17	Urban Area	Kuwana		505BG	not avail	32	0
274	July 16/17	Urban Area	Hiratsuka	314BW		137	129*	0
274	July 16/17	Urban Area	Hiratsuka		19BG	33	not avail	
274	July 16/17	Urban Area	Hiratsuka		29BG	not avail	33	
274	July 16/17	Urban Area	Hiratsuka		39BG	34	33	0
274	July 16/17	Urban Area	Hiratsuka		330BG	not avail	35	
275	July 17/18	Mining Mission 36	See Below	313BW		30	28	0
275	July 17/18	Mine Field Mike	Moji		6BG	10	not avail	0
275	July 17/18	Mine Field Seishin	Seishin, Korea		6BG	6	not avail	0
275	July 19	Mine Field Nan	Nanao Bay		6BG	14	not avail	0
276	July 19	Mining Mission 37	See Below		6BG	31	29	1
276	July 19	Mine Field Able	Shimonoseki		6BG	10	not avail	
276	July 19	Mine Field Uncle/Nan	Niigata		6BG	7	not avail	
276	July 19	Mine Field Zebra	Maizura/Miyama		6BG	9	not avail	
276	July 19	Mine Field Genzan — Konan	Korea		6BG	5	not avail	
277	July 19/20	Urban Area	Fukui	58BW		130	127	0
277	July 19/20	Urban Area	Fukui		40BG	32	31	0
277	July 19/20	Urban Area	Fukui		444BG	32	31	0
277	July 19/20	Urban Area	Fukui		462BG	33	32	0
277	July 19/20	Target of Opportunity	Seto		462BG	in above	1	0
277	July 19/20	Urban Area	Fukui		468BG	33	33	0

VII. The XXI Bomber Command

MSN	Date	Target	Location	Wings	Grps	No. Up	Bombed	Lost
278	July 19/20	Hitachi Urban Area	Hitachi	73BW		133	127	2
278	July 19/20	Hitachi Urban Area	Hitachi		497BG	32	30	0
278	July 19/20	Hitachi Urban Area	Hitachi		498BG	33	32	0
278	July 19/20	Hitachi Urban Area	Hitachi		499BG	not avail	33	1
278	July 19/20	Hitachi Urban Area	Hitachi		500BG	not avail	32	1
279	July 19/20	Choshi Urban Area	Choshi	313BW		97	91	0
279	July 19/20	Choshi Urban Area	Choshi		9BG	34	30	0
279	July 19/20	Choshi Urban Area	Choshi		504BG	33	30	0
279	July 19/20	Choshi Urban Area	Choshi		505BG	not avail	not avail	
280	July 19/20	Okazaki Urban Area	Okazaki	314BW		131	126	0
280	July 19/20	Okazaki Urban Area	Okazaki		19BG	not avail	not avail	
280	July 19/20	Okazaki Urban Area	Okazaki		29BG	not avail	32	
280	July 19/20	Okazaki Urban Area	Okazaki		39BG	34	33	0
280	July 19/20	Target of Opportunity	Nakii		39BG	in above	1	0
280	July 19/20	Okazaki Urban Area	Okazaki		330BG	32	31	0
281	July 19/20	Amagasaki — Nihon Oil Refinery	Amagasaki	315BW		84	83	0
281	July 19/20	Amagasaki — Nihon Oil Refinery	Amagasaki		16BG	30	not avail	not avail
281	July 19/20	Amagasaki — Nihon Oil Refinery	Amagasaki		501BG	not avail	29	0
	July 20	Pumpkin Mission 1	Koriyama		509CG	3	3	0
	July 20	Pumpkin Mission 2	Fukushima		509CG	2	2	0
	July 20	Pumpkin Mission 3	Nagaoka		509CG	2	2	0
	July 20	Pumpkin Mission 4	Toyama		509CG	3	3	0
282	July 22	Mining Mission 38	See Below	313BW	6BG	30	26	0
282	July 22	Mine Field Mike/Love	Shimonoseki		6BG	15	not avail	0
282	July 22	Mine Field Rashin	Korea		6BG	8	not avail	0
282	July 23/24	Mine Field Fusan/Musan	Korea		6BG	3/4	not avail	0
283	July 23/24	Coal Liquidification Plant	Ube	315BW		82	82	0
283	July 23/24	Coal Liquidification Plant	Ube		16BG	30	not avail	0
283	July 23/24	Coal Liquidification Plant	Ube		not avail	23	not avail	0
283	July 23/24	Coal Liquidification Plant	Ube		501BG	29	23	0
284	July 23/24	Sumitomo Metal Factory	Kochi	58BW		90	82	1
284	July 23/24	Sumitomo Metal Factory	Suzaki		40BG	43	40	0
284	July 23/24	Target of Opportunity	Kochi		40BG	in above	1	0
284	July 23/24	Target of Opportunity	Saimodo		40BG	in above	1	0
284	July 23/24	Sumitomo Metal Factory	Shimizu		444BG	47	42	1
284	July 23/24	Target of Opportunity	Saimoda		444BG	in above	1	0
284	July 23/24	Target of Opportunity	Shimizu		444BG	in above	1	0
284	July 23/24	Urban Area	Kuwana	73BW		81	71	0
284	July 23/24	Urban Area	Kuwana		497BG	38	35	0
284	July 23/24	Urban Area	Kuwana		498BG	43	36	0
284	July 23/24	Target of Opportunity	Kuwana		498BG	in above	6	0
285	July 24	Kawanishi Aircraft Factory	Kurakawa	58BW		88	81	0
285	July 24	Kawanishi Aircraft Factory	Tokushima		462BG	42	39	0
285	July 24	Target of Opportunity	Kurakowa		462BG	in above	1	0
285	July 24	Target of Opportunity	Tokushima		462BG	in above	1	0
285	July 24	Kawanishi Aircraft Factory	Takarazuka		468BG	46	42	0
285	July 24	Target of Opportunity	Kawanishi, Osaka		468BG	in above	1	0
286	July 24	Armory & Kuwana — Factory	Osaka	58BW		not avail	43	0
286	July 24	Armory & Kuwana — Factory	Osaka		444BG	not avail	43	0
286	July 24	Armory & Kuwana — Factory	Osaka	73BW		173	154	0
286	July 24	Armory & Kuwana — Factory	Osaka		497BG	38	35	0
286	July 24	Armory & Kuwana — Factory	Osaka		498BG	43	36	0
286	July 24	Armory & Kuwana — Factory	Osaka		498BG	in above	6	0
286	July 24	Armory & Kuwana — Factory	Osaka		499BG	not avail	35	0
286	July 24	Armory & Kuwana — Factory	Osaka		500BG	not avail	42	0
286	July 24	Target of Opportunity	Osaka		500BG	not avail	1	0
287	July 24	Aichi Aircraft Factory	Etitoku	313BW		74	71 plus	0
287	July 24	Aichi Aircraft Factory	Etitoku		504BG	38	33	0
287	July 24	Aichi Aircraft Factory	Etitoku		505BG	not avail	33	0
288	July 24	City of Tsu	Tsu		9BG	41	38	0
288	July 24	Target of Opportunity	Heki		9BG	in above	1	0
288	July 24	Target of Opportunity	Shingu		9BG	in above	1	0
289	July 24	City of Tsu	Tsu	314BW		not avail	75	0
289	July 24	City of Tsu	Tsu		19BG	not avail	34	0
289	July 24	City of Tsu	Tsu		330BG	not avail	41	0

MSN	Date	Target	Location	Wings	Grps	No. Up	Bombed	Lost
290	July 24	Nakajima Aircraft Plant	Handa		WING	81+	111	0
290	July 24	Nakajima Aircraft Plant	Handa		29BG	40	40	
290	July 24	Nakajima Aircraft Plant	Handa		330BG	36	32	
290	July 24	Target of Opportunity	Shingu Urban		39BG	41	39	0
290	July 24	Target of Opportunity	Kawasaki		39BG	in above	1	0
	July 24	Pumpkin Mission #5	Niihama		509CG	3	3	0
	July 24	Pumpkin Mission #6	Kobe		509CG	4	4	0
	July 24	Pumpkin Mission #7	Osaka		509CG	3	3	0
291	July 25/26	Oil Refinery	Korea	315BW		85	75	1
291	July 25/26	Oil Refinery	Kawasaki		16BG	32		
291	July 25/26	Oil Refinery	Kawasaki		Unk	not avail	not avail	
291	July 25/26	Oil Refinery	Kawasaki		501BG	not avail	29	0
292	July 25/26	Mining Mission 39	See Below	313BW		30	29	0
292	July 25/26	Mine Field Seishin	Seishin, Korea		504BG	6	5	0
292	July 25/26	Mine Field Fusan	Fusan, Korea		504BG	7	7	0
292	July 25/26	Mine Field Nan	Nanao Bay		504BG	11	11	0
292	July 25/26	Mine Field Zebra	Tsuruga Bay		504BG	6	6	0
293	July 25/26	Urban Area	Matsuyama	73BW		132	128	0
293	July 25/26	Urban Area	Matsuyama		497BG	32	32	0
293	July 25/26	Urban Area	Matsuyama		498BG	33	33	0
293	July 25/26	Urban Area	Matsuyama		499BG	not avail	31	0
293	July 25/26	Urban Area	Matsuyama		500BG	not avail	32	0
294	July 26/27	Urban Area	Tokuyama	313BW		102	97	0
294	July 26/27	Urban Area	Tokuyama		6BG	not avail	36	
294	July 26/27	Urban Area	Tokuyama		9BG	33	30	0
294	July 26/27	Urban Area	Tokuyama		9BG	in above	1	0
294	July 26/27	Urban Area	Omuta		505BG	not avail	32	0
295	July 26/27	Urban Area	Omuta	314BW		130	124	1
295	July 26/27	Urban Area	Omuta		19BG	32	not avail	
295	July 26/27	Urban Area	Omuta		29BG	not avail	32	
295	July 26/27	Urban Area	Omuta		39BG	33	31	0
295	July 26/27	Urban Area	Omuta		330BG	33	31	0
296	July 27/28	Mining Mission 40	See Below	313BW		31	24	2
296	July 26/27	Mine Field Mike	Shimonoseki		504BG	10	27	3
296	July 26/27	Mine Field Uncle	Niigata		504BG	8	in above	
296	July 26/27	Mine Field Zebra	Maizura		504BG	10	in above	
296	July 26/27	Mine Field X-Ray	Senzaki		504BG	3	in above	
	July 26	Pumpkin Mission #8	Nagaoka		509CG	4	4	0
	Jily 26	Pumpkin Mission #9	Toyama		509CG	6	6	0
297	July 27/28	Urban Area	Tsu/Aomori	58BW		82	80	0
297	July 27/28	Urban Area	Tsu		40BG	40	39	0
297	July 27/28	Urban Area	Tsu		468BG	42	41	0
298	July 28/29	Urban Area	Aomori		WING	68	67	0
298	July 28/29	Urban Area	Aomori		444BG	35	34	0
298	July 28/29	Target of Opportunity	Taira		444BG	in above	1	0
298	July 28/29	Urban Area	Aomori		462BG	33	33	0
298	July 28/29	Target of Opportunity	Taira		462BG	in above	2	0
298	July 28/29	Target of Opportunity	Morioka		462BG	in above	1	0
299	July 28/29	Urban Area	Ichinomiya	73BW		136	133	
299	July 28/29	Urban Area	Ichinomiya		497BG	29	26	0
299	July 28/29	Urban Area	Ichinomiya		498BG	34	33	0
299	July 28/29	Urban Area	Ichinomiya		499BG	30	41	0
299	July 28/29	Urban Area	Ichinomiya		500BG	34	33	0
300	July 28/29	Urban Area	Ujiyamada	313BW		99	93	
300	July 28/29	Urban Area	Ujiyamada		6BG	30	29	
300	July 28/29	Urban Area	Ujiyamada		9BG	33	32	0
300	July 28/29	Urban Area	Ujiyamada		505BG	37	32	
301	July 28/29	Urban Area	Ogaki	314BW		97	90*	0
301	July 28/29	Urban Area	Ogaki		29BG		33	
301	July 28/29	Urban Area	Ogaki		39BG	33	29	0
301	July 28/29	Urban Area	Ogaki		330BG	33	32	
302	July 28/29	Urban Area	Ogaki		19BG	32	29	0
303	July 28/29	Shimotsu	Maruzen Refinery	315BW		82	76	0
303	July 28/29	Shimotsu	Maruzen Refinery		16BG	21	not avail	
303	July 28/29	Shimotsu	Maruzen Refinery					

VII. The XXI Bomber Command

MSN	Date	Target	Location	Wings	Grps	No. Up	Bombed	Lost
303	July 28/29	Shimotsu	Maruzen Refinery		501BG	not avail	25	0
	July 29	Pumpkin Mission #10	Ube		509CG	3	3	0
	July 29	Pumpkin Mission #11	Koriyama		509CG	3	3	0
	July 29	Pumpkin Mission #12	Yokkaichi		509CG	2	2	0
304	July 29/30	Mining Mission 41	See Below	313BW		30	27	0
304	July 29/30	Mine Field Rashin	Raishin, Korea		504BG	14	in above	0
304	July 29/30	Mine Field Mike	Shimonoseki		504BG	10	in above	0
304	July 29/30	Mine Field Charlie	Fukuoka Bay		504BG	6	in above	0
305	August 1/2	Mining Mission 42	See Below		WING	45	38	0
305	August 1/2	Mine Field Rashin	Raishin, Korea		504BG	9	in above	0
305	August 1/2	Mine Field Seishin	Seishin, Korea		504BG	7	in above	0
305	August 1/2	Mine Field Mike	Moji Area		504BG	17	in above	0
305	August 1/2	Mine Field George	Hamada,		504BG	7	in above	0
305	August 1/2	Mine Field X-Ray	Hagi, Oura		504BG	5	in above	0
306	August 1/2	Urban Area	See Below	MISSION		794	750	2
306	August 1/2	Urban Area	Hachiogi	58BW		180	169	2
306	August 1/2	Urban Area	Hachiogi		40BG	44	42	1
306	August 1/2	Target of Opportunity	Hahajima		40BG	in above	1	0
306	August 1/2	Urban Area	Hachiogi		444BG	45	41	0
306	August 1/2	Target of Opportunity	Hahajima		444BG	in above	1	0
306	August 1/2	Target of Opportunity	Shimazu		444BG	in above	1	0
306	August 1/2	Urban Area	Shimoda		462BG	46	42	1
306	August 1/2	Target of Opportunity	Hahajima		462BG	in above	1	0
306	August 1/2	Urban Area	Shimoda		468BG	45	44	0
306	August 1/2	Target of Opportunity	Yokosuka		468BG	in above	1	0
307	August 1/2	Urban & Industrial Areas	Toyame	73BW		184	174	0
307	August 1/2	Urban & Industrial Areas	Toyame		497BG	47	44	0
307	August 1/2	Urban & Industrial Areas	Toyame		498BG	46	44	0
307	August 1/2	Urban & Industrial Areas	Toyame		499BG	not avail	41	0
307	August 1/2	Urban & Industrial Areas	Toyame		500BG	not avail	41	0
308	August 1/2	Urban & Industrial Areas	Nagaoka	313BW		136	125	0
308	August 1/2	Urban & Industrial Areas	Nagaoka		6BG	not avail	45	
308	August 1/2	Urban & Industrial Areas	Nagaoka		9BG	47	43	0
308	August 1/2	Urban & Industrial Areas	Tanabe		9BG	in above	1	0
308	August 1/2	Urban & Industrial Areas	Fukui		9BG	in above	1	0
308	August 1/2	Urban & Industrial Areas	Toyama		9BG	in above	1	0
308	August 1/2	Urban & Industrial Areas	Hamamatsu		9BG	in above	1	0
308	August 1/2	Urban & Industrial Areas	Nagaoka		504BG	not avail	not avail	
308	August 1/2	Urban & Industrial Areas	Nagaoka		505BG	not avail	38	0
309	August 1/2	Urban & Industrial Areas	Mito	314BW		168	161	0
309	August 1/2	Urban & Industrial Areas	Mito		19BG	45	43	0
309	August 1/2	Urban & Industrial Areas	Mito		29BG	44	40	0
309	August 1/2	Urban & Industrial Areas	Mito		39BG	38	37	0
309	August 1/2	Target of Opportunity	Choshi		39BG	in above	1	0
309	August 1/2	Urban & Industrial Areas	Kawasaki		330BG	41	41	0
310	August 1/2	Mitsubishi Havana Oil Refinery	Kawasaki	315BG		128	120	1
310	August 1/2	Mitsubishi Havana Oil Refinery	Kawasaki		16BG	36	not avail	
310	August 1/2	Mitsubishi Havana Oil Refinery	Kawasaki		331BG	not avail	not avail	
310	August 1/2	Mitsubishi Havana Oil Refinery	Kawasaki		502BG	not avail	not avail	
310	August 1/2	Mitsubishi Havana Oil Refinery	Kawasaki		501BG	not avail	34	0
311	August 5/6	Mining Mission 43	See Below	313BW	504BG	30	27	0
311	August 5/6	Mine Field X-Ray	Hagi Harbor		504BG	6	in above	0
311	August 5/6	Mine Field Zebra	Tsuruga Harbor		504BG	10	in above	0
311	August 5/6	Mine Field Geijitsu	Geijitsu, Korea		504BG	5	in above	0
311	August 5/6	Mine Field Rashin	Rashin, Korea		504BG	9	in above	0
313	August 5/6	Urban & Industrial Areas	Maebashi	313BW		102	92	0
313	August 5/6	Urban & Industrial Areas	Maebashi		WING	in above	4	0
313	August 5/6	Urban & Industrial Areas	Maebashi		6BG	37	31*	0
313	August 5/6	Urban & Industrial Areas	Tateyama		9BG	35	29	0
313	August 5/6	Urban & Industrial Areas	Maebashi		9BG	in above	2	0
313	August 5/6	Urban & Industrial Areas	Maebashi		505BG	30	30	0
314	August 5/6	Urban & Industrial Areas	Nishinomiya	MISSION		268	259	1
314	August 5/6	Target of Opportunity	Nishinomiya	WINGS		in above	3	0
314	August 5/6	Urban & Industrial Areas	Nishinomiya	73BW		133	130	1
314	August 5/6	Urban & Industrial Areas	Nishinomiya		497BG	33	33	0

MSN	Date	Target	Location	Wings	Grps	No. Up	Bombed	Lost
314	August 5/6	Urban & Industrial Areas	Nishinomiya		498BG	33	32	1
314	August 5/6	Urban & Industrial Areas	Nishinomiya		499BG	not avail	30	0
314	August 5/6	Urban & Industrial Areas	Nishinomiya		500BG	not avail	35	0
314	August 5/6	Urban & Industrial Areas	Nishinomiya	314BW		135	129	0
314	August 5/6	Urban & Industrial Areas	Nishinomiya		19BG	34	30	0
314	August 5/6	Urban & Industrial Areas	Nishinomiya		29BG	34	32	0
314	August 5/6	Urban & Industrial Areas	Nishinomiya		39BG	33	33	0
314	August 5/6	Urban & Industrial Areas	Nishinomiya		330BG	34	34	0
315	August 5/6	Ube Coal Liquidification Co.	Ube	315BW		110	100	0
315	August 5/6	Ube Coal Liquidification Co.	Ube		WING	in above	2	0
315	August 5/6	Ube Coal Liquidification Co.	Ube		16BG	30	not avail	0
315	August 5/6	Ube Coal Liquidification Co.	Ube		331BG	not avail	not avail	0
315	August 5/6	Ube Coal Liquidification Co.	Ube		501BG	not avail	23	0
315	August 5/6	Ube Coal Liquidification Co.	Ube		502BG	not avail	not avail	0
312	August 5/6	Urban & Industrial Areas	Saga/Imabari	58BW		133	127	0
312	August 5/6	Urban & Industrial Areas	Saga		WING	66	63	1
312	August 5/6	Urban & Industrial Areas	Saga		462BG	32	31	0
312	August 5/6	Urban & Industrial Areas	Saga		468BG	34	32	1
316	August 5/6	Urban & Industrial Areas	Imabari		WING	67	64	0
316	August 5/6	Urban & Industrial Areas	Imabari		40BG	33	33	0
316	August 5/6	Urban & Industrial Areas	Imabari		444BG	34	31	0
	August 6	LITTLE BOY BOMB	Hiroshima		509CG	7	1	0
317	August 7	Toyokawa Naval Arsenal	Toyokawa	MISSION		131	127	1
317	August 7	Toyokawa Naval Arsenal	Toyokawa	58BW		32	30	0
317	August 7	Toyokawa Naval Arsenal	Toyokawa		40BG	11	11	0
317	August 7	Toyokawa Naval Arsenal	Toyokawa		444BG	10	9	0
317	August 7	Toyokawa Naval Arsenal	Toyokawa		462BG	11	10	0
317	August 7	Toyokawa Naval Arsenal	Toyokawa	73BW		34	29	0
317	August 7	Toyokawa Naval Arsenal	Toyokawa		498BG	11	11	0
317	August 7	Toyokawa Naval Arsenal	Toyokawa		499BG	15	10	0
317	August 7	Toyokawa Naval Arsenal	Toyokawa		500BG	8	8	0
317	August 7	Toyokawa Naval Arsenal	Toyokawa	313BW		36	35	1
317	August 7	Toyokawa Naval Arsenal	Toyokawa		6BG	12	12	0
317	August 7	Toyokawa Naval Arsenal	Toyokawa		9BG	12	9	1
317	August 7	Toyokawa Naval Arsenal	Toyokawa		505BG	12	12	0
317	August 7	Toyokawa Naval Arsenal	Toyokawa	314BW		33	31	2
317	August 7	Toyokawa Naval Arsenal	Toyokawa		19BG	12	12	0
317	August 7	Toyokawa Naval Arsenal	Toyokawa		29BG	9	7	2
317	August 7	Toyokawa Naval Arsenal	Toyokawa		39BG	12	12	0
318	August 7/8	Mining Mission 44	See Below	313BW	504BG	27	27	0
318	August 7/8	Mine Field Rashin	Raishin, Korea		504BG	8	8	0
318	August 7/8	Mine Field Mike/Love	Shimonoseki		504BG	5	5	0
318	August 7/8	Mine Field Zebra	Maizuru		504BG	8	8	0
318	August 7/8	Mine Field Yoke	Sakai		504BG	6	6	0
319	August 7/8	Urban & Industrial Area	Yawata	MISSION		314	304	4
319	August 7/8	Urban & Industrial Area	Yawata Area		WINGS	in above	6	0
319	August 7/8	Urban & Industrial Area	Yawata	58BW		35	30	2
319	August 7/8	Urban & Industrial Area	Yawata		40BG	11	11	0
319	August 7/8	Urban & Industrial Area	Yawata		444BG	10	9	1
319	August 7/8	Urban & Industrial Area	Yawata		462BG	11	10	1
319	August 7/8	Urban & Industrial Area	Yawata	73BW		122	108	0
319	August 7/8	Urban & Industrial Area	Yawata		WING	in above	3	0
319	August 7/8	Urban & Industrial Area	Yawata		497BG	33	30	0
319	August 7/8	Urban & Industrial Area	Yawata		498BG	31	26	0
319	August 7/8	Urban & Industrial Area	Yawata		498BG	in above	2	0
319	August 7/8	Urban & Industrial Area	Yawata		499BG	not avail	26	0
319	August 7/8	Urban & Industrial Area	Yawata		500BG	not avail	26	0
319	August 7/8	Urban & Industrial Area	Yawata	313BW		90	84	1
319	August 7/8	Urban & Industrial Area	Yawata		6BG	29	27	0
319	August 7/8	Urban & Industrial Area	Yawata		9BG	30	29	1
319	August 7/8	Urban & Industrial Area	Yawata		505BG	31	28	1
319	August 7/8	Urban & Industrial Area	Yawata	314BW		69	67	3
319	August 7/8	Urban & Industrial Area	Yawata		29BG	33	33	1
319	August 7/8	Urban & Industrial Area	Yawata		330BG	36	34	0
	August 8	Pumpkin Mission #14	Osaka		509CG	4	4	0

VII. The XXI Bomber Command

MSN	Date	Target	Location	Wings	Grps	No. Up	Bombed	Lost
	August 8	Pumpkin Mission #15	Yokkaichi		509CG	2	2	0
321	August 8/9	Urban Area	Fukuyama	58BW		98	91	0
321	August 8/9	Urban Area	Fukuyama		40BG	33	31	0
321	August 8/9	Urban Area	Fukuyama		462BG	32	29	0
321	August 8/9	Urban Area	Fukuyama		468BG	33	31	0
322	August 8/9	Nippon Oil Refinery	Amagasaki	315BW		107	95	0
322	August 8/9	Target of Opportunity	Amagasaki Area		WING	in above	2	0
322	August 8/9	Nippon Oil Refinery	Amagasaki		16BG	29	not avail	0
322	August 8/9	Nippon Oil Refinery	Amagasaki		331BG	not avail	not avail	0
322	August 8/9	Nippon Oil Refinery	Amagasaki		501BG	not avail	23	0
322	August 8/9	Nippon Oil Refinery	Amagasaki		502BG	not avail	not avail	0
	August 9	FAT MAN BOMB	Nagasaki		509CG	6	1	0
323	August 9/10	Nakajima Aircraft Factory	Tokyo Arsenal	314BW		78	72	2
323	August 9/10	Target of Opportunity	Tokyo Arsenal		WING	in above	3	0
323	August 9/10	Nakajima Aircraft Factory	Tokyo Arsenal		19BG	40	39	0
323	August 9/10	Urban Area	Kawaguchi		39BG	38	33	2
323	August 9/10	Target of Opportunity	Yakaichiba A/F		39BG	in above	2	0
323	August 9/10	Target of Opportunity	Hachijo Jima		39BG	in above	1	0
324	August 9/10	Mining Mission 45	See Below	313BW		32	32	0
324	August 9/10	Mine Field Rashin	Raishin, Korea		504BG	7	7	0
324	August 9/10	Mine Field Seishin	Seishin, Korea		504BG	7	7	0
324	August 9/10	Mine Field Mike	Shimonoseki		504BG	11	11	0
324	August 9/10	Mine Field Genzan	Genzan, Korea		504BG	7	7	0
325	August 9/10	Hikari Naval Arsenal	Hikari	58BW		190	157	0
325	August 9/10	Hikari Naval Arsenal	Hikari		40BG	44	42	0
325	August 9/10	Hikari Naval Arsenal	Hikari		444BG	58	37	0
325	August 9/10	Hikari Naval Arsenal	Hikari		462BG	40	35	0
325	August 9/10	Hikari Naval Arsenal	Hikari		468BG	48	43	0
	August 14	Pumpkin Mission #17	Nagoya		509CG	4	4	0
	August 14	Pumpkin Mission #18	Koroma		509GC	3	3	0
326	August 14	Army Arsenal	Osaka	73BW		165	145	0
326	August 14	Target of Opportunity	Osaka Area		WING	in above	2	0
326	August 14	Army Arsenal	Osaka		497BG	40	34	0
326	August 14	Army Arsenal	Osaka		498BG	40	38	0
326	August 14	Secondary Target	Osaka		498BG	in above	1	0
326	August 14	Army Arsenal	Osaka		499BG	not avail	38	0
326	August 14	Army Arsenal	Osaka		500BG	not avail	35	0
327	August 14	Marshalling Yards	Marifu	313BW		116	109	0
327	August 14	Target of Opportunity	Marifu Area		WING	in above	2	0
327	August 14	Marshalling Yards	Marifu		6BG	43	39	0
327	August 14	Marshalling Yards	Marifu		9BG	41	37	0
327	August 14	Marshalling Yards	Marifu		505BG	32	31	0
328	August 14	Nippon Oil Refinery	Tsuchisaki	315BW		141	132	0
328	August 14	Nippon Oil Refinery	Tsuchisaki		16BG	38	not avail	0
328	August 14	Nippon Oil Refinery	Tsuchisaki		331BG	not avail	not avail	0
328	August 14	Nippon Oil Refinery	Tsuchisaki		501BG	not avail	31	0
328	August 14	Nippon Oil Refinery	Tsuchisaki		502BG	not avail	not avail	0
329	August 14	Kumagaya/Isezaki	See below	MISSION		751	723	
329	August 14	Urban Area	Kumagaya	WINGS		198	183	0
329	August 14	Urban Area	Kumagaya	313BW		116	111	0
329	August 14	Urban Area	Kumagaya		9BG	4	3	0
329	August 14	Urban Area	Kumagaya		505BG	5	5	0
329	August 14	Urban Area		314BW		164	155	0
329	August 14	Urban Area	Kumagaya	WING		76	71	0
329	August 14	Urban Area	Kumagaya		29BG	38	37	0
329	August 14	Urban Area	Kumagaya		330BG	38	34	0
330	August 14	Urban Area	Isesake	MISSION		95	90	0
330	August 14	Target of Opportunity	Isesake	WINGS		in above	2	0
330	August 14	Urban Area	Isesake	73BW		10	9	0
330	August 14	Urban Area	Isesake		497BG	9	9	0
330	August 14	Urban Area	Isesake	314BW		88	84	1
330	August 14	Target of Opportunity	Unknown		Unk	in above	2	0
330	August 14	Urban Area	Isesake		19BG	44	41	0
330	August 14	Urban Area	Isesake		39BG	44	43	1
330	August 14	Target of Opportunity	Choshi Point		39BG	in above	1	0

MSN	Date	Target	Location	Wings	Grps	No. Up	Bombed	Lost
331	August 14	Mining Mission 46	See Below	313BW		41	38	0
331	August 14	Mine Field Nan	Nanao Bay		504BG	8	in above	0
331	August 14	Mine Field Mike	Shimonoseki		504BG	10	in above	0
331	August 14	Mine Field Zebra	Tsuruga Bay		504BG	17	in above	0
331	August 14	Mine Field George	Hamada Harbor		504BG	6	in above	0

Bomb Group Mission Lists

Here are the data from the composite XXI Bomber Command mission list, broken down by bomb group. Following these lists is a group-by-group listing of the planes assigned to each bomb group.

40thBG/58thBW

MSN	Date	Target	Location	Number Up	Bombed	Lost
146	May 5	Hiro Naval Aircraft Factory	Kure	31	27	0
146	May 5	Target of Opportunity	Nageoka	in above	1	0
166	May 10	Naval Oil Storage	Oshima	31	28	0
166	May 10	Target of Opportunity	Sukumo Seaplane	in above	1	0
174	May 14	Urban Area	Nagoya	35	32	0
174	May 14	Target of Opportunity	Unknown	in above	1	0
176	May 16/17	Urban Area	Nagoya	32	29	0
176	May 16/17	Target of Opportunity	Matsuyaka	in above	1	0
178	May 19	Aircraft factory	Tachikawa	24	23	1
178	May 19	Target of Opportunity	Kofu	in above	1	0
181	May 23/24	Urban Area	Tokyo	32	30	1
181	May 23/24	Target of Opportunity	Sagara	in above	1	0
183	May 25/26	Urban Area	Tokyo	33	30	2
186	May 29	Urban Area	Yokohama	31	30	0
187	June 1	Urban Area	Osaka	29	27	0
188	June 5	Urban Area	Kobe	29	24	0
188	June 5	Target of Opportunity	Shizuki	in above	1	0
189	June 7	Urban Area	Osaka	32	31	0
195	June 10	Kasumigahara Seaplane Base	Kasumiguara	29	23	0
195	June 10	Target of Opportunity	Hachido Jima	in above	1	0
195	June 10	Target of Opportunity	Shinoda	in above	1	0
203	June 15	Urban Area	Osaka	33	29	0
203	June 15	Target of Opportunity	Kimomoto	in above	1	0
203	June 15	Target of Opportunity	Kobe	in above	1	0
207	June 17/18	Urban Area	Omuta	32	30	0
207	June 17/18	Target of Opportunity	Nagazu	in above	1	0
210	June 19/20	Urban Area	Toyohashi	35	34	0
217	June 22	Himeji — Kawanishi Aircraft Factory	Tokushima	28	24	0
217	June 22	Target of Opportunity	Wadashima	in above	1	0
217	June 22	Target of Opportunity	Tokoshima	in above	1	0
225	June 26	Kawasaki Aircraft Factory	Ahashi	38	31	1
225	June 26	Secondary Target	Tsu	in above	4	0
228	June 26	Mitsubishi Aircraft Factory	Kagamigahara	38	33	1
234	June 28/29	Okayama Urban Area	Okiyama	36	34	0
240	July 1/2	Kure Urban Area	Kure	41	39	0
247	July 3/4	Urban Area	Takamatsu	31	28	0
251	July 6/7	Chiba Urban Area	Chiba	32	30	0
251	July 6/7	Target of Opportunity	Unknown	in above	1	0
257	July 9/10	Urban Area	Sendai	33	31	0
263	July 12	Urban Area	Utsonomiya	33	29	0
271	July 15/16	Urban Area	Namazu	40	26	0
277	July 19/20	Urban Area	Fukui	32	31	0

284	July 23/24	Sumitomo Metal Factory	Suzaki	43	40	0
284	July 23/24	Target of Opportunity	Kochi	in above	1	0
284	July 23/24	Target of Opportunity	Saimodo	in above	1	0
297	July 27/28	Urban Area	Tsu	40	39	0
306	August 1/2	Urban Area	Hachiogi	44	42	1
306	August 1/2	Target of Opportunity	Hahajima	in above	1	0
316	August 5/6	Urban & Industrial Areas	Imabari	33	33	0
317	August 7	Toyokawa Naval Arsenal	Toyokawa	11	11	0
319	August 7/8	Ubrban & Industrial Area	Yawata	11	11	0
321	August 8/9	Urban Area	Fukuyama	33	31	0
325	August 9/10	Hikari Naval Arsenal	Hikari	44	42	0

444thBG/58thBW

MSN	Date	Target	Location	Number Up	Bombed	Lost
166	May 10	Naval Oil Storage	Oshima	24	20	0
166	May 10	Target of Opportunity — Ship	32-57N, 132-58E	in above	1	0
166	May 10	Target of Opportunity	Otake Oil Refinery	in above	2	0
172	May 11	Kawanishi Aircraft Factory	Kobe	13	12	0
172	May 11	Target of Opportunity	Sakimahara	in above	1	0
174	May 14	Urban Area	Nagoya	32	28	0
176	May 16/17	Urban Area	Nagoya	32	26	0
176	May 16/17	Target of Opportunity	Koshimoto	in above	1	0
178	May 19	Aircraft Factory	Tachikawa	21	17	
178	May 19	Target of Opportunity	Matsugami	in above	1	
178	May 19	Target of Opportunity	Shimata	in above	1	
181	May 23/24	Urban Area	Tokyo	34	32	0
183	May 25/26	Urban Area	Tokyo	34	33	5
186	May 29	Urban Area	Yokohama	29	26	
187	June 1	Urban Area	Osaka	27	22	3
187	June 1	Target of Opportunity	Matsuke	in above	1	0
188	June 5	Urban Area	Kobe	27	25	1
189	June 7	Urban Area	Osaka	26	24	0
189	June 7	Target of Opportunity	Shingo	in above	1	0
191	June 9	Kawanishi Aircraft Factory	Osaka	29	28	0
191	June 9	Target of Opportunity	Aichi Plant-Nagoya	in above	1	0
203	June 15	Urban Area	Osaka	35	32	0
203	June 15	Target of Opportunity	Kushimoto	in above	1	0
207	June 17/18	Urban Area	Omuta	35	32	0
207	June 17/18	Target of Opportunity	Nokeoka	in above	1	0
210	June 19/20	Urban Area	Toyohashi	35	34	0
217	June 22	Himeji — Kawanishi Aircraft Factory	Tomashima	30	28	0
217	June 22	Target of Opportunity	Tomashima	in above	2	0
225	June 26	Kawasaki Aircraft Factory	Ahashi	41	27	1
225	June 26	Secondary Target	Tsu	in above	11	0
225	June 26	Target of Opportunity	Kochi A/F	in above	2	0
225	June 26	Target of Opportunity	Banzai	in above	1	0
228	June 26	Mitsubishi Aircraft Factory	Kagamigahara	41	27	0
228	June 26	Target of Opportunity	Komatsushima	in above	2	0
228	June 26	Target of Opportunity	Yokkaichi	in above	2	0
234	June 28/29	Okayama Urban Area	Okiyama	35	35	0
240	July 1/2	Kure Urban Area	Kure	39	38	0
247	July 3/4	Urban Area	Takamatsu	32	28	0
251	July 6/7	Chiba Urban Area	Chiba	31	30	0
257	July 9/10	Urban Area	Sendai	33	32	0
263	July 12	Urban Area	Utsonomiya	33	29	0
271	July 15/16	Urban Area	Namazu	40	31	0
277	July 19/20	Urban Area	Fukui	32	31	0
284	July 23/24	Sumitomo Metal Factory	Shimizu	47	42	1
284	July 23/24	Target of Opportunity	Saimoda	in above	1	0
284	July 23/24	Target of Opportunity	Shimizu	in above	1	0
286	July 24	Armory & Kuwana — Factory	Osaka	not avail	43	0
298	July 28/29	Urban Area	Aomori	35	34	0
298	July 28/29	Target of Opportunity	Taira	in above	1	0
306	August 1/2	Urban Area	Hachiogi	45	41	0
306	August 1/2	Target of Opportunity	Hahajima	in above	1	0

MSN	Date	Target	Location	Number Up	Bombed	Lost
306	August 1/2	Target of Opportunity	Shimazu	in above	1	0
316	August 5/6	Urban & Industrial Areas	Imabari	34	31	0
317	August 7	Toyokawa Naval Arsenal	Toyokawa	10	9	0
319	August 7/8	Ubrban & Industrial Area	Yawata	10	9	1
325	August 9/10	Hikari Naval Arsenal	Hikari	58	37	0

462ndBG/58thBW

MSN	Date	Target	Location	Number Up	Bombed	Lost
146	May 5	Hiro Naval Aircraft Factory	Kure	11	10	0
166	May 10	Naval Oil Storage	Oshima	33	32	0
174	May 14	Urban Area	Nagoya	36	33	0
176	May 16/17	Urban Area	Nagoya	37	33	0
178	May 19	Aircraft Factory	Tachikawa	22	20	0
178	May 19	Target of Opportunity	Tokyo	in above	1	0
181	May 23/24	Urban Area	Tokyo	30	27	1
183	May 25/26	Urban Area	Tokyo	31	30	1
186	May 29	Urban Area	Yokohama	34	30	0
186	May 29	Target of Opportunity	Hachijo Jima	in above	1	0
187	June 1	Urban Area	Osaka	28	26	0
188	June 5	Urban Area	Kobe	31	26	4
188	June 5	Target of Opportunity	Wakayama	in above	1	0
189	June 7	Urban Area	Osaka	26	25	0
189	June 7	Target of Opportunity	Shirahama	in above	1	0
191	June 9	Kawanishi Aircraft Factory	Naruo	17	16	0
203	June 15	Urban Area	Amagasaki	30	25	0
203	June 15	Target of Opportunity	Shirahama	in above	1	0
207	June 17/18	Urban Area	Omuta	22	20	0
210	June 19/20	Urban Area	Toyohashi	33	33	0
215	June 22	Kure Arsenal	Kure	31	27	1
223	June 26	Sumitomo Light Metal Industry	Osaka	35	34	0
234	June 28/29	Okayama Urban Area	Okiyama	34	33	0
240	July 1/2	Kure Urban Area	Kure	40	37	0
247	July 3/4	Urban Area	Takamatsu	32	31	1
247	July 3/4	Target of Opportunity	Nobeoka	in above	1	0
251	July 6/7	Chiba Urban Area	Chiba	33	33	0
257	July 9/10	Urban Area	Sendai	33	29	1
263	July 12	Urban Area	Utsonomiya	32	27	0
271	July 15/16	Urban Area	Namazu	26	21	0
277	July 19/20	Urban Area	Fukui	33	32	0
277	July 19/20	Target of Opportunity	Seto	in above	1	0
285	July 24	Kawanishi Aircraft Factory	Tokushima	42	39	0
285	July 24	Target of Opportunity	Kurakowa	in above	1	0
285	July 24	Target of Opportunity	Tokushima	in above	1	0
298	July 28/29	Urban Area	Aomori	33	33	0
298	July 28/29	Target of Opportunity	Taira	in above	2	0
298	July 28/29	Target of Opportunity	Morioka	in above	1	0
306	August 1/2	Urban Area	Shimoda	46	42	1
306	August 1/2	Target of Opportunity	Hahajima	in above	1	0
312	August 5/6	Urban & Industrial Areas	Saga	32	31	0
317	August 7	Toyokawa Naval Arsenal	Toyokawa	11	10	0
319	August 7/8	Ubrban & Industrial Area	Yawata	11	10	1
321	August 8/9	Urban Area	Fukuyama	32	29	0
325	August 9/10	Hikari Naval Arsenal	Hikari	40	35	0

468thBG/58thBW

MSN	Date	Target	Location	Number Up	Bombed	Lost
172	May 11	Kawanishi Aircraft Factory	Kobe	25	23	0
174	May 14	Urban Area	Nagoya	38	31	
174	May 14	Target of Opportunity	Neve (sp?) Lake	in above	1	
176	May 16/17	Urban Area	Nagoya	39	34	0
176	May 16/17	Target of Opportunity	Yokosuka	in above	1	0

VII. The XXI Bomber Command

176	May 16/17	Target of Opportunity	Osaka	in above	1	0
176	May 16/17	Target of Opportunity	Okura	in above	1	0
181	May 23/24	Urban Area	Tokyo	32	29	1
181	May 23/24	Target of Opportunity	Shizuoka	in above	1	0
183	May 25/26	Urban Area	Tokyo	35	26	1
183	May 25/26	Target of Opportunity	Hamamatsu	in above	1	0
183	May 25/26	Target of Opportunity	Tateyama Hoto	in above	1	0
186	May 29	Urban Area	Yokohama	38	36	
186	May 29	Target of Opportunity	Hachijo Jima	in above	1	
187	June 1	Urban Area	Osaka	35	32	1
187	June 1	Target of Opportunity	Shingu	in above	1	0
188	June 5	Urban Area	Kobe	33	32	0
188	June 5	Target of Opportunity	Komatsushima	in above	1	0
189	June 7	Urban Area	Osaka	31	27	0
189	June 7	Target of Opportunity	Chichi Jima	in above	11	0
189	June 7	Target of Opportunity	Tokushima	in above	1	0
189	June 7	Target of Opportunity	None'	in above	1	0
196	June 10	Tomioko Engine Plant	Tomioka	33	32	0
196	June 10	Target of Opportunity	Shimizu	in above	1	0
203	June 15	Urban Area	Osaka	40	33	0
203	June 15	Target of Opportunity	Kushimoto	in above	1	0
203	June 15	Target of Opportunity	Katsaura	in above	1	0
203	June 15	Target of Opportunity	Tokushima	in above	1	0
203	June 15	Target of Opportunity	24:15N; 138:15E	in above	1	0
207	June 17/18	Urban Area	Omuta	37	34	0
207	June 17/18	Target of Opportunity	Hatamara	in above	1	0
210	June 19/20	Urban Area	Toyohashi	38	35	0
215	June 22	Kure Arsenal	Uwajima	33	19	0
215	June 22	Target of Opportunity	Kure	in above	1	0
215	June 22	Target of Opportunity	Nobeoka	in above	1	0
215	June 22	Target of Opportunity	Sukumo N.B	in above	1	0
215	June 22	Target of Opportunity	Hoseshima	in above	1	0
215	June 22	Target of Opportunity	Oita A/F	in above	1	0
215	June 22	Target of Opportunity	Sagonoseki	in above	1	0
215	June 22	Target of Opportunity	Sadei	in above	1	0
215	June 22	Target of Opportunity	Ochimura	in above	1	0
215	June 22	Target of Opportunity	Shimazu RR Brdg	in above	1	0
215	June 22	Target of Opportunity	Meisho	in above	1	0
215	June 22	Target of Opportunity	Okina Shima	in above	1	0
215	June 22	Target of Opportunity	Tomitaki A/F	in above	1	0
223	June 26	Sumitomo Light Metal Industry	Osaka	36	18	0
223	June 26	Target of Opportunity	Kushimoto	in above	2	0
223	June 26	Target of Opportunity	Oshimo	in above	1	0
223	June 26	Target of Opportunity	Tokushima	in above	3	0
223	June 26	Target of Opportunity	Ujiyamada	in above	6	0
234	June 28/29	Okayama Urban Area	Okiyama	36	36	1
240	July 1/2	Kure Urban Area	Kure	40	38	0
247	July 3/4	Urban Area	Takamatsu	33	29	2
247	July 3/4	Target of Opportunity	Chichi Jima	in above	1	0
247	July 3/4	Target of Opportunity	Sukumo	in above	1	0
251	July 6/7	Chiba Urban Area	Akashi	33	31	0
257	July 9/10	Urban Area	Sendai	33	32	0
263	July 12	Urban Area	Utsonomiya	33	30	1
271	July 15/16	Urban Area	Namazu	42	21	0
277	July 19/20	Urban Area	Fukui	33	33	0
285	July 24	Kawanishi Aircraft Factory	Takarazuka	46	42	0
285	July 24	Target of Opportunity	Kawanishi, Osaka	in above	1	0
297	July 27/28	Urban Area	Tsu	42	41	0
306	August 1/2	Urban Area	Shimoda	45	44	0
306	August 1/2	Target of Opportunity	Yokosuka	in above	1	0
312	August 5/6	Urban & Industrial Areas	Saga	34	32	1
321	August 8/9	Urban Area	Fukuyama	33	31	0
325	August 9/10	Hikari Naval Arsenal	Hikari	48	43	0

497thBG/73rdBW

MSN	Date	Target	Location	Number Up	Bombed	Lost
1	Oct 28, 1944	Submarine Base	Truk	9	7	0
2	October 30	Submarine Base	Truk	9	9	0
3	November 2	Submarine Base	Truk	11	9	0
4	November 5	Airfields	Iwo Jima	18	16	0
5	November 8	Airfields	Iwo Jima	7	6	0
5	November 10	Recon (Aborted — weather)	Tokyo	0	n/a	0
7	November 24	Docks & Urban Area — Secondary	Tokyo	28	20	1
8	November 27	Musashino — Bombed Secondary	Tokyo	26	14	0
9	November 29/30	Industrial Area	Tokyo	9	7	0
10	December 3	Musashino — Nakajima AC Eng	Tokyo	24	11	1
11	December 8	Airfields	Iwo Jima	16	12	0
12	December 13	Mitsubishi Aircraft Engine Plant	Nagoya	18	16	0
13	December 18	Mitsubishi Aircraft Factory	Nagoya	18	14	1
14	December 22	Mitsubishi Aircraft Engine Plant	Nagoya	14	7	1
15	December 24	Navigation mission	Iwo Jima	3	n/a	0
16	December 27	Nakajima AC Engine Plant	Tokyo	21	14	0
17	Jan 3, 1945	Urban Area	Nagoya	21	17	1
18	January 9	Musashino — Nakajima AC EngineTokyo	Tokyo	17	10	3
19	January 14	Mitsubishi Aircraft Factory	Nagoya	21	17	2
20	January 19	Kawasaki Aircraft Factory	Akashi	19	13	0
22	January 23	Mitsubishi Aircraft Engine Plant	Nagoya	17	11	1
24	January 27	Musashi — Nakajima Aircraft Factory	Tokyo	19	13	5
26	February 4	Urban Area	Kobe	12	10	0
29	February 10	Nakajima AircraftAssy Plant	Ota	11	9	0
34	February 15	Mitsubishi Aircraft Engine Factory	Nagoya	20	6	0
34	February 15	Target Last Resort	Nagoya City	in above	11	0
37	February 19	Nakajima Aircraft Engine Factory	Tokyo	23	20	1
38	February 25	Urban Area	Tokyo	24	19	2
39	March 4	Secondary Target	Tokyo Docks	34	32	1
40	March 9/10	Urban Area	Tokyo	37	34	0
41	March 11/12	Urban Area	Nagoya	34	34	0
42	March 13/14	Urban Area	Osaka	37	35	0
43	March 16/17	Urban Area	Kobe	38	38	0
44	March 18/19	Urban Area	Nagoya	39	37	0
45	March 24/25	Mitsubishi Aircraft Engine Plant	Nagoya	30	27	1
46	March 27	Tachiari Airport	Kyushu	30	29	0
50	March 31	Tachiari Machine Works	Kyushu	31	29	0
51	April 1/2	Nakajima Aircraft Engine Plant	Tokyo	26	25	1
57	April 3/4	Aircraft Frame Plant	Tachikawa	26	17	1
57	April 3/4	Secondary — Urban Area	Tachikawa	in above	7	0
58	April 7	Nakajima Aircraft Factory	Tokyo	35	29	0
63	April 12	Nakajima Aircraft Factory	Tokyo	30	24	0
67	April 13/14	Tokyo Arsenal	Tokyo	30	29	1
69	April 15/16	Urban Area	Tokyo	28	23	0
71	April 17	Tachiari Airfield	Kyushu	11	11	0
76	April 18	Tachiari Airfield	Kyushu	11	10	2
85	April 21	Usa Airfield	Kyushu	22	20	0
94	April 22	Tomitaka Airfield	Kyushu	11	8	0
96	April 24	Hitachi Aircraft Factory	Tachikawa	11	11	0
97	April 26	Usa Airfield	Kyushu	22	18	0
97	April 26	Target of Opportunity	Tomitaki A/F	in above	1	0
97	April 26	Target of Opportunity	Hi-Saki A/F	in above	1	0
109	April 27	Miyazaki Airfield	Kyushu	11	10	0
120	April 29	Miyazaki Airfield	Kyushu	22	19	2
126	April 30	Tachikawa Air Arsenal	Hamamatsu	11	9	0
146	May 5	Hiro Naval Aircraft Factory	Kure	33	25	1
163	May 10	Naval Fueling Station	Tokuyama	33	28	0
172	May 11	Kawanishi Aircraft Factory	Kobe	11	7	0
174	May 14	Urban Area	Nagoya	44	36	0
176	May 16/17	South Nagoya Urban	Nagoya	38	35	0
178	May 19	Tachikawa Arsenal	Hamamatsu	22	22	0
181	May 23/24	South Tokyo Urban	Tokyo	41	35	2
183	May 25/26	Tokyo Palace Area	Tokyo	35	30	1

VII. The XXI Bomber Command

186	May 29	Urban Area	Yokohama	39	34	0
187	June 1	Urban Area	Osaka	34	32	2
188	June 5	Urban Area	Kobe	36	31	1
189	June 7	Urban Area	Osaka	30	24	0
197	June 10	Hitachi Engine Works (2ndary)	Kaigan	30	29	0
203	June 15	Osaka/Amagasaki Urban Area	Osaka/Amagasaki	40	36	0
208	June 17/18	Urban Area	Hamamatsu	36	35	0
211	June 19/20	Urban Area	Fukuoka	37	34	0
215	June 22	Kure Arsenal	Kure	30	24	0
224	June 26	Osaka Arsenal	Oaska	30	24	0
235	June 28/29	Sasebo Urban Area	Sasebo	42	41	0
241	July 1/2	Urban Area	Kumamoto	40	38	0
248	July 3/4	Urban Area	Kochi	33	30	1
252	July 6/7	Urban Area	Akashi	33	30	0
258	July 9/10	Urban Area	Sakai	33	29	0
264	July 12/13	Urban Area	Ichinomiya	33	33	0
272	July 16/17	Urban Area	Oita	33	30	0
278	July 19/20	Hitachi Urban Area	Hitachi	32	30	0
284	July 23/24	Urban Area	Kuwana	38	35	0
286	July 24	Armory & Kuwana — Factory	Osaka	38	35	0
293	July 25/26	Urban Area	Matsuyama	32	32	0
299	July 28/29	Urban Area	Ichinomiya	29	26	0
307	August 1/2	Urban & Industrial Areas	Toyame	47	44	0
314	August 5/6	Urban & Industrial Areas	Nishinomiya	33	33	0
319	August 7/8	Ubrban & Industrial Area	Yawata	33	30	0
326	August 14	Army Arsenal	Osaka	40	34	0
330	August 14	Urban Area	Isesake	9	9	0

498thBG/73rdBW

MSN	Date	Target	Location	Number Up	Bombed	Lost
1	Oct 28, 1944	Submarine Base	Truk	9	2	0
2	October 30	Submarine Base	Truk	9	8	0
3	November 2	Submarine Base	Truk	9	8	0
4	November 5	Airfields	Iwo Jima	18	9	0
5	November 8	Airfields	Iwo Jima	10	0	1
7	November 24	Docks & Urban Area — Secondary	Tokyo	29	24	0
8	November 27	Docks Area — Secondary	Tokyo	27	20	0
9	Nov 29/30	Light Industrial & Dock Area	Tokyo	10	8	0
10	December 3	Nakajima AC Engine Plant	Tokyo	24	20	3
11	December 8	Airfields	Iwo Jima	23	18	0
12	December 13	Mitsubishi Aircraft Engine Plant	Nagoya	25	22	2
13	December 18	Mitsubishi Aircraft Factory	Nagoya	29	24	1
14	December 22	Mitsubishi Aircraft Engine Plant	Nagoya	25	13	0
14	December 22	Secondary Target	Nagoya	in above	4	0
15	December 24	Airfields	Iwo Jima	2	2	0
16	December 27	Nakajima AC Engine Plant	Tokyo	23	15	3
16	December 27	Secondary Target	Tokyo	in above	1	0
17	Jan 3, 1945	Urban Area	Nagoya	24	13	1
17	Jan 3, 1945	Secondary Target	Unreported	in above	1	0
18	January 9	Musashino — Nakajima AC EngineTokyo	Tokyo	18	2	0
18	January 9	Secondary Targets	Tokyo	in above	6	0
19	January 14	Mitsubishi Aircraft Factory	Nagoya	22	9	1
20	January 19	Kawasaki Aircraft Factory	Akashi	19	16	0
22	January 23	Secondary Target	Nagoya City	22	16	0
24	January 27	Docks, urban area — secondary	Tokyo	21	18	2
26	February 4	Urban Area	Kobe	23	19	1
26	February 4	Secondary Target	Kobe	in above	1	0
29	February 10	Nakajima AircraftAssy Plant	Ota	24	16	2
29	February 10	Secondary target	Ota	in above	6	0
34	February 15	Mitsubishi Aircraft Engine Factory	Nagoya	24	10	0
37	February 19	Secondary Target	Tokyo Docks	33	27	0
38	February 25	Urban area & Docks	Tokyo	37	27	0
39	March 4	Secondary Target	Tokyo Docks	37	30	0
40	March 9/10	Urban Area	Tokyo	42	34	1

40	March 9/10	Secondary Target	Tokyo	in above	3	0	
41	March 11/12	Urban Area	Nagoya	39	38	0	
42	March 13/14	Urban Area	Osaka	30	28	1	
43	March 16/17	Urban Area	Kobe	41	39	0	
44	March 18/19	Urban Area	Nagoya	38	34	0	
45	March 24/25	Mitsubishi Aircraft Engine Plant	Nagoya	30	26	1	
46	March 27	Tachiari Airport	Kyushu	30	29	0	
50	March 31	Tachiari Machine Works	Kyushu	30	27	1	
51	April 1/2	Nakajima Aircraft Engine Plant	Tokyo	31	29	3	
57	April 3/4	Aircraft Frame Plant	Tachikawa	29	13	0	
57	April 3/4	Secondary Target	Tachikawa	in above	14	0	
58	April 7	Nakajima Aircraft Factory	Tokyo	32	31	1	
63	April 12	Nakajima Aircraft Factory	Tokyo	28	17	0	
67	April 13/14	Tokyo Arsenal	Tokyo	32	30	0	
69	April 15/16	Urban Area	Tokyo	33	30	0	
71	April 17	Tachiari Airfield	Kyushu	10	10	0	
76	April 18	Tachiari Airfield	Kyushu	11	10	0	
88	April 21	Tachiari Airfield	Kyushu	21	17	0	
88	April 21	Target of Opportunity	Oyodo A/F	in above	1	0	
94	April 22	Tomitaka Airfield	Kyushu	10	10	0	
96	April 24	Hitachi AircraftEngine Plant	Tachikawa	11	9	0	
100	April 26	Tomitaka Airfield	Kyushu	22	21	0	
101	April 26	Target of Opportunity	Usa A/F	22	21	0	
109	April 27	Miyazaki Airfield	Kyushu	11	11	0	
121	April 29	Miyakonojo Airfield	Kyushu	23	22	0	
126	April 30	Tachikawa Air Arsenal	Hamamatsu	11	11	0	
146	May 5	Hiro Naval Aircraft Factory	Kure	36	33	1	
163	May 10	Naval Fueling Station	Tokuyama	34	29	0	
174	May 14	Urban Area	Nagoya	31	29	0	
176	May 16/17	Urban Area	Nagoya	40	38	0	
178	May 19	Tachikawa Arsenal	Hamamatsu	23	20	0	
181	May 23/24	South Tokyo Urban	Tokyo	44	44	4	
183	May 25/26	Urban Area	Tokyo	33	30	1	
186	May 29	Urban Area	Yokohama	41	34	0	
187	June 1	Urban Area	Osaka	39	37	1	
188	June 5	Urban Area	Kobe	37	32	1	
189	June 7	Urban Area	Osaka	33	30	0	
197	June 10	Hitachi Engine Works (2ndary)	Kaigan	33	32	0	
203	June 15	Osaka/Amagasaki Urban Area	Osaka/Amagasaki	36	34	0	
208	June 17/18	Urban Area	Hamamatsu	31	31	0	
211	June 19/20	Urban Area	Fukuoka	34	29	0	
215	June 22	Kure Arsenal	Kure	30	27	1	
215	June 22	Target of Opportunity	Kure	in above	1	0	
224	June 26	Osaka Arsenal	Oaska	31	28	0	
235	June 28/29	Sasebo Urban Area	Sasebo	31	28	0	
241	July 1/2	Urban Area	Kumamoto	41	39	0	
248	July 3/4	Urban Area	Kochi	33	31	0	
252	July 6/7	Urban Area	Akashi	33	30	0	
252	July 6/7	Target of Opportunity	Akashi	in above	1	0	
258	July 9/10	Urban Area	Sakai	30	30	0	
264	July 12/13	Urban Area	Ichinomiya	33	31	0	
272	July 16/17	Urban Area	Oita	35	34	0	
278	July 19/20	Hitachi Urban Area	Hitachi	33	32	0	
284	July 23/24	Urban Area	Kuwana	43	36	0	
284	July 23/24	Target of Opportunity	Kuwana	in above	6	0	
286	July 24	Armory & Kuwana — Factory	Osaka	43	36	0	
286	July 24	Armory & Kuwana — Factory	Osaka	in above	6	0	
293	July 25/26	Urban Area	Matsuyama	33	33	0	
299	July 28/29	Urban Area	Ichinomiya	34	33	0	
307	August 1/2	Urban & Industrial Areas	Toyame	46	44	0	
314	August 5/6	Urban & Industrial Areas	Nishinomiya	33	32	1	
317	August 7	Toyokawa Naval Arsenal	Toyokawa	11	11	0	
319	August 7/8	Ubrban & Industrial Area	Yawata	31	26	0	
319	August 7/8	Ubrban & Industrial Area	Yawata	in above	2	0	
326	August 14	Army Arsenal	Osaka	40	38	0	
326	August 14	Secondary Target	Osaka	in above	1	0	

499thBG/73rdBW

MSN	Date	Target	Location	Number Up	Bombed	Lost
7	November 24	Docks & Urban Area — Secondary	Tokyo	not avail	28	1
8	November 27	Docks Area — Secondary	Tokyo	in above	13	0
9	Nov 29/30	Light Industrial & Dock Area	Tokyo	not avail	5	0
10	December 3	Nakajima AC Engine Plant	Tokyo	not avail	not avail	0
11	December 8	Airfields	Iwo Jima	not avail	17	0
12	December 13	Mitsubishi Aircraft Engine Plant	Nagoya	not avail	21	1
13	December 18	Mitsubishi Aircraft Factory	Nagoya	not avail	21	1
14	December 22	Mitsubishi Aircraft Engine Plant	Nagoya	not avail	32	1
15	December 24	Airfields	Iwo Jima	not avail	12	1
16	December 27	Nakajima AC Engine Plant	Tokyo	not avail	17	0
17	Jan 3, 1945	Urban Area	Nagoya	not avail	28	0
18	January 9	Musashino — Nakajima AC Engine	Tokyo	not avail	6	2
19	January 14	Mitsubishi Aircraft Factory	Nagoya	not avail	28	1
20	January 19	Kawasaki Aircraft Factory	Akashi	not avail	24	0
22	January 23	Mitsubishi aircraft engine plant	Nagoya	not avail	20	1
24	January 27	Musashi — Nakajima Aircraft Factory	Tokyo	not avail	21	1
26	February 4	Urban Area	Kobe	not avail	22	0
29	February 10	Nakajima Aircraft Assy Plant	Ota	not avail	19	1
34	February 15	Mitsubishi Aircraft Engine Factory	Nagoya	not avail	not avail	1
37	February 19	Secondary Target	Tokyo Docks	not avail	22	1
38	February 25	Urban area & Docks	Tokyo	not avail	29	
39	March 4	Secondary Target	Tokyo Urban	not avail	15	0
40	March 9/10	Urban Area	Tokyo	not avail	44	0
41	March 11/12	Urban Area	Nagoya	19	40	1
42	March 13/14	Urban Area	Osaka	not avail	not avail	1
43	March 16/17	Urban Area	Kobe	32	28	0
44	March 18/19	Urban Area	Nagoya	31	28	0
45	March 24/25	Mitsubishi Aircraft Engine Plant	Nagoya	not avail	28	1
46	March 27	Oita Airfield	Kyushu	not avail	not avail	0
50	March 31	Tachiari Machine Works	Kyushu	not avail	not avail	0
51	April 1/2	Nakajima Aircraft Engine Plant	Tokyo	not avail	27	1
57	April 3/4	Kawasaki Aircraft Factory	Tachikawa	not avail	14	0
58	April 7	Nakajima Aircraft Factory	Tokyo	not avail	not avail	2
63	April 12	Nakajima Aircraft Factory	Tokyo	not avail	not avail	0
67	April 13/14	Tokyo Arsenal	Tokyo	not avail	28	2
69	April 15/16	Urban Area	Tokyo	not avail	26	1
70	April 17	Izumi Airfield	Kyushu	not avail	9	0
79	April 18	Izumi Airfield	Kyushu	not avail	10	0
82	April 21	Oita Airfield	Kyushu	not avail	not avail	0
91	April 22	Izumi Airfield	Kyushu	not avail	9	0
96	April 24	Hitachi Aircraft Factory	Tachikawa	not avail	no avail	1
98	April 26	Oita Airfield	Kyushu	22	19	0
98	April 26	Target of Opportunity	Tomitaki A/f	in above	2	0
108	April 27	Izumi Airfield	Kyushu	11	10	0
115	April 28	Miyazaki Airfield	Kyushu	21	20	1
126	April 30	Tachikawa Air Arsenal	Hamamatsu	not avail	15	0
146	May 5	Hiro Naval Aircraft Factory	Kure	not avail	24	0
164	May 10	Naval Coal Yard/Briquette Plant	Tokuyama	not avail	28	0
174	May 14	Urban Area	Nagoya	Not avail	40	1
176	May 16/17	Urban Area	Nagoya	not avail	25	
178	May 19	Tachikawa Arsenal	Hamamatsu	23	20	0
181	May 23/24	Urban Area	Tokyo	Not avail	37	0
183	May 25/26	Urban Area	Tokyo	not avail	35	1
186	May 29	Urban Area	Yokohama	not avail	26	
187	June 1	Urban Area	Osaka	not avail	36	1
188	June 5	Urban Area	Kobe	not avail	35	0
189	June 7	Urban Area	Osaka	not avail	31	0
197	June 10	Musashi — Nakajima Aircraft Factory	Tokyo	24	0	
203	June 15	Osaka/Amagasaki Urban Area	Osaka/Amagasaki	35	30	0
208	June 17/18	Urban Area	Hamamatsu	34	31	0
211	June 19/20	Urban Area	Fukuoka	not avail	38	0
215	June 22	Kure Arsenal	Kure	not avail	29	0
224	June 26	Osaka Arsenal	Oaska	28	1	

MSN	Date	Target	Location	Number Up	Bombed	Lost
235	June 28/29	Sasebo Urban Area	Sasebo	not avail	38	0
241	July 1/2	Urban Area	Kumamoto	not avail	36	0
248	July 3/4	Urban Area	Kochi	32	0	
252	July 6/7	Urban Area	Akashi	not avail	not avail	
258	July 9/10	Urban Area	Sakai	not avail	26	0
264	July 12/13	Urban Area	Ichinomiya	not avail	30	0
272	July 16/17	Urban Area	Oita	29	31	0
278	July 19/20	Hitachi Urban Area	Hitachi	not avail	33	1
286	July 24	Armory & Kuwana — Factory	Osaka	not avail	35	0
293	July 25/26	Urban Area	Matsuyama	not avail	31	0
299	July 28/29	Urban Area	Ichinomiya	30	41	0
307	August 1/2	Urban & Industrial Areas	Toyame	not avail	41	0
314	August 5/6	Urban & Industrial Areas	Nishinomiya	not avail	30	0
317	August 7	Toyokawa Naval Arsenal	Toyokawa	15	10	0
319	August 7/8	Ubrban & Industrial Area	Yawata	not avail	26	0
326	August 14	Army Arsenal	Osaka	not avail	38	0

500thBG/73rdBW

MSN	Date	Target	Location	Number Up	Bombed	Lost
6	November 11	Submarine Base	Truk	9	8	0
7	November 24	Docks & Urban Area — Secondary	Tokyo	not avail	16	0
8	November 27	Docks Area — Secondary	Tokyo	in above	15	1
9	Nov 29/30	Light Industrial & Dock Area	Tokyo	not avail	6	1
10	December 3	Musashino — Nakajima AC Eng	Tokyo	not avail	19	1
11	December 8	Airfields	Iwo Jima	not avail	18	0
12	December 13	Mitsubishi Aircraft Engine Plant	Nagoya	not avail	21	1
13	December 18	Mitsubishi Aircraft Factory	Nagoya	not avail	14	1
14	December 22	Mitsubishi Aircraft Engine Plant	Nagoya	not avail	6	0
15	December 24	Airfields	Iwo Jima	not avail	9	0
16	December 27	Nakajima AC Engine Plant	Tokyo	not avail	5	0
17	Jan 3, 1945	Urban Area	Nagoya	not avail	19	2
18	January 9	Target of Opportunity	Tokyo Area	not avail	10	1
19	January 14	Mitsubishi Aircraft Factory	Nagoya	not avail	8	0
20	January 19	Kawasaki Aircraft Factory	Akashi	not avail	16	0
22	January 23	Mitsubishi Aircraft Engine Plant	Nagoya	not avail	13	1
24	January 27	Docks, urban area — secondary	Tokyo	not avail	10	0
26	February 4	Matsuzaka	Kobe	not avail	15	0
29	February 10	Nakajima AircraftAssy Plant	Ota	not avail	18	1
34	February 15	Mitsubishi Aircraft Engine Factory	Nagoya	not avail	7	0
37	February 19	Industrial Area	Tokyo	not avail	17	2
38	February 25	Urban area & Docks	Tokyo	not avail	27	0
39	March 4	Secondary Target	Tokyo Urban	not avail	19	0
40	March 9/10	Urban Area	Tokyo	42	31	1
41	March 11/12	Urban Area	Nagoya	39	36	
42	March 13/14	Urban Area	Osaka	32	32	0
43	March 16/17	Urban Area	Kobe	39	37	1
44	March 18/19	Urban Area	Nagoya	32	29	0
45	March 24/25	Mitsubishi Aircraft Engine Plant	Nagoya	not avail	25	0
46	March 27	Oita Airfield	Kyushu	39	30	0
50	March 31	Tachiari Machine Works	Kyushu	not avail	24	0
51	April 1/2	Nakajima Aircraft Engine Plant	Tokyo	not avail	34	1
57	April 3/4	Kawasaki Aircraft Factory	Tachikawa	not avail	30	1
58	April 7	Nakajima Aircraft Factory	Tokyo	not avail	9	1
60	April 8	Kanoya Airfield	Kyushu	in above	3	1
61	April 8	Kagoshima	Kyushu	not avail	28	
63	April 12	Nakajima Aircraft Factory	Tokyo	not avail	29	0
67	April 13/14	Tokyo Arsenal	Tokyo	not avail	28	0
69	April 15/16	Urban Area	Tokyo	34	30	1
70	April 17	Izumi Airfield	Kyushu	not avail	11	0
79	April 18	Izumi Airfield	Kyushu	not avail	11	0
82	April 21	Oita Airfield	Kyushu	not avail	9	0
85	April 21	Usa Airfield	Kyushu	9	9	0
91	April22	Izumi Airfield	Kyushu	not avail	10	0
96	April 24	Hitachi Aircraft Factory	Tachikawa	11	9	0

VII. The XXI Bomber Command

99	April 26	Saeki Airfield	Kyushu	23	19	0
108	April 27	Izumi Airfield	Kyushu	11	11	
114	April 28	Izumi Airfield	Kyushu	25	23	0
126	April 30	Tachikawa Air Arsenal	Hamamatsu	not avail	20	0
146	May 5	Hiro Naval Aircraft Factory	Kure	not avail	32	0
164	May 10	Naval Coal Yard/Briquette Plant	Tokuyama	not avail	28	0
172	May 11	Kawanishi Aircraft Factory	Kobe	not avail	10	0
174	May 14	Urban Area	Nagoya	not avail	41	
176	May 16/17	Urban Area	Nagoya	not avail	40	
178	May 19	Tachikawa Aircraft Compamnu	Hamamatsu	23	23	
181	May 23/24	Urban Area	South Tokyo	not avail	42	2
183	May 25/26	Urban Area	Tokyo	not avail	40	1
186	May 29	Urban Area	Yokohama	not avail	42	
187	June 1	Urban Area	Osaka	not avail	35	
188	June 5	Urban Area	Kobe	not avail	41	0
189	June 7	Urban Area	Osaka	not avail	35	0
197	June 10	Musashi — Nakajima Aircraft Factory	Tokyo	33	0	
203	June 15	Osaka/Amagasaki Urban Area	Osaka/Amagasaki	33	32	0
208	June 17/18	Urban Area	Hamamatsu	37	33	0
211	June 19/20	Urban Area	Fukuoka	not avail	32	0
215	June 22	Kure Arsenal	Kure	not avail	26	0
224	June 26	Osaka Arsenal	Oaska	29	0	
235	June 28/29	Sasebo Urban Area	Sasebo	not avail	34	0
241	July 1/2	Urban Area	Kumamoto	not avail	41	0
248	July 3/4	Urban Area	Kochi	32	0	
252	July 6/7	Urban Area	Shimizu	not avail	3	
258	July 9/10	Urban Area	Sakai	no avail	30	
264	July 12/13	Urban Area	Ichinomiya	not avail	31	0
272	July 16/17	Urban Area	Oita	32	32	0
278	July 19/20	Hitachi Urban Area	Hitachi	not avail	32	1
286	July 24	Armory & Kuwana — Factory	Osaka	not avail	42	0
286	July 24	Target of Opportunity	Osaka	not avail	1	0
293	July 25/26	Urban Area	Matsuyama	not avail	32	0
299	July 28/29	Urban Area	Ichinomiya	34	33	0
307	August 1/2	Urban & Industrial Areas	Toyame	not avail	41	0
314	August 5/6	Urban & Industrial Areas	Nishinomiya	not avail	35	0
317	August 7	Toyokawa Naval Arsenal	Toyokawa	8	8	0
319	August 7/8	Ubrban & Industrial Area	Yawata	not avail	26	0
326	August 14	Army Arsenal	Osaka	not avail	35	0

6thBG/313thBW

MSN	Date	Target	Location	Number Up	Bombed	Lost
28	February 9	Moen Island Airfield	Truk	30	30	0
30	February 11	Sea Search	9	8	0	
32	February 12	Sea Search	10	8	0	
33	February 14	Sea Search	6	5	1	
36	February 18	Moen Airfield #2	Truk	19	19	0
37	February 19	Urban Area	Tokyo	3		
38	February 25	Urban Area	Tokyo	not avail	21	0
39	March 4	Secondary Target	Tokyo Urban	20	20	0
40	March 9/10	Urban Area	Tokyo	not avail	32	0
41	March 11/12	Urban Area	Nagoya	not avail	32	
42	March 13/14	Urban Area	Osaka	not avail	30	0
43	March 16/17	Urban Area	Kobe	34	33	0
44	March 18/19	Urban Area	Nagoya	34	32	0
45	March 24/25	Mitsubishi Aircraft Engine Plant	Nagoya	not avail	18	0
47	March 27/28	Mine Field Mike	Shimonoseki	not avail	30	0
49	March 30/31	Mine Field Item	Kure	23	23	0
49	March 30/31	Mine Field Roger	Sasebo Approach	2	2	0
56	April 3/4	Nakajima Aircraft Plant	Koizama	not avail	19	0
53	April 7	Mitsubishi Aircraft Engine Plant	Nagoya	30	30	1
59	April 7	Mitsubishi Aircraft Engine Plant	Nagoya	not avail	30	0
61	April 8	Kanoya East Airfield	Kyushu	10	10	0
62	April 9/10	Shimonoseki Mining	Shimonoseki	10	10	0

64	April 12	Hodagaya Chemicak Works	Koriyama	not avail	20	0
67	April 13/14	Tokyo Arsenal	Tokyo	29	29	0
68	April 15/16	Urban Area	Kawasaki	24	24	0
73	April 17	Kanoya East Airfield	Kyushu	21	10	0
77	April 18	Kushira Airfield	Kyushu	not avail	10	0
83	April 21	Kanoya East Airfield	Kyushu	23	22	0
95	April 22	Kanoya Airfield	Kyushu	21	16	0
96	April 24	Hitachi Aircraft Factory	Tachikawa	12	9	0
101	April 26	Matsuyama Airfield	Kyushu	not avail	18	0
111	April 27	Miyakonojo Airfield	Kyushu	not avail	6	0
117	April 28	Miyakonojo Airfield	Kyushu	19	18	0
117	April 28	Target of Opportunity	Tanega Shima AF	in above	1	0
126	April 30	Tachikawa Air Arsenal	Hamamatsu	not avail	7	0
139	May 3/4	Mine Field Love	Moji Area	14	14	0
139	May 3/4	Mine Field Love	Suo Nada	20	19	0
150	May 5/6	Mine Field Baker	Aki/Nada	5	in above	0
150	May 5/6	Mine Field Fox	Bingo Nada	20	in above	0
151	May 5/6	Kanoya Airfield	Kyushu	10	10	0
152	May 7	Ibusuki Airfield	Kyushu	11	10	0
160	May 10	Usa Airfield	Kyushu	22	15	0
160	May 10	Target of Opportunity	Matsuyama A/F	in above	2	0
160	May 10	Target of Opportunity	Nakamura	in above	1	0
160	May 10	Target of Opportunity	Nobeoka	in above	1	0
160	May 10	Target of Opportunity	Susaki Docks	in above	1	0
169	May 11	Nittagahara Airfield	Kyushu	11	5	0
169	May 11	Targets of Opportunity	Hoso Shima NB	in above	6	0
169	May 11	Targets of Opportunity	Kochi A/F	in above	1	0
169	May 11	Targets of Opportunity	Kubokawa	in above	1	0
169	May 11	Targets of Opportunity	Suma	in above	1	0
169	May 11	Targets of Opportunity	Shibushi	in above	1	0
169	May 11	Targets of Opportunity	Tanega Shima	in above	1	0
174	May 14	Urban Area	Nagoya	32	25	0
176	May 16/17	Urban Area	Nagoya	33	33	0
178	May 19	Tachikawa Arsenal	Hamamatsu	30	25	
181	May 23/24	Urban Area	Tokyo	35	31	3
183	May 25/26	Urban Area	Tokyo	not avail	24	1
186	May 29	Urban Area	Yokohama	27	25	
187	June 1	Urban Area	Osaka	27	27	1
188	June 5	Urban Area	Kobe	29	28	
189	June 7	Urban Area	Osaka	27	26	
192	June 9	Kawasaki Aircraft Factory	Akashi	26	24	
203	June 15	Urban Area	Osaka/Amagasaki	35	35	0
209	June 17/18	Urban Area	Yokkaichi	30	28	0
211	June 19/20	Urban Area	Fukuoka	29	29	0
216	June 22	Akashi — Kawasaki Aircraft Factory	Akashi	not avail	29	
219	June 22	Kagoshima Aircraft Factory	Akashi	not avail	29	0
219	June 22	Target of Opportunity	Kagamigahara	in above	1	0
219	June 22	Target of Opportunity	Hamamatsu	in above	1	0
219	June 22	Target of Opportunity	Kawasaki AC	in above	1	0
219	June 22	Target of Opportunity	Hachionan	in above	1	0
236	June 28/29	Urban Area	Moji	not avail	30	0
242	July 1/2	Secondary Target	Ube	39	35	0
249	July 3/4	Urban Area	Himeji	35		
	July 5	Marcus Island	Tokushima	3	3	0
253	July 6/7	Urban Area	Shimizu	not avail	36	
256	July 9/10	Mine Field Mike	Shimonoseki	14	not avail	not avail
256	July 9/10	Mine Field Uncle	Niigata/Nanao	9	not avail	not avail
256	July 9/10	Mine Field Nan	Fushiki	7	not avail	not avail
262	July 11	Mine Field Mike	Shimonoseki	7	not avail	0
262	July 11	Mine Field Rashin	Korea	6	not avail	0
262	July 11	Mine Field Fusan	Korea	9	not avail	0
262	July 11	Mine Field Zebra	Maizuru	8	not avail	0
268	July 13	Mine Field Mike	Shimonoseki	9	not avail	0
268	July 13	Mine Field Seishin	Korea	6	not avail	0
268	July 13	Mine Field Masan-Reisu	Korea	13	not avail	0
268	July 15/16	Mine Field Charlie	Fukuoka	3	not avail	0
269	July 15/16	Mining Mission 35	See Below	28	26	0

VII. The XXI Bomber Command

MSN	Date	Target	Location	Number Up	Bombed	Lost
269	July 15/16	Mine Field Uncle/Nan	Niigata/Naoetsu	9	not avail	0
269	July 15/16	Mine Field Rashin	Rashin, Korea	8	not avail	0
269	July 15/16	Mine Field Genzen-Konan	Genzen, Korea	4	not avail	0
269	July 15/16	Mine Field Fusan	Fusan, E1079Korea	7	not avail	0
275	July 17/18	Mine Field Mike	Moji	10	not avail	0
275	July 17/18	Mine Field Seishin	Seishin, Korea	6	not avail	0
275	July 19	Mine Field Nan	Nanao Bay	14	not avail	0
276	July 19	Mining Mission 37	See Below	31	29	1
276	July 19	Mine Field Able	Shimonoseki	10	not avail	
276	July 19	Mine Field Uncle/Nan	Niigata	7	not avail	
276	July 19	Mine Field Zebra	Maizura/Miyama	9	not avail	
276	July 19	Mine Field Genzan — Konan	Korea	5	not avail	
282	July 22	Mining Mission 38	See Below	30	26	0
282	July 22	Mine Field Mike/Love	Shimonoseki	15	not avail	0
282	July 22	Mine Field Rashin	Korea	8	not avail	0
282	July 23/24	Mine Field Fusan/Musan	Korea	3/4	not avail	0
294	July 26/27	Urban Area	Tokuyama	not avail	36	
300	July 28/29	Urban Area	Ujiyamada	30	29	
308	August 1/2	Urban & Industrial Areas	Nagaoka	not avail	45	
313	August 5/6	Urban & Industrial Areas	Maebashi	37	31*	0
317	August 7	Toyokawa Naval Arsenal	Toyokawa	12	12	0
319	August 7/8	Ubrban & Industrial Area	Yawata	29	27	0
327	August 14	Marshalling Yards	Marifu	43	39	0

9thBG/313rdBW

MSN	Date	Target	Location	Number Up	Bombed	Lost
28	February 9	Moen Island Airfield	Truk	30	29	0
31	February 12	Antiaircraft	Iwo Jima	21	21	0
36	February 18	Moen Airfield #2	Truk	18	18	0
38	February 25	Urban Area	Tokyo	32	23	0
38	February 25	Targets of Opportunity	Tokyo Area	in above	5	0
39	March 4	Secondary Target	Tokyo Urban	23	20	0
39	March 4	Target of Opportunity	Shizuoka	in above	1	0
39	March 4	Target of Opportunity	Maug	in above	1	0
40	March 9/10	Urban Area	Tokyo	32	29	2
40	March 9/10	Target of Opportunity	Unknown	in above	1	0
41	March 11/12	Urban Area	Nagoya	31	27	2
42	March 13/14	Urban Area	Osaka	33	33	0
43	March 16/17	Urban Area	Kobe	34	30	1
44	March 18/19	Urban Area	Nagoya	31	30	0
45	March 24/25	Mitsubishi Aircraft Engine Plant	Nagoya	22	19	1
47	March 27/28	Mine Field Love	Shimonoseki	31	31	0
49	March 30/31	Mine Field Item	Kure	8	in above	0
49	March 30/31	Mine Field Jig	Hiroshima Entry	11	10	1
49	March 30/31	Mine Field Charlie	Hiroshima	1	1	0
49	March 30/31	Mine Field Roger	Sasebo Approach	4	4	0
52	April 1/2	Mining Mission 3	Kure	6	6	0
56	April 3/4	Nakajima Aircraft Plant	Koizama	28	22	0
56	April 3/4	Secondary Targets	Koizama	in above	3	0
53	April 7	Mitsubishi Aircraft Engine Plant	Nagoya	39	31	0
59	April 7	Mitsubishi Aircraft Engine Plant	Nagoya	39	31	0
64	April 12	Hodagaya Chemicak Works	Koriyama	23	20	0
66	April 12/13	Mining Mission 7	Shimonoseki	6	5	0
67	April 13/14	Tokyo Arsenal	Tokyo	28	28	0
68	April 15/16	Urban Area	Kawasaki	33	26	4
68	April 15/16	Targets of Opportunity	Kawasaki	in above	4	0
80	April 18	Kokubu Airfield	Kyushu	22	18	0
80	April 18	Target of Opportunity	Kanoya East A/F	in above	1	0
80	April 18	Target of Opportunity	Izumi A/F	in above	1	0
83	April 21	Kanoya Airfield	Kyushu	10	7	0
83	April 21	Secondary Target	Kyushu	in above	1	0
83	April 21	Last Resort Target	Kyushu	in above	1	0
86	April 21	Kokubu Airfield	Kyushu	12	12	0
92	April 22	Kushira Airfield	Kyushu	9	7	0

95	April 22	Kanoya Airfield	Kyushu	4	3	0
96	April 24	Hitachi Aircraft Factory	Tachikawa	12	1	0
96	April 24	Last Resort — Urban Area	Tachikawa	in above	9	0
103	April 26	Miyazaki Airfield	Kyushu	21	19	0
103	April 26	Secondary Target	Kokubu A/F	in above	1	0
110	April 27	Kokubu Airfield	Kyushu	10	9	0
116	April 28	Kokubu Airfield	Kyushu	20	17	1
126	April 30	Tachikawa Air Arsenal	Hamamatsu	17	15	0
139	May 3/4	Mine Field Mike	Shimonoseki	24	19	0
139	May 3/4	Secondary Target	Shimonoseki	in above	1	0
150	May 5/6	Mine Field Dog	Shoda Shima	28	in above	0
157	May 8	Oita Airfield	Kyushu	12	11	0
158	May 8	Matsuyama West Airfield	Shikoku	22	16	0
159	May 8	Matsuyama West Airfield	Shikoku	23	15	0
159	May 8	Target of Opportunity	Kumamota A/F	in above	2	0
	May 10	Kyushu Airfields	Unknown	23	16	0
	May 10	Target of Opportunity	Unknown	in above	2	0
170	May 11	Miyazaki Airfield	Kyushu	12	11	0
171	May 11	Miyakonojo Airfield	Kyushu	11	10	0
173	May 13/14	Mining Mission 10	See Below	12	12	0
173	May 13/14	Mine Field Mike	Shimonoseki	3	3	0
173	May 13/14	Mine Field Love	Shimonoseki	5	5	0
173	May 13/14	Mine Field Uncle	Niigata	4	4	0
175	May 16/17	Mine Field Mike	Shimonoseki	24	20	0
175	May 16/17	Mine Field Zebra	Maizuru	6	5	0
177	May 18/19	Mining Mission 12	See Below	34	30	0
177	May 18/19	Mine Field Love	Shimonoseki	22	18	0
177	May 18/19	Mine Field Zebra	Tsuruga	12	12	0
179	May 20/21	Mining-Mission 13	Shimonoseki	32	30	3
179	May 20/21	Mine field Mike	Shimonoseki	23	22	0
179	May 20/21	Mine Field Love	Shimonoseki	5	4	0
179	May 20/21	Mine Field Zebra	Maizura Bay	4	4	0
180	May 22/23	Mining Mission 14	Shimonoseki	32	30	1
180	May 22/23	Mine Field Mike	Shimonoseki	9	8	0
180	May 22/23	Mine Field Mike	Shimonoseki	9	8	0
180	May 22/23	Mine Field Love	Shimonoseki	14	14	0
182	May 24/25	Mining Mission 15	Fushiki/Niigata	30	27	0
182	May 24/25	Secondary Targets	Fushiki/Niigata	in above	2	0
182	May 24/25	Mine Field Uncle	Niigata	11	9	0
182	May 24/25	Mine field Nan	Fushiki	10	9	0
182	May 24/25	Mine Field Mike	Shimonoseki	9	7	0
184	May 25/26	Mine Field Mike	Shimonoseki	2	2	0
184	May 25/26	Mine Field Nan	Fushiki	6	6	0
184	May 25/26	Mine Field Love	Shimonoseki	7	7	0
184	May 25/26	Mine Field Charlie	Fukuoka	15	14	0
185	May 27	Mining Mission 17	See Below	11	9	1
185	May 27	Mine Field Charlie	Fukuoka Bay	1	1	0
185	May 27	Mine Field Love	Shimonoseki	10	8	0
187	June 1	Urban Area	Osaka	36	31	0
187	June 1	Target of Opportunity	Osaka	in above	1	0
188	June 5	Urban Area	Kobe	35	32	0
189	June 7	Urban Area	Osaka	32	32	0
193	June 9	Aichi Aircraft Engine Plant	Atsuta	26	24	0
193	June 9	Target of Opportunity	Unknown	in above	1	0
203	June 15	Urban Area	Osaka/Amagasaki	35	33	0
203	June 15	Target of Opportunity	Osaka/Amagasaki	in above	2	0
209	June 17/18	Urban Area	Yokkaichi	38	36	0
211	June 19/20	Urban Area	Fukuoka	37	34	0
211	June 19/20	Target of Opportunity	Fukuoka	in above	1	0
216	June 22	Utsube Oil refinery	Unknown	28	17	0
216	June 22	Secondary target	Himeji	in above	6	0
218	June 22	Kagashima Aircraft Factory	Akashi	37	32	0
218	June 22	Secondary Target	Unknown	in above	3	0
229	June 26	Aichi Aircraft Works	Eitoku	37	32	0
236	June 28/29	Urban Area	Moji	36	32	0
236	June 28/29	Target of Opportunity	Moji	in above	1	0
242	July 1/2	Secondary Target	Ube	38	30	0

VII. The XXI Bomber Command

249	July 3/4	Urban Area	Himeji	36	36	0
253	July 6/7	Urban Area	Shimizu	34	34	0
259	July 9/10	Urban Area	Wakayama	36	36	0
265	July 12/13	Urban Area	Tsuruga	34	30	0
265	July 12/13	Target of Opportunity	Yamada	in above	1	0
265	July 12/13	Target of Opportunity	Shingu	in above	1	0
273	July 16/17	Urban Area	Kuwana	33	32	0
273	July 16/17	Target of Opportunity	Shingu	in above	1	0
279	July 19/20	Choshi Urban Area	Choshi	34	30	0
288	July 24	City of Tsu	Tsu	41	38	0
288	July 24	Target of Opportunity	Heki	in above	1	0
288	July 24	Target of Opportunity	Shingu	in above	1	0
294	July 26/27	Urban Area	Tokuyama	33	30	0
294	July 26/27	Urban Area	Tokuyama	in above	1	0
300	July 28/29	Urban Area	Ujiyamada	33	32	0
308	August 1/2	Urban & Industrial Areas	Nagaoka	47	43	0
308	August 1/2	Urban & Industrial Areas	Tanabe	in above	1	0
308	August 1/2	Urban & Industrial Areas	Fukui	in above	1	0
308	August 1/2	Urban & Industrial Areas	Toyama	in above	1	0
308	August 1/2	Urban & Industrial Areas	Hamamatsu	in above	1	0
313	August 5/6	Urban & Industrial Areas	Tateyama	35	29	0
313	August 5/6	Urban & Industrial Areas	Maebashi	in above	2	0
317	August 7	Toyokawa Naval Arsenal	Toyokawa	12	9	1
319	August 7/8	Ubrban & Industrial Area	Yawata	30	29	1
327	August 14	Marshalling Yards	Marifu	41	37	0
329	August 14	Urban Area	Kumagaya	4	3	0

504thBG/313thBW

MSN	Date	Target	Location	Number Up	Bombed	Lost
26	February 4	Urban Area	Kobe	14	12	0
29	February 10	Nakajima AircraftAssy Plant	Ota	14	11	1
34	February 15	Mitsubishi Aircraft Engine Factory	Nagoya	13	0	1
37	February 19	Secondary Target	Tokyo Urban	18	0	0
38	February 25	Urban Area	Tokyo	14	8	0
39	March 4	Secondary Target	Tokyo Urban	11	10	0
40	March 9/10	Tokyo Urban Area	Tokyo	21	18	0
41	March 11/12	Urban Area	Nagoya	21	17	
42	March 13/14	Urban Area	Osaka	21	20	0
43	March 16/17	Urban Area	Kobe	24	23	1
44	March 18/19	Urban Area	Nagoya	23	21	0
45	March 24/25	Mitsubishi Aircraft Engine Plant	Nagoya	14	13	1
47	March 27/28	Mine Field Mike	Shimonoseki	49	20	1
49	March 30/31	Mine Field Love	East Shimonoseki	14	14	0
49	March 30/31	Mine Field Roger	Sasebo Approach	3	3	0
53	April 2/3	Mine Field Jig- Mission 4	Hiroshima	10	9	0
56	April 3/4	Nakajima Aircraft Plant	Koizama	11	6	0
53	April 7	Mitsubishi Aircraft Works	Nagoya	31	28	0
59	April 7	Mitsubishi Aircraft Works	Nagoya	not avail	not avail	
64	April 12	Hodagaya Chemicak Works	Koriyama	18	18	0
67	April 13/14	Tokyo Arsenal	Tokyo	26	24	0
68	April 15/16	Urban Area	Kawasaki	22	20	1
73	April 17	Kanoya East Airfield	Kyushu	in above	in above	0
77	April 18	Kanoya East Airfield	Kyushu	5	5	0
	April 20	Izumi Airfield	Kyushu	6	5	0
89	April 21	Izumi Airfield	Kyushu	16	13	0
89	April 21	Target of Opportunity	Miyazake A/F	in above	2	0
89	April 21	Target of Opportunity	Kanoya East A/F	in above	1	0
96	April 24	Hitachi Aircraft Factory	Tachikawa	11	9	2
96	April 24	Secondary Target	Sukumo A/F	16	14	0
96	April 24	Secondary Target	Kochi A/F	in above	in above	0
107	April 26	Miyakonojo Airfield	Kyushu	9	8	0
111	April 27	Miyakonojo Airfield	Kyushu	not avail	8	0
123	April 29	Kanoya East Airfield	Kyushu	15	14	0
126	April 30	Tachikawa Air Arsenal	Hamamatsu	5	0	0

126	April 30	Secondary Target	Unknown	in above	5	0	
139	May 3/4	Mine Field Able	Kobe/Osaka	16	16	0	
150	May 5/6	Mine Field Easy	Tokyo & Ise Bays	10	in above	0	
150	May 5/6	Mine Field Jig/Item	Hiroshima/Kure	7	in above	0	
150	May 5/6	Mine Field Oboe	Tokyo	4	in above	0	
155	May 7	Kanoya Airfield	Kyushu	21	17	0	
155	May 7	Target of Opportunity	Omura City	in above	1	0	
156	May 8	Miyakonojo Airfield	Kyushu	12	12	0	
167	May 10	Oita Airfield	Kyushu	20	17	0	
167	May 10	Target of Opportunity	Miyachi	in above	1	0	
174	May 14	Urban Area	Nagoya	17	13	0	
174	May 14	Secondary Target	Nagoya	in above	2	0-	
176	May 16/17	Urban Area	Nagoya	not avail	not avail		
181	May 23/24	Urban Area	Tokyo	31	29	2	
183	May 25/26	Urban Area	Tokyo	26	26	2	
186	May 29	Urban Area	Yokohama	25	22	2	
187	June 1	Urban Area	Osaka	24	20	0	
188	June 5	Urban Area	Kobe	20	18		
188	June 5	Secondary Target	Kobe	in above	2	0	
189	June 7	Urban Area	Osaka	23	19	1	
189	June 7	Secondary Target	Osaka	in above	1	0	
192	June 9	Kawasaki Aircraft Factory	Akashi	20	19	0	
203	June 15	Urban Area	Osaka/Amagasaki	24	20	0	
203	June 15	Secondary Target	Osaka/Amagasaki	in above	1	0	
209	June 17/18	Urban Area	Yokkaichi	25	23	0	
211	June 19/20	Urban Area	Fukuoka	27	27	0	
218	June 22	Mitsubishi Aircraft Factory	Kagamigahara	22	20	1	
229	June 26	Aichi Aircraft Works	Eitoku	30	18	2	
229	June 26	Secondary Target	Eitoku	in above	11	0	
236	June 28/29	Urban Area	Moji	35	32	0	
242	July 1/2	Secondary Target	Ube	35	33	0	
249	July 3/4	Urban Area	Himeji	35	35	0	
253	July 6/7	Urban Area	Shimizu	32	30	0	
259	July 9/10	Urban Area	Wakayama	37	37	0	
265	July 12/13	Urban Area	Tsuruga	33	30	0	
273	July 16/17	Urban Area	Kuwana	34	31	0	
279	July 19/20	Choshi Urban Area	Choshi	33	30	0	
287	July 24	Aichi Aircraft Factory	Etitoku	38	33	0	
292	July 25/26	Mine Field Seishin	Seishin, Korea	6	5	0	
292	July 25/26	Mine Field Fusan	Fusan, Korea	7	7	0	
292	July 25/26	Mine Field Nan	Nanao Bay	11	11	0	
292	July 25/26	Mine Field Zebra	Tsuruga Bay	6	6	0	
296	July 26/27	Mine Field Mike	Shimonoseki	10	27	3	
296	July 26/27	Mine Field Uncle	Niigata	8	in above		
296	July 26/27	Mine Field Zebra	Maizura	10	in above		
296	July 26/27	Mine Field X-Ray	Senzaki	3	in above		
304	July 29/30	Mine Field Rashin	Raishin, Korea	14	in above	0	
304	July 29/30	Mine Field Mike	Shimonoseki	10	in above	0	
304	July 29/30	Mine Field Charlie	Fukuoka Bay	6	in above	0	
305	August 1/2	Mine Field Rashin	Raishin, Korea	9	in above	0	
305	August 1/2	Mine Field Seishin	Seishin, Korea	7	in above	0	
305	August 1/2	Mine Field Mike	Moji Area	17	in above	0	
305	August 1/2	Mine Field George	Hamada,	7	in above	0	
305	August 1/2	Mine Field X-Ray	Hagi, Oura	5	in above	0	
308	August 1/2	Urban & Industrial Areas	Nagaoka	not avail	not avail		
311	August 5/6	Mining Mission 43	See Below	30	27	0	
311	August 5/6	Mine Field X-Ray	Hagi Harbor	6	in above	0	
311	August 5/6	Mine Field Zebra	Tsuruga Harbor	10	in above	0	
311	August 5/6	Mine Field Geijitsu	Geijitsu,Korea	5	in above	0	
311	August 5/6	Mine Field Rashin	Rashin, Korea	9	in above	0	
318	August 7/8	Mining Mission 44	See Below	27	27	0	
318	August 7/8	Mine Field Rashin	Raishin, Korea	8	8	0	
318	August 7/8	Mine Field Mike/Love	Shimonoseki	5	5	0	
318	August 7/8	Mine Field Zebra	Maizuru	8	8	0	
318	August 7/8	Mine Field Yoke	Sakai	6	6	0	
324	August 9/10	Mine Field Rashin	Raishin, Korea	7	7	0	
324	August 9/10	Mine Field Seishin	Seishin, Korea	7	7	0	

VII. The XXI Bomber Command

MSN	Date	Target	Location			
324	August 9/10	Mine Field Mike	Shimonoseki	11	11	0
324	August 9/10	Mine Field Genzan	Genzan, Korea	7	7	0
331	August 14	Mine Field Nan	Nanao Bay	8	in above	0
331	August 14	Mine Field Mike	Shimonoseki	10	in above	0
331	August 14	Mine Field Zebra	Tsuruga Bay	17	in above	0
331	August 14	Mine Field George	Hamada Harbor	6	in above	0

505thBG/313thBW

MSN	Date	Target	Location	WINGS	GRPS	Number Up	Bombed	Lost
26	February 4	Urban Area	Kobe	313BW	505BG	not avail	20	1
29	February 10	Nakajima AircraftAssy Plant	Ota	313BW	505BG	21	18	8
34	February 15	Mitsubishi Aircraft Engine Factory	Nagoya	313BW	505BG	15	4	0
34	February 15	Last Resort Target	Hachijo Jima	313BW	505BG	in above	Unk	0
34	February 15	Last Resort Target	Hamamatsu	313BW	505BG	in above	Unk	0
34	February 15	Last Resort Target	Maisaka	313BW	505BG	in above	Unk	0
34	February 15	Last Resort Target	Fukude	313BW	505BG	in above	Unk	0
37	February 19	Nakajima Aircraft Engine Factory	Tokyo	313BW	505BG	0	0	0
37	February 19	Port of Tokyo — Secondary	Tokyo	313BW	505BG	20	17	0
37	February 19	Last Resort Target	Kawasaki	313BW	505BG	in above	1	0
37	February 19	Last Resort Target	Toyohashi	313BW	505BG	in above	1	0
37	February 19	Last Resort Target	Shizuoka	313BW	505BG	in above	1	0
38	February 25	Urban Area	Tokyo	313BW	505BG	15	8	0
38	February 25	Last Resort Target	Toyohashi	313BW	505BG	in above	2*	0
38	February 25	Last Resort Target	Hamamatsu	313BW	505BG	in above	2*	0
38	February 25	Last Resort Target	Numazu	313BW	505BG	in above	2*	0
38	February 25	Last Resort Target	Kega	313BW	505BG	in above	1*	0
39	March 4	Tokyo Port & Urban — (Second	Tokyo	313BW	505BG	21	19	0
39	March 4	Hamamatsu (Last Resort)	Hamamatsu	313BW	505BG	in above	2*	0
40	March 9/10	Tokyo Urban Area	Tokyo	313BW	505BG	not avail	20	1
40	March 9/10	Target of Last Resort	Kasumigoura AF	313BW	505BG	in above	1	0
40	March 9/10	Target of Last Resort	Haha Jima	313BW	505BG	in above	1	0
40	March 9/10	Target of Last Resort	Chichi Jima	313BW	505BG	in above	1	0
40	March 9/10	Target of Last Resort	Guguan	313BW	505BG	in above	1	0
41	March 11/12	Urban Area	Nagoya	313BW	505BG	25	25	1
42	March 13/14	Urban Area	Osaka	313BW	505BG	not avail	26	0
43	March 16/17	Urban Area	Kobe	313BW	505BG	36	32	0
44	March 18/19	Urban Area	Nagoya	313BW	505BG	33	31	1
45	March 24/25	Mitsubishi Aircraft Engine Plant	Nagoya	313BW	505BG	not avail	21	0
47	March 27/28	Mine Field Mike	Shimonoseki Strait	313BW	505BG	24	24	2
49	March 30/31	Mine Field Love	East Shimonoseki	313BW	505BG	29	in above	0
54	April 3/4	Mine Field Item — Mission 5	Hiroshima	313BW	505BG	9	9	0
56	April 3/4	Nakajima Aircraft Plant	Koizama	313BW	505BG	11	4	0
56	April 3/4	Secondary Target	Tokyo Urban	313BW	505BG	in above	16	0
53	April 7	Mitsubishi Aircraft Works	Nagoya	313BW	505BG	32	21	0
53	April 7	Last Resort Target	Hamamatsu	313BW	505BG	in above	1	0
53	April 7	Last Resort Target	Ujiyamada	313BW	505BG	in above	1	0
53	April 7	Last Resort Target	Akazaki	313BW	505BG	in above	1	0
53	April 7	Last Resort Target	Kushimoto	313BW	505BG	in above	1	0
53	April 7	Last Resort Target	Owase	313BW	505BG	in above	1	0
59	April 7	Mitsubishi Aircraft Works	Nagoya	313BW	505BG	32	21	0
59	April 7	Last Resort Target	Hamamatsu	313BW	505BG	in above	1	0
59	April 7	Last Resort Target	Ujiyamada	313BW	505BG	in above	1	0
59	April 7	Last Resort Target	Akazaki	313BW	505BG	in above	1	0
59	April 7	Last Resort Target	Kushimoto	313BW	505BG	in above	1	0
59	April 7	Last Resort Target	Owase	313BW	505BG	in above	1	0
62	April 9/10	Mine Field Mike	Shomonoseki	313BW	505BG	10	10	0
64	April 12	Hodagaya Chemicak Works	Koriyama	313BW	505BG	not avail	not avail	0
67	April 13/14	Tokyo Arsenal	Tokyo	313BW	505BG	not avail	35	1
68	April 15/16	Urban Area	Kawasaki	313BW	505BG	32	25	1
72	April 17	Kokubu Airfield	Kyushu	313BW	505BG	24	20	0
72	April 17	Target of Opportunity	Nittigahara A/F	313BW	505BG	in above	2	0
86	April 21	Kokubu Airfield	Kyushu	313BW	505BG	23	22	0
92	April 22	Target of Opportunity	Kanoya A/F	313BW	505BG	in above	5	0
92	April 22	Kushira Airfield	Kyushu	313BW	505BG	not avail	2	0

95	April 22	Kanoya Airfield	Kyushu	313BW	505BG	5	5	
96	April 24	Hitachi Aircraft Factory	Tachikawa	313BW	505BG	10	10	0
101	April 26	Matsuyama Airfield	Kyushu	313BW	505BG	not avail	not avail	0
102	April 26	Nittagahara Airfield	Kyushu	313BW	505BG	18	18	0
102	April 26	Targets of Opportunity	Kaimon Dake AF	313BW	505BG	in above	1	0
102	April 26	Targets of Opportunity	Miyazaki A/F	313BW	505BG	in above	1	0
110	April 27	Kokubu Airfield	Kyushu	313BW	505BG	12	8	0
122	April 29	Kokubu Airfield	Kyushu	313BW	505BG	22	22	0
126	April 30	Tachikawa Air Arsenal	Hamamatsu	313BW	505BG	not avail	7	0
139	May 3/4	Mine Field Able	Kobe/Osaka	313BW	505BG	22	19	0
139	May 3/4	Mine Field Mike	Shimonoseki	313BW	505BG	4	4	0
150	May 5	Mine Field Able	Kobe-Osaka	313BW	505BG	12	in above	0
150	May 5/6	Mine Field King	Tokuyama	313BW	505BG	8	in above	0
150	May 5/6	Mine Field Tare	Nagoya	313BW	505BG	5	in above	0
153	May 7	Oita Airfield	Kyushu	313BW	505BG	10	10	2
154	May 7	Usa Airfield	Kyushu	313BW	505BG	11	11	1
161	May 10	Miyazaki Airfield	Kyushu	313BW	505BG	12	7	0
161	May 10	Target of Opportunity	Tanoga Naval	313BW	505BG	in above	1	0
162	May 10	Kanoya Airfield	Kyushu	313BW	505BG	12	4	0
162	May 10	Target of Opportunity	Kyushu	313BW	505BG	in above	6	0
168	May 11	Saeki Airfield	Kyushu	313BW	505BG	11	7	0
168	May 11	Target of Opportunity	shimo Kawaguchi	313BW	505BG	in above	1	0
174	May 14	Urban Area	Nagoya	313BW	505BG	not avail	not avail	2
181	May 23/24	Urban Area	Tokyo	313BW	505BG	not avail	35	
183	May 25/26	Urban Area	Tokyo	313BW	505BG	not avail	not avail	4
187	June 1	Urban Area	Osaka	313BW	505BG	31	33	0
188	June 5	Urban Area	Kobe	313BW	505BG	not avail	not avail	
190	June 7/8	Mine Field Mike	Shimonoseki	313BW	505BG	10	9	0
190	June 7/8	Mine Field Love	Shimonoseki	313BW	505BG	10	7	0
190	June 7/8	Mine Field Charlie	Fukuoka Bay	313BW	505BG	5	4	0
190	June 7/8	Mine Field Mike	Shimonoseki	313BW	505BG	10	6	0
194	June 9/10	Mine Field Mike	Shimonoseki	313BW	505BG	20	19	0
194	June 9/10	Mine Field Love	Shimonoseki	313BW	505BG	8	7	0
201	June 11/12	Mine Field Mike	Shimonoseki	313BW	505BG	15	15	0
201	June 11/12	Mine Field Zebra	Tsuruga	313BW	505BG	11	11	0
202	June 13/14	Mine Field Love	Shimonoseki	313BW	505BG	17	17	0
202	June 13/14	Mine Field Uncle	Niigata	313BW	505BG	12	12	0
204	June 15/16	Mine Field Mike	Shimonoseki	313BW	505BG	13	13	0
204	June 15/16	Mine Field Nan	Fushiki	313BW	505BG	9	9	0
204	June 15/16	Mine Field Charlie	Fukuoka	313BW	505BG	8	8	0
205	June 17/18	Mine Field Mike	Shimonoseki	313BW	505BG	18	18	0
205	June 17/18	Mine Field Able	Kobe/Osaka	313BW	505BG	7	7	0
213	June 19/20	Mine Field Mike	Shimonoseki	313BW	505BG	10	10	0
213	June 19/20	Mine Field Uncle	Niigata	313BW	505BG	8	8	0
213	June 19/20	Mine Field Zebra	Miazuru/Miyazu	313BW	505BG	10	10	0
214	June 21/22	Mine Field X-Ray	Senzaki	313BW	505BG	9	9	0
214	June 21/22	Mine Field Nan	Fushiki	313BW	505BG	10	10	0
214	June 21/22	Mime Field Able	Kobe/Osaka	313BW	505BG	6	6	0
221	June 23/24	Mine Field Charlie	Fukuoka	313BW	505BG	9	9	
221	June 23/24	Mine Field Yoke	Sakai	313BW	505BG	8	8	
221	June 23/24	Mine Field Uncle	Niigata	313BW	505BG	9	9	
222	June 25/26	Mine Field Mike	Shimonoseki	313BW	505BG	11	11	0
222	June 25/26	Mine Field Love	Shimonoseki	313BW	505BG	6	6	0
222	June 25/26	Mine Field Zebra	Tsuruga Bay	313BW	505BG	9	9	0
233	June 27/28	Mine Field X-Ray	Hagi	313BW	505BG	14	14	0
233	June 27/28	Mine Field Able	Kobe/Osaka	313BW	505BG	8	8	0
233	June 27/28	Mine Field Uncle	Niigata	313BW	505BG	7	7	0
239	June 29/30	Mine Field Mike	Yawata	313BW	505BG	12	12	0
239	June 29/30	Mine Field Zebra	Maizyuru	313BW	505BG	4	4	0
239	June 29/30	Mine Field Uncle	Miyahoura	313BW	505BG	9	9	0
244	July 1/2	Mine Field Love	Shimonoseki City	313BW	505BG	9	2	0
244	July 1/2	Mine Field Nan	Fushiki, Nanao	313BW	505BG	15	22	0
246	July 3/4	Mine Field William	Funakoshi	313BW	505BG	6	10	0
246	July 3/4	Mine Field Zebra	Maizuru	313BW	505BG	4	3	0
246	July 3/4	Mine Field Mike	Shimonoseki	313BW	505BG	16	13	0
253	July 6/7	Urban Area	Kofu	313BW	505BG	not avail	33	0
259	July 9/10	Urban Area	Wakayama	313BW	505BG	not avail	35	0

VII. The XXI Bomber Command

MSN	Date	Target	Location			Number Up	Bombed	Lost
265	July 12/13	Urban Area	Tsuruga	313BW	505BG	not avail	not avail	
273	July 16/17	Urban Area	Kuwana	313BW	505BG	not avail	32	0
279	July 19/20	Choshi Urban Area	Choshi	313BW	505BG	not avail	not avail	
287	July 24	Aichi Aircraft Factory	Etitoku	313BW	505BG	not avail	33	0
294	July 26/27	Urban Area	Omuta	313BW	505BG	not avail	32	0
300	July 28/29	Urban Area	Ujiyamada	313BW	505BG	37	32	
308	August 1/2	Urban & Industrial Areas	Nagaoka	313BW	505BG	not avail	38	0
313	August 5/6	Urban & Industrial Areas	Maebashi	313BW	505BG	30	30	0
317	August 7	Toyokawa Naval Arsenal	Toyokawa	313BW	505BG	12	12	0
319	August 7/8	Urbran & Industrial Area	Yawata	313BW	505BG	31	28	1
327	August 14	Marshalling Yards	Marifu	313BW	505BG	32	31	0
329	August 14	Urban Area	Kumagaya	313BW	505BG	5	5	0

19thBG/314thBW

MSN	Date	Target	Location	Number Up	Bombed	Lost
38	February 25	NE Tokyo	NE Tokyo	11	10	
38	February 25	Target of opportunity	Nagoya	in above	1	
39	March 4	Nakajima Aurcraft Factory	Tokyo	9	not avail	
40	March 9/10	Urban Area	Tokyo	28	27	3
41	March 11/12	Urban Area	Nagoya	21	20	0
42	March 13/14	Urban Area	Osaka	22	21	0
43	March 16/17	Urban Area	Kobe	26	24	0
44	March 18/19	Urban Area	Nagoya	26	25	0
45	March 24/25	Mitsubishi Aircraft Engine Plant	Nagoya	22	22	0
46	March 27	Omura Airfield	Omura	22	21	0
46	March 27	Target of Opportunity	Kami Kushita	in above	1	0
48	March 30/31	Mitsubishi Aircraft Engine Plant	Nagoya	14	12	0
48	March 30/31	Target of Opportunity	Seto	in above	1	0
50	March 31	Omura Airfield	Kyushu	12	10	0
55	April 3/4	Aircraft Engine Factory	Shizuoka	25	24	0
59	April 7	Mitsubishi Aircraft Engine Plant	Nagoya	32	29	0
59	April 7	Target of Opportunity	Shingu	in above	1	0
65	April 12	Hodagaya Chemicak Works	Koriyama	21	22	0
67	April 13/14	Tokyo Arsenal	Tokyo	35	34	1
68	April 15/16	Urban Area	Kawasaki	33	33	3
74	April 17	Nittagahara Airfield	Kyushu	10	6	0
81	April 18	Nittagahara Airfield	Kyushu	11	11	0
90	April 21	Nittagahara Airfield	Kyushu	23	22	0
93	April 22	Miyazaki Airfield	Kyushu	11	11	
96	April 24	Hitachi Aircraft Factory	Tachikawa	11	11	0
104	April 26	Kanoya Airfield	Kyushu	22	19	0
104	April 26	Targets of Opportunity	Sakamoto	in above	1	0
104	April 26	Targets of Opportunity	Kaimon Dace	in above	1	0
112	April 27	Kanoya Airfield	Kyushu	11	10	0
118	April 28	Kanoya Airfield	Kyushu	11	11	
124	April 28	Kanoya Airfield	Kyushu	11	9	0
127	April 30	Kanoya Airfield	Kyushu	11	10	0
128	April 30	Kanoya East Airfield	Kyushu	11	10	0
136	April 30	Kanoya Airfield	Kyushu	11	8	0
136	April 30	Target of Opportunity	Kokubu Airfield	in above	2	0
136	April 30	Target of Opportunity	Tanega Shima AF	in above	1	0
137	May 3	Kanoya East Airfield	Kyushu	11	11	1
138	May 3	Kokubu Airfield	Kyushu	11	9	1
138	May 3	Target of Opportunity	Tanega Shima AF	in above	1	0
147	May 5	Kanoya Airfield	Kyushu	11	10	1
148	May 5	Chiran Airfield	Kyushu	11	8	0
149	May 5	Ibusuki Airfield	Kyushu	10	10	0
165	May 10	Oil Refinery	Otake	not avail	33	
172	May 11	Kawanishi Aircraft Factory	Fukae	not avail	33	0
174	May 14	Urban Area	Nagoya	not avail	39	
176	May 16/17	South Nagoya Urban	Nagoya	not avail	37	
178	May 19	City of Hamamatsu	Hamamatsu	not avail	22	0
181	May 23/24	South Tokyo Urban	Tokyo	43	43	0
183	May 25/26	South Central Tokyo Urban Area	Tokyo	not avail	34	

MSN	Date	Target	Location	Number Up	Bombed	Lost
186	May 29	Urban Area	Yokohama	not avail	39	
187	June 1	Urban Area	Osaka	36	32	0
188	June 5	Urban Area	Kobe	33	32	0
189	June 7	Urban Area	Osaka	32	31	1
199	June 10	Kasumiguara Air Base	Kasumiguara	33	26	1
203	June 15	Osaka/Amagasaki Urban Area	Osaka/Amagasaki	34	27	0
206	June 17/18	Urban Area	Kagoshima	32	32	
212	June 19/20	Urban Area	Shizuoka	35	33	1
216	June 22	Mitsubishi Aircraft Factory	Tamashima	33	29	0
225	June 26	Kawasaki Aircraft Factory	Akashi	35	33	0
231	June 26	Kita-Shioya Aircraft Factory	Kagamihara	35	25	2
231	June 26	Target of Opportunity	Matsusaka	in above	2	
231	June 26	Target of Opportunity	Toyohashi	in above	1	
231	June 26	Target of Opportunity	Katada	in above	1	
231	June 26	Target of Opportunity	Kita Shioya	in above	1	
243	July 1/2	Urban Area	Shimonoseki City	34	25	
250	July 3/4	Urban Area	Tokushima	31	29	0
254	July 6/7	Urban Area	Kofu	not avail	30	
260	July 9/10	Urban Area	Gifu	32	not avail	
266	July 12/13	Urban Area	Uwajima	not avail	26	
274	July 16/17	Urban Area	Hiratsuka	33	not avail	not avail
280	July 19/20	Okazaki Urban Area	Okazaki	not avail	not avail	
289	July 24	City of Tsu	Tsu	not avail	34	0
295	July 26/27	Urban Area	Omuta	32	not avail	
302	July 28/29	Urban Area	Ogaki	32	29	0
309	August 1/2	Urban & Industrial Areas	Mito	45	43	0
314	August 5/6	Urban & Industrial Areas	Nishinomiya	34	30	0
317	August 7	Toyokawa Naval Arsenal	Toyokawa	12	12	0
323	August 9/10	Nakajima Aircraft Factory	Tokyo Arsenal	40	39	0
330	August 14	Urban Area	Isesake	44	41	0

29thBG/314thBW

MSN	Date	Target	Location	Number Up	Bombed	Lost
40	March 9/10	Urban Area	Tokyo	26	22	5
41	March 11/12	Urban Area	Nagoya	21	19	0
42	March 13/14	Urban Area	Osaka	23	22	0
43	March 16/17	Urban Area	Kobe	26	23	0
44	March 18/19	Urban Area	Nagoya	23	23	0
45	March 24/25	Mitsubishi Aircraft Engine Plant	Nagoya	28	26	0
46	March 27	Omura Airfield	Omura	22	22	0
46	March 27	Target of Opportunity	Chichi Jima	in above	1	0
46	March 27	Target of Opportunity	Uchinoura	in above	1	0
50	March 31	Omura Airfield	Kyushu	22	22	0
55	April 3/4	Aircraft Engine Factory	Shizuoka	24	24	0
59	April 7	Mitsubishi Aircraft Engine Plant	Nagoya	31	28	2
59	April 7	Target of Opportunity	Oitsu	in above	1	0
59	April 7	Target of Opportunity	Shingu	in above	1	0
59	April 7	Target of Opportunity	Tamaski A/F	in above	1	0
59	April 7	Target of Opportunity	Chichi Jima	in above	1	0
59	April 7	Target of Opportunity	Nakiri	in above	1	0
65	April 12	Hodagaya Chemicak Works	Koriyama	22	22	0
67	April 13/14	Tokyo Arsenal	Tokyo	38	36	1
68	April 15/16	Urban Area	Kawasaki	32	30	2
75	April 17	Kanoya Airfield	Kyushu	12	9	0
78	April 18	Kanoya Airfield	Kyushu	not avail	not avail	
84	April 21	Kanoya Airfield	Kyushu	11	10	0
87	April 21	Kushira Airfield	Kyushu	9	8	0
93	April 22	Miyazaki Airfield	Kyushu	11	11	
96	April 24	Hitachi Aircraft Factory	Tachikawa	11	11	0
105	April 26	Kushira Airfield	Kyushu	22	13	0
105	April 26	Target of Opportunity	Kokubu A/F	in above	5	0
105	April 26	Target of Opportunity	Chiran A/F	in above	1	0
105	April 26	Target of Opportunity	Kanoya A/F	in above	1	0
105	April 26	Target of Opportunity	Tanaga Shima	in above	2	0

VII. The XXI Bomber Command

MSN	Date	Target	Location	Number Up	Bombed	Lost
113	April 27	Kushira Airfield	Kyushu	7	4	0
119	April 28	Kushira Airfield	Kyushu	11	10	
125	April 29	Kushira Airfield	Kyushu	9	9	0
130	April 30	Oita Airfield	Kyushu	11	10	0
132	April 30	Saeki Airfield	Kyushu	11	11	0
133	April 30	Tachiarai Airfield	Kyushu	11	9	0
133	April 30	Target of Opportunity	Kochi A/F	in above	1	0
134	April 30	Miyazaki Airfield	Kyushu	11	11	0
135	April 30	Miyakonojo Airfield	Kyushu	11	11	0
144	May 4	Oita Airfield	Kyushu	17	17	0
145	May 4	Tachiari A/F	Kyushu	11	10	2
165	May 10	Oil Refinery	Otake	not avail	31	0
172	May 11	Kawanishi Aircraft Factory	Fukae	not avail	9	0
174	May 14	Urban Area	Nagoya	Not avail	38	
176	May 16/17	South Nagoya Urban	Nagoya	not avail	30	
178	May 19	City of Hamamatsu	Hamamatsu	not avail	22	
181	May 23/24	Urban Area	Tokyo	not avail	44	
181	May 23/24	South Tokyo Urban	Tokyo	not avail	44	0
183	May 25/26	South Central Tokyo Urban Area	Tokyo	not avail	40	
186	May 29	Urban Area	Yokohama	not avail	37	
187	June 1	Urban Area	Osaka	30	29	0
188	June 5	Urban Area	Kobe	38	35	1
189	June 7	Urban Area	Osaka	31	28	0
200	June 10	Tachikawa Air Depot-Parts Factory	Tachikawa	34	32	0
203	June 15	Osaka/Amagasaki Urban Area	Osaka/Amagasaki	37	34	1
206	June 17/18	Urban Area	Kagoshima	33	32	
212	June 19/20	Urban Area	Shizuoka	37	32	0
216	June 22	Mitsubishi Aircraft Factory	Tamashima	28	28	0
226	June 26	Chigura Factory, Nagoya Arsenal	Nagoya	35	31	0
237	June 28/29	Urban Area	Nobeoka	28	27	0
243	July 1/2	Urban Area	Shimonoseki City	32	34	
250	July 3/4	Urban Area	Tokushima	35	0	
254	July 6/7	Urban Area	Kofu	not avail	35	0
260	July 9/10	Urban Area	Gifu	not avail	34	
266	July 12/13	Urban Area	Uwajima	not avail	32	0
274	July 16/17	Urban Area	Hiratsuka	not avail	33	
280	July 19/20	Okazaki Urban Area	Okazaki	not avail	32	
290	July 24	Nakajima Aircraft Plant	Handa	40	40	
295	July 26/27	Urban Area	Omuta	not avail	32	
301	July 28/29	Urban Area	Ogaki	33		
309	August 1/2	Urban & Industrial Areas	Mito	44	40	0
314	August 5/6	Urban & Industrial Areas	Nishinomiya	34	32	0
317	August 7	Toyokawa Naval Arsenal	Toyokawa	9	7	2
319	August 7/8	Ubrban & Industrial Area	Yawata	33	33	1
329	August 14	Urban Area	Kumagaya	38	37	0

39thBG/314thBW

MSN	Date	Target	Location	Number Up	Bombed	Lost
65	April 12	Hodagaya Chemicak Works	Koriyama	22	20	0
67	April 13/14	Tokyo Arsenal	Tokyo	24	21	0
68	April 15/16	Urban Area	Kawasaki	23	23	1
75	April 17	Kanoya Airfield	Kyushu	11	10	0
78	April 18	Kanoya Airfield	Kyushu	11	7	0
78	April 18	Kanoya East with 313BW	Kyushu	in above	2	0
84	April 21	Kanoya Airfield	Kyushu	11	9	0
84	April 21	Simon DakeA/F	Kyushu	in above	1	0
87	April 21	Kushira Airfield	Kyushu	11	9	0
96	April 24	Bombed Tof O with 313BW	Unknown	in above	3	0
96	April 24	Target of Opportunity	Unknown	in above	1	0
96	April 24	Hitachi Aircraft Factory	Tachikawa	10	6	0
106	April 26	Kokubu Airfield	Kyushu	22	17	0
106	April 26	Targets of Opportunity	Miyakonojo A/F	in above	2	0
106	April 26	Targets of Opportunity	Nagasaki Docks	in above	1	0
106	April 26	Targets of Opportunity	Shibuski A/F	in above	1	0

113	April 27	Kushira Airfield	Kyushu	12	11	1
119	April 28	Kushira Airfield	Kyushu	12	11	0
125	April 29	Kushira Airfield	Kyushu	11	7	0
129	April 30	Kokubu Airfield	Kyushu	10	7	0
140	May 3/4	Oita Airfield	Kyushu	22	11	1
142	May 4	Saeki Airfield	Kyushu	9	9	0
165	May 10	Oil Refinery	Otake	33	29	0
172	May 11	Kawanishi Aircraft Factory	Fukae	11	10	0
174	May 14	Urban Area	North Nagoya	33	32	1
178	May 19	City of Hamamatsu	Hamamatsu	22	16	0
178	May 19	Target of Opportunity	Shizuiko Urban	in above	1	0
178	May 19	Target of Opportunity	Shimoda A/F	in above	1	0
181	May 23/24	South Tokyo Urban	Tokyo	38	35	1
183	May 25/26	South Central Tokyo Urban Area	Tokyo	36	33	1
186	May 29	Urban Area	Yokohama	31	25	2
186	May 29	Target of Opportunity	Sagara	in above	1	0
186	May 29	Target of Opportunity	Shimoda A/F	in above	1	0
187	June 1	Urban Area	Osaka	15	13	1
187	June 1	Target of Opportunity	Wakayama	in above	1	0
188	June 5	Urban Area	Kobe	27	24	0
189	June 7	Urban Area	Osaka	29	21	0
193	June 9	Hitachi Aircraft Plant	Chiba	27	26	0
198	June 10	Hitachi — Aircraft Parts	Chiba	32	26	0
203	June 15	Urban Area	Osaka/Amagasaki	31	23	0
203	June 15	Target of Opportunity	Uwano	in above	1	0
203	June 15	Target of Opportunity	Kushimoto	in above	2	0
206	June 17/18	Urban Area	Kagoshima	27	26	0
212	June 19/20	Urban Area	Shizuoka	32	28	1
216	June 22	Mitsubishi Aircraft Factory	Tamashima	29	27	0
216	June 22	Target of Opportunity	Tamashima	in above	1	0
216	June 22	Target of Opportunity	Kushimoto	in above	2	0
227	June 26	Atsuta Factory Nagoya Arsenal	Nagoya	32	16	1
227	June 26	Secondary target	City of Tsu	in above	9	0
227	June 26	Target of Opportunity	City of Tsu	in above	2	0
227	June 26	Target of Opportunity	Hamamatsu	in above	4	0
227	June 26	Target of Opportunity	Shiroki	in above	1	0
227	June 26	Target of Opportunity	Shingu	in above	2	0
237	June 28/29	Urban Area	Nobeoka	29	28	0
243	July 1/2	Urban Area	Shimonoseki City	38	32	0
243	July 1/2	Target of Opportunity	Shimonoseki City	in above	1	0
243	July 1/2	Target of Opportunity	Kumamoto	in above	1	0
250	July 3/4	Urban Area	Tokushima	34	32	0
254	July 6/7	Urban Area	Kofu	not avail	not avail	0
260	July 9/10	Urban Area	Gifu	34	33	1
266	July 12/13	Urban Area	Uwajima	33	30	0
266	July 12/13	Target of Opportunity	Sukumo	in above	1	0
274	July 16/17	Urban Area	Hiratsuka	34	33	0
280	July 19/20	Okazaki Urban Area	Okazaki	34	33	0
280	July 19/20	Target of Opportunity	Nakii	in above	1	0
290	July 24	Target of Opportunity	Shingu Urban	41	39	0
290	July 24	Target of Opportunity	Kawasaki	in above	1	0
295	July 26/27	Urban Area	Omuta	33	31	0
301	July 28/29	Urban Area	Ogaki	33	29	0
309	August 1/2	Urban & Industrial Areas	Mito	38	37	0
309	August 1/2	Target of Opportunity	Choshi	in above	1	0
314	August 5/6	Urban & Industrial Areas	Nishinomiya	33	33	0
317	August 7	Toyokawa Naval Arsenal	Toyokawa	12	12	0
323	August 9/10	Urban Area	Kawaguchi	38	33	2
323	August 9/10	Target of Opportunity	Yakaichiba A/F	in above	2	0
323	August 9/10	Target of Opportunity	Hachijo Jima	in above	1	0
330	August 14	Urban Area	Isesake	44	43	1
330	August 14	Target of Opportunity	Choshi Point	in above	1	0

330thBG/314thBW

MSN	Date	Target	Location	Number Up	Bombed	Lost
65	April 12	Hodagaya Chemicak Works	Koriyama	20	17	3
65	April 12	Target of Opportunity	Onehame	in above	1	0
67	April 13/14	Tokyo Arsenal	Tokyo	16	11	1
67	April 13/14	T of O, Targets 213, 214	Tokyo	in above	1	0
67	April 13/14	T of O — Matsuda Airfield	Hachijo Jima	in above	2	0
68	April 15/16	Urban Area	Kawasaki	20	20	0
75	April 17	Kanoya Airfield	Kyushu	11	11	0
78	April 18	Kanoya Airfield	Kyushu	11	10	0
84	April 21	Kanoya Airfield	Kyushu	11	11	0
87	April 21	Kushira Airfield	Kyushu	11	11	0
96	April 24	Hitachi Aircraft Factory	Tachikawa	10	10	1
107	April 26	Miyakonojo Air Field	Kyushu	20	2	0
107	April 26	Secondary Target	Miyazaki AF	in above	15	0
107	April 26	Target of Opportunity	Yanikawa	in above	1	0
107	April 26	Target of Opportunity	Makurazaki	in above	1	0
112	April 27	Kanoya Airfield	Kyushu	10	10	0
118	April 28	Kanoya Airfield	Kyushu	12	10	0
124	April 28	Kanoya	Kyushu	9	9	0
131	April 30	Tomitaka Airfield	Kyushu	10	10	0
141	May 4	Omura Airfield	Kyushu	10	10	0
141	May 4	Target of Opportunity	Kushikino A/F	in above	1	0
143	May 4	Matsuyama Naval Air Station	Kyushu	19	17	0
143	May 4	Target of Opportunity	Lomuta	in above	1	0
143	May 4	Target of Opportunity	Wakagama	in above	1	0
165	May 10	Oil Refinery	Otake	33	33	0
172	May 11	Kawanishi Aircraft Factory	Fukae	11	11	0
174	May 14	Urban Area	North Nagoya	32	32	0
176	May 16/17	South Nagoya Urban	Nagoya	32	32	1
178	May 19	City of Hamamatsu	Hamamatsu	22	22	0
181	May 23/24	South Tokyo Urban	Tokyo	35	35	1
183	May 25/26	South Central Tokyo Urban Area	Tokyo	28	18	0
186	May 29	Urban Area	Yokohama	38	34	0
187	June 1	Urban Area	Osaka	37	34	0
188	June 5	Urban Area	Kobe	31	25	1
189	June 7	Urban Area	Osaka	27	27	0
195	June 10	Kasumigahara Seaplane Base	Kasumiguara	32	25	0
195	June 10	Secondary Target	Gifu	in above	1	0
199	June 10	Ogikubu — Nakajima Aircraft Factory	Ogikubu	32	25	0
203	June 15	Osaka/Amagasaki Urban Area	Osaka/Amagasaki	33	24	0
206	June 17/18	Urban Area	Kagoshima	28	28	0
212	June 19/20	Urban Area	Shizuoka	33	30	0
216	June 22	Mitsubishi Aircraft Factory	Gobo	34	30	1
230	June 26	Sumitome Light Metals	Nagoya	33	29	0
230	June 26	Target of Opportunity	Owase	in above	1	0
230	June 26	Target of Opportunity	Hachijo Jima	in above	1	0
230	June 26	Target of Opportunity	Hamamatsu A/F	in above	1	0
237	June 28/29	Urban Area	Nobeoka	35	32	0
243	July 1/2	Urban Area	Shimonoseki	38	37	0
250	July 3/4	Urban Area	Tokushima	35	35	0
254	July 6/7	Urban Area	Kofu	33	32	0
260	July 9/10	Urban Area	Gifu	34	31	0
266	July 12/13	Urban Area	Uwajima	33	33	
274	July 16/17	Urban Area	Hiratsuka	not avail	35	
280	July 19/20	Okazaki Urban Area	Okazaki	32	31	0
289	July 24	City of Tsu	Tsu	not avail	41	0
290	July 24	Nakajima Aircraft Plant	Handa	36	32	
295	July 26/27	Urban Area	Omuta	33	31	0
301	July 28/29	Urban Area	Ogaki	33	32	
309	August 1/2	Urban & Industrial Areas	Kawasaki	41	41	0
314	August 5/6	Urban & Industrial Areas	Nishinomiya	34	34	0
319	August 7/8	Ubrban & Industrial Area	Yawata	36	34	
329	August 14	Urban Area	Kumagaya	38	34	0

16thBG/315thBW

MSN	Date	Target	Location	Number Up	Bombed	Lost
232	June 26/27	Utsube Oil Refinery	Yokkaichi	15	15	0
238	June 29/30	Nippon Oil Refinery	Kadamatsu	18	16	0
245	July 2/3	Maruzan Oil Refirnery	Shimotsu	19	19	0
255	July 6/7	Maruzen Oil Refinery	Shimotsu	31	30	0
261	July 9/10	Utsube Oil refinery	Yokkaichi	30	33	0
267	July 12/13	Petroleum Oil Center	Kawasaki	28	28	
270	July 15/16	Kudamatsu Oil Refinery	Kudamatsu	30	30	0
281	July 19/20	Amagasaki — Nihon Oil Refinery	Amagasaki	30	not avail	not avail
283	July 23/24	Coal Liquidification Plant	Ube	30	not avail	0
291	July 25/26	Oil Refinery	Kawasaki	32		
303	July 28/29	Shimotsu	Maruzen Refinery	21	not avail	
310	August 1/2	Mitsubishi Havana Oil Refinery	Kawasaki	36	not avail	
315	August 5/6	Ube Coal Liquidification Co.	Ube	30	not avail	0
322	August 8/9	Nippon Oil Refinery	Amagasaki	29	not avail	0
328	August 14	Nippon Oil Refinery	Tsuchisaki	38	not avail	0

331stBG/315thBW

MSN	Date	Target	Location	Number Up	Bombed	Lost
310	August 1/2	Mitsubishi Havana Oil Refinery	Kawasaki	not avail		
315	August 5/6	Ube Coal Liquidification Co.	Ube	not avail	not avail	
322	August 8/9	Nippon Oil Refinery	Amagasaki	not avail	not avail	
328	August 14	Nippon Oil Refinery	Tsuchisaki	not avail	not avail	

501stBG/315thBW

MSN	Date	Target	Location	Number Up	Bombed	Lost
232	June 26/27	Utsube Oil Refinery	Yokkaichi	19	17	0
238	June 29/30	Nippon Oil Refinery	Kadamatsu	18	16	0
245	July 2/3	Maruzan Oil Refirnery	Shimotsu	21	20	0
255	July 6/7	Maruzen Oil Refinery	Shimotsu	28	28	0
261	July 9/10	Utsube Oil refinery	Yokkaichi	34	28	0
267	July 12/13	Petroleum Oil Center	Kawasaki	34	25	
270	July 15/16	Kudamatsu Oil Refinery	Kudamatsu	39	27	0
281	July 19/20	Amagasaki — Nihon Oil Refinery	Amagasaki	not avail	29	0
283	July 23/24	Coal Liquidification Plant	Ube	29	23	0
291	July 25/26	Oil Refinery	Kawasaki	not avail	29	0
303	July 28/29	Shimotsu	Maruzen Refinery	not avail	25	0
310	August 1/2	Mitsubishi Havana Oil Refinery	Kawasaki	not avail	34	0
315	August 5/6	Ube Coal Liquidification Co.	Ube	not avail	23	0
322	August 8/9	Nippon Oil Refinery	Amagasaki	not avail	23	0
328	August 14	Nippon Oil Refinery	Tsuchisaki	not avail	31	0

502ndBG/315thBW

MSN	Date	Target	Location	Number Up	Bombed	Lost
310	August 1/2	Mitsubishi Havana Oil Refinery	Kawasaki	not avail		
315	August 5/6	Ube Coal Liquidification Co.	Ube	not avail	not avail	0
322	August 8/9	Nippon Oil Refinery	Amagasaki	not avail	not avail	
328	August 14	Nippon Oil Refinery	Tsuchisaki	not avail	not avail	0

VII. The XXI Bomber Command

BOMB GROUP PLANE ROSTERS

This section is a group-by-group listing of planes known to have been assigned to the individual groups when the groups were assigned to the XXI Bomber Command 1944–45. The planes are identified by serial number only.

In addition, the Names List earlier in the book includes a subsection of names with known group assignments but unknown serial numbers. That subsection lists by name more than 300 airplanes assigned to specific groups. Some redundancy is certain to exist between those planes and the planes listed below by serial number in the individual group rosters.

40thBG/58thBW/XXI BC

42-24492	42-24738	42-24886	42-63455	42-63580	44-61542	44-61656	44-69698	44-70139	44-87635
42-24541	42-24739	42-24888	42-63462	42-65233	44-61548	44-61694	44-69828	44-70151	44-87660
42-24579	42-24740	42-24894	42-63498	42-65269	44-61554	44-61746	44-69906	44-84054	
42-24620	42-24752	42-24908	42-63505	42-65271	44-61556	44-61747	44-69992	44-86261	
42-24685	42-24757	42-24914	42-63527	42-65274	44-61565	44-61812	44-70015	44-86535	
42-24714	42-24795	42-24915	42-63538	42-65328	44-61587	44-61848	44-70085	44-87632	
42-24718	42-24846	42-63396	42-63542	42-93859	44-61634	44-69659	44-70094	44-87623	
42-24729	42-24860	42-63420	42-63555	44-61529	44-61651	44-69668	44-70100	44-87633	

444thBG/58thBW/XXI BC

42-24462	42-24580	42-24873	42-63464	42-65228	44-27307	44-61670	44-70002	44-70131	44-87662
42-24464	42-24584	42-24891	42-63496	42-65268	44-61525	44-61776	44-70102	44-70132	
42-24472	42-24720	42-24897	42-63530	42-65270	44-61526	44-61821	44-70108	44-70137	
42-24485	42-24723	42-24899	42-63533	42-65273	44-61543	44-61845	44-70120	44-70143	
42-24492	42-24724	42-63378	42-63537	42-65277	44-61546	44-69922	44-70123	44-70149	
42-24507	42-24730	42-63411	42-63548	42-65327	44-61553	44-69963	44-70127	44-84060	
42-24524	42-24731	42-63422	42-63557	42-65337	44-61564	44-69981	44-70128	44-84068	
42-24525	42-24732	42-63446	42-63559	42-93857	44-61646	44-69982	44-70129	44-86247	
42-24538	42-24861	42-63451	42-65227	42-93884	44-61653	44-69988	44-70130	44-87630	

462ndBG/58thBW/XXI BC

42-24590	42-24865	42-63448	42-63476	42-63560	42-93873	44-61569	44-61823	44-69989	44-86344
42-24711	42-24896	42-63450	42-63480	42-65230	42-93984	44-61628	44-61836	44-70076	44-87636
42-24728	42-24898	42-63454	42-63488	42-65232	44-27340	44-61639	44-69661	44-70086	44-87640
42-24786	42-24904	42-63457	42-63502	42-65252	44-61515	44-61650	44-69694	44-70104	44-87643
42-24800	42-24919	42-63459	42-63503	42-65299	44-61535	44-61652	44-69730	44-70110	44-87658
42-24801	42-63503	42-63472	42-63521	42-65329	44-61555	44-61657	44-69734	44-70142	44-87734
42-24838	42-65336	42-63473	42-63531	42-65336	44-61559	44-61745	44-69965	44-70148	
42-24848	42-63378	42-63474	42-63540	42-65352	44-61560	44-61786	44-69966	44-84068	

468thBG/58thBW/XXI BC

42-24471	42-24719	42-24895	42-63500	42-65227	44-61517	44-61677	44-69660	44-70140	44-87668
42-24486	42-24734	42-24909	42-63529	42-65272	44-61521	44-61681	44-69663	44-70146	44-87671
42-24487	42-24792	42-63415	42-63530	42-65275	44-61562	44-61695	44-69665	44-87644	44-87674
42-24525	42-24855	42-63417	42-63532	42-65276	44-61568	44-61702	44-69701	44-87647	
42-24542	42-24858	42-63424	42-63534	42-65279	44-61573	44-61703	44-69977	44-87656	
42-24691	42-24879	42-63445	42-63536	42-65315	44-61640	44-61708	44-70014	44-87659	
42-24703	42-24892	42-63456	42-63552	42-93877	44-61672	44-61807	44-70042	44-87661	
42-24714	42-24893	42-63460	42-63561	44-61516	44-61674	44-61816	44-70084	44-87666	

497thBG/73rdBW/XXI BC

42-24583	42-24599	42-24628	42-24772	42-63414	42-63466	42-63594	42-65348	44-61809	44-69899
42-24591	42-24604	42-24641	42-24774	42-63418	42-63467	42-65231	42-93858	44-69724	44-69912
42-24592	42-24615	42-24648	42-24807	42-63423	42-63471	42-65246	42-93883	44-69731	44-69913
42-24593	42-24616	42-24655	42-24808	42-63425	42-63485	42-65282	42-93885	44-69732	44-69929
42-24594	42-24619	42-24717	42-24845	42-63426	42-63492	42-65292	42-94050	44-69743	44-69932
42-24595	42-24622	42-24733	42-24855	42-63427	42-63519	42-65293	44-61663	44-69745	44-69998
42-24596	42-24623	42-24741	42-24857	42-63431	42-63523	42-65309	44-61709	44-69750	44-70007
42-24597	42-24626	42-24745	42-63412	42-63461	42-63626	42-65338	44-61716	44-69863	44-87665
42-24598	42-24627	42-24756	42-63413	42-63463	42-63541	42-65343	44-61785	44-69898	

498thBG/73rdBW/XXI BC

42-24468	42-24613	42-24695	42-24767	42-63468	42-65212	42-93999	44-61711	44-69749	44-70001
42-24544	42-24614	42-24727	42-24771	42-63469	42-65232	42-94014	44-61734	44-69752	44-70075
42-24601	42-24624	42-24735	42-24777	42-63475	42-65248	42-94027	44-61757	44-69765	44-70082
42-24603	42-24625	42-24742	42-24794	42-63478	42-65331	42-94030	44-61766	44-69772	44-70135
42-24605	42-24629	42-24748	42-63416	42-63501	42-65332	42-94046	44-61775	44-69777	44-70141
42-24606	42-24642	42-24749	42-63428	42-63510	42-65345	42-94095	44-69699	44-69848	44-70145
42-24607	42-24645	42-24750	42-63430	42-63522	42-93852	44-61623	44-69706	44-69852	44-70147
42-24608	42-24646	42-24751	42-63432	42-63524	42-93881	44-61655	44-69722	44-69932	44-84014
42-24609	42-24649	42-24755	42-63433	42-63554	42-93944	44-61661	44-69728	44-69969	44-84046
42-24610	42-24654	42-24760	42-63444	42-65210	42-93945	44-61666	44-69729	44-69987	
42-24611	42-24681	42-24763	42-63445	42-65211	42-93959	44-61678	44-69747	44-69993	

499thBG/73rdBW/XXI BC

42-24450	42-24661	42-24682	42-24758	42-63409	42-63483	42-65245	42-94005	44-69879	44-70099
42-24550	42-24665	42-24683	42-24765	42-63438	42-63491	42-65256	42-94006	44-69880	44-70119
42-24633	42-24666	42-24684	42-24769	42-63439	42-63492	42-65330	44-61690	44-69887	44-70134
42-24638	42-24667	42-24688	42-24773	42-63440	42-63493	42-65333	44-69655	44-69896	44-70152
42-24644	42-24669	42-24693	42-24775	42-63442	42-63495	42-65335	44-69735	44-69926	44-70153
42-24647	42-24670	42-24698	42-24782	42-63447	42-63513	42-65340	44-69740	44-69957	44-87649
42-24650	42-24673	42-24699	42-24833	42-63453	42-63550	42-65344	44-69832	44-70014	44-87651
42-24651	42-24674	42-24700	42-24887	42-63465	42-65220	42-93850	44-69833	44-70081	44-87652
42-24658	42-24677	42-24753	42-63369	42-63477	42-65222	42-93897	44-69842	44-70083	44-87657
42-24659	42-24679	42-24754	42-63381	42-63481	42-65224	42-93941	44-69843	44-70096	44-87669

500thBG/73rdBW/XXI BC

42-24436	42-24671	42-24696	42-24792	42-63489	42-65296	44-61530	44-61797	44-69775	44-70144
42-24600	42-24672	42-24700	42-24813	42-63490	42-65341	44-61658	44-69657	44-69829	44-84001
42-24643	42-24675	42-24714	42-24849	42-63494	42-65346	44-61669	44-69666	44-69848	44-84006
42-24652	42-24676	42-24721	42-24850	42-63497	42-93875	44-61684	44-69712	44-69868	
42-24653	42-24680	42-24743	42-63366	42-65218	42-93876	44-61685	44-69721	44-69878	
42-24656	42-24686	42-24744	42-63429	42-65219	42-93878	44-61692	44-69725	44-69885	
42-24657	42-24687	42-24761	42-63435	42-65221	42-93889	44-61693	44-69742	44-69944	
42-24660	42-24688	42-24762	42-63436	42-65245	42-93947	44-61699	44-69746	44-70101	
42-24662	42-24689	42-24766	42-63441	42-65247	42-94001	44-61737	44-69751	44-70113	
42-24664	42-24692	42-24770	42-63486	42-65249	42-94049	44-61782	44-69755	44-70117	
42-24668	42-24694	42-24785	42-63487	42-65251	42-94094	44-61789	44-69761	44-70136	

6thBG/313thBW/XXI BC

42-6396	42-24843	42-63410	42-63558	42-93901	44-27316	44-61836	44-69831	44-69979	44-84008
42-24540	42-24866	42-63444	42-65207	42-93902	44-61538	44-61905	44-69838	44-69980	44-84058
42-24547	42-24868	42-6351?	42-65229	42-93906	44-61549	44-62038	44-69839	44-70006	44-84081
42-24759	42-24870	42-63514	42-65281	42-93911	44-61558	44-69667	44-69840	44-70069	44-86303
42-24776	42-24872	42-63515	42-65339	42-93939	44-61635	44-69672	44-69841	44-70114	44-87650
42-24783	42-24874	42-63516	42-65347	42-93951	44-61671	44-69675	44-69847	44-70116	44-87672
42-24788	42-24878	42-63518	42-65381	42-94008	44-61675	44-69736	44-69849	44-70118	44-87734
42-24825	42-24880	42-63533	42-93842	42-94022	44-61679	44-69744	44-69850	44-70124	44-87650
42-24826	42-24884	42-63535	42-93843	42-94042	44-61686	44-69753	44-69854	44-70126	44-87672
42-24830	42-24885	42-63548	42-93842	42-94058	44-61688	44-69757	44-69855	44-70150	44-87734
42-24835	42-24890	42-63551	42-93843	42-94063	44-61705	44-69802	44-69857	44-83950	
42-24836	42-24901	42-63552	42-93887	42-94103	44-61784	44-69825	44-69865	44-84002	
42-24842	42-24916	42-63553	42-93898	44-27296	44-61803	44-69830	44-69884	44-84003	

9thBG/313thBW/XXI BC

42-24789	42-24853	42-63511	42-65278	42-93893	42-94025	44-61760	44-69811	44-69994	44-87737
42-24791	42-24856	42-63512	42-65283	42-93896	42-94041	44-61797	44-69834	44-70011	
42-24796	42-24859	42-63544	42-65284	42-93907	42-94043	44-61800	44-69849	44-70070	
42-24820	42-24875	42-63545	42-65285	42-93915	42-94045	44-61840	44-69874	44-70072	
42-24822	42-24876	42-63546	42-65286	42-93926	42-94067	44-69733	44-69883	44-70112	
42-24831	42-24900	42-63556	42-65298	42-93956	42-94119	44-69748	44-69920	44-70121	
42-24835	42-24907	42-63561	42-93886	42-93962	44+61648	44-69754	44-69934	44-83893	
42-24840	42-24913	42-63571	42-93888	42-94007	44-61689	44-69760	44-69975	44-86343	
42-24847	42-63509	42-63574	42-93892	42-94010	44-61691	44-69764	44-69985	44-87641	

VII. The XXI Bomber Command

504thBG/313thBW/XXI BC

42-2447X	42-24632	42-24816	42-24881	42-63507	42-65244	42-94056	44-69771	44-69952	44-87654
42-24489	42-24779	42-24821	42-24882	42-63522	42-65266	42-94108	44-69844	44-69953	
42-24491	42-24780	42-24826	42-24918	42-63523	42-65280	44-61520	44-69846	44-69970	
42-24493	42-24781	42-24834	42-63364	42-63525	42-65349	44-61550	44-69859	44-69978	
42-24501	42-24788	42-24837	42-63384	42-63536	42-93875	44-61551	44-69894	44-70018	
42-24512	42-24790	42-24851	42-63385	42-65216	42-93888	44-61660	44-69918	44-70106	
42-24519	42-24799	42-24852	42-63387	42-65217	42-93992	44-61792	44-69919	44-70109	
42-24551	42-24806	42-24854	42-63481	42-65235	42-93996	44-69727	44-69936	44-70111	
42-24555	42-24812	42-24863	42-63499	42-65241	42-94002	44-69739	44-69938	44-70118	
42-24557	42-24814	42-24865	42-63504	42-65242	42-94041	44-69758	44-69939	44-70118	

505thBG/313thBW/XXI BC

42-24437	42-24784	42-24827	42-24898	42-63524	42-65334	44-61561	44-69674	44-69941	44-70115
42-24448	42-24787	42-24828	42-24904	42-63525	42-93846	44-61563	44-69739	44-69958	44-70122
42-24477	42-24793	42-24839	42-63373	42-63549	42-93878	44-61572	44-69767	44-69961	44-70133
42-24495	42-24794	42-24844	42-63391	42-63614	42-93882	44-61664	44-69809	44-69964	44-70138
42-24502	42-24797	42-24848	42-63482	42-63619	42-93890	44-61665	44-69813	44-69997	44-86343
42-24532	42-24802	42-24860	42-63484	42-65243	42-93894	44-61667	44-69820	44-69999	44-87667
42-24534	42-24809	42-24860	42-63486	42-65250	42-93907	44-61715	44-69821	44-70004	
42-24536	42-24811	42-24862	42-63508	42-65253	42-93909	44-61728	44-69823	44-70005	
42-24634	42-24815	42-24864	42-63510	42-65255	42-93912	44-61778	44-69824	44-70010	
42-24764	42-24818	42-24867	42-63513	42-65257	42-93968	44-61788	44-69835	44-70071	
42-24778	42-24823	42-24889	42-63517	42-65265	42-93997	44-69662	44-69853	44-70078	
42-24781	42-24824	42-24890	42-63520	42-65272	42-94019	44-69664	44-69900	44-70111	

19thBG/314thBW/XXI BC

42-24883	42-65300	42-93917	42-94039	44-69673	44-69689	44-69856	44-69968	44-70095	44-87639
42-24903	42-65303	42-93918	42-94098	44-69678	44-69696	44-69860	44-69984	44-70098	44-87653
42-24906	42-65304	42-93923	44-61638	44-69679	44-69703	44-69862	44-69990	44-70102	44-87655
42-63547	42-65307	42-93967	44-61642	44-69680	44-69707	44-69871	44-69996	44-70103	
42-63563	42-65308	42-93984	44-61647	44-69681	44-69804	44-69872	44-70000	44-83959	
42-63567	42-62	42-93989	44-61680	44-69682	44-69815	44-69873	44-70003	44-83967	
42-63569	42-65310	42-94003	44-61700	44-69684	44-69818	44-69893	44-70013	44-83974	
42-63573	42-65342	42-94026	44-61751	44-69685	44-69843	44-69959	44-70064	44-83988	
42-65297	42-93913	42-94028	44-61768	44-69686	44-69845	44-69967	44-70092	44-84012	

29thBG/314thBW/XXI BC

42-6395	42-63562	42-65302	42-93913	42-93930	42-94061	44-69676	44-69847	44-69890	44-70105
42-6428	42-63564	42-65305	42-93916	42-93958	42-94066	44-69677	44-69851	44-69891	44-70107
42-24433	42-63565	42-65306	42-93918	42-93998	44-61658	44-69691	44-69869	44-69962	44-70XX9
42-24554	42-63566	42-65311	42-93920	42-94009	44-61676	44-69693	44-69875	44-69983	44-84009
42-24571	42-63568	42-65312	42-93925	42-94013	44-61706	44-69728	44-69876	44-69991	44-87662
42-24572	42-63570	42-65350	42-93925	42-94023	44-61721	44-69762	44-69877	44-70008	44-87663
42-24636	42-63571	42-93900	42-93927	42-94033	44-61805	44-69772	44-69881	44-70009	44-87664
42-24912	42-63579	42-93905	42-93928	42-94034	44-69669	44-69776	44-69882	44-70080	
42-24917	42-65301	42-93912	42-93929	42-94035	44-69670	44-69797	44-69886	44-70093	

39thBG/314thBW/XXI BC

42-24578	42-65368	42-94021	44-61643	44-69763	44-69788	44-69810	44-69901	44-69974	44-84036
42-65361	42-65369	42-94053	44-61795	44-69768	44-69791	44-69844	44-69902	44-70077	44-84037
42-65362	42-65373	42-94079	44-69681	44-69769	44-69792	44-69867	44-69903	44-70079	44-84059
42-65364	42-93966	44-27346	44-69688	44-69773	44-69794	44-69870	44-69907	44-84015	44-84061
42-65365	42-93974	44-61524	44-69716	44-69779	44-69796	44-69888	44-69908	44-84016	44-84062
42-65366	42-93975	44-61524	44-69756	44-69783	44-69799	44-69889	44-69910	44-84028	44-87642
42-65367	42-93979	44-61524	44-69757	44-69785	44-69810	44-69895	44-69914	44-84029	

330thBG/314thBW/XXI BC

42-24917	42-93910	42-93957	42-93979	42-94032	44-61537	44-69766	44-69801	44-69935	44-84024
42-63517	42-93912	42-93961	42-93980	42-94037	44-61539	44-69774	44-69814	44-69971	44-84040
42-63539	42-93935	42-93964	42-93982	42-94040	44-61664	44-69786	44-69817	44-69995	
42-65363	42-93937	42-93969	42-93995	42-94047	44-61670	44-69790	44-69857	44-69997	
42-65370	42-93943	42-93970	42-93996	42-94052	44-61696	44-69795	44-69897	44-70118	
42-65371	42-93953	42-93971	42-94016	42-94059	44-61756	44-69797	44-69897	44-70010	
42-93837	42-93954	42-93976	42-94024	42-94062	44-69696	44-69799	44-69911	44-70016	
42-93908	42-93955	42-93978	42-94029	42-94071	44-69741	44-69800	44-69928	44-70101	

16thBG/315thBW/XXI BC

42-63589	42-63621	42-63649	42-63661	42-63692	42-63708	42-63739	42-63746	44-83930
42-63603	42-63626	42-63651	42-63666	42-63697	42-63712	42-63740	44-83899	44-83931
42-63605	42-63628	42-63653	42-63667	42-63699	42-63715	42-63741	44-83906	44-83932
42-63608	42-63630	42-63655	42-63673	42-63703	42-63734	42-63742	44-83913	44-83933
42-63613	42-63644	42-63657	42-63688	42-63704	42-63736	42-63743	44-83915	44-84077
42-63617	42-63646	42-63659	42-63691	42-63707	42-63738	42-63745	44-83918	

331stBG/315thBW/XXI BC

42-63593	42-63631	42-63700	44-83893	44-83904	44-83929	44-83952	44-83992	44-84069
42-63606	42-63658	42-93862	44-83895	44-83905	44-83937	44-83954	44-83993	44-84087
42-63609	42-63676	44-61776	44-83896	44-83907	44-83939	44-83956	44-84031	44-84127
42-63610	42-63679	44-61815	44-83897	44-83919	44-83941	44-83958	44-84056	44-84129
42-63611	42-63681	44-83890	44-83898	44-83924	44-83943	44-83961	44-84057	44-84131
42-63612	42-63690	44-83891	44-83902	44-83925	44-83944	44-83981	44-84066	
42-63618	42-63698	44-83892	44-83903	44-83928	44-83946	44-83985	44-84067	

501stBG/315thBW/XXI BC

42-63599	42-63627	42-63652	42-63668	42-63675	42-63684	42-63694	42-63710	42-63720	42-63726
42-63600	42-63639	42-63654	42-63669	42-63677	42-63685	42-63696	42-63714	42-63721	42-63749
42-63601	42-63640	42-63663	42-63670	42-63680	42-63686	42-63702	42-63717	42-63723	44-83986
42-63607	42-63641	42-63664	42-63671	42-63682	42-63687	42-63705	42-63718	42-63724	44-83987
42-63615	42-63650	42-63665	42-63674	42-63683	42-63689	42-63709	42-63719	42-63725	

502ndBG/315thBW/XXI BC

42-63614	42-63620	42-63656	42-63695	42-63713	42-63748	44-83909	44-83942
42-63616	42-63623	42-63662	42-63701	42-63722	42-93835	44-83910	44-83948
42-63619	42-63624	42-63678	42-63711	42-63747	44-83901	44-83912	

Targets Listed by Name

Target	Target #	Location
33rd Infantry Regiment and Arsenal	1214	Hisai — Tsu
Aeronautical Experimental Laboratory	1246	Tokyo
Aeronautical Experimental Laboratory	1393	Tokyo
Aeronautical Instruments	368	Tokyo
Aichi Aircraft Eitoku Plant	1077	Fukushima
Aichi Aircraft Works,	1101	Aomoi
Aichi Aircraft Works, Atsuta Plant	198	Nagoya
Aichi Aircraft Works, Bomimaru Plant	1730	Koromo
Aichi Aircraft Works, Eitoku Plant	1729	Nagoya
Aichi Aircraft Works, Mizuho Branch	199	Nagoya
Aichi Aircraft Works, Tsukiji Plant	1828	Nagoya
Aikoku Petroleum Refinery	147	Kawasaki
Ainoura Steam Power Plant	1835	Ainoura
Aircraft Engine Plant	1393	Tokyo
Akabane Powder Storage, Tokyo Arsenal	1463	Kasumguara
Akabane Railroad Bridge	1368	Tokyo
Akano Military Aviation School & Field	1129	Nagoya
Akano Military Aviation School & Field	1213	Uji Yamada
Akashi Industrial — Aircraft Factory	1464	Choshi City
Akashi Urban Area	1496	Chichibu
Akunoura Engine Works	542	Nagasaki
Amagasaki Oil Refinery	1671	Unknown

VII. The XXI Bomber Command

Target	Target #	Location
Amagasaki Steam Power Plant	536	Amagasaki
Amagasaki Steel Works	1769	Amagasaki
Amagasaki Transformer Station	1629	Amagasaki
Amagasaki Urban Area	1675	Yonago
Amatsuji Steel Ball Mfg Co	661	Osaka
Ammunition Storage	1339	Yokohama
Anju Seaplane Station	1729	Nagoya
Aomori Urban Area	1943	Higashi-Iwase
Army Arsenal	206	Oji Ward Tokyo
Army Branch Powder Co	208	Oji Ward Tokyo
Army Central Clothing Depot, Tokyo Arsenal	201	Tokyo
Army Central Clothing Depot, Tokyo Arsenal	202	Oji Ward Tokyo
Army Clothing Depot	737	Hiroshima
Army Division Headquarters	748	Hiroshima
Army Food Depot	737	Ujina Island
Army Ordnance Depot	736	Hirishima
Army Provisioning Depot	2010	Nagoya
Army Provisions Depot	355	Tokyo
Army Transportation Base	735	Ujina Island
Asada Aluminum Plants	1648	Shihama
Asahi Bemberg Cupraammonium Plant	1314	Nobeoka
Asahi Bemberg Gunpowder Plant	1310	Nobeoka
Asahi Electrochemical Company	202	Tokyo
Asahi Electrochemical Company	212	Tokyo
Asahi Glass Company	134	Yokohama
Asahi Glass Company	1760	Amagasaki
Asahi Glass Company	567	Tobata
Asahi Oil Refinery	40	Hiroshima
Asahi Petroleum Co	130	Kawasaki
Asahi Pottery Works		
Asahi Precision Work Co.	1716	Sakai
Asahi Substatiion	102	Yokohama
Asano Cement Company	1874	Nogota
Asano Cement Company	475	Kawasaki
Asano Dockyard	70	Yokohama
Asano Iron Works	51	Kawasaki
Asano Karite Co Ltd	1334	Yokohama
Asano Portland Cement Co	1725	Osaka
Asano Portland Cement Co	39	Moji
Atsuta Factory, Nagoya Arsenal	197	Nagoya
Atsuta Factory, Nagoya Arsenal	208	Tokyo
Bridgestone Tire Company	1263	Kurume
Bridgestone Tire Company	1431	Totsuka
Briquette Plant	212	Tokyo
Central Meteorlogical Observatory	1424	
Chiba Urban Area	228	Tokyo
Chichibu Cement Company	1496	Chichibu
Chichibu Cement Company	2470	Nagoya
Chigiri Shima Copper Smelter	927b	Innoshima Is
Chiran Airfield	304	Futtsu Point
Chofu Airport	1412	Tokyo
Chosi Army Airdrome	335	Tokyo
Chosi Urban Area	4011	Tokyo
Chugoku Steam Power Plant	1284	Okayama
Cooperative Auto Works	135	Kawasaki
Dai Niehi Electric Wire Co	1743	Amagasaki
Dai Nipon Cellophane Plant	1827	Nagoya
Dai Nippon Celluloid Co	1281	Abashi
Dai Nippon Celluloid Co	383	Sakai
Daido Machinery Co Showa Plant	1800	Nagoya
Daido Machinery, Showa Plant	1800	Machinery
Daido Steel Mfg Co	1766	Amagasaki
Dainoura Mine	1274	Nogata
Dejima Wharf	485	Unknown
Diado Electric Steel Arsuta Plant	247b	Nagoya

Target	Target #	Location
Diado Electric Steel Minami Plant	247d	Nagoya
Diado Electric Steel Oe Plant	247a	Nagoya
Diado Electric Steel Tsukiji Plant	247c	Nagoya
Diado Electric Steel Tsukiji Plant	674	Tokuyama
East Asia Development Co Warehouses	1446	Yokohama
East Hiroshima Rail Road Station	740	Hiroshima
Eastern Petroleum Refinery	89	Yokohama
Edogawa Petroleum Refinery	360	Tokyo
Eitoku Industrial/Aircraft Factory	810a	Tsudo
Eitoku Plant	810b	Honzan
Electric Instrument Plant	810c	Moto
Electro-Chemical Industry	1100	Aomi
Electro-Chemical Industry	811	Kiyotaki
Electro-Chemical Industry Co	1246	
Electro-Chemical Industry, Aomi Plant	826	Tokyo
Fiji Spinning Mills	836	Nagasaki
Ford Motor Company	72	Yokohama
Fuji Elec Works	136	Kawasaki
Fuji Spinning Mills	485	Kawasaki
Fuji Steel Works Of Imperial Iron and Steel	56	Kawasaki
Fujikoshi Steel Products, Higashi-Iwase Plant	1943	Higashi-Iwase
Fujikoshi Steel Products, Toyama Plant	941	Toyama
Fujikura Electric Cable Works	867	Fushiki
Fujinagata Shipbuilding Co	273a	Osaka
Fujinagata Shipbuilding Co	273b	Osaka
Fukagawa Pier	1434	Tokyo
Fukayama Urban Area		Akashi
Fukuda Light Aeroplane Co	793	Amagasaki
Fukui Urban Area		Akashi
Fukuoka Air Station	663	Fukuoka
Fukuoka Urban Area		Amagasaki
Fukuoka Wireless Station	1423	
Fukushima Railroad Yards	1077	Fukushima
Fukuszaki Steam Power Plant	324	Osaka
Fukuyama Urban Area		Aomori
Funabashi Naval Transmitting Station	1420	Funabashi
Furakawa Electric Co., Plant No. 1	75	yokohama
Furakawa Electrical Co	1739	Amagasaki
Furakawa Electrical Co	1740	Osaka
Furuawa Copper Refinery	811	Kiyotaki
Furukawa Copper Ore Dressing Plant	810a	Tsudo
Furukawa Machine Shops	810c	Mato
Furukawa Smelter (code name: Treadle)	810b	Kyushu
Fushiki Harbor	867	Fushiki
Futase Mines	1276	Nogata
Futtsu Point Forts	304	Fukayama
Gifu Urban Area		Fukui
Goto Railroad Shops	1675	Yonago
Government Aircraft Parts Plant	1671	Sendai
Government Power Station, Tokyo Arsenal		Gifu
Government Steam Power Station	228	Tokyo
Ha Shima Colliery	843	Ha Shima
Haba Shipyard	927a	Innoshima Is
Hachioji Junctions and Bridge	1375	Tokyo
Hachioji Urban Area		Hachioji
Haiki Railroad Terminal	939	Haiki
Hakata Railroad Yards and Station	1270	Fukuoka
Hakate Harbor	1255	Fukuoka
Hako Point Oil Tanks	297	Yokosuka
Hamamatsu Urban Area		Hamamatsu
Hanao Wake Oil Storage	669	O Shima
Haneda Airport	337	Tokyo
Harima Shipyard	1296	Oo
Harley-Davidsen Motorcycle Co	1340	Tokyo
Hatsudoki Engine Works	1191	Osaka

VII. The XXI Bomber Command

Target	Target #	Location
Hattori Company	913	Tokyo
Hayama Petroleum Refinery	127	Kawasaki
Hayashi Commercial Company Engine Works	828	Nagasaki
Higashi Yokohame RR Yards	68	Yokohama
Higashi-Iwase Harbor	862	Higashi-Iwase
Hikari Arsenal (code name: Thunderhead)	794	Tokuyama
Hikari Naval Arsenal	671	Tokuyama
Himeji Urban Area		Himeji
Hinode-Cho Pier	886	Tokyo
Hiraidzuma Drug Depot	1447	Yokohama
Hirano Loom Works	1831	Nagoya
Hiratsuka Urban Area		Hiratsuka
Hiro Arsenal	794	Hiro
Hiro Naval Aircraft Factory (code name: Thunderhead)	660	Kure
Hiro Naval Engine and Turbines Factory	730	Hiro
Hiro Naval Shipyard, Dock & Arsenal	660	Hiro
Hiroshima Urban Area		Hiroshima
Hitachi Aircraft Company, Chiba Plant		Chiba
Hitachi Aircraft Engine Company (code name: Catcall)		Tachikawa
Hitachi Aircraft Plant	2009	Tokyo
Hitachi Copper Refineery	812	Sakagawa
Hitachi Copper Smelter	1492	Sakagawa
Hitachi Engine Works	1191	
Hitachi Engineering Co., Kaigan Plant	1476	Sakegawa
Hitachi Engineering Kaigan Plant	1467	Hatachi
Hitachi Mfg Co Kasado Plant	825	Kudamatsu
Hitachi mfg co. Ube plant	1700	ube
Hitachi Urban Area		Hitachi
Hitachi Works	1802	Kuwana
Hitachi+A91 Engineering Kameido Plant	916	Tokyo
Hitachi-Solex Aircraft	1390	Totsaka
Hitonose Oil Storage	659	Yeta Shima
Hodagaya Substation	62	Yokohama
Hodogaya Chemical Industries (code name: Butterball)	2025	Koriyama
Hodogaya Soda	1399	Yokohama
Hodogaya Soda	1400	Tokyo
Hokada Airdrome	1487	
Hokoku Machinery Co., Atsuta Plant	1809	Nagoya
Hokoku Machinery Kasadera Plant	1799	Nagoya
Horai Hydro Plant	1607	Abukuma River
Horai Hydro Plant	1667	Horai
I.G.R.Shops	538	Taketori
Ibagawa Electro-Chemical Co., Kido Plant	1822	Ogaki
Ibagawa Electro-Chemical Co., Kita Kiriishi Plant	1832	Ogaki
Ibusuki Air Field (code name: Infirmary)	2507	Kyushu
Ichinomiya Urban Area		Ichinomiya
Ikegai Motor Plant	883	Tokyo
Imabari Urban Area		Imabari
Imperial Dye Works	1931	Fukuyama
Imperial Government Railways Shops	1230	Hamamatsu
Imperial Government Railways Shops	894	Omiya
Inaka Hydro-Electric Plant	1604	Kambara
Inasawa Shunting Yards	1133	Inaswaw
Inawashiro Hydro Plant #1	881	Nipasshi River
Inawashiro Hydro Plant #2	900	Nipasshi River
Inawashiro Transformer Station	211	Tokyo
Inflammable Storage Docks	1338	Yokohama
Inuyama Transformer Station	1638	Inuyama
Isahaya Railroad Junctions	838	Isahaya
Isesaki Urban Area		Isesaki
Ishihara Smelter & Refinery	1776	Yokkaichi
Ishii Iron Works	1360	Tokyo
Ishikawajima Dockyard Of Ishikawajima Shipbuild	330	Tokyo
Ishikawajima Engine Plant	1391	Tomioka
Itabashi Powder Plant, Tokyo Arsenal	205	Tokyo

Target	Target #	Location
Ito Aircraft Plant	1519	Ito
Itozaki Oil Storage	934	Itozaki
Iwakura Transformer Station	1143	Iwakura
Iwatsaki Wireless Station	1425	Tokyo
Iwo Jima Air Fields (code name: Sledgehammer)		Iwo Jima
Iwo Jima Air Fields 1 & 2 (code name: Rock Crusher)		Iwo Jima
Iwo Jima Air Fields 1 & 2, 1/24/45 (code name: Starlit)		Iwo Jima
Izumi Air Field (code name: Bullish)	2512	Kyushu
Japan (Showa) Elctro-Chemical Company	521	Yokohama
Japan Aircraft Company, Tomioka Plant		Tomioka
Japan American Oil Company	88	Yokohama
Japan Artificial Fertilizer	1401	Tokyo
Japan Artificial Fertilizer	204	Tokyo
Japan Artificial Fertilizer	479	Kawasaki
Japan Bakelite Company	1365	Tokyo
Japan Dyestuff Mfg Co	1734	Osaka
Japan Dyestuffs Co	1733	Osaka
Japan Explosives Mfg	1875	Nogata
Japan Iron Works	1765	Osaka
Japan Iron Works, Hirohato Plant	1290	Hirohato
Japan Iron Works, Tobata Plant	29	Tobata
Japan Iron Works, Yawata Plant	28	Yawata
Japan Light Metals Aluminum Plant	1176	Shimizu
Japan Light Metals Aluminum Plant	1177	Kamabara
Japan Light Metals Company	1003	Niigata
Japan Machine Industry	1342	Tokyo
Japan Military Goods Co., Explosives Plant	898	Tomioka
Japan Military Goods Company	899	Tomioka
Japan Musical Instrument Co Propeller Plant	1219	Hamamatsu
Japan Nitrogen Fertilizer Co	1386	Minamata
Japan Paraffin Mfg Co	670	O Shima
Japan Physico-Chemical Co	1357	Tokyo
Japan Porcelain Company	1153	Nagoya
Japan Powder Mfg Co Asa Plant	820	Asa
Japan Radio and Phonograph Company	523	Yokohama
Japan Refining Company	1088	Koriyama
Japan Refining Works	1397	Tokyo
Japan Service Company	1347	Kawasaki
Japan Special Steel Company	336	Tokyo
Japan Special Steel Works	1349	Tokyo
Japan Steel Bearing Company	1355	Tokyo
Japan Steel Co., Hiroshima Plant	1891	Hiroshima
Japan Streel Bearing Company Fujisawa Plant	1356	Fujisawa
Japanese Steel Tube Company	52	Kawasaki
Junction	1376	Ofuna
Kabota Iron & Machinery	1778	
Kadamatsu Industrial/Oil Refinery		Kadamatsu
Kagamigahara Machinery Works	1812	Gifu
Kagamigahara Military Airport	249	Gifu
Kagoshima Freight Yards	1518	Kagoshima
Kagoshima Railroad Repair Shops	1517	Kagoshima
Kagoshima Urban Area		Kagoshima
Kakigi Dry Dock	1454	Yokosuka
Kakowa Mfg Company	1645	Utsonomiya
Kakowa Mfg Company — Kumagaya Parts Plant	1650	Kumagaya
Kakusa Kagyo Foundry	554	Wakamatsu
Kakuwa Mfg Company	1645	Utsonomiya
Kamagawa Steam Power Plant	63	Yokohama
Kameyama Rr Junction	1217	Kameyama
Kamioka Lead & Zinc Mines	1484	Kamioka
Kamizaki Heavy Industry	169	
Kanagafuchi Spinning Mill Company	1394	Tokyo
Kanagawa Gasoline Tanks	92	Yokohama
Kanazawa Heavy Industry	171	
Kanazawa Railroad Shops	871	Matsuto-Machi

VII. The XXI Bomber Command

Target	Target #	Location
Kanda Market	1449	Tokyo
Kanegafuchi Soda Industry	785	Kobe
Kanidera Hydro Electric Plant	874	Jintsu River
Kanishi Photo Works	880	Tokyo
Kanokawa Oil Storage	1907	Nishi Nomi Jima
Kanose Hydro Plant	1590	Agano River
Kanoya Air Field (code name: Checkbook)	1378	Kyushu
Kanoya East Air Field (code name: Famish)	2516	Kanoya
Kansi Kyodo Steam Power Plant No 1	540a	Amagasaki
Kansi Kyodo Steam Power Plant No 2	540b	Amagasaki
Kanson Point Fortification	1494	Uraga
Kasado Dock Co	668	Kasado Is
Kasagi Hydro-Electric Plant	1615	Kiso River
Kashima Naval Air Station	1465	Kasumiguara
Kasugade No 1 Steam Power Plant	322	Osaka
Kasugade No 2 Steam Power Plant	323	Osaka
Kasumigaura Seaplane Station		Kasumiguara
Kasumiguara Naval Air Station	1466	Kasumiguara
Kasumiguara Naval Air Station	1491	Kasumiguara
Kawanami Industry Co Shipyard	860	Koyagi Shima
Kawanishi Aircraft Co., Kagamigahara Plant		Kagamigahara
Kawanishi Aircraft Co., Takarazuka Plant		Takarazuka
Kawanishi Aircraft Company (code name: Leafstalk)	1702	Kobe
Kawanishi Aircraft Company, Akashi		Akashi
Kawanishi Aircraft Company, Fukae		Fukae
Kawanishi Aircraft Company, Himeji Plant		Himeji
Kawanishi Aircraft Company, Narua Plant		Narua
Kawanishi Machine Shops	1745	Kobe
Kawasahe Elec Pwr Plant For Govt Railroad	111	Kawasaki
Kawasaki Aircraft Co	1724	Akashi
Kawasaki Aircraft Plant (code name: Fruitcake)	1547	Akashi
Kawasaki Aircraft Works, Kagamigahara Plant	240	Gifu
Kawasaki Heavy Industry	1762	Kobe
Kawasaki Heavy Industry	1775	Kobe
Kawasaki Heavy Industry	7	Kobe
Kawasaki Heavy Industry Co.	171	Kobe
Kawasaki Industrial Area		Kawasaki
Kawasaki Locomotive & Car Co.	11	Kobe
Kawasaki Naval Wireless Sta	132	Kawasaki
Kawasaki Petroleum Center		Kawasaki
Kawasaki Substation #1	105	Kawasaki
Kawasaki Urban Area (code name: Brisket)		Kawasaki
Kayaba Engineering Company	1331	Tokyo
Kenda Market	1449	Tokyo
Kineshima Mine	1271	Saga
Kinugawa Steam Power Station	213	Tokyo
Kioroshi Army Airdrome	1488	Kioroshi
Kisarazu Fleet Fueling Base	1477	Tokyo
Kisarazu Naval Air Station	373	Tokyo
Kitashi Electrical Engineering Company	1362	Tokyo
Kizu Transformer Station	1632	Kizu
Kizugawa Steam Power Plant	325	Osaka
Koa Machinery Company	1813	Tomita
Kobe Harbor District No 1	22	Kobe
Kobe Harbor District No 2	34	Kobe
Kobe Industrial/Aircraft Factory		Kobe
Kobe Steel Works	1768	Kobe
Kobe Steel Works	1215	Toba
Kobe Steel Works	5	Kobe
Kobe Steel Works	6	Kobe
Kobe Steel Works — Moji Plant	46	Moji
Kobe Steel Works Nagoya Plant	1753	Nagoya
Kobe Urban Area (code name: Middleman)		Kobe
Kobokuro Machinery Works	1257	Nogata
Kochi Urban Area		Kochi

Target	Target #	Location
Kofu Urban Area		Kofu
Koibota Iron & Maainery Works	1778	Osaka
Kokubu Air Field (code name: Barranca)	2520	Kyushu
Kokura Arsenal	168	Kokura
Kokura Railroad Shops	184	Kokura
Kokura Steam Power Plant #2	184	Kikura
Kokura Steel Works	165	Kokura
Kokusan Machinery Company	1361	Tokyo
Komaki Hydro Electric Plant	873	Sho River
Koreha Textile Mill		
Koriyama Chemical Plant (code name: Lunchroom)	6129	Koriyama
Koriyama Railroad Shops	1655	Koriyama
Koriyama Urban Area		Koriyama
Koto Market	1450	Tokyo
Koyagi Shima Shipyard, Matsu Iron Works	860	Nagasaki
Kozaki Point Oil Storage	545	Nagasaki
Kubato Iron Machinery Works	1777	Sakai
Kubato Iron Machinery Works	688	Osaka
Kubato Iron Machinery Works	701	Osaka
Kubota Iron & Machinery Works	1197	Amagasaki
Kuje Power Plant	1616	Osaka
Kumagaya Air Base	1644	Kumagaya
Kumagaya Urban Area		Kumagaya
Kumamoto Military Zone	1248	Kumamoto
Kumamoto Urban Area		Kumamoto
Kurada Iron Works	498	Yokohama
Kure Naval Air Station	656	Hiro
Kure Naval Arsenal	657a	Kure
Kure Naval Shipyard	657b	Kure
Kure Provisions and Clothing Depot	657d	Kure
Kure Railroad Station	798	Kure
Kure Submarine Base	658	Kure
Kure Torpedo Boat Base & Mine Depot	657c	Kure
Kure Urban Area		Kure
Kureha Textile Mill	196	Nagoya
Kurimoto Iron Works	1763	Osaka
Kurimoto Iron Works	685	Osaka
Kurobe #2 Hydro Electric Plant	877	Kurobe River
Kurobe #3 Hydro Electric Plant	1608	Kurobe River
Kurume Military Zone	1247	Kurume
Kurume Railroad Station	1269	Kurume
Kushira Air Field (code name: Aeroscope)	2534	Kyushu
Kuwana Urban Area		Kuwana
Kwoyo Precision Works Co.	1714	Kokubu
Kwoyo Precision Works Co.	1715	Osaka
Kwoyo Precision Works Co.	1718	Osaka
Kwoyo Precision Works Co.	1720	Osaka
Kyoritsu Warehouse	1438	Yokohama
Kyushu Chemical Industry	569	Yawata
Kyushu Coal and SS Co. Colliery	1843	Sekito Shima
Kyushu Ordnance Plant	1239	Zasshonoguma
Lighthouse Bureau Oil Tanks	92	Yokohama
Maebashi Urban Area		Maebashi
Maibara Railroad Yards	1160	Maibara
Maizura Naval Base	1041	Maizura
Manda Mine	1275	Omuta
Manda Mine	1278	Omuta
Mansushita Wireless Company	1744	Moriguchi
Marifu Marshalling Yards		Marifu
Marunoichi Telephone Exchange	1418	Tokyo
Maruzen Industrial/Oil Refinery		Maruzen
Maruzen Oil Refinery	1764	Shimatsu
Maruzen Oil Refinery	257	Osaka
Matsido Airport	1416	Tokyo
Matsukata Oil Storage	94	Yokohama

VII. The XXI Bomber Command

Target	TGT #	Location
Matsuo Engine Works	803	Nagasaki
Matsuo Engine Works	828	Unknown
Matsushita Dry Battery Co	1750	Moriguchi
Matsushita Dry Battery Co	1751	Osaka
Matsuyama Air Field (code name: Mopish)	2777	Kyushu
Matsuzaka Junction and Round House	1218	Matsuzaka
Megami Point Oil Storage	832	Nagasaki
Meideisha Elec Mfg Co Nishi Biwajima Plant	1803	Heisaka
Meiji Sugar Factory	487	Kawasaki
Migata Engineering Works, S. Tokyo	573	Tokyo
Miike Coal Yards	1267	Miike
Miike Dyestuffs	1243	Omuta
Miike Harbor	1254	Omuta
Miike Machinery Works	1256	Omuta
Military Gunpowder Works, Tokyo Arsenal	207	Tokyo
Military Works	209	Oji Ward Tokyo
Minakata Hydro-Electric Plant	1652	Tenryu River
Minatogawa Steam Power Plant	4	Kobe
Mine Field Able		Kobe/Osaka
Mine Field Baker		Aki Nada
Mine Field Charlie		Fukuoka
Mine Field Dog		Shoda Shima
Mine Field Easy		Bisan Seto
Mine Field Fox		Bingo Nada
Mine Field Fusan		Korea
Mine Field Geijitsu		Korea
Mine Field Genzan		Korea
Mine Field George		Hamada
Mine Field How		Kure
Mine Field Item		Kure
Mine Field Jig		Hiroshima
Mine Field King		Tokuyama
Mine Field Love		Shimonoseki
Mine Field Masan		Korea
Mine Field Mike		Shimonoseki
Mine Field Nan		Fushiki
Mine Field Oboe		Tokyo Bay
Mine Field Pusan		Korea
Mine Field Rashin		Korea
Mine Field Roger		Sasebo
Mine Field Seishin		Korea
Mine Field Tare		Nagoya
Mine Field Uncle		Niigata
Mine Field William		Fungkoshi
Mine Field X-Ray		Senzaki, Hagi, Oura
Mine Field Yoke		Sakai
Mine Field Zebra		Tsuruga, Miyazu
Minoshima Industrial/Oil Refinery		Minoshima
Minsukasa Coke Plant	24	Wakamatsu
Mita Leather	1761	
Mitani Railroad Workshops	1268	Mitani
Mito Army Airdrome	1468	Mito
Mito Railroad Junction and Station	1478	Mito
Mito Urban Area		Mito
Mitsibishi Dock	543	Nagasaki
Mitsubishi Aircft Co., Kagamigahara Plant		Kagamigahara
Mitsubishi Aircraft Co., Mishima Plant		Tamashima
Mitsubishi Aircraft Engine Plant (code name: Eradicate)		Nagoya
Mitsubishi Aircraft Engine Works (code name: Memphis)	193	Nagoya
Mitsubishi Aircraft Shibaura Plant	327	Tokyo
Mitsubishi Aircraft Works	802	Unknown
Mitsubishi Aircraft Works, Kagamigahara Plant	1833	Gifu
Mitsubishi Aircraft Works, Nagoya (code name: Hesitation)	194	Nagoya
Mitsubishi Aircraft, Oimochi Plant	799	Tokyo
Mitsubishi Arms Works	546	Nagasaki

Target	Target #	Location
Mitsubishi Coal Depot	191	Kawasaki
Mitsubishi Copper Refining	697	Osaka
Mitsubishi Copper Smelter/Zinc Refinery	1297	Nao Island
Mitsubishi Dockyard	543	Nagasaki
Mitsubishi Electrical Mfg Co	829	Nagasaki
Mitsubishi Electrical Mfg Co	1803	Heisaka
Mitsubishi Electrical Mfg Co	1885	Hiroshima
Mitsubishi Electrical Mfg Co	254	Nagoya
Mitsubishi Heavy Industry	169	Kobe
Mitsubishi Oil Company & Chiyoda Oil Tanks	117	Yokohama
Mitsubishi Oil Refinery	116	Kawasaki
Mitsubishi Oil Storage	258a	Osaka
Mitsubishi Oil Storage	258b	Osaka
Mitsubishi Piece Good Wharf	1437	Kawasaki
Mitsubishi Steel and Arms Works	546	Nagasaki
Mitsubishi Steel Rolling Mill	1795	Nagasaki
Mitsubishi Warehouse	1440	Yokohama
Mitsui Coal Liquidification Plant	1262	Omuta
Mitsui Electrolytic Zinc Refinery	1261	Omuta
Mitsui Rr Workshops	1268	Omuta
Mitsui Tame Shipyard	1295	Tamashima
Mitsui Zinc Distilling Plant	1260	Omuta
Mittan Leather Belt Company	1761	Osaka
Miyakonojo Air Field (code name: Dripper)	2527	Kyushu
Miyata Engineering Works	573	Tokyo
Miyata Motor Works	1241	Zasshonoguma
Miyazaki Air Field (code name: Neckcloth)	2529	Kyushu
Miyazu Steam Power Plant	1949	Miyazu
Mizaguchi Gear Works	1631	Amagasaki
Mizaguchi Gear Works	1634	Amagasaki
Mizuno Sporting Goods Co	1706	Osaka
Moen Island Air Fields (code name: Panhandle)		Truk
Moji Central Wharf	44a	Moji
Moji Coaling Station	44c	Moji
Moji Ordnance Storage	49	Moji
Moji Southern Wharf	44b	Moji
Moji Urban Area		Moji
Momoyama Hydro Electric Plant and Substation	1499	Kiso River
Montenaga Inlet Oil Tanks	1457	Yokosuka
Mushishino Aircraft Plant	357	Tokyo
Nagamachi Marshalling Yards	1104	Sendai
Nagano Government Railroad Shops	1098	Nagano
Nagaoka Urban Area (code name: Microscope)		Nagaoka
Nagasaki Spinning Mill	851	Nagasaki
Nagasaki Steam Power Plant	847	Nagasaki
Nagasaki Urban Area		Nagasaki
Nagasaki Wharves	1842	Nagasaki
Nagasaki Wharves	835	Nagasaki
Nagoya #1 Steam Power Plant	195	Nagoya
Nagoya #2 Steam Power Plant	1598	Nagoya
Nagoya Arenal, Atsuda Plant	197	Nagoya
Nagoya Arsenal, Chigusa Factory	196	Nagoya
Nagoya Freight Yards	250a	Nagoya
Nagoya Harbor Facilities	251	Nagoya
Nagoya Industrial Area		Nagoya
Nagoya Plant		Nagoya
Nagoya Repair Shops	250b	Nagoya
Nagoya Station	250c	Nagoya
Nagoya Urban Area		Nagoya
Najima Seaplane Base	1237	Fukuoka
Najima Steam Power Plant	664	Fukuoka
Nakagama Steel Mfg Company	1755	Osaka
Nakajima Aircraft Company	1544	New Ota
Nakajima Aircraft Company (code name: Enkindle)	357	Musashino
Nakajima Aircraft Company		Handa

VII. The XXI Bomber Command

Target	Target #	Location
Nakajima Aircraft Company		Ogikubu
Nakajima Aircraft Company, Musashino Plant (code name: Enkindle)	357	Tokyo
Nakajima Aircraft Company, New Ota Plant (code name: Furious)	1545	Koizumi
Nakajima Aircraft Company-Kumagaya Parts Plant	1650	Kumagaya
Nakajima Aircraft Company-Maebashi Parts Plant	1546	Maebashi
Nakajima Aircraft Company-Ota Parts Plant	789	Ota
Nakajima Aircraft Ogikubo Plant	356	Tokyo
Nakajima Aircraft Plant (code name: Fraction)	1544	Ota
Nakajima Aircraft Plant	1834	Koromo
Nakajima Aircraft Tanishi Foundry Plant	359	Tokyo
Nakajima Aircraft Works	1635	Handa
Nakajima Seaplane Works	332	Tokyo
Nakatsugawa Hydro-Electric Plant	1611	Nakatsugawa
Nakayama Steel Mfg Co	1755	Osaka
Nakayama Steel Mfg Co	1771	Amagasaki
Nakayama Steel Mfg Co	713	Osaka
Namazu Urban Area		Namazu
Nanao Harbor	870	Nanao
Naniwa Warehouse	1451	Yokohama
Narashino Army Airdrome	1409	Funabashi
National Silk Conditioning House	1443	Yokhama
Naval Gunpowder Works, Tokyo Arsenal	1335	Oji Ward Tokyo
Navy Arsenal	1336	Hiratsuka
Navy Department Radio Towers	1419	Tokyo
Nesshin Spinning Mill	912	Tokyo
Nichiman Aluminum Company	861	Higashi-Iwase
Niigata Harbor Facilities	1670	Niigata
Niigata Iron Works	1353	Niigata
Niigata Iron Works	918	Niigata
Niigata Iron Works Plant #1	998	Niigata
Niigata Iron Works Plant #2	999	Niigata
Niigata Iron Works Plant Kashiwazaki Plant	1659	Kashiwazaki
Niigata Iron Works Plant Kashiwazaki Plant	1668	Kashiwazaki
Niigata Iron Works Plant Nagaoka Plant	1660	Nagaoka
Niigata Oil Storage #1	1012	Niigata
Niigata Oil Storage #2	1013	Niigata
Niigata Railroad Yards	1023	Niigata
Niigata Sulphuric Acid Company	80	Kawasaki
Niihama Copper Concentrating Mill	925	Niihama
Niihama Steam Power Plant	931	Niihama
Nipon Magnesium Company	1942	Sasazu
Nipon Soda K.K.	1642	Nihongo
Nippon Air Brake Co	1719	Kobe
Nippon Aluminum Works	1708	Osaka
Nippon Carbon Company, Factory #1	499	Yokohama
Nippon Dunlop Ruber Co	14	Kobe
Nippon Dye Works	1317	Tsurusaka
Nippon Electric Battery Company, Plant #1	1677	Kyoto
Nippon Electric Battery Company, Plant #2	1678	Kyoto
Nippon Electric Factory #2	326	Tokyo
Nippon Electric Power Transforming Station	104	Yokohama
Nippon Electric Wire & Cable Company	489	Kawasaki
Nippon Electric Wire and Cable	1430	Tokyo
Nippon Electric, Factory #1	497	Kawasaki
Nippon Electric, Factory #2	326	Tokyo
Nippon Explosives Works	1138	Taketoya
Nippon Insulator Company	1171	Nagoya
Nippon Iron Sand Industry	1770	Takasago
Nippon Iron Sand Industry, Shikama Plant	1921	Shikama
Nippon Magnesium Co	1942	Kamioka
Nippon Mining Co	1387	Tagame
Nippon Motor Oil Company	1880	Ube
Nippon Nitrogen Explosives Nobeoka Plant	1311	Kawashima
Nippon Nitrogen Explosives Raiken Plant	1312	Nobeoka
Nippon Oil Co; Kudamatsu Plant Refinery	672	Kudamatsu

Target	Target #	Location
Nippon Oil Company	1366	Tokyo
Nippon Oil Refinery	1000	Niigata
Nippon Oil Refinery & Tank Farm	1203	Amagasaki
Nippon Oil Refinery Kashiwazaki	1649	Kashiwazaki
Nippon Oil Storage	261	Osaka
Nippon Rubber Company	1264	Kurume
Nippon Soda Company, Aluminum Plant	866	Takaoka
Nippon Soda Company, Magnesium Plant	1944	Higashi-Iwase
Nippon Soda, Odera Refinery	1666	Odera
Nippon Stainless Steel Naoetsu Plant	1651	Naoetsu
Nippon Super Fuel Co	129	Yokohama
Nippon Synthetic Chemicals Plany	1826	Ogaki
Nippon Typewriter Co	1333	Tokyo
Nippon Vehicle Manufacturing Company	241	Nagoya
Nirada Firecracker Co	78	Yokohama
Nishinomiya Urban Area		Nishinomiya
Nissan Auto Company	522	Yokohama
Nissan Chemical Industry Co, Onoda Plant	819	Onoda
Nissan Chemical Plant	253	Nagoya
Nissan Chemical Plant #1	936	Sasakura
Nissan Liquid Fuel Company	1123	Unknown
Nisshin Spinning Mill	912	Tokyo
Nisso Steel Mfg Company	334	Tokyo
Nitta Leather Belt Mfg Co	1761	Osaka
Nittigahara Air Field (code name: Bushing)	2531	Kyushu
Nobeoka Urban Area		Nobeoka
Northeast Aluminum Plant	1658	Koriyama
Numazu Railroad Yards	1181	Numazu
Numazu Urban Area		Numazu
Oaska Machinery Company Plant #1	1661	
Ogaki Iron Works	1810	Ogaki
Ogaki Iron Works	1811	Gifu
Ogaki Urban Area		Ogaki
Ogura Oil Company	87	Yokohama
Ogura Oil Company	911	Tokyo
Oi Hydro-Electric Plant	1505	Kiso River
Oi Railroad Works	370	Tokyo
Oigana Hydro-Electric Plant	1609	Oi River
Oita Air Field (code name: Camlet)	1307	Kyushu
Oita Air Field (code name: Fearless)	1308	Oita
Oita Railroad Yards	1329	Oita
Oita Urban Area		Oita
Oji Army Arsenal	206	Tokyo
Okamoto Aircraft Works, Kosagara Plant	1129	Nagoya
Okamoto Aircraft Works, Showa Plant	242	Nagoya
Okamoto Aircraft Works, Tarui Plant	1736	Tarui
Okayama Urba Area		Okayama
Okazaki Urban Area		Okazaki
Oki Electric Company #1	1746	Osaka
Oki Electric Company #1	1747	Kobe
Oki Electric Company #1	888	Tokyo
Okuma Iron Works, Asahi Plant	1807	Seto
Okuma Iron Works, Hagino Plant	1146	Nagoya
Okuma Iron Works, Kachikawa Plant	1806	Kachikawa
Okuma Iron Works, Kami Iida Cho Plant	1797	Nagoya
Okuma Iron Works, Kusonoki Plant	1805	Kachikawa
Okuma Iron Works, Nunoike Plant	1147	Nagoya
Okuma Iron Works, Ozone Plant	461	Nagoya
Omura Aircraft Plant (code name: Fearless)	1627	Omura
Omura Airfield		Omura
Omura Naval Air Station (code name: Vamoose)	849	Omura
Omuta #1 Steam Power Plant	1249	Omuta
Omuta #2 Steam Power Plant	1595	Omuta
Omuta #3 Steam Power Plant	1596	Omuta
Omuta Mill Kanegafuchi Spinning Company	1273	Omuta

VII.   The XXI Bomber Command

Target	Target #	Location
Omuta Plant, Kyushu Power Company	1249	Omuta
Omuta Urban Area		Omuta
Onada Cement Co	822	Onada
Onada Steam Power Plant	1600	Onada
Onagohata Power Station	1520	Hida
Oppama Naval Air Station	298	Yokosuka
Ordnance Storehouse Amunition	201	Oji Ward Tokyo
Ordnance Storehouse, Tokyo Arsenal	218	Oji Ward Tokyo
Ordnance Supply Depot, Tokyo Arsenal	203	Oji Ward Tokyo
Oriental Babcock Company	1354	Yokohama
Oriental Cable Company	1756	Kaji
Oriental Can Mfg Co	1785	Osaka
Oriental Can Mfg Co	1787	Oaska
Oriental High Tension Company, Plant A	1244	Tokyo
Oriental High Tension Company, Plant B	1245	Tokyo
Oriental Oil Company	90	Yokohama
Oriental Otis Elevator, S.Tokyo	541	Tokyo
Oriental Steel Products Co	57	Kawasaki
Oriental Weaving Company	914	Tokyo
Oriental Weaving Company	915	Tokyo
Osaka Army Arsenal Hirakata Branch	1723	Hirakata
Osaka Arsenal	382	Osaka
Osaka Ceramic Industry Cement Co	1726	Oaska
Osaka Chain and Machinery Works	1773	Ibarigi
Osaka Electric Machinery Co	1752	Osaka
Osaka Gas Company Coke Oven Plant	1712	Osaka
Osaka Harbor	53	Osaka
Osaka Iron Works	430b	Hiroshima
Osaka Iron Works, Unit No 1	272	Osaka
Osaka Iron Works, Unit No 2	699	Osaka
Osaka Machinery Co	1780	Itami
Osaka Machinery Co	1781	Osaka
Osaka Machinery Company	388	Osaka
Osaka Machinery Company Plant #1	1661	Nagaoka
Osaka Machinery Company Plant #2	1662	Nagaoka
Osaka Machinery Works	1797	Nagoya
Osaka Machinery Works	1798	Nagoya
Osaka Metal Industry	1782	Sakai
Osaka Metals Industries Company	686	Osaka
Osaka Railroad Station	1206	Osaka
Osaka Ship Building Works	1711	Osaka
Osaka Steel Mfg Co	268	Osaka
Osaka Transformer Station	1630	Osaka
Osaka Urban Area (code name: Peachblow)		Osaka
Osaka Wakayama Iron Works	1784	Tsuda
Osaka Wakayama Iron Works	687	Osaka
Oshima	1884	Oshima
Oshima Naval Oil Storage (code name: Anaphase)	1884	Oshima
Ota Industrial/Aircraft Engines		Ota
Otake Oil Refinery (code name: Fainter)	2121	Otake
Otaki Hydro-Electric Plant	1498	Kiso River
Otawa Bay Airport	1470	Otawa Bay
Oyama Point Ammo and Powder Stores	1455	Yokosuka
Petroleum Center	128	Kawasaki
Physico Chemical Industries	1364	Tokyo
Port Of Tobata	33	Tobata
Powder Factory, Tokyo Arsenal	217	Oji Ward Tokyo
Powder Magazine	516	Yokohama
Powder Magazine, Tokyo Arsenal	219	Oji Ward Tokyo
Radio Station Joak	363	Tokyo
Railroad Shop	561	Wakamatsu
Railroad Transformer Station	106	Kawasaki
Rasa Industry Co.	1732	Osaka
Riken Alunite Plant	1821	Nagoya
Riken Heavy Industries Co	1358	Tokyo

Target	Target #	Location
Riken Heavy Industry, Kashiwazaki Plant	1668	Kashiwazaki
Riken Heavy Industry, Nagaoka Plant	1669	Nagaoka
Riken Industrial, Maebashi	1646	Shin-Maebashi
Riken Industrial, Takasaki	1647	Takasaki
Riken Metal Company, Ube Plant	922	Ube
Riken Vacuum Tube Plant	1520	Mohara
Rikuo Engine Company	1341	Tokyo
Rising Sun Petroleum Oil Storage	17	Kobe
Rising Sun Petroleum Oil Storage	260	Osaka
Rising Sun Petroleum Terminal	910	Tokyo
Rolling Stock Company	1332	Tokyo
Rolling Stock Mfg Co	548	Osaka
Saeki Air Field (code name: Cockcrow)	1306	Kyushu
Saga Urban Area		Saga
Sagami (Bangu) Rr Bridge	1377	Hiratsuka
Saganoseki Copper Works	1328	Saganoseki
Saijo Steam Power Plant	1602	Saijo
Saito Zaki Petroleum Storage	665	Fukuoka
Saka Steam Power Plant	796	Sakai
Sakai Hydro-Electric Plant	1059	Shibukawa
Sakai Urban Area		Sakai
Saku Hydro-Electric Plant	1059	Shibukawa
Sakurada Engineering Works	904	Tokyo
Sangyo Cement Co	1242	
Sankyo Co	1741	Osaka
Sannomiya Railroad Station	157	Kobe
Sanyo Steam Power Plant No 3	1283	Shikawa
Sanyo Steam Power Plants No 1 and 2	1286	Shikawa
Sanyo Steel Co Shikawa Plant	1920	Shikawa
Sasasu Transformer Station	1633	Sasazu
Sasebo Aircraft Factory	834	Sasebo
Sasebo Coal Yard	844	Sasebo
Sasebo Fuel & Munition Storage	762	Sasebo
Sasebo Mine & Torpedo Storage	757	Sasebo
Sasebo Naval Air Station	754	Sasebo
Sasebo Naval Arsenal	758	Sasebo
Sasebo Naval Dockyard	752	Sasebo
Sasebo Oil Storage	755	Sasebo
Sasebo Outfitting Wharf	845	Sasebo
Sasebo Provisions Wharf	845	Sasebo
Sasebo Railroad Station	759	Sasebo
Sasebo Urban Area		Sasebo
Sayama Airport	1408	Tokyo
Sendai Urban Area		Sendai
Senju Hydro Plant	1593	Shinano River
Senju Steam Power Plant	230	Tokyo
Senkari Reservoir Dam	153	Sendai
Seto Hydro-Electric Station	1144	Shimohara
Shannosho Shipyard	927b	Innoshima Is
Shibaura Engineering Works	133	Yokohama
Shibaura Machine Tool Company	354	Tokyo
Shibaura Wharf	887	Tokyo
Shimagawa Mfg Plant	1665	
Shimizu Urban Area		Shimizu
Shimonoseki Port	42	Shimonoseki
Shimonoseki Straits (code name: Starvation)		Shimonoseki
Shimonoseki-Maeda Power Plant	99	Shimonoseki
Shimoshizu Army Airdrome	1470	Shimoshizu
Shimotaki Hydro-Electric Plant	1613	Shimotaki
Shimotsu Oil Refinery		Shimotsu
Shinagawa Airport	1413	Tokyo
Shinagawa Mfg Plant	1665	Fukushima
Shinagawa Railroad Yards	364	Tokyo
Shinano Hydro Plant	1592	Shinano River
Shinetsu Nitrogen Fertilizer, Naoetsu Plant	1667	Naoetsu

VII. The XXI Bomber Command

Target	Target #	Location
Shingagawa Shipyard	1462	Tokyo
Shingo Hydro Plant	1588	Agano River
Shingo Hydro Plant	1592	
Shinjuku Railroad Station & Rr Junction	1374	Tokyo
Shinohara Machinery Works	1522	
Shiodome Freight Yards	365	Tokyo
Shizuoka Aircraft Engine Works		Shizuoka
Shizuoka Aircraft Plant (code name: Upcast)	2011	Shizuoka
Shizuoka Urban Area		Shizuoka
Shoda Aircraft	1395	Tokyo
Shoun Engineering Co	1363	Ofuna
Showa Aircraft	791	Tokyo
Showa Airport	1407	Tokyo
Showa Chemical & Fertilizer Plant	1521	
Showa Electrode Co	1748	Naruo
Showa Fertilizer	137	Kawasaki
Showa Fertilizer, Kanose Plant	1536	Kanose
Showa Iron Works	1872	Fukuoka
Showa Soda Plant	467	Nagoya
Showa Wire Cable Company	484	Kawasaki
Showa-Denko Aluminum	1100	Omachi
Sobu Railroad Bridge.Shinkoiwa Yard	1373	Tokyo
South Manchurian Wharf & Warehouse	1485	Kawasaki
Soyama Hydro Electric Plant	875	Sho River
Spinning and Weaving Mill	859	Nagasaki
Standard Vacuum Oil Storage	261	Osaka
Steam Engineering & Rolling Stock Company	366	Tokyo
Steam Power Plant #2	1595	Omuta
Steel Plants Ne Joto Ward	1352	Tokyo
Strong Engineering Works	1704	Ibaragi
Strong Engineering Works	1705	Osaka
Suita Railroad Yars and Shops	1209	Suita
Sumida River Rr Yard/ Bridge Of Joban Line	1370	Tokyo
Sumidar Bridge Sobr Rr/Ryogoku Sta & Rr Yadr	1372	Tokyo
Sumitoma Aluminum Company	1709	Yao
Sumitoma Aluminum Company	263b	Osaka
Sumitoma Duraluminum Company, Nagoya Plant		Nagoya
Sumitoma Duraluminum Mill Kuwona Plant	1639	Kuwana
Sumitoma Electric Industries Co	262	Osaka
Sumitomo Aluminum Plant	1657	Niihawa
Sumitomo Aluminum Reduction Plant	924	Niihawa
Sumitomo Chemical Co	923	Niihama
Sumitomo Copper Refinery	815	Niihawa
Sumitomo Copper Smelter	814	Mino Is
Sumitomo Light Metal Plant	X14065	Nagoya
Sumitomo Light Metals Plant	2024	Tokyo
Sumitomo Machinery Works	932	Niihawa
Sumitomo Metal Industry	1195	Amagasaki
Sumitomo Metal Industry	263a	Osaka
Sumitomo Metal Industry	264	Osaka
Sunamachi Airport	1405	Tokyo
Susaki Airport	1411	Tokyo
Susaki Dock Yard Of Ishikawajima Shipbuilding	330	Tokyo
Suzuka Naval Air Station	1130	Karbe
Suzuki Loom Works	1227	Hamamatsu
Suzuki Loom Works Takatsuka Plant	1929	Hamamatsu
Tabata — Nippori Railroad Yard	224	Tokyo
Tachiari Air Field (code name: Fearless)	1236	Tachiari
Tachiari Machine Works	1870	Tachiari
Tachikawa Air Base	1404	Tachikawa
Tachikawa Air Depot (code name: Blockhouse)	2008	Tachikawa
Tachikawa Aircraft Plant (code name: Modeller)	792	Tachikawa
Tadotsu Railroad Yards and Repair Shops	1303	Tadotsu
Tagawa Mines	1275	Nogata
Tagi Fertilizer Company	1713	Behu

Target	Target #	Location
Taka Shima Colliery	842	Taka Shima
Takada Aluminum Mfg Co	1709	Sakai
Takahagi Airport	1415	Tokyo
Takamatsu Urban Area		Takamatsu
Takasaki Railroad Yards	1051	Takasaki
Take Hara Electrolytic Coper Refinery	1936	Takehars
Takeda Drug Co	1628	Osaka
Takeshiba Pier	885	Tokyo
Takeshina Freight Yards	66	Yokohama
Tama River Bridges & Tokado/Keihin Elec Lines	1367	Kawasaki
Tama River Bridges & Tokado/Keihin Elec Lines	1369	Kawasaki
Tanaka Piston Ring Co	1721	Osaka
Tankiji Markey and Warehouse	1448	Tokyo
Tatara Machinery Works	1873	Fukuoka
Tategame Shipyard	544	Nagasaki
Tateyama Naval Air Station	371	Airports
Tatigami Shipyard	544	Nagasaki
Tatigami Urban Arera		Nagasaki
Teikoku Mining Co/Hibi Refinery	1922	Hibi
Three Powder Magazine	315	Yokohama
Titanium Industry Co	1879	Ube
Tkinogawa Army Arsenal	209	Tokyo
Toa Oil Company; Shimotsu Refinery		Shimotsu
Toba Dockyard	1216	Tobata
Tobata Foundry Company	30	Tobata
Tobata Steam Power Plant	1594	Unknown
Tobato Port	33	Tobata
Tochigi Shipbuilding Yard	558	Wakamatsu
Togami Electric Works	1259	Saga
Togi Fertilizer Co	1713	Behu
Toho Gas Works	1821	Nagoya
Toho Gas Works Synthetic Oil Plant	456	Nagoya
Toho Heavy Industry	1738	Yokkaichi
Toho Steel Foundry Co	1348	Tokyo
Tokada Aluminum Mfg Company	1710	Sakai
Tokai Electrical Co Nagoya #1 Plant	1823	Nagoya
Tokai Electrical Co Nagoya #2 Plant	1824	Nagoya
Tokai Electrical Co Nagoya #3 Plant	1825	Nagoya
Tokai Electrode Co	1539	Tanoura
Tokai Steel Works	555	Wakamatsu
Tokorazawa Airport	1406	Tokyo
Tokoshu Light Metals-Atsuta Plant	1818	Nagoya
Tokoyama Naval Arsenal		Tokuyama
Tokushima Urban Area		Tokushima
Tokushu Light Metals Atsuta Plant	1818	Nagoya
Tokushu Light Metals Inasawa Plant	1820	Inasawa
Tokushu Light Metals Ogaki Plant	1819	Ogaki
Tokuyama Coal Yard (code name: Rotative)	674	Tokuyama
Tokuyama Naval Fueling Station	673	Tokuyama
Tokuyama Naval Fueling Station (code name: Indices)	73	Tokuyama
Tokuyama Sheet Iron Co	1881	O Shima
Tokuyama Soda Co	675	Tokuyama
Tokuyama Urban Area		Tokuyama
Tokyo (code name: San Antonio)		Tokyo
Tokyo Armory	214	Tokyo
Tokyo Army Arsenal & Engineering School	352	Tokyo
Tokyo Arsenal (code name: Perdition)	3600	Tokyo
Tokyo Central Station	367	Tokyo
Tokyo Elec Power Generating Sta, Tsurumi Plant	110	Kawasaki
Tokyo Electric Power Station	470	Kawasaki
Tokyo Electric Wire & Mfg Company	494	Kawasaki
Tokyo Gas & Electric Engineering, S. Tokyo	321	Tokyo
Tokyo Gas Company	907	Tokyo
Tokyo Gas Company, Omari Branch	1398	Tokyo
Tokyo Gas Company, Sunamachi Branch	907	Tokyo

VII. The XXI Bomber Command

Target	Target #	Location
Tokyo Gas Company, Tsurumi Branch	481	Kawasaki
Tokyo Industrial — Aircraft Engines (code name: Brooklyn)		Tokyo
Tokyo Industrial Area		Tokyo
Tokyo Kazai Company	1350	Tokyo
Tokyo Measuring Instrument Works	919	Tokyo
Tokyo Municipality Steram Plant	1345	Tokyo
Tokyo Nakayama Iron Works	1351	Kawasaki
Tokyo Rolling Works	1767	Amagasaki
Tokyo Shibaura Electric Company Factory #1	496	Tokyo
Tokyo Shibaura Electric Company Factory #2	488	Tokyo
Tokyo Shibaura Electric Company Oguni Plant	1664	Oguni
Tokyo Shibaura Ellectrical Mfg Co	1814	Kuwana
Tokyo Special Chrome Steel Works	58	Kawasaki
Tokyo Special Machinery Mfg Company	1357	Tokyo
Tokyo Steam Power	493	Kawasaki
Tokyo Urban Area (code name: Arrange)		Tokyo
Tokyo Urban Area (code name: Meetinghouse)		Tokyo
Tomioka Seaplane Base	1403	Yokohama
Tomitaki Air Field (code name: Skewer)	2536	Kyushu
Tomobe Navy Airdraome	1489	Kasumiguara
Toride Railroad Bridge	1479	Mito
Torimatsu Factory, Nagoya Arsenal	200	Nagoya
Toshin & Shinko Warehouses	1493	Yokohama
Tosu Railroad Yards and Repair Shops	1871	Tosu
Totsuka Air Base	82	Totsuka
Toyama Urban Area		Toyama
Toyami Hydro Plant	1589	
Toyo Aluminum Co	1877	Miike
Toyo Bearing Co, Dojima Plant	1815	Kuwana
Toyo Bearing Co., Uchibori Plant	1816	Kuwana
Toyo Bearing Mfg Co	1198	Takarazuka
Toyo Industry	1890	Hiroshima
Toyo Metal and Wood Co	1707	Itami
Toyo Oil Refinery	257	Osaka
Toyo Soda Compay	1882	Tonda
Toyo Steel Plate Co	1883	Kudamatsu
Toyoda Auto Works	1140	Kariya
Toyoda Auto Works, Koromo Plant	1139	Koromo
Toyoda Machinery Mfg Co	430	Nagoya
Toyoda Steel Plant	1830	Yokosura
Toyohashi Army Base and Arsenal	1224	Toyohashi
Toyohashi Urban Area		Toyohashi
Toyokawa Arsenal	1653	Toyokawa
Toyomi Hydro Plant	1589	Agano River
Toyowa Heavy Industries Co.	1703	Osaka
Toyowa Heavy Industries, Nishi Plant	1735	Nagoya
Toyowa Heavy Industries, Shinkawa Plant	1141	Nagoya
Truk Airfield		Truk Atoll
Truk Submarine Base		Truk Atoll
Tsu Urban Area		Tsu
Tsuchiura Navy Airdraome	1473	Kasumiguara
Tsuchiya Tabi Co	1266	Kurume
Tsuchizaki Oil Refinery		Tsuchizaki
Tsudanuma Airdrome	1472	
Tsudanuma Airdrome	1490	
Tsugami-Atagi Mfg Company	1656	Nagaoka
Tsukiji Market/Wholesale Warehouses	1448	Tokyo
Tsukishima East Wharf	1436	Tokyo
Tsukishima Electrical Engineering Company	350	Tokyo
Tsukishima Pier	1435	Tokyo
Tsukuba Navy Airdrome	1472	Kasumiguara
Tsurama Soda	1396	Yokohama
Tsuruga Harbor Facilities	1950	Tsuruga
Tsuruga Railroad Yards	1676	Tsuruga
Tsuruga Urban Area		Tsuruga

Target	Target #	Location
Tsurumi Army and Navy Stores	1337	Yokohama
Tsurumi Shunting Yards	112	Tsurumi
Tsurumi Steel and Shipbuilding Company	122	Yokohama
Ube Cement Co	822	Ube
Ube Coal Liquidification Company	1841	Ube
Ube Industrial Area		Ube
Ube Iron Works	1878	Ube
Ube Nitrogen Fertilizer Co	818	Ube
Ube Soda Co	1844	Ube
Ube Steam Power Plant #1	827	Ube
Ube Steam Power Plant #2	1626	Ube
Ube Urban Area		Ube
Ueno Railroad Station	369	Tokyo
Uibar Oil Company	359	Tokyo
Uji Hydro Electric Plant	1155	Uji River
Uji Powder Factory	1169	Uji
Ujina Shipbuilding	1889	Ujina Island
Ujiyamada Urban Area		Ujiyamada
Umekoje Freight Yards	1159	Kyoto
Unidentified	X13058	Yokohama
Unidentified	X13069	Yokohama
Unidentified	X13070	Kawasaki
Unidentified	X13075	Yokohama
Unidentified Arms Plant	142	Kawasaki
Unidentified Bldg — Tokyo Arsenal	3006	Tokyo
Unidentified Bldg — Tokyo Arsenal	3007	Tokyo
Unidentified Bldg — Tokyo Arsenal	3109	Tokyo
Unidentified Bldg — Tokyo Arsenal	X13004	Tokyo
Unidentified Bldg — Tokyo Arsenal	X13008	Tokyo
Unidentified Bldg — Tokyo Arsenal	X13062	Tokyo
Uno Harbor	1289	Uno
Uraga Dockyard #1	1460	Yokosuka
Uraga Dockyard #2	1461	Yokosuka
Uraga Dockyard #3	71	Yokohama
Urawa Airport	1414	Tokyo
Usa Air Field (code name: Blowzy)	1307	Kyushu
Utsonomiya Air Training School & Arsenal	1643	Utsonomiya
Utsonomiya Urban Area		Utsonomiya
Utsube River Oil Refinery		
Uwajima Urban Area		Uwajima
Wakamatsu Port	32	Wakamatsu
Wakamatsu Railroad Shop	561	Wakamatsu
Wakayama Urban Area		Wakayama
Watanabe Aircraft Plant	662	Zasshonoguma
Watanabe Iron Works Plant No 2	1238	Fukuoka
Watanabe Steel Works	328	Tokyo
Yahagi Electro-Chemical Plant	255	Nagoya
Yahagi Steel Plant	1829	Nagoya
Yamagawara Hydro Electric Plant	876	Kurobe River
Yamamoto Heavy Industries	1779	Yokkaichi
Yamamoto Heavy Industry	1779	Kuwana
Yamato Steel Works	1774	Osaka
Yanagawara Hydro-Electric Plant	876	Kurobe River
Yao Transformer Station	1631	Yao
Yasuoka Hydro-Electric Plant	1591	Tenryu River
Yatabe Navy Airdrome	1471	Kasumiguara
Yatsuyama Mine	1229	Unknown
Yatzuzawa Hydro-Electric Plant	1515	Sagomi River
Yawata Steel Works		Yawata
Yawata Urban Area		Yawata
Yokkaichi Harbor	1737	Yokkaichi
Yokkaichi Industrial/Oil Refinery		Yokkaichi
Yokkaichi Urban Area		Yokkaichi
Yokohama Central Wholesale Warehouse	1439	Yokohama
Yokohama Dockyard Of Mitsubishi Heavy Industry	69	Yokohama

Target	Target #	Location
Yokohama Electric Light Co Power Plant	1346	yokohama
Yokohama Harbor Coop Purchasing Warehouse	1445	Yokohama
Yokohama Harbor Passenger Wharves	1432	Yokohama
Yokohama Harbor Timber Basin	1444	Yokohama
Yokohama Main Pier	1433	Yokohama
Yokohama Naval Air Station	1402	Yokohama
Yokohama Station, Junction, Overpass	67	Yokohama
Yokohama Urban Area		Yokohama
Yokohama Warehouse	1441	N. Yokohama
Yokohama Warehouse	1442	S. Yokohama
Yokohama Rubber Company	141	Unknown
Yokose Oil Storage	1835	Yokose
Yokosuka Arsenal	282	Yokosuka
Yokosuka Base Headquarters	1452	Yokosuka
Yokosuka Experimental Labs/Ordnance Plants	276	Yokosuka
Yokosuka Gunnery and Navigation Schools	1453	Yokosuka
Yokosuka Mine and Aircraft Stores	295	Yokosuka
Yokosuka Naval Aircraft Factory	1392	Yokosuka
Yokosuka Naval Barracks	278	Yokosuka
Yokosuka Naval Base Ship Yard	274	Yokosuka
Yokosuka Naval Radio	1458	Yokosuka
Yokosuka Old Aircraft Center	1486	Yokosuka
Yokosuka Railroad Station Oil Tanks	295	Yokosuka
Yokosuka Refittings Berths — Battleships/Cruiser	277	Yokosuka
Yokosuka Torpedo and Wireless Schools	1456	Yokosuka
Yomikaki Hydro-Electric Plant	1502	Kiso River
Yotsuyama Mine	1279	Omuta
Yuasa Storage Battery Mfg	1717	Takasuki

Targets Listed by Number

Target #	Target	Location
4	Minatogawa Steam Power Plant	Kobe
5	Kobe Steel Works	Kobe
6	Kobe Steel Works	Kobe
7	Kawasaki Heavy Industry	Kobe
11	Kawasaki Locomotive & Car Co.	Kobe
14	Nippon Dunlop Rubber Co	Kobe
17	Rising Sun Petroleum Oil Storage	Kobe
22	Kobe Harbor District No 1	Kobe
24	Minsukasa Coke Plant	Wakamatsu
28	Japan Iron Works, Yawata Plant	Yawata
29	Japan Iron Works, Tobata Plant	Tobata
30	Tobata Foundry Company	Tobata
32	Wakamatsu Port	Wakamatsu
33	Port Of Tobata	Tobata
34	Kobe Harbor District No 2	Kobe
39	Asano Portland Cement Co	Moji
40	Asahi Oil Refinery	Hiroshima
42	Shimonoseki Port	Shimonoseki
44a	Moji Central Wharf	Moji
44b	Moji Southern Wharf	Moji
44c	Moji Coaling Station	Moji
46	Kobe Steel Works — Moji Plant	Moji
49	Moji Ordnance Storage	Moji
51	Asano Iron Works	Kawasaki
52	Japanese Steel Tube Company	Kawasaki
53	Osaka Harbor	Osaka
56	Fuji Steel Works Of Imperial Iron And Steel	Kawasaki
57	Oriental Steel Products Co	Kawasaki

Target #	Target	Location
58	Tokyo Special Chrome Steel Works	Kawasaki
62	Hodagaya Substation	Yokohama
63	Kamagawa Steam Power Plant	Yokohama
66	Takeshina Freight Yards	Yokohama
67	Yokohama Station, Junction, Overpass	Yokohama
68	Higashi Yokohame RR Yards	Yokohama
69	Yokohama Dockyard Of Mitsubishi Heavy Industr	Yokohama
70	Asano Dockyard	Yokohama
71	Uraga Dockyard #3	Yokohama
72	Ford Motor Company	Yokohama
73	Tokuyama Naval Fueling Station (code name: Indices)	Tokuyama
75	Furakawa Electric Co., Plant No 1	Yokohama
78	Nirada Firecracker Co	Yokohama
80	Niigata Sulphuric Acid Company	Kawasaki
82	Totsuka Air Base	Totsuka
87	Ogura Oil Company	Yokohama
88	Japan American Oil Company	Yokohama
89	Eastern Petroleum Refinery	Yokohama
90	Oriental Oil Company	Yokohama
92	Kanagawa Gasoline Tanks	Yokohama
92	Lighthouse Bureau Oil Tanks	Yokohama
94	Matsukata Oil Storage	Yokohama
99	Shimonoseki-Maeda Power Plant	Shimonoseki
102	Asahi Substatiion	Yokohama
104	Nippon Electric Power Transforming Station	Yokohama
105	Kawasaki Substation #1	Kawasaki
106	Railroad Transformer Station	Kawasaki
110	Tokyo Elec Power Generating Sta, Tsurumi Plant	Kawasaki
111	Kawasahe Elec Pwr Plant For Govt Railroad	Kawasaki
112	Tsurumi Shunting Yards	Tsurumi
116	Mitsubishi Oil Refinery	Kawasaki
117	Mitsubishi Oil Company & Chiyoda Oil Tanks	Yokohama
122	Tsurumi Steel And Shipbuilding Company	Yokohama
127	Hayama Petroleum Refinery	Kawasaki
128	Petroleum Center	Kawasaki
129	Nippon Super Fuel Co	Yokohama
130	Asahi Petroleum Co	Kawasaki
132	Kawasaki Naval Wireless Sta	Kawasaki
133	Shibaura Engineering Works	Yokohama
134	Asahi Glass Company	Yokohama
135	Cooperative Auto Works	Kawasaki
136	Fuji Elec Works	Kawasaki
137	Showa Fertilizer	Kawasaki
141	Yokohame Rubber Company	Unknown
142	Unidentified Arms Plant	Kawasaki
147	Aikoku Petroleum Refinery	Kawasaki
153	Senkari Reservoir Dam	Sendai
157	Sannomiya Railroad Station	Kobe
165	Kokura Steel Works	Kokura
168	Kokura Arsenal	Kokura
169	Kamizaki Heavy Industry	
169	Mitsubishi Heavy Industry	Kobe
171	Kanazawa Heavy Industry	
184	Kokura Railroad Shops	Kokura
184	Kokura Steam Power Plant #2	Kokura
191	Mitsubishi Coal Depot	Kawasaki
193	Mitsubishi Aircraft Engine Works (code name: Memphis)	Nagoya
194	Mitsubishi Aircraft Works, Nagoya (code name: Hesitation)	Nagoya
195	Nagoya #1 Steam Power Plant	Nagoya
196	Kureha Textile Mill	Nagoya
196	Nagoya Arsenal, Chigusa Factory	Nagoya
197	Atsuta Factory, Nagoya Arsenal	Nagoya
197	Nagoya Arenal, Atsuda Plant	Nagoya
198	Aichi Aircraft Works, Atsuta Plant	Nagoya
199	Aichi Aircraft Works, Mizuho Branch	Nagoya

VII. The XXI Bomber Command

Target #	Target	Location
273a	Fujinagata Shipbuilding Co	Osaka
273b	Fujinagata Shipbuilding Co	Osaka
200	Torimatsu Factory, Nagoya Arsenal	Nagoya
201	Army Central Clothing Depot, Tokyo Arsenal	Tokyo
201	Ordnance Storehouse Amunition	Oji Ward Tokyo
202	Army Central Clothing Depot, Tokyo Arsenal	Oji Ward Tokyo
202	Asahi Electrochemical Company	Tokyo
203	Ordnance Supply Depot, Tokyo Arsenal	Oji Ward Tokyo
204	Japan Artificial Fertilizer	Tokyo
205	Itabashi Powder Plant, Tokyo Arsenal	Tokyo
206	Oji Army Arsenal	Oji Ward Tokyo
207	Military Gunpowder Works, Tokyo Arsenal	Tokyo
208	Army Branch Powder Co	Oji Ward Tokyo
208	Atsuta Factory, Nagoya Arsenal	Tokyo
209	Military Works	Oji Ward Tokyo
209	Tkinogawa Army Arsenal	Tokyo
211	Inawashiro Transformer Station	Tokyo
212	Asahi Electrochemical Company	Tokyo
212	Briquette Plant	Tokyo
213	Kinugawa Steam Power Station	Tokyo
214	Tokyo Armory	Tokyo
217	Powder Factory, Tokyo Arsenal	Oji Ward Tokyo
218	Ordnance Storehouse, Tokyo Arsenal	Oji Ward Tokyo
219	Powder Magazine, Tokyo Arsenal	Oji Ward Tokyo
224	Tabata - Nippori Railroad Yard	Tokyo
228	Chiba Urban Area	Tokyo
228	Government Steam Power Station	Tokyo
230	Senju Steam Power Plant	Tokyo
240	Kawasaki Aircraft Works, Kagamigahara Plant	Gifu
241	Nippon Vehicle Manufacturing Company	Nagoya
242	Okamoto Aircraft Works, Showa Plant	Nagoya
247a	Diado Electric Steel Oe Plant	Nagoya
247b	Diado Electric Steel Arsuta Plant	Nagoya
247c	Diado Electric Steel Tsukiji Plant	Nagoya
247d	Diado Electric Steel Minami Plant	Nagoya
249	Kagamigahara Military Airport	Gifu
250a	Nagoya Freight Yards	Nagoya
250b	Nagoya Repair Shops	Nagoya
250c	Nagoya Station	Nagoya
251	Nagoya Harbor Facilities	Nagoya
253	Nissan Chemical Plant	Nagoya
254	Mitsubishi Electrical Mfg Co	Nagoya
255	Yahagi Electro-Chemical Plant	Nagoya
257	Maruzen Oil Refinery	Osaka
257	Toyo Oil Refinery	Osaka
258a	Mitsubishi Oil Storage	Osaka
258b	Mitsubishi Oil Storage	Osaka
260	Rising Sun Petroleum Oil Storage	Osaka
261	Nippon Oil Storage	Osaka
261	Standard Vacuum Oil Storage	Osaka
262	Sumitoma Electric Industries Co	Osaka
263a	Sumitoma Metal Industry	Osaka
263b	Sumitoma Aluminum Company	Osaka
264	Sumitoma Metal Industry	Osaka
268	Osaka Steel Mfg Co	Osaka
272	Osaka Iron Works, Unit No 1	Osaka
274	Yokosuka Naval Base Ship Yard	Yokosuka
276	Yokosuka Experimental Labs/Ordnance Plants	Yokosuka
277	Yokosuka Refittings Berths - Battleships/Cruiser	Yokosuka
278	Yokosuka Naval Barracks	Yokosuka
282	Yokosuka Arsenal	Yokosuka
295	Yokosuka Mine And Aircraft Stores	Yokosuka
295	Yokosuka Railroad Station Oil Tanks	Yokosuka
297	Hako Point Oil Tanks	Yokosuka
298	Oppama Naval Air Station	Yokosuka

Target #	Target	Location
304	Chiran Airfield	Futtsu Point
304	Futtsu Point Forts	Fukayama
315	Three Powder Magazine	Yokohama
321	Tokyo Gas & Electric Engineering, S. Tokyo	Tokyo
322	Kasugade No 1 Steam Power Plant	Osaka
323	Kasugade No 2 Steam Power Plant	Osaka
324	Fukuszaki Steam Power Plant	Osaka
325	Kizugawa Steam Power Plant	Osaka
326	Nippon Electric Factory #2	Tokyo
327	Mitsubishi Aircraft Shibaura Plant	Tokyo
328	Watanabe Steel Works	Tokyo
330	Ishikawajima Dockyard Of Ishikawajima Shipbuild	Tokyo
330	Susaki Dock Yard Of Ishikawajima Shipbuilding	Tokyo
332	Nakajima Seaplane Works	Tokyo
334	Nisso Steel Mfg Company	Tokyo
335	Chosi Army Airdrome	Tokyo
336	Japan Special Steel Company	Tokyo
337	Haneda Airport	Tokyo
350	Tsukishima Electrical Engineering Company	Tokyo
352	Tokyo Army Arsenal & Engineering School	Tokyo
354	Shibaura Machine Tool Company	Tokyo
355	Army Provisions Depot	Tokyo
356	Nakajima Aircraft Ogikubo Plant	Tokyo
357	Mushishino Aircraft Plant	Tokyo
357	Nakajima Aircraft Company (code name: Enkindle)	Musashino
357	Nakajima Aircraft Company, Musashino Plant (code name: Enkindle)	Tokyo
359	Uibar Oil Company	Tokyo
360	Edogawa Petroleum Refinery	Tokyo
363	Radio Station Joak	Tokyo
364	Shinagawa Railroad Yards	Tokyo
365	Shiodome Freight Yards	Tokyo
366	Steam Engineering & Rolling Stock Company	Tokyo
367	Tokyo Central Station	Tokyo
368	Aeronautical Instruments	Tokyo
369	Ueno Railroad Station	Tokyo
370	Oi Railroad Works	Tokyo
371	Tateyama Naval Air Station	Airports
373	Kisarazu Naval Air Station	Tokyo
382	Osaka Arsenal	Osaka
383	Dai Nippon Celluloid Co	Sakai
388	Osaka Machinery Company	Osaka
430	Toyoda Machinery Mfg Co	Nagoya
430b	Osaka Iron Works	Hiroshima
456	Toho Gas Works Synthetic Oil Plant	Nagoya
461	Okuma Iron Works, Ozone Plant	Nagoya
467	Showa Soda Plant	Nagoya
470	Tokyo Electric Power Station	Kawasaki
475	Asano Cement Company	Kawasaki
479	Japan Artificial Fertilizer	Kawasaki
481	Tokyo Gas Company, Tsurumi Branch	Kawasaki
484	Showa Wire Cable Company	Kawasaki
485	Dejima Wharf	Unknown
485	Fuji Spinning Mills	Kawasaki
487	Meiji Sugar Factory	Kawasaki
488	Tokyo Shibaura Electric Company Factory #2	Tokyo
489	Nippon Electric Wire & Cable Company	Kawasaki
493	Tokyo Steam Power	Kawasaki
494	Tokyo Electric Wire & Mfg Company	Kawasaki
496	Tokyo Shibaura Electric Company Factory #1	Tokyo
497	Nippon Electric, Factory #1	Kawasaki
498	Kurada Iron Works	Yokohama
499	Nippon Carbon Company, Factory #1	Yokohama
554	Kakusa Kagyo Foundry	Wakamatsu
555	Tokai Steel Works	Wakamatsu
558	Tochigi Shipbuilding Yard	Wakamatsu

VII. The XXI Bomber Command

Target #	Target	Location
561	Railroad Shop	Wakamatsu
359	Nakajima Aircraft Tanishi Foundry Plant	Tokyo
516	Powder Magazine	Yokohama
521	Japan (Showa) Elctro-Chemical Company	Yokohama
522	Nissan Auto Company	Yokohama
523	Japan Radio And Phonograph Company	Yokohama
536	Amagasaki Steam Power Plant	Amagasaki
538	I.G.R. Shops	Taketori
540a	Kansi Kyodo Steam Power Plant No 1	Amagasaki
540b	Kansi Kyodo Steam Power Plant No 2	Amagasaki
541	Oriental Otis Elevator, S. Tokyo	Tokyo
542	Akunoura Engine Works	Nagasaki
543	Mitsibishi Dock	Nagasaki
543	Mitsubishi Dockyard	Nagasaki
544	Tatigami Shipyard	Nagasaki
545	Kozaki Point Oil Storage	Nagasaki
546	Mitsibishi Steel And Arms Works	Nagasaki
548	Rolling Stock Mfg Co	Osaka
561	Wakamatsu Railroad Shop	Wakamatsu
567	Asahi Glass Company	Tobata
569	Kyushu Chemical Industry	Yawata
573	Migata Engineering Works, S. Tokyo	Tokyo
573	Miyata Engineering Works	Tokyo
660	Hiro Naval Aircraft Factory (code name: Thunderhead)	Kure
656	Kure Naval Air Station	Hiro
657a	Kure Naval Arsenal	Kure
657b	Kure Naval Shipyard	Kure
657c	Kure Torpedo Boat Base & Mine Depot	Kure
657d	Kure Provisions And Clothing Depot	Kure
658	Kure Submarine Base	Kure
659	Hitonose Oil Storage	Yeta Shima
660	Hiro Naval Shipyard, Dock & Arsenal	Hiro
661	Amatsuji Steel Ball Mfg Co	Osaka
662	Watanabe Aircraft Plant	Zasshonoguma
663	Fukuoka Air Station	Fukuoka
664	Najima Steam Power Plant	Fukuoka
665	Saito Zaki Petroleum Storage	Fukuoka
668	Kasado Dock Co	Kasado Is
669	Hanao Wake Oil Storage	O Shima
670	Japan Paraffin Mfg Co	O Shima
671	Hikari Naval Arsenal	Tokuyama
672	Nippon Oil Co; Kudamatsu Plant Refinery	Kudamatsu
673	Tokuyama Naval Fueling Station	Tokuyama
674	Diado Electric Steel Tsukiji Plant	Tokuyama
674	Tokuyama Coal Yard (code name: Rotative)	Tokuyama
675	Tokuyama Soda Co	Tokuyama
685	Kurimoto Iron Works	Osaka
686	Osaka Metals Industries Company	Osaka
687	Osaka Wakayama Iron Works	Osaka
688	Kubato Iron Machinery Works	Osaka
697	Mitsubishi Copper Refining	Osaka
699	Osaka Iron Works, Unit No 2	Osaka
752	Sasebo Naval Dockyard	Sasebo
754	Sasebo Naval Air Station	Sasebo
755	Sasebo Oil Storage	Sasebo
757	Sasebo Mine & Torpedo Storage	Sasebo
758	Sasebo Naval Arsenal	Sasebo
759	Sasebo Railroad Station	Sasebo
762	Sasebo Fuel & Munition Storage	Sasebo
701	Kubato Iron Machinery Works	Osaka
713	Nakayama Steel Mfg Co	Osaka
730	Hiro Naval Engine And Turbines Factory	Hiro
735	Army Transportation Base	Ujina Island
736	Army Ordnance Depot	Hirishima
737	Army Clothing Depot	Hiroshima

Target #	Target	Location
737	Army Food Depot	Ujina Island
740	East Hiroshima Rail Road Station	Hiroshima
748	Army Division Headquarters	Hiroshima
785	Kanegafuchi Soda Industry	Kobe
789	Nakajima Aircraft Company-Ota Parts Plant	Ota
791	Showa Aircraft	Tokyo
792	Tachikawa Aircraft Plant (code name: Modeller)	Tachikawa
793	Fukuda Light Aeroplane Co	Amagasaki
794	Hikari Arsenal (code name: Thunderhead)	Tokuyama
794	Hiro Arsenal	Hiro
796	Saka Steam Power Plant	Sakai
798	Kure Railroad Station	Kure
799	Mitsubishi Aircraft, Oimochi Plant	Tokyo
803	Matsuo Engine Works	Nagasaki
829	Mitsubishi Electrical Mfg Co	Nagasaki
834	Sasebo Aircraft Factory	Sasebo
844	Sasebo Coal Yard	Sasebo
845	Sasebo Outfitting Wharf	Sasebo
859	Spinning And Weaving Mill	Nagasaki
860	Koyagi Shima Shipyard, Matsu Iron Works	Nagasaki
802	Mitsubishi Aircraft Works	Unknown
810a	Eitoku Industrial/Aircraft Factory	Tsudo
810a	Furukawa Copper Ore Dressing Plant	Tsudo
810b	Eitoku Plant	Honzan
810b	Furukawa Smelter (code name: Treadle)	Kyushu
810c	Electric Instrument Plant	Moto
810c	Furukawa Machine Shops	Mato
811	Electro-Chemical Industry	Kiyotaki
811	Furuawa Copper Refinery	Kiyotaki
812	Hitachi Copper Refineery	Sakagawa
814	Sumitomo Copper Smelter	Mino Is
815	Sumitomo Copper Refinery	Niihawa
818	Ube Nitrogen Fertilizer Co	Ube
819	Nissan Chemical Industry Co, Onoda Plant	Onoda
820	Japan Powder Mfg Co Asa Plant	Asa
822	Onada Cement Co	Onada
822	Ube Cement Co	Ube
825	Hitachi Mfg Co Kasado Plant	Kudamatsu
826	Electro-Chemical Industry, Aomi Plant	Tokyo
827	Ube Steam Power Plant #1	Ube
828	Hayashi Commercial Company Engine Works	Nagasaki
828	Matsuo Engine Works	Unknown
832	Megami Point Oil Storage	Nagasaki
835	Nagasaki Wharves	Nagasaki
836	Fiji Spinning Mills	Nagasaki
838	Isahaya Railroad Junctions	Isahaya
842	Taka Shima Colliery	Taka Shima
843	Ha Shima Colliery	Ha Shima
845	Sasebo Provisions Wharf	Sasebo
847	Nagasaki Steam Power Plant	Nagasaki
849	Omura Naval Air Station (code name: Vamoose)	Omura
851	Nagasaki Spinning Mill	Nagasaki
860	Kawanami Industry Co Shipyard	Koyagi Shima
861	Nichiman Aluminum Company	Higashi-Iwase
862	Higashi-Iwase Harbor	Higashi-Iwase
866	Nippon Soda Company, Aluminum Plant	Takaoka
867	Fujikura Electric Cable Works	Fushiki
867	Fushiki Harbor	Fushiki
870	Nanao Harbor	Nanao
871	Kanazawa Railroad Shops	Matsuto-Machi
873	Komaki Hydro Electric Plant	Sho River
874	Kanidera Hydro Electric Plant	Jintsu River
875	Soyama Hydro Electric Plant	Sho River
876	Yanagawara Hydro-Electric Plant	Kurobe River
877	Kurobe #2 Hydro Electric Plant	Kurobe River

VII. The XXI Bomber Command

Target #	Target	Location
880	Kanishi Photo Works	Tokyo
881	Inawashiro Hydro Plant #1	Nipasshi River
883	Ikegai Motor Plant	Tokyo
885	Takeshiba Pier	Tokyo
886	Hinode-Cho Pier	Tokyo
887	Shibaura Wharf	Tokyo
888	Oki Electric Company #1	Tokyo
894	Imperial Government Railways Shops	Omiya
898	Japan Military Goods Co., Explosives Plant	Tomioka
899	Japan Military Goods Company	Tomioka
1943	Fujikoshi Steel Products, Higashi-Iwase Plant	Higashi-Iwase
900	Inawashiro Hydro Plant #2	Nipasshi River
904	Sakurada Engineering Works	Tokyo
907	Tokyo Gas Company, Sunamachi Branch	Tokyo
910	Rising Sun Petroleum Terminal	Tokyo
911	Ogura Oil Company	Tokyo
912	Nesshin Spinning Mill	Tokyo
913	Hattori Company	Tokyo
914	Oriental Weaving Company	Tokyo
916	Hitachi+A91 Engineering Kameido Plant	Tokyo
918	Niigata Iron Works	Niigata
919	Tokyo Measuring Instrument Works	Tokyo
922	Riken Metal Company, Ube Plant	Ube
923	Sumitomo Chemical Co	Niihama
924	Sumitomo Aluminum Reduction Plant	Niihawa
925	Niihama Copper Concentrating Mill	Niihama
927a	Haba Shipyard	Innoshima Is
927b	Chigiri Shima Copper Smelter	Innoshima Is
927b	Shannosho Shipyard	Innoshima Is
931	Niihama Steam Power Plant	Niihama
932	Sumitomo Machinery Works	Niihawa
934	Itozaki Oil Storage	Itozaki
936	Nissan Chemical Plant #1	Sasakura
939	Haiki Railroad Terminal	Haiki
941	Fujikoshi Steel Products, Toyama Plant	Toyama
998	Niigata Iron Works Plant #1	Niigata
999	Niigata Iron Works Plant #2	Niigata
1000	Nippon Oil Refinery	Niigata
1003	Japan Light Metals Company	Niigata
1012	Niigata Oil Storage #1	Niigata
1013	Niigata Oil Storage #2	Niigata
1023	Niigata Railroad Yards	Niigata
1041	Maizura Naval Base	Maizura
1051	Takasaki Railroad Yards	Takasaki
1059	Sakai Hydro-Electric Plant	Shibukawa
1077	Aichi Aircraft Eitoku Plant	Fukushima
1077	Fukushima Railroad Yards	Fukushima
1088	Japan Refining Company	Koriyama
1098	Nagano Government Railroad Shops	Nagano
1123	Nissan Liquid Fuel Company	Unknown
1100	Electro-Chemical Industry	Aomi
1100	Showa-Denko Aluminum	Omachi
1101	Aichi Aircraft Works,	Aomoi
1104	Nagamachi Marshalling Yards	Sendai
1129	Akano Military Aviation School & Field	Nagoya
1129	Okamoto Aircraft Works, Kosagara Plant	Nagoya
1130	Suzuka Naval Air Station	Karbe
1133	Inasawa Shunting Yards	Inaswaw
1138	Nippon Explosives Works	Taketoya
1139	Toyoda Auto Works, Koromo Plant	Koromo
1140	Toyoda Auto Works	Kariya
1141	Toyowa Heavy Industries, Shinkawa Plant	Nagoya
1143	Iwakura Transformer Station	Iwakura
1144	Seto Hydro-Electric Station	Shimohara
1146	Okuma Iron Works, Hagino Plant	Nagoya

Target #	Target	Location
1147	Okuma Iron Works, Nunoike Plant	Nagoya
1153	Japan Porcelain Company	Nagoya
1155	Uji Hydro Electric Plant	Uji River
1159	Umekoje Freight Yards	Kyoto
1160	Maibara Railroad Yards	Maibara
1169	Uji Powder Factory	Uji
1171	Nippon Insulator Company	Nagoya
1176	Japan Light Metals Aluminum Plant	Shimizu
1177	Japan Light Metals Aluminum Plant	Kamabara
1181	Numazu Railroad Yards	Numazu
1191	Hatsudoki Engine Works	Osaka
1191	Hitachi Engine Works	
1195	Sumitomo Metal Industry	Amagasaki
1197	Kubota Iron & Machinery Works	Amagasaki
1198	Toyo Bearing Mfg Co	Takarazuka
1229	Yatsuyama Mine	Unknown
1243	Miike Dyestuffs	Omuta
1244	Oriental High Tension Company, Plant A	Tokyo
1245	Oriental High Tension Company, Plant B	Tokyo
1246	Aeronautical Experimental Laboratory	Tokyo
1249	Omuta Plant, Kyushu Power Company	Omuta
1254	Miike Harbor	Omuta
1260	Mitsui Zinc Distilling Plant	Omuta
1261	Mitsui Electrolytic Zinc Refinery	Unknown
1262	Mitsui Coal Liquidification Plant	Omuta
1267	Miike Coal Yards	Miike
1268	Mitani Railroad Workshops	Mitani
1273	Omuta Mill Kanegafuchi Spinning Company	Omuta
1275	Manda Mine	Omuta
1203	Nippon Oil Refinery & Tank Farm	Amagasaki
1206	Osaka Railroad Station	Osaka
1209	Suita Railroad Yars And Shops	Suita
1213	Akano Military Aviation School & Field	Uji Yamada
1214	33rd Infantry Regiment And Arsenal	Hisai - Tsu
1215	Kobe Steel Works	Toba
1216	Toba Dockyard	Tobata
1217	Kameyama Rr Junction	Kameyama
1218	Matsuzaka Junction And Round House	Matsuzaka
1219	Japan Musical Instrument Co Propeller Plant	Hamamatsu
1224	Toyohashi Army Base And Arsenal	Toyohashi
1227	Suzuki Loom Works	Hamamatsu
1230	Imperial Government Railways Shops	Hamamatsu
1236	Tachiari Air Field (code name: Fearless)	Tachiari
1237	Najima Seaplane Base	Fukuoka
1238	Watanabe Iron Works Plant No 2	Fukuoka
1239	Kyushu Ordnance Plant	Zasshonoguma
1241	Miyata Motor Works	Zasshonoguma
1242	Sangyo Cement Co	
1246	Electro-Chemical Industry Co	
1247	Kurume Military Zone	Kurume
1248	Kumamoto Military Zone	Kumamoto
1249	Omuta #1 Steam Power Plant	Omuta
1255	Hakate Harbor	Fukuoka
1257	Kobokuro Machinery Works	Nogata
1259	Togami Electric Works	Saga
1261	Mitsui Electrolytic Zinc Refinery	Omuta
1263	Bridgestone Tire Company	Kurume
1264	Nippon Rubber Company	Kurume
1266	Tsuchiya Tabi Co	Kurume
1268	Mitsui Rr Workshops	Omuta
1269	Kurume Railroad Station	Kurume
1270	Hakata Railroad Yards And Station	Fukuoka
1271	Kineshima Mine	Saga
1274	Dainoura Mine	Nogata
1275	Tagawa Mines	Nogata

VII. The XXI Bomber Command

Target #	Target	Location
1276	Futase Mines	Nogata
1278	Manda Mine	Omuta
1279	Yotsuyama Mine	Omuta
1281	Dai Nippon Celluloid Co	Abashi
1283	Sanyo Steam Power Plant No 3	Shikawa
1284	Chugoku Steam Power Plant	Okayama
1286	Sanyo Steam Power Plants No 1 And 2	Shikawa
1289	Uno Harbor	Uno
1290	Japan Iron Works, Hirohato Plant	Hirohato
1295	Mitsui Tame Shipyard	Tamashima
1296	Harima Shipyard	Oo
1297	Mitsubishi Copper Smelter/Zinc Refinery	Nao Island
1303	Tadotsu Railroad Yards And Repair Shops	Tadotsu
1306	Saeki Air Field (code name: Cockrow)	Kyushu
1307	Oita Air Field (code name: Camlet)	Kyushu
1307	Usa Air Field (code name: Blowzy)	Kyushu
1308	Oita Air Field (code name: Fearless)	Oita
1310	Asahi Bemberg Gunpowder Plant	Nobeoka
1311	Nippon Nitrogen Explosives Nobeoka Plant	Kawashima
1312	Nippon Nitrogen Explosives Raiken Plant	Nobeoka
1314	Asahi Bemberg Cupraammonium Plant	Nobeoka
1317	Nippon Dye Works	Tsurusaka
1328	Saganoseki Copper Works	Saganoseki
1329	Oita Railroad Yards	Oita
1331	Kayaba Engineering Company	Tokyo
1332	Rolling Stock Company	Tokyo
1333	Nippon Typewriter Co	Tokyo
1334	Asano Karite Co Ltd	Yokohama
1335	Naval Gunpowder Works, Tokyo Arsenal	Oji Ward Tokyo
1336	Navy Arsenal	Hiratsuka
1337	Tsurumi Army And Navy Stores	Yokohama
1338	Inflammable Storage Docks	Yokohama
1339	Ammunition Storage	Yokohama
1340	Harley-Davidsen Motorcycle Co	Tokyo
1341	Rikuo Engine Company	Tokyo
1342	Japan Machine Industry	Tokyo
1345	Tokyo Municipality Steram Plant	Tokyo
1346	Yokohama Electric Light Co Power Plant	Yokohama
1347	Japan Service Company	Kawasaki
1348	Toho Steel Foundry Co	Tokyo
1349	Japan Special Steel Works	Tokyo
1350	Tokyo Kazai Company	Tokyo
1351	Tokyo Nakayama Iron Works	Kawasaki
1352	Steel Plants Ne Joto Ward	Tokyo
1353	Niigata Iron Works	Niigata
1354	Oriental Babcock Company	Yokohama
1355	Japan Steel Bearing Company	Tokyo
1356	Japan Streel Bearing Company Fujisawa Plant	Fujisawa
1357	Japan Physico-Chemical Co	Tokyo
1357	Tokyo Special Machinery Mfg Company	Tokyo
1358	Riken Heavy Industries Co	Tokyo
1360	Ishii Iron Works	Tokyo
1361	Kokusan Machinery Company	Tokyo
1362	Kitashi Electrical Engineering Company	Tokyo
1363	Shoun Engineering Co	Ofuna
1364	Physico Chemical Industries	Tokyo
1365	Japan Bakelite Company	Tokyo
1366	Nippon Oil Company	Tokyo
1367	Tama River Bridges & Tokado/Keihin Elec Lines	Kawasaki
1368	Akabane Railroad Bridge	Tokyo
1369	Tama River Bridges & Tokado/Keihin Elec Lines	Kawasaki
1370	Sumida River Rr Yard/ Bridge Of Joban Line	Tokyo
1372	Sumidar Bridge Sobr Rr/Ryogoku Sta & Rr Yadr	Tokyo
1373	Sobu Railroad Bridge.Shinkoiwa Yard	Tokyo
1374	Shinjuku Railroad Station & Rr Junction	Tokyo

Target #	Target	Location
1375	Hachioji Junctions And Bridge	Tokyo
1376	Junction	Ofuna
1377	Sagami (Bangu) Rr Bridge	Hiratsuka
1378	Kanoya Air Field (code name: Checkbook)	Kyushu
1386	Japan Nitrogen Fertilizer Co	Minamata
1387	Nippon Mining Co	Tagame
1390	Hitachi-Solex Aircraft	Totsaka
1391	Ishikawajima Engine Plant	Tomioka
1392	Yokosuka Naval Aircraft Factory	Yokosuka
1393	Aeronautical Experimental Laboratory	Tokyo
1393	Aircraft Engine Plant	Tokyo
1394	Kanagafuchi Spinning Mill Company	Tokyo
1395	Shoda Aircraft	Tokyo
1396	Tsurama Soda	Yokohama
1397	Japan Refining Works	Tokyo
1398	Tokyo Gas Company, Omari Branch	Tokyo
1399	Hodogaya Soda	Yokohama
1400	Hodogaya Soda	Tokyo
1401	Japan Artificial Fertilizer	Tokyo
1402	Yokohama Naval Air Station	Yokohama
1403	Tomioka Seaplane Base	Yokohama
1404	Tachikawa Air Base	Tachikawa
1405	Sunamachi Airport	Tokyo
1406	Tokorazawa Airport	Tokyo
1407	Showa Airport	Tokyo
1408	Sayama Airport	Tokyo
1409	Narashino Army Airdrome	Funabashi
1411	Susaki Airport	Tokyo
1412	Chofu Airport	Tokyo
1413	Shinagawa Airport	Tokyo
1414	Urawa Airport	Tokyo
1415	Takahagi Airport	Tokyo
1416	Matsido Airport	Tokyo
1418	Marunoichi Telephone Exchange	Tokyo
1419	Navy Department Radio Towers	Tokyo
1420	Funabashi Naval Transmitting Station	Funabashi
1423	Fukuoka Wireless Station	
1424	Central Meteorlogical Observatory	
1425	Iwatsaki Wireless Station	Tokyo
1430	Nippon Electric Wire And Cable	Tokyo
1431	Bridgestone Tire Company	Totsuka
1432	Yokohama Harbor Passenger Wharves	Yokohama
1433	Yokohama Main Pier	Yokohama
1434	Fukagawa Pier	Tokyo
1435	Tsukishima Pier	Tokyo
1436	Tsukishima East Wharf	Tokyo
1437	Mitsubishi Piece Good Wharf	Kawasaki
1438	Kyoritsu Warehouse	Yokohama
1439	Yokohama Central Wholesale Warehouse	Yokohama
1440	Mitsubishi Warehouse	Yokohama
1441	Yokohama Warehouse	N. Yokohama
1442	Yokohama Warehouse	S. Yokohama
1443	National Silk Conditioning House	Yokhama
1444	Yokohama Harbor Timber Basin	Yokohama
1445	Yokohama Harbor Coop Purchasing Warehouse	Yokohama
1446	East Asia Development Co Warehouses	Yokohama
1447	Hiraidzuma Drug Depot	Yokohama
1448	Tsukiji Market/Wholesale Warehouses	Tokyo
1449	Kenda Market	Tokyo
1450	Koto Market	Tokyo
1451	Naniwa Warehouse	Yokohama
1452	Yokosuka Base Headquarters	Yokosuka
1453	Yokosuka Gunnery And Navigation Schools	Yokosuka
1454	Kakigi Dry Dock	Yokosuka
1455	Oyama Point Ammo And Powder Stores	Yokosuka

VII.   The XXI Bomber Command

Target #	Target	Location
1456	Yokosuka Torpedo And Wireless Schools	Yokosuka
1457	Montenaga Inlet Oil Tanks	Yokosuka
1458	Yokosuka Naval Radio	Yokosuka
1460	Uraga Dockyard #1	Yokosuka
1461	Uraga Dockyard #2	Yokosuka
1462	Shingagawa Shipyard	Tokyo
1463	Akabane Powder Storage, Tokyo Arsenal	Kasumguara
1464	Akashi Industrial - Aircraft Factory	Choshi City
1465	Kashima Naval Air Station	Kasumiguara
1466	Kasumiguara Naval Air Station	Kasumiguara
1467	Hitachi Engineering Kaigan Plant	Hatachi
1468	Mito Army Airdrome	Mito
1470	Otawa Bay Airport	Otawa Bay
1470	Shimoshizu Army Airdrome	Shimoshizu
1471	Yatabe Navy Airdrome	Kasumiguara
1472	Tsudanuma Airdrome	Kasumiguara
1472	Tsukuba Navy Airdrome	Kasumiguara
1473	Tsuchiura Navy Airdraome	Kasumiguara
1476	Hitachi Engineering Co., Kaigan Plant	Sakegawa
1477	Kisarazu Fleet Fueling Base	Tokyo
1478	Mito Railroad Junction And Station	Mito
1479	Toride Railroad Bridge	Mito
1484	Kamioka Lead & Zinc Mines	Kamioka
1485	South Manchurian Wharf & Warehouse	Kawasaki
1486	Yokosuka Old Aircraft Center	Yokosuka
1487	Hokada Airdrome	Hokada
1488	Kioroshi Army Airdrome	Kioroshi
1489	Tomobe Navy Airdraome	Kasumiguara
1490	Tsudanuma Airdrome	Tsudanuma
1491	Kasumiguara Naval Air Station	Kasumiguara
1492	Hitachi Copper Smelter	Sakagawa
1493	Toshin & Shinko Warehouses	Yokohama
1494	Kanson Point Fortification	Uraga
1496	Akashi Urban Area	Chichibu
1496	Chichibu Cement Company	Chichibu
1498	Otaki Hydro-Electric Plant	Kiso River
1499	Momoyama Hydro Electric Plant And Substation	Kiso River
1594	Tobata Steam Power Plant	Unknown
1595	Steam Power Plant #2	Omuta
1502	Yomikaki Hydro-Electric Plant	Kiso River
1505	Oi Hydro-Electric Plant	Kiso River
1515	Yatzuzawa Hydro-Electric Plant	Sagomi River
1517	Kagoshima Railroad Repair Shops	Kagoshima
1518	Kagoshima Freight Yards	Kagoshima
1519	Ito Aircraft Plant	Ito
1520	Onagohata Power Station	Hida
1520	Riken Vacuum Tube Plant	Mohara
1521	Showa Chemical & Fertilizer Plant	
1522	Shinohara Machinery Works	
1536	Showa Fertilizer, Kanose Plant	Kanose
1539	Tokai Electrode Co	Tanoura
1544	Nakajima Aircraft Company	New Ota
1544	Nakajima Aircraft Plant (code name: Fraction)	Ota
1545	Nakajima Aircraft Company, New Ota Plant (code name: Furious)	Koizumi
1546	Nakajima Aircraft Company-Maebashi Parts Plant	Maebashi
1547	Kawasaki Aircraft Plant (code name: Fruitcake)	Akashi
1588	Shingo Hydro Plant	Agano River
1589	Toyomi Hydro Plant	Agano River
1590	Kanose Hydro Plant	Agano River
1591	Yasuoka Hydro-Electric Plant	Tenryu River
1592	Shinano Hydro Plant	Shinano River
1592	Shingo Hydro Plant	
1593	Senju Hydro Plant	Shinano River
1595	Omuta #2 Steam Power Plant	Omuta
1596	Omuta #3 Steam Power Plant	Omuta

Target #	Target	Location
1598	Nagoya #2 Steam Power Plant	Nagoya
1600	Onada Steam Power Plant	Onada
1602	Saijo Steam Power Plant	Saijo
1604	Inaka Hydro-Electric Plant	Kambara
1607	Horai Hydro Plant	Abukuma River
1608	Kurobe #3 Hydro Electric Plant	Kurobe River
1609	Oigana Hydro-Electric Plant	Oi River
1611	Nakatsugawa Hydro-Electric Plant	Nakatsugawa
1613	Shimotaki Hydro-Electric Plant	Shimotaki
1615	Kasagi Hydro-Electric Plant	Kiso River
1616	Kuje Power Plant	Osaka
1626	Ube Steam Power Plant #2	Ube
1627	Omura Aircraft Plant (code name: Fearless)	Omura
1628	Takeda Drug Co	Osaka
1629	Amagasaki Transformer Station	Amagasaki
1630	Osaka Transformer Station	Osaka
1631	Mizaguchi Gear Works	Amagasaki
1631	Yao Transformer Station	Yao
1632	Kizu Transformer Station	Kizu
1633	Sasasu Transformer Station	Sasazu
1634	Mizaguchi Gear Works	Amagasaki
1635	Nakajima Aircraft Works	Handa
1638	Inuyama Transformer Station	Inuyama
1639	Sumitoma Duraluminum Mill Kuwona Plant	Kuwana
1642	Nipon Soda K.K.	Nihongo
1643	Utsonomiya Air Training School & Arsenal	Utsonomiya
1644	Kumagaya Air Base	Kumagaya
1645	Kakowa Mfg Company	Utsonomiya
1646	Riken Industrial, Maebashi	Shin-Maebashi
1647	Riken Industrial, Takasaki	Takasaki
1648	Asada Aluminum Plants	Shihama
1649	Nippon Oil Refinery Kashiwazaki	Kashiwazaki
1650	Kakowa Mfg Company - Kumagaya Parts Plant	Kumagaya
1650	Nakajima Aircraft Company-Kumagaya Parts Plant	Kumagaya
1651	Nippon Stainless Steel Naoetsu Plant	Naoetsu
1652	Minakata Hydro-Electric Plant	Tenryu River
1653	Toyokawa Arsenal	Toyokawa
1655	Koriyama Railroad Shops	Koriyama
1656	Tsugami-Atagi Mfg Company	Nagaoka
1657	Sumitomo Aluminum Plant	Niihawa
1658	Northeast Aluminum Plant	Koriyama
1659	Niigata Iron Works Plant Kashiwazaki Plant	Kashiwazaki
1660	Niigata Iron Works Plant Nagaoka Plant	Nagaoka
1661	Osaka Machinery Company Plant #1	Nagaoka
1662	Osaka Machinery Company Plant #2	Nagaoka
1664	Tokyo Shibaura Electric Company Oguni Plant	Oguni
1665	Shinagawa Mfg Plant	Fukushima
1666	Nippon Soda, Odera Refinery	Odera
1667	Horai Hydro Plant	Horai
1667	Shinetsu Nitrogen Fertilizer, Naoetsu Plant	Naoetsu
1668	Niigata Iron Works Plant Kashiwazaki Plant	Kashiwazaki
1668	Riken Heavy Industry, Kashiwazaki Plant	Kashiwazaki
1669	Riken Heavy Industry, Nagaoka Plant	Nagaoka
1670	Niigata Harbor Facilities	Niigata
1671	Amagasaki Oil Refinery	Unknown
1671	Government Aircraft Parts Plant	Sendai
1675	Amagasaki Urban Area	Yonago
1675	Goto Railroad Shops	Yonago
1676	Tsuruga Railroad Yards	Tsuruga
1677	Nippon Electric Battery Company, Plant #1	Kyoto
1678	Nippon Electric Battery Company, Plant #2	Kyoto
1700	Hitachi Mfg Co. Ube Plant	Ube
1702	Kawanishi Aircraft Company (code name: Leafstalk)	Kobe
1703	Toyowa Heavy Industries Co.	Osaka
1704	Strong Engineering Works	Ibaragi

VII. The XXI Bomber Command

Target #	Target	Location
1705	Strong Engineering Works	Osaka
1706	Mizuno Sporting Goods Co	Osaka
1707	Toyo Metal And Wood Co	Itami
1708	Nippon Aluminum Works	Osaka
1709	Sumitoma Aluminum Company	Yao
1709	Takada Aluminum Mfg Co	Sakai
1710	Tokada Aluminum Mfg Company	Sakai
1711	Osaka Ship Building Works	Osaka
1712	Osaka Gas Company Coke Oven Plant	Osaka
1713	Tagi Fertilizer Company	Behu
1714	Kwoyo Precision Works Co.	Kokubu
1715	Kwoyo Precision Works Co.	Osaka
1716	Asahi Precision Work Co.	Sakai
1718	Kwoyo Precision Works Co.	Osaka
1719	Nippon Air Brake Co	Kobe
1720	Kwoyo Precision Works Co.	Osaka
1721	Tanaka Piston Ring Co	Osaka
1723	Osaka Army Arsenal Hirakata Branch	Hirakata
1724	Kawasaki Aircraft Co	Akashi
1725	Asano Portland Cement Co	Osaka
1726	Osaka Ceramic Industry Cement Co	Oaska
1729	Aichi Aircraft Works, Eitoku Plant	Nagoya
1729	Anju Seaplane Station	Nagoya
1730	Aichi Aircraft Works, Bomimaru Plant	Koromo
1733	Japan Dyestuffs Co	Osaka
1734	Japan Dyestuff Mfg Co	Osaka
1735	Toyowa Heavy Industries, Nishi Plant	Nagoya
1736	Okamoto Aircraft Works, Tarui Plant	Tarui
1737	Yokkaichi Harbor	Yokkaichi
1738	Toho Heavy Industry	Yokkaichi
1739	Furakawa Electrical Co	Amagasaki
1740	Furakawa Electrical Co	Osaka
1741	Sankyo Co	Osaka
1743	Dai Niehi Electric Wire Co	Amagasaki
1744	Mansushita Wireless Company	Moriguchi
1745	Kawanishi Machine Shops	Kobe
1746	Oki Electric Company #1	Osaka
1747	Oki Electric Company #1	Kobe
1750	Matsushita Dry Battery Co	Moriguchi
1751	Matsushita Dry Battery Co	Osaka
1752	Osaka Electric Machinery Co	Osaka
1753	Kobe Steel Works Nagoya Plant	Nagoya
1755	Nakagama Steel Mfg Company	Osaka
1756	Oriental Cable Company	Kaji
1760	Asahi Glass Company	Amagasaki
1761	Mita Leather	
1761	Mittan Leather Belt Company	Osaka
1761	Nitta Leather Belt Mfg Co	Osaka
1762	Kawasaki Heavy Industry	Kobe
1763	Kurimoto Iron Works	Osaka
1764	Maruzen Oil Refinery	Shimatsu
1765	Japan Iron Works	Osaka
1766	Daido Steel Mfg Co	Amagasaki
1767	Tokyo Rolling Works	Amagasaki
1768	Kobe Steel Works	Kobe
1769	Amagasaki Steel Works	Amagasaki
1770	Nippon Iron Sand Industry	Takasago
1773	Osaka Chain And Machinery Works	Ibarigi
1774	Yamato Steel Works	Osaka
1775	Kawasaki Heavy Industry	Kobe
1776	Ishihara Smelter & Refinery	Yokkaichi
1777	Kubato Iron Machinery Works	Sakai
1778	Kabota Iron & Machinery	
1779	Yamamoto Heavy Industries	Yokkaichi
1779	Yamamoto Heavy Industry	Kuwana

Target #	Target	Location
1780	Osaka Machinery Co	Itami
1781	Osaka Machinery Co	Osaka
1782	Osaka Metal Industry	Sakai
1784	Osaka Wakayama Iron Works	Tsuda
1785	Oriental Can Mfg Co	Osaka
1787	Oriental Can Mfg Co	Oaska
1795	Mitsubishi Steel Rolling Mill	Nagasaki
1797	Okuma Iron Works, Kami Iida Cho Plant	Nagoya
1797	Osaka Machinery Works	Nagoya
1798	Osaka Machinery Works	Nagoya
1799	Hokoku Machinery Kasadera Plant	Nagoya
1883	Toyo Steel Plate Co	Kudamatsu
1884	Oshima Naval Oil Storage (code name: Anaphase)	Oshima
1884	Oshima	Oshima
1717	Yuasa Storage Battery Mfg	Takasuki
1732	Rasa industry co.	osaka
1748	Showa Electrode Co	Naruo
1771	Nakayama Steel Mfg Co	Amagasaki
1800	Daido Machinery Co Showa Plant	Nagoya
1802	Hitachi Works	Kuwana
1803	Meideisha Elec Mfg Co Nishi Biwajima Plant	Heisaka
1803	Mitsubishi Electrical Mfg Co	Heisaka
1805	Okuma Iron Works, Kusonoki Plant	Kachikawa
1806	Okuma Iron Works, Kachikawa Plant	Kachikawa
1807	Okuma Iron Works, Asahi Plant	Seto
1809	Hokoku Machinery Co., Atsuta Plant	Nagoya
1810	Ogaki Iron Works	Ogaki
1811	Ogaki Iron Works	Gifu
1812	Kagamigahara Machinery Works	Gifu
1813	Koa Machinery Company	Tomita
1814	Tokyo Shibaura Electrical Mfg Co	Kuwana
1815	Toyo Bearing Co, Dojima Plant	Kuwana
1816	Toyo Bearing Co., Uchibori Plant	Kuwana
1818	Tokoshu Light Metals-Atsuta Plant	Nagoya
1818	Tokushu Light Metals Atsuta Plant	Nagoya
1819	Tokushu Light Metals Ogaki Plant	Ogaki
1820	Tokushu Light Metals Inasawa Plant	Inasawa
1821	Riken Alunite Plant	Nagoya
1821	Toho Gas Works	Nagoya
1822	Ibagawa Electro-Chemical Co., Kido Plant	Ogaki
1823	Tokai Electrical Co Nagoya #1 Plant	Nagoya
1824	Tokai Electrical Co Nagoya #2 Plant	Nagoya
1825	Tokai Electrical Co Nagoya #3 Plant	Nagoya
1826	Nippon Synthetic Chemicals Plant	Ogaki
1827	Dai Nipon Cellophane Plant	Nagoya
1828	Aichi Aircraft Works, Tsukiji Plant	Nagoya
1829	Yahagi Steel Plant	Nagoya
1830	Toyoda Steel Plant	Yokosura
1831	Hirano Loom Works	Nagoya
1832	Ibagawa Electro-Chemical Co., Kita Kiriishi Plant	Ogaki
1833	Mitsubishi Aircraft Works, Kagamigahara Plant	Gifu
1834	Nakajima Aircraft Plant	Koromo
1835	Ainoura Steam Power Plant	Ainoura
1835	Yokose Oil Storage	Yokose
1841	Ube Coal Liquidification Company	Ube
1842	Nagasaki Wharves	Nagasaki
1843	Kyushu Coal and SS Co. Colliery	Sekito Shima
1844	Ube Soda Co	Ube
1870	Tachiari Machine Works	Tachiari
1871	Tosu Railroad Yards And Repair Shops	Tosu
1872	Showa Iron Works	Fukuoka
1873	Tatara Machinery Works	Fukuoka
1874	Asano Cement Company	Nogota
1875	Japan Explosives Mfg	Nogata
1877	Toyo Aluminum Co	Miike

VII. The XXI Bomber Command

Target #	Target	Location
1878	Ube Iron Works	Ube
1879	Titanium Industry Co	Ube
1880	Nippon Motor Oil Company	Ube
1881	Tokuyama Sheet Iron Co	O Shima
1882	Toyo Soda Compay	Tonda
1885	Mitsubishi Electrical Mfg Co	Hiroshima
1889	Ujina Shipbuilding	Ujina Island
1890	Toyo Industry	Hiroshima
1891	Japan Steel Co., Hiroshima Plant	Hiroshima
1907	Kanokawa Oil Storage	Nishi Nomi Jima
1920	Sanyo Steel Co Shikawa Plant	Shikawa
1921	Nippon Iron Sand Industry, Shikama Plant	Shikama
1922	Teikoku Mining Co/Hibi Refinery	Hibi
1929	Suzuki Loom Works Takatsuka Plant	Hamamatsu
1931	Imperial Dye Works	Fukuyama
1936	Take Hara Electrolytic Coper Refinery	Takehars
1942	Nipon Magnesium Company	Sasazu
1942	Nippon Magnesium Co	Kamioka
1943	Aomori Urban Area	Higashi-Iwase
1944	Nippon Soda Company, Magnesium Plant	Higashi-Iwase
1949	Miyazu Steam Power Plant	Miyazu
1950	Tsuruga Harbor Facilities	Tsuruga
2008	Tachikawa Air Depot (code name: Blockhouse)	Tachikawa
2009	Hitachi Aircraft Plant	Tokyo
2010	Army Provisioning Depot	Nagoya
2011	Shizuoka Aircraft Plant (code name: Upcast)	Shizuoka
2024	Sumitomo Light Metals Plant	Tokyo
2025	Hodagoya Chemical Industries (code name: Butterball)	Koriyama
2121	Otake Oil Refinery (code name: Fainter)	Otake
2470	Chichibu Cement Company	Nagoya
2507	Ibusuki Air Field (code name: Infirmary)	Kyushu
2512	Izumi Air Field (code name: Bullish)	Kyushu
2516	Kanoya East Air Field (code name: Famish)	Kanoya
2520	Kokubu Air Field (code name: Barranca)	Kyushu
2527	Miyakonojo Air Field (code name: Dripper)	Kyushu
2529	Miyazaki Air Field (code name: Neckcloth)	Kyushu
2531	Nittigahara Air Field (code name: Bushing)	Kyushu
2534	Kushira Air Field (code name: Aeroscope)	Kyushu
2536	Tomitaki Air Field (;code name: Skewer)	Kyushu
2777	Matsuyama Air Field (code name: Mopish)	Kyushu
3006	Unidentified Bldg - Tokyo Arsenal	Tokyo
3007	Unidentified Bldg - Tokyo Arsenal	Tokyo
3109	Unidentified Bldg - Tokyo Arsenal	Tokyo
3600	Tokyo Arsenal (code name: Peridition)	Tokyo
4011	Chosi Urban Area	Tokyo
6129	Koriyama Chemical Plant (code name: Lunchroom)	Koriyama
XX13004	Unidentified Bldg - Tokyo Arsenal	Tokyo
XX13008	Unidentified Bldg - Tokyo Arsenal	Tokyo
XX13058	Unidentified	Yokohama
XX13062	Unidentified Bldg - Tokyo Arsenal	Tokyo
XX13069	Unidentified	Yokohama
XX13070	Unidentified	Kawasaki
XX13075	Unidentified	Yokohama
XX14065	Sumitomo Light Metal Plant	Nagoya

THE FIRE RAIDS ON JAPAN

In January 1945, General Curtis LeMay arrived in the Marianas and assumed command of the XXI Bomber Command. His appointment came directly from General Arnold, who was far from happy with the results, or more precisely, the lack of results, from the missions flown to date.

General Hansell, who had commanded the XXIst, was a

proponent of daylight precision bombing and was sending the B-29s over Japanese targets at altitudes above 30,000 feet. The vaunted "pickle barrel" accuracy of the Norden bombsight was not present over Japan.

An unanticipated factor was that at 30,000 feet the planes were in the jet stream. If they made their bomb run from west to east, with the wind, their true ground speed was so high as to be off the scales of the bombsight. Coming the other way, upwind, sometimes resulted in a true ground speed of zero.

Hansell started lowering the bombing altitude, but he never intended to go below 20,000 feet, as that would put the B-29s in reach of Japanese anti-aircraft guns. This became academic when LeMay took over. He let the missions continue as before while analyzing the bombing results to date. On January 3rd he had the 73rd Bomb Wing drop a mixed load of high explosives and incendiaries on Nagoya. A month later the same mission profile was used against Kobe. The results in both cases were unimpressive; the dropping of non-aerodynamic incendiaries from 25,000 feet did not assure hits in the target area.

However, LeMay saw enough in the burned-out parts of Nagoya and Kobe to start thinking of low-level incendiary bombing. His faith in the abilities of the B-29s to survive at the lower altitudes was bolstered when a plane of the 19th Bomb Group, dispatched to get radarscope pictures of the Kobe area, flew over the city for two hours at an altitude of 5000 feet.

LeMay ordered a strike on Tokyo for the night of March 9, returning March 10. Bombing altitude would be 5000 feet. Guns would be removed from the aircraft to save weight for more incendiary clusters. The lead squadron would drop M-47 napalm bombs with the planes intervalometers set to have the bombs hit 100 feet apart. The rest of the planes would drop M-69 incendiary cluster bombs set for fifty foot intervals.

Results were far beyond expectations. Nearly 16 square miles of the city were burnt out. Japanese casualty figures, more than likely low, placed the number of dead at 78,660. In the next few months, the B-29s of the XXI Bomber command would bring such destruction to the cities of Japan that by July of 1945 there were no strategic targets left in the Home Islands.

For an explanation of the columns, see page 141.

MSN	Date	Target	Location	Wings	Grps	No. Up	Bombed	Lost
17	Jan 3, 1945	Urban Area	Nagoya	73BW	WING	97	57	5
17	Jan 3, 1945	Targets of Opportunity	Nagoya Area		WING	in above	21	0
17	Jan 3, 1945	Urban Area	Nagoya		497BG	21	17	1
17	Jan 3, 1945	Urban Area	Nagoya		498BG	24	13	1
17	Jan 3, 1945	Secondary Target	Unreported		498BG	in above	1	0
17	Jan 3, 1945	Urban Area	Nagoya		499BG	not avail	28	0
17	Jan 3, 1945	Urban Area	Nagoya		500BG	not avail	19	2
26	February 4	Urban Area	Kobe	MISSION	WINGS	110	69	2
26	February 4	Urban Area	Kobe	73BW	WING	74	37	2
26	February 4	Secondary Target			WING	in above	29	0
26	February 4	Target of Opportunity			WING	in above	1	0
26	February 4	Urban Area	Kobe		497BG	12	10	0
26	February 4	Urban Area	Kobe		498BG	23	19	1
26	February 4	Secondary Target	Kobe		498BG	in above	1	0
26	February 4	Urban Area	Kobe		499BG	not avail	22	0
26	February 4	Matsuzaka	Kobe		500BG	not avail	15	0
26	February 4	Urban Area	Kobe	313BW	Wing	36	32	1
26	February 4	Urban Area	Kobe		504BG	14	12	0
26	February 4	Urban Area	Kobe		505BG	22	20	1
38	February 25	Urban Area	Tokyo	MISSION	WINGS	229	172	3
38	February 25	Last Resort Target	Tokyo		WINGS	in above	28	
38	February 25	Target of Opportunity	Tokyo		WINGS	in above	1	
38	February 25	Urban Area	Tokyo	73BW	WING	116	94	3
38	February 25	Targets of Opportunity	Tokyo		WING	in above	8	0
38	February 25	Urban Area	Tokyo		497BG	24	19	2
38	February 25	Urban Area & Docks	Tokyo		498BG	37	27	0
38	February 25	Urban Area & Docks	Tokyo		499BG	not avail	29	
38	February 25	Urban Area & Docks	Tokyo		500BG	not avail	27	0
38	February 25	Urban Area	Tokyo	313BW	WING	not avail	60	0
38	February 25	Urban Area	Tokyo		6BG	not avail	21	0
38	February 25	Urban Area	Tokyo		9BG	32	23	0
38	February 25	Targets of Opportunity	Tokyo Area		9BG	in above	5	0
38	February 25	Urban Area	Tokyo		504BG	14	8	0
38	February 25	Urban Area	Tokyo		505BG	15	8	0
38	February 25	Last Resort Target	Toyohashi		505BG	in above	2	0
38	February 25	Last Resort Target	Hamamatsu		505BG	in above	2	0

VII. The XXI Bomber Command

MSN	Date	Target	Location	Wings	Grps	No. Up	Bombed	Lost
38	February 25	Last Resort Target	Numazu		505BG	in above	2	0
38	February 25	Last Resort Target	Kega		505BG	in above	1	0
38	February 25	NE Tokyo	NE Tokyo	314BW	19BG	11	10	0
38	February 25	Target of Opportunity	Nagoya		19BG	in above	1	0
40	March 9/10	Urban Area	Tokyo	MISSION	WINGS	325	279	14
40	March 9/10	Urban Area	Tokyo	73BW	WING	165	137	1
40	March 9/10	Target of Opportunity	Tokyo		WING	in above	9	0
40	March 9/10	Urban Area	Tokyo		497BG	37	34	0
40	March 9/10	Urban Area	Tokyo		498BG	42	34	1
40	March 9/10	Secondary Target	Tokyo		498BG	in above	3	0
40	March 9/10	Urban Area	Tokyo		499BG	not avail	44	0
40	March 9/10	Urban Area	Tokyo		500BG	42	31	1
40	March 9/10	Urban Area	Tokyo	313BW	WING	106	99	3
40	March 9/10	Urban Area	Tokyo		6BG	not avail	32	0
40	March 9/10	Urban Area	Tokyo		9BG	32	29	2
40	March 9/10	Target of Opportunity	Unknown		9BG	in above	1	0
40	March 9/10	Tokyo Urban Area	Tokyo		504BG	21	18	0
40	March 9/10	Tokyo Urban Area	Tokyo		505BG	not avail	20	1
40	March 9/10	Target of Last Resort	Kasumigoura A	505BG	in above	1	0	
40	March 9/10	Target of Last Resort	Haha Jima		505BG	in above	1	0
40	March 9/10	Target of Last Resort	Chichi Jima		505BG	in above	1	0
40	March 9/10	Target of Last Resort	Guguan		505BG	in above	1	0
40	March 9/10	Urban Area	Tokyo	314BW	WING	54	49	8
40	March 9/10	Urban Area	Tokyo		19BG	28	27	3
40	March 9/10	Urban Area	Tokyo		29BG	26	22	5
41	March 11/12	Urban Area	Nagoya	MISSION	WINGS	310	285	1
41	March 11/12	Urban Area	Nagoya	73BW	WING	160	145	1
41	March 11/12	Target of Opportunity	Nagoya		WING	in above	3	0
41	March 11/12	Urban Area	Nagoya		497BG	34	34	0
41	March 11/12	Urban Area	Nagoya		498BG	39	38	0
41	March 11/12	Urban Area	Nagoya		499BG	19	40	1
41	March 11/12	Urban Area	Nagoya		500BG	39	36	0
41	March 11/12	Urban Area	Nagoya	313BW	WING	108	101	3
41	March 11/12	Urban Area	Nagoya		6BG	31	30	0
41	March 11/12	Urban Area	Nagoya		9BG	31	27	2
41	March 11/12	Urban Area	Nagoya		504BG	21	17	0
41	March 11/12	Urban Area	Nagoya		505BG	25	25	1
41	March 11/12	Urban Area	Nagoya	314BW	WING	42	39	0
41	March 11/12	Urban Area	Nagoya		19BG	21	20	0
41	March 11/12	Urban Area	Nagoya		29BG	21	19	0
42	March 13/14	Urban Area	Osaka	MISSION	WINGS	298	274	2
42	March 13/14	Urban Area	Osaka	73BW	WING	141	125	2
42	March 13/14	Target of Opportunity	Osaka		WING	in above	2	0
42	March 13/14	Urban Area	Osaka		497BG	37	35	0
42	March 13/14	Urban Area	Osaka		498BG	30	28	1
42	March 13/14	Urban Area	Osaka		499BG	not avail	not avail	1
42	March 13/14	Urban Area	Osaka		500BG	32	32	0
42	March 13/14	Urban Area	Osaka	313BW	WING	106	106	0
42	March 13/14	Urban Area	Osaka		6BG	not avail	30	0
42	March 13/14	Urban Area	Osaka		9BG	33	33	0
42	March 13/14	Urban Area	Osaka		504BG	21	20	0
42	March 13/14	Urban Area	Osaka		505BG	not avail	26	0
42	March 13/14	Urban Area	Osaka	314BW	WING	45	43	0
42	March 13/14	Urban Area	Osaka		19BG	22	21	0
42	March 13/14	Urban Area	Osaka		29BG	23	22	0
43	March 16/17	Urban Area	Kobe	MISSION	WINGS	334	306	3
43	March 16/17	Urban Area	Kobe	73BW	WING	154	142	1
43	March 16/17	Target of Opportunity	Kobe		WING	in above	2	0
43	March 16/17	Urban Area	Kobe		497BG	38	38	0
43	March 16/17	Urban Area	Kobe		498BG	41	39	0
43	March 16/17	Urban Area	Kobe		499BG	32	28	0
43	March 16/17	Urban Area	Kobe		500BG	39	37	1
43	March 16/17	Urban Area	Kobe	313BW	WING	128	117	2
43	March 16/17	Urban Area	Kobe		6BG	34	33	0
43	March 16/17	Urban Area	Kobe		9BG	34	30	1

MSN	Date	Target	Location	Wings	Grps	No. Up	Bombed	Lost
43	March 16/17	Urban Area	Kobe		504BG	24	23	1
43	March 16/17	Urban Area	Kobe		505BG	36	32	0
43	March 16/17	Urban Area	Kobe	314BW	WING	52	47	0
43	March 16/17	Urban Area	Kobe		19BG	26	24	0
43	March 16/17	Urban Area	Kobe		29BG	26	23	0
44	March 18/19	Urban Area	Nagoya	MISSION	WINGS	312	290	2
44	March 18/19	Urban Area	Nagoya	73BW	WING	142	128	1
44	March 18/19	Urban Area	Nagoya		497BG	39	37	0
44	March 18/19	Urban Area	Nagoya		498BG	38	34	0
44	March 18/19	Urban Area	Nagoya		499BG	31	28	0
44	March 18/19	Urban Area	Nagoya		500BG	32	29	0
44	March 18/19	Urban Area	Nagoya	313BW	WING	121	114	1
44	March 18/19	Urban Area	Nagoya		6BG	34	32	0
44	March 18/19	Urban Area	Nagoya		9BG	31	30	0
44	March 18/19	Urban Area	Nagoya		504BG	23	21	0
44	March 18/19	Urban Area	Nagoya		505BG	33	31	1
44	March 18/19	Urban Area	Nagoya	314BW	WING	49	48	0
44	March 18/19	Urban Area	Nagoya		19BG	26	25	0
44	March 18/19	Urban Area	Nagoya		29BG	23	23	0
68	April 15/16	Urban Area	Kawasaki	MISSION	WINGS	219	197	12
68	April 15/16	Urban Area	Kawasaki	313BW	WING	111	95	6
68	April 15/16	Urban Area	Kawasaki		6BG	24	24	0
68	April 15/16	Urban Area	Kawasaki		9BG	33	26	4
68	April 15/16	Targets of Opportunity	Kawasaki		9BG	in above	4	0
68	April 15/16	Urban Area	Kawasaki		504BG	22	20	1
68	April 15/16	Urban Area	Kawasaki		505BG	32	25	1
68	April 15/16	Urban Area	Kawasaki	314BW	WING	108	102	6
68	April 15/16	Urban Area	Kawasaki		19BG	33	33	3
68	April 15/16	Urban Area	Kawasaki		29BG	32	30	2
68	April 15/16	Urban Area	Kawasaki		39BG	23	23	1
68	April 15/16	Urban Area	Kawasaki		330BG	20	20	0
69	April 15/16	Tokyo Urban Area	Tokyo	73BW	WING	120	109	1
69	April 15/16	Urban Area	Tokyo		497BG	28	23	0
69	April 15/16	Urban Area	Tokyo		498BG	33	30	0
69	April 15/16	Urban Area	Tokyo		499BG	not avail	26	1
69	April 15/16	Urban Area	Tokyo		500BG	34	30	1
174	May 14	Urban Area	Nagoya	MISSION	XXIBC	524	472	11
174	May 14	Urban Area	Nagoya	58BW	WING	141	124	2
174	May 14	Urban Area	Nagoya		40BG	35	32	
174	May 14	Target of Opportunity	Unknown		40BG	in above	1	
174	May 14	Urban Area	Nagoya		444BG	32	28	
174	May 14	Urban Area	Nagoya		462BG	36	33	
174	May 14	Urban Area	Nagoya		468BG	38	31	
174	May 14	Target of Opportunity	Neve (sp?) Lake		468BG	in above	1	
174	May 14	Urban Area	Nagoya	73BW	WING	162	145	2
174	May 14	Urban Area	Nagoya		497BG	44	36	1
174	May 14	Urban Area	Nagoya		498BG	31	29	0
174	May 14	Urban Area	Nagoya		499BG	Not avail	40	1
174	May 14	Urban Area	Nagoya		500BG	not avail	41	
174	May 14	Urban Area	Nagoya		6BG	32	25	0
174	May 14	Urban Area	Nagoya		504BG	17	13	0
174	May 14	Secondary Target	Nagoya		504BG	in above	2	0-
174	May 14	Urban Area	Nagoya		505BG	not avail	25	2
174	May 14	North Urban Area	Nagoya	314BW	WING	144	135	4
174	May 14	Urban Area	Nagoya		19BG	not avail	39	
174	May 14	Urban Area	Nagoya		29BG	Not avail	38	
174	May 14	Urban Area	North Nagoya		39BG	33	32	1
174	May 14	Urban Area	North Nagoya		330BG	32	32	0
176	May 16/17	South Urban Area	Nagoya	MISSION	WINGS	516	439	3
176	May 16/17	Urban Area	Nagoya	58BW	WING	140	122	7
176	May 16/17	Urban Area	Nagoya		40BG	32	29	0
176	May 16/17	Target of Opportunity	Matsuyaka		40BG	in above	1	0
176	May 16/17	Urban Area	Nagoya		444BG	32	26	0
176	May 16/17	Target of Opportunity	Koshimoto		444BG	in above	1	0
176	May 16/17	Urban Area	Nagoya		462BG	37	33	0

VII. The XXI Bomber Command

MSN	Date	Target	Location	Wings	Grps	No. Up	Bombed	Lost
176	May 16/17	Urban Area	Nagoya		468BG	39	34	0
176	May 16/17	Target of Opportunity	Yokosuka		468BG	in above	1	0
176	May 16/17	Target of Opportunity	Osaka		468BG	in above	1	0
176	May 16/17	Target of Opportunity	Okura		468BG	in above	1	0
176	May 16/17	Urban Area	Nagoya	73BW	WING	154	137	0
176	May 16/17	South Nagoya Urban	Nagoya		497BG	38	35	0
176	May 16/17	Urban Area	Nagoya		498BG	40	38	0
176	May 16/17	Urban Area	Nagoya		499BG	not avail	25	
176	May 16/17	Urban Area	Nagoya		500BG	not avail	40	
176	May 16/17	Urban Area	Nagoya		6BG	33	33	0
176	May 16/17	Urban Area	Nagoya		504BG	not avail	not avail	
176	May 16/17	South Nagoya Urban	Nagoya	314BW	WING	131	127	0
176	May 16/17	South Nagoya Urban	Nagoya		19BG	not avail	37	
176	May 16/17	South Nagoya Urban	Nagoya		29BG	not avail	30	
176	May 16/17	South Nagoya Urban	Nagoya		39BG`	32	28	0
176	May 16/17	South Nagoya Urban	Nagoya		330BG	32	32	1
181	May 23/24	Urban Area	Tokyo	MISSION	WINGS	558	526	17
181	May 23/24	Urban Area	Tokyo	58BW	WING	127	118	4
181	May 23/24	Urban Area	Tokyo		40BG	32	30	1
181	May 23/24	Target of Opportunity	Sagara		40BG	in above	1	0
181	May 23/24	Urban Area	Tokyo		444BG	34	32	0
181	May 23/24	Urban Area	Tokyo		462BG	30	27	1
181	May 23/24	Urban Area	Tokyo		468BG	32	29	1
181	May 23/24	Target of Opportunity	Shizuoka		468BG	in above	1	0
181	May 23/24	South Tokyo Urban	Tokyo	73BW	WING	170	156	6
181	May 23/24	South Tokyo Urban	Tokyo		497BG	41	35	2
181	May 23/24	South Tokyo Urban	Tokyo		498BG	44	44	4
181	May 23/24	Urban Area	Tokyo		499BG	Not avail	37	0
181	May 23/24	Urban Area	South Tokyo		500BG	not avail	42	2
181	May 23/24	Urban Area	Tokyo	313BW	WING	103	95	5
181	May 23/24	Urban Area	Tokyo		6BG	35	31	3
181	May 23/24	Urban Area	Tokyo		504BG	31	29	2
181	May 23/24	Urban Area	Tokyo		505BG	not avail	35	
181	May 23/24	South Tokyo Urban	Tokyo	314BW	WING	not avail	157	2
181	May 23/24	South Tokyo Urban	Tokyo		19BG	43	43	0
181	May 23/24	South Tokyo Urban	Tokyo		29BG	not avail	44	0
181	May 23/24	South Tokyo Urban	Tokyo		39BG	38	35	1
181	May 23/24	South Tokyo Urban	Tokyo		330BG	35	35	1
183	May 25/26	Urban Area	Tokyo	MISSION	WING	498	464	26
183	May 25/26	Urban Area	Tokyo	58BW	WING	133	119	11
183	May 25/26	Urban Area	Tokyo		40BG	33	30	2
183	May 25/26	Urban Area	Tokyo		444BG	34	33	5
183	May 25/26	Urban Area	Tokyo		462BG	31	30	1
183	May 25/26	Urban Area	Tokyo		468BG	35	26	1
183	May 25/26	Target of Opportunity	Hamamatsu		468BG	in above	1	0
183	May 25/26	Target of Opportunity	Tateyama Hoto		468BG	in above	1	0
183	May 25/26	Urban Area	Tokyo	73BW	WING	144	132	4
183	May 25/26	Tokyo Palace Area	Tokyo		497BG	35	30	1
183	May 25/26	Urban Area	Tokyo		498BG	33	30	1
183	May 25/26	Urban Area	Tokyo		499BG	not avail	35	1
183	May 25/26	Urban Area	Tokyo		500BG	not avail	40	1
183	May 25/26	Urban Area	Tokyo		6BG	not avail	24	1
183	May 25/26	Urban Area	Tokyo		504BG	26	26	2
183	May 25/26	Urban Area	Tokyo		505BG	not avail	not avail	4
183	May 25/26	South Central Tokyo Urban Area	Tokyo	314BW	WING	138	125	2
183	May 25/26	South Central Tokyo Urban Area	Tokyo		19BG	not avail	34	
183	May 25/26	South Central Tokyo Urban Area	Tokyo		29BG	not avail	40	
183	May 25/26	South Central Tokyo Urban Area	Tokyo		39BG	36	33	1
183	May 25/26	South Central Tokyo Urban Area	Tokyo		330BG	28	18	0
186	May 29	Urban Area	Yokohama	MISSION	WINGS	510	454	7
186	May 29	Urban Area	Yokohama	58BW	WING	132	122	2
186	May 29	Urban Area	Yokohama		40BG	31	30	
186	May 29	Urban Area	Yokohama		444BG	29	26	
186	May 29	Urban Area	Yokohama		462BG	34	30	
186	May 29	Target of Opportunity	Hachijo Jima		462BG	in above	1	

MSN	Date	Target	Location	Wings	Grps	No. Up	Bombed	Lost
186	May 29	Urban Area	Yokohama		468BG	38	36	
186	May 29	Target of Opportunity	Hachijo Jima		468BG	in above	1	
186	May 29	Urban Area	Yokohama	73BW	WING	152	136	0
186	May 29	Urban Area	Yokohama		497BG	39	34	0
186	May 29	Urban Area	Yokohama		498BG	41	34	
186	May 29	Urban Area	Yokohama		499BG	not avail	26	
186	May 29	Urban Area	Yokohama		500BG	not avail	42	
186	May 29	Urban Area	Yokohama	313BW	WING	83	65	2
186	May 29	Urban Area	Yokohama		6BG	27	25	
186	May 29	Urban Area	Yokohama		504BG	25	22	2
186	May 29	Urban Area	Yokohama	314BW	WING	146	131	3
186	May 29	Target of Opportunity	Shimoda A/F	WING	in above	2		
186	May 29	Target of Opportunity	Nii Shima		WING	in above	1	
186	May 29	Target of Opportunity	Hamamatsu		WING	in above	1	
186	May 29	Target of Opportunity	Sagara		WING	in above	1	
186	May 29	Target of Opportunity	Kisarazu Nav Sta		WING	in above	1	
186	May 29	Urban Area	Yokohama		29BG	not avail	37	
186	May 29	Urban Area	Yokohama		19BG	not avail	39	
186	May 29	Urban Area	Yokohama		39BG	31	25	2
186	May 29	Target of Opportunity	Sagara		39BG	in above	1	0
186	May 29	Target of Opportunity	Shimoda A/F	39BG		in above	1	0
186	May 29	Urban Area	Yokohama		330BG	38	34	
187	June 1	Urban Area	Osaka	MISSION	WINGS	509	487	10
187	June 1	Urban Area	Osaka	58BW	WING	119	107	4
187	June 1	Target of Opportunity	Katsuuri		WING	in above	1	
187	June 1	Target of Opportunity	Shingu		WING	in above	1	
187	June 1	Urban Area	Osaka		40BG	29	27	0
187	June 1	Urban Area	Osaka		444BG	27	22	3
187	June 1	Target of Opportunity	Matsuke		444BG	in above	1	0
187	June 1	Urban Area	Osaka		462BG	28	26	0
187	June 1	Urban Area	Osaka		468BG	35	32	1
187	June 1	Target of Opportunity	Shingu		468BG	in above	1	0
187	June 1	Urban Area	Osaka	73BW	WING	158	140	4
187	June 1	Target of Opportunity	Chichi Jima		WING	in above	2	0
187	June 1	Target of Opportunity	Muki Tomika		WING	in above	1	0
187	June 1	Target of Opportunity	Nara Kushimoto		WING	in above	1	0
187	June 1	Target of Opportunity	Tanabe		WING	in above	1	0
187	June 1	Urban Area	Osaka		497BG	34	32	2
187	June 1	Urban Area	Osaka		498BG	39	37	1
187	June 1	Urban Area	Osaka		499BG	not avail	36	1
187	June 1	Urban Area	Osaka		500BG	not avail	35	
187	June 1	Urban Area	Osaka	314BG	WING	118	103	2
187	June 1	Target of Opportunity	Takashima		WING	in above	1	
187	June 1	Target of Opportunity	Kushimoto A/F	WING	in above	1		
187	June 1	Urban Area	Osaka		29BG	30	29	0
187	June 1	Urban Area	Osaka		19BG	36	32	0
187	June 1	Urban Area	Osaka		39BG	15	13	1
187	June 1	Target of Opportunity	Wakayama		39BG	in above	1	0
187	June 1	Urban Area	Osaka		330BG	37	34	0
187	June 1	Urban Area	Osaka	313BW	WING	119	108	1
187	June 1	Target of Opportunity	Nagashima		WING	in above	1	
187	June 1	Target of Opportunity	Kochi		WING	in above	1	
187	June 1	Target of Opportunity	Shiona Misaki		WING	in above	1	
187	June 1	Urban Area	Osaka		6BG	27	27	1
187	June 1	Urban Area	Osaka		9BG	36	31	0
187	June 1	Target of Opportunity	Osaka		9BG	in above	1	0
187	June 1	Urban Area	Osaka		504BG	24	20	0
187	June 1	Urban Area	Osaka		505BG	31	33	0
188	June 5	Urban Area	Kobe	MISSION	WINGS	524	494	12
188	June 5	Urban Area	Kobe	58BW	WING	120	107	5
188	June 5	Urban Area	Kobe		40BG	29	24	0
188	June 5	Target of Opportunity	Shizuki		40BG	in above	1	
188	June 5	Urban Area	Kobe		444BG	27	25	1
188	June 5	Urban Area	Kobe		462BG	31	26	4
188	June 5	Target of Opportunity	Wakayama		462BG	in above	1	0

VII. The XXI Bomber Command

MSN	Date	Target	Location	Wings	Grps	No. Up	Bombed	Lost
188	June 5	Urban Area	Kobe		468BG	33	32	0
188	June 5	Target of Opportunity	Komatsushima		468BG	in above	1	0
188	June 5	Urban Area	Kobe	73BW	WING	154	139	2
188	June 5	Urban Area	Kobe		497BG	36	31	1
188	June 5	Urban Area	Kobe		498BG	37	32	1
188	June 5	Urban Area	Kobe		499BG	not avail	35	0
188	June 5	Urban Area	Kobe		500BG	not avail	41	0
188	June 5	Urban Area	Kobe	313BW	WING	124	113	2
188	June 5	Target of Opportunity	Shingu		WING	in above	1	
188	June 5	Target of Opportunity	Komatushima		WING	in above	1	
188	June 5	Urban Area	Kobe		6BG	29	28	
188	June 5	Urban Area	Kobe		9BG	35	32	0
188	June 5	Urban Area	Kobe		504BG	20	18	
188	June 5	Secondary Target	Kobe		504BG	in above	2	0
188	June 5	Urban Area	Kobe		505BG	not avail	not avail	
188	June 5	Urban Area	Kobe	314BW	WING	129	120	2
188	June 5	Target of Opportunity	Futaminora A/F	WING	in above	1		
188	June 5	Target of Opportunity	Saki Nohana		WING	in above	1	
188	June 5	Target of Opportunity	Chichi Jima		WING	in above	1	
188	June 5	Urban Area	Kobe		29BG	38	35	1
188	June 5	Urban Area	Kobe		19BG	33	32	0
188	June 5	Urban Area	Kobe		39BG	27	24	0
188	June 5	Urban Area	Kobe		330BG	31	25	1
189	June 7	Urban Area	Osaka	MISSION	WINGS	449	409	2
189	June 7	Urban Area	Osaka	58BW	WING	115	107	0
189	June 7	Urban Area	Osaka		40BG	32	31	0
189	June 7	Target of Opportunity	Shingo		444BG	26	24	0
189	June 7	Urban Area	Osaka		444BG	in above	1	0
189	June 7	Target of Opportunity	Shirahama		462BG	26	25	0
189	June 7	Urban Area	Osaka		462BG	in above	1	0
189	June 7	Target of Opportunity	Chichi Jima		468BG	31	27	0
189	June 7	Target of Opportunity	Tokushima		468BG	in above	11	0
189	June 7	Target of Opportunity	None'		468BG	in above	1	0
189	June 7	Urban Area	Osaka	73BW	WING	133	120	0
189	June 7	Urban Area	Osaka		497BG	30	24	0
189	June 7	Urban Area	Osaka		498BG	33	30	0
189	June 7	Urban Area	Osaka		499BG	not avail	31	0
189	June 7	Urban Area	Osaka		500BG	not avail	35	0
189	June 7	Urban Area	Osaka	313BW	WING	83	83	
189	June 7	Urban Area	Osaka		6BG	27	26	
189	June 7	Urban Area	Osaka		9BG	32	32	0
189	June 7	Urban Area	Osaka		504BG	23	19	1
189	June 7	Secondary Target	Osaka		504BG	in above	1	0
189	June 7	Urban Area	Osaka	314BW	WING	119	107	1
189	June 7	Urban Area	Osaka		29BG	31	28	0
189	June 7	Urban Area	Osaka		19BG	32	31	1
189	June 7	Urban Area	Osaka		39BG	29	21	0
189	June 7	Urban Area	Osaka		330BG	27	27	0
203	June 15	Osaka/Amagasaki Urban Area	Osaka/Amagasaki	MISSION	WINGS	511	476	2
203	June 15	Osaka/Amagasaki Urban Area	Osaka/Amagasaki	58BW	WING	138	119	1
203	June 15	Urban Area	Osaka		40BG	33	29	0
203	June 15	Target of Opportunity	Kimomoto		40BG	in above	1	0
203	June 15	Target of Opportunity	Kobe		40BG	in above	1	0
203	June 15	Urban Area	Osaka		444BG	35	32	0
203	June 15	Target of Opportunity	Kushimoto		444BG	in above	1	0
203	June 15	Urban Area	Amagasaki		462BG	30	25	0
203	June 15	Target of Opportunity	Shirahama		462BG	in above	1	0
203	June 15	Urban Area	Osaka		468BG	40	33	0
203	June 15	Target of Opportunity	Kushimoto		468BG	in above	1	0
203	June 15	Target of Opportunity	Katsaura		468BG	in above	1	0
203	June 15	Target of Opportunity	Tokushima		468BG	in above	1	0
203	June 15	Target of Opportunity	24:15N; 138:15E		468BG	in above	1	0
203	June 15	Osaka/Amagasaki Urban Area	Osaka/Amagasaki	73BW	WING	145	132	0
203	June 15	Target of Opportunity	Wakayama		WING	in above	1	0

MSN	Date	Target	Location	Wings	Grps	No. Up	Bombed	Lost
203	June 15	Osaka/Amagasaki Urban Area	Osaka/Amagasaki		497BG	40	36	0
203	June 15	Osaka/Amagasaki Urban Area	Osaka/Amagasaki		498BG	36	34	0
203	June 15	Osaka/Amagasaki Urban Area	Osaka/Amagasaki		499BG	35	30	0
203	June 15	Osaka/Amagasaki Urban Area	Osaka/Amagasaki		500BG	33	32	0
203	June 15	Osaka/Amagasaki Urban Area	Osaka/Amagasaki	314BW	WING	135	108	1
203	June 15	Target of Opportunity	Wakayama		WING	in above	1	0
203	June 15	Target of Opportunity	Mugi		WING	in above	1	0
203	June 15	Target of Opportunity	Tokushima		WING	in above	1	0
203	June 15	Target of Opportunity	Shiono Misaki		WING	in above	1	0
203	June 15	Target of Opportunity	Chichi Jima		WING	in above	2	0
203	June 15	Target of Opportunity	Matsune		WING	in above	1	0
203	June 15	Target of Opportunity	Katsuura		WING	in above	1	0
203	June 15	Osaka/Amagasaki Urban Area	Osaka/Amagasaki		19BG	34	27	0
203	June 15	Osaka/Amagasaki Urban Area	Osaka/Amagasaki		29BG	37	34	1
203	June 15	Urban Area	Osaka/Amagasaki		39BG	31	23	0
203	June 15	Target of Opportunity	Uwano		39BG	in above	1	0
203	June 15	Target of Opportunity	Kushimoto		39BG	in above	2	0
203	June 15	Osaka/Amagasaki Urban Area	Osaka/Amagasaki		330BG	33	24	0
203	June 15	Osaka/Amagasaki Urban Area	Osaka/Amagasaki	313BW	WING	94	85	0
203	June 15	Target of Opportunity	Ujiyamada		WING	in above	1	0
203	June 15	Target of Opportunity	Fukido		WING	in above	1	0
203	June 15	Target of Opportunity	Wakayama		WING	in above	1	0
203	June 15	Target of Opportunity	Tokushima		WING	in above	1	0
203	June 15	Target of Opportunity	Shingu		WING	in above	1	0
203	June 15	Target of Opportunity	Kochi		WING	in above	1	0
203	June 15	Urban Area	Osaka/Amagasaki		6BG	35	35	0
203	June 15	Urban Area	Osaka/Amagasaki		9BG	35	33	0
203	June 15	Target of Opportunity	Osaka/Amagasaki		9BG	in above	2	0
203	June 15	Urban Area	Osaka/Amagasaki		504BG	24	20	0
203	June 15	Secondary Target	Osaka/Amagasaki		504BG	in above	1	0
211	June 19/20	Urban Area	Fukuoka	MISSION	WINGS	237	223	0
211	June 19/20	Urban Area	Fukuoka	73BW	WING	142	131	0
211	June 19/20	Target of Opportunity	Tomitaka		WING	in above	1	0
211	June 19/20	Target of Opportunity	Chichi Jima		WING	in above	1	0
211	June 19/20	Urban Area	Fukuoka		497BG	37	34	0
211	June 19/20	Urban Area	Fukuoka		498BG	34	29	0
211	June 19/20	Urban Area	Fukuoka		499BG	not avail	38	0
211	June 19/20	Urban Area	Fukuoka		500BG	not avail	32	0
211	June 19/20	Urban Area	Fukuoka	313BW	WING	95	92	0
211	June 19/20	Target of Opportunity	Miyazaki		WING	in above	1	0
211	June 19/20	Urban Area	Fukuoka		6BG	29	29	0
211	June 19/20	Urban Area	Fukuoka		9BG	37	34	0
211	June 19/20	Target of Opportunity	Fukuoka		9BG	in above	1	0
211	June 19/20	Urban Area	Fukuoka		504BG	27	27	0
212	June 19/20	Urban Area	Shizuoka	314BW	WING	137	123	2
212	June 19/20	Target of Opportunity	Hachijo Jima		WING	in above	1	0
212	June 19/20	Target of Opportunity	Chichi Jima		WING	in above	1	0
212	June 19/20	Urban Area	Shizuoka		19BG	35	33	1
212	June 19/20	Urban Area	Shizuoka		29BG	37	32	0
212	June 19/20	Urban Area	Shizuoka		39BG	32	28	1
212	June 19/20	Urban Area	Shizuoka		330BG	33	30	0

Missions Against Kyushu Airfields

In late March 1945, the XXI Bomber Command received a request for assistance from the U. S. Navy. The invasion of Okinawa was well under way and the Navy had a very large number of ships close in shore for support.

The Japanese were sending a major part of their aircraft reserves south from Kyushu airfields to attack the American ships. Many of these aircraft were the so-called Kamikazes — pilots so dedicated to sinking American ships that they were willing to give their lives to do so.

The numbers of attacking planes were so great that they

VII. The XXI Bomber Command

overwhelmed the anti-aircraft fire volume of the Navy ships, allowing Kamikazes to reach their targets. The XXI Bomber Command was asked to strike at the home airfields of the Japanese attackers with two objectives: destroy as many Japanese planes on the ground as possible, and damage the airfields themselves enough to reduce their ability to launch planes.

Starting on March 27, 161 planes, 117 of the veteran 73rd Bomb Wing, and 44 from the 314th Bomb Wing, attacked Tachiari, Oita, Kanoya and Omura airfields in southern Kyushu. Serious airfield attacks resumed on April 17 and continued until May 11. In that period eighteen bases were attacked numerous times, and attacks by Japanese planes on the Navy at Okinawa were reduced.

For an explanation of the columns, see page 141.

MSN	Date	Target	Location	Wings	Grps	No. Up	Bombed	Lost
46	March 27	Kyushu Airfields	Kyushu	MISSION	WINGS	161	146	0
46	March 27	Kyushu Airfields	Kyushu	73BW	WING	117	112	0
46	March 27	Tachiari Airport	Kyushu	497BG	30	29	0	
46	March 27	Tachiari Airport	Kyushu	498BG	30	29	0	
46	March 27	Oita Airfield	Kyushu	499BG	not avail	not avail	0	
46	March 27	Oita Airfield	Kyushu	500BG	39	30	0	
46	March 27	Kanoya Airfield	Kyushu	314BW	WING	44	0	0
46	March 27	Omura Airfield (Secondary)	Omura		WING	in above	40	0
46	March 27	Omura Airfield	Omura	19BG	22	21	0	
46	March 27	Omura Airfield	Omura	29BG	22	22	0	
46	March 27	Target of Opportunity	Kami Kushita	19BG	in above	1	0	
46	March 27	Target of Opportunity	Chichi Jima	29BG	in above	1	0	
46	March 27	Target of Opportunity	Uchinoura	29BG	in above	1	0	
70	April 17	Izumi Airfield	Kyushu	73BW	WING	22	20	0
70	April 17	Target of Opportunity	Chichi Jima	WING	in above	1	0	
70	April 17	Target of Opportunity	Pagan Island	WING	in above	1	0	
70	April 17	Izumi Airfield	Kyushu	499BG	not avail	9	0	
70	April 17	Izumi Airfield	Kyushu	500BG	not avail	11	0	
71	April 17	Tachiari Airfield	Kyushu	73BW	WING	21	21	0
71	April 17	Tachiari Airfield	Kyushu	497BG	11	11	0	
71	April 17	Tachiari Airfield	Kyushu	498BG	10	10	0	
72	April 17	Kokubu Airfield	Kyushu	313BW	505BG	24	20	0
72	April 17	Target of Opportunity	Nittigahara A/F	505BG	in above	2	0	
73	April 17	Kanoya East Airfield	Kyushu	6BG	21	10	0	
73	April 17	Kanoya East Airfield	Kyushu	504BG	in above	in above	0	
74	April 17	Nittagahara Airfield	Kyushu	314BW	19BG	10	6	0
75	April 17	Kanoya Airfield	Kyushu	314BW	WING	34	30	0
75	April 17	Kanoya Airfield	Kyushu	29BG	12	9	0	
75	April 17	Kanoya Airfield	Kyushu	330BG	11	11	0	
75	April 17	Kanoya Airfield	Kyushu	39BG	11	10	0	
76	April 18	Tachiari Airfield	Kyushu	73BW	WING	22	20	2
76	April 18	Target of Opportunity	Miyazaki A/F	WING	in above	1	0	
76	April 18	Tachiari Airfield	Kyushu	497BG	11	10	2	
76	April 18	Tachiari Airfield	Kyushu	498BG	11	10	0	
77	April 18	Kanoya East Airfield	Kyushu	313BW	WING	20	19	0
77	April 18	Target of Opportunity	Kushira A/F	WING	in above	7	0	
77	April 18	Kushira Airfield	Kyushu	6BG	not avail	10	0	
77	April 18	Kanoya East Airfield	Kyushu	504BG	5	5	0	
78	April 18	Kanoya Airfield	Kyushu	314BW	WING	33	30	0
78	April 18	Kanoya Airfield	Kyushu	29BG	not avail	not avail	0	
78	April 18	Kanoya Airfield	Kyushu	39BG	11	7	0	
78	April 18	Kanoya East with 313BW	Kyushu	39BG	in above	2	0	
78	April 18	Kanoya Airfield	Kyushu	330BG	11	10	0	
79	April 18	Izumi Airfield	Kyushu	73BW	WING	23	21	0
79	April 18	Target of Opportunity	Kaimon Dake A/F	WING	in above	1	0	
79	April 18	Target of Opportunity	Unknown	WING	in above	1	0	
79	April 18	Izumi Airfield	Kyushu	499BG	not avail	10	0	
79	April 18	Izumi Airfield	Kyushu	500BG	not avail	11	0	
80	April 18	Kokubu Airfield	Kyushu	313BW	9BG	22	18	0
80	April 18	Target of Opportunity	Kanoya East A/F	9BG	in above	1	0	
80	April 18	Target of Opportunity	Izumi A/F	9BG	in above	1	0	
81	April 18	Nittagahara Airfield	Kyushu	314BW	19BG	11	11	0
	April 20	Izumi Airfield	Kyushu	313BW	504BG	6	5	0
82	April 21	Oita Airfield	Kyushu	73BW	WING	30	17	0
82	April 21	Target of Opportunity	Kagoshima	WING	in above	11	0	

MSN	Date	Target	Location	Wings	Grps	Number Up	Bombed	Lost
82	April 21	Target of Opportunity	Tomitaki A/F	WING	in above	1	0	
82	April 21	Oita Airfield	Kyushu	499BG	not avail	not avail	0	
82	April 21	Oita Airfield	Kyushu	500BG	not avail	9	0	
83	April 21	Kanoya East Airfield	Kyushu	313BW	WING	33	27	0
83	April 21	Target of Opportunity	Kanoya	WING	in above	2	0	
83	April 21	Target of Opportunity	Kokubu A/F	WING	in above	1	0	
83	April 21	Target of Opportunity	Yamakawa Harbor	WING	in above	1	0	
83	April 21	Kanoya East Airfield	Kyushu	6BG	23	22	0	
83	April 21	Kanoya Airfield	Kyushu	9BG	10	7	0	
83	April 21	Secondary Target	Kyushu	9BG	in above	1	0	
83	April 21	Last Resort Target	Kyushu	9BG	in above	1	0	
84	April 21	Kanoya Airfield	Kyushu	314BW	WING	33	27	0
84	April 21	Target of Opportunity	Kaimon Dake A/F	WING	in above	1	0	
84	April 21	Kanoya Airfield	Kyushu	29BG	11	10	0	
84	April 21	Kanoya Airfield	Kyushu	39BG	11	9	0	
84	April 21	Simon Dake A/F	Kyushu	39BG	in above	1	0	
84	April 21	Kanoya Airfield	Kyushu	330BG	11	11	0	
85	April 21	USA Airfield	Kyushu	73BW	WING	30	29	0
85	April 21	USA Airfield	Kyushu	497BG	22	20	0	
85	April 21	USA Airfield	Kyushu	500BG	9	9	0	
86	April 21	Kokubu Airfield	Kyushu	313BW	WING	35	34	0
86	April 21	Kokubu Airfield	Kyushu	505BG	23	22	0	
86	April 21	Kokubu Airfield	Kyushu	9BG	12	12	0	
87	April 21	Kushira Airfield	Kyushu	314BW	WING	31	27	0
87	April 21	Kushira Airfield	Kyushu	29BG	9	8	0	
87	April 21	Kushira Airfield	Kyushu	39BG	11	9	0	
87	April 21	Kushira Airfield	Kyushu	330BG	11	11	0	
88	April 21	Tachiari Airfield	Kyushu	73BW	WING	21	17	0
88	April 21	Tachiari Airfield	Kyushu	498BG	21	17	0	
88	April 21	Target of Opportunity	Oyodo A/F	498BG	in above	1	0	
89	April 21	Izumi Airfield	Kyushu	313BW	504BG	16	13	0
89	April 21	Target of Opportunity	Miyazake A/F	504BG	in above	2	0	
89	April 21	Target of Opportunity	Kanoya East A/F	504BG	in above	1	0	
90	April 21	Nittagahara Airfield	Kyushu	314BW	19BG	23	22	0
91	April 22	Izumi Airfield	Kyushu	73BW	WING	22	19	0
91	April 22	Target of Opportunity	Shimizu	WING	in above	1	0	
91	April 22	Izumi Airfield	Kyushu	499BG	not avail	9	0	
91	April 22	Izumi Airfield	Kyushu	500BG	not avail	10	0	
92	April 22	Kushira Airfield	Kyushu	313BW	WING	18	10	0
92	April 22	Target of Opportunity	Kanoya A/F	505BG	in above	5	0	
92	April 22	Kushira Airfield	Kyushu	9BG	9	7	0	
92	April 22	Kushira Airfield	Kyushu	505BG	not avail	2	0	
93	April 22	Miyazaki Airfield	Kyushu	314BW	WING	22	21	0
93	April 22	Miyazaki Airfield	Kyushu	19BG	11	11	0	
93	April 22	Miyazaki Airfield	Kyushu	29BG	11	11	0	
94	April 22	Tomitaka Airfield	Kyushu	73BW	WING	19	18	0
94	April 22	Tomitaka Airfield	Kyushu	497BG	11	8	0	
94	April 22	Tomitaka Airfield	Kyushu	498BG	10	10	0	
95	April 22	Kanoya Airfield	Kyushu	313BW	WING	36	28	1
95	April 22	Kanoya Airfield	Kyushu	9BG	4	3	0	
95	April 22	Kanoya Airfield	Kyushu	6BG	21	16	0	
95	April 22	Kanoya Airfield	Kyushu	505BG	5	5	0	
97	April 26	Usa Airfield	Kyushu	73BW	497BG	22	18	0
97	April 26	Target of Opportunity	Tomitaki A/F	497BG	in above	1	0	
97	April 26	Target of Opportunity	Hi-Saki A/F	497BG	in above	1	0	
98	April 26	Oita Airfield	Kyushu	73BW	WING	22	19	0
98	April 26	Oita Airfield	Kyushu	499BG	22	19	0	
98	April 26	Target of Opportunity	Tomitaki A/f	499BG	in above	2	0	
99	April 26	Saeki Airfield	Kyushu	73BW	500BG	23	19	0
100	April 26	Tomitaka Airfield	Kyushu	73BW	WING	22	21	0
100	April 26	Tomitaka Airfield	Kyushu	498BG	22	21	0	
101	April 26	Target of Opportunity	Usa A/F	498BG	22	21	0	
101	April 26	Matsuyama Airfield	Kyushu	313BW	WING	37	18 plus	0
101	April 26	Targets of Opportunity	Kochi A/F	WING	in above	16	0	
101	April 26	Targets of Opportunity	Susaki	WING	in above	1	0	

VII. The XXI Bomber Command

239

MSN	Date	Target	Location	Wings	Grps	Number Up	Bombed	Lost
101	April 26	Targets of Opportunity	Masaki	WING	in above	1	0	
101	April 26	Targets of Opportunity	Susaki A/F	WING	in above	1	0	
101	April 26	Targets of Opportunity	Sakumo	WING	in above	1	0	
101	April 26	Matsuyama Airfield	Kyushu	6BG	not avail	18	0	
101	April 26	Matsuyama Airfield	Kyushu	505BG	not avail	not avail	0	
102	April 26	Nittagahara Airfield	Kyushu	313BW	WING	18	18	0
102	April 26	Nittagahara Airfield	Kyushu	505BG	18	18	0	
102	April 26	Targets of Opportunity	Kaimon Dake AF	505BG	in above	1	0	
102	April 26	Targets of Opportunity	Miyazaki A/F	505BG	in above	1	0	
103	April 26	Miyazaki Airfield	Kyushu	9BG	21	19	0	
103	April 26	Secondary Target	Kokubu A/F	9BG	in above	1	0	
104	April 26	Kanoya Airfield	Kyushu	314BW	19BG	22	19	0
104	April 26	Targets of Opportunity	Sakamoto	19BG	in above	1	0	
104	April 26	Targets of Opportunity	Kaimon Dace	19BG	in above	1	0	
105	April 26	Kushira Airfield	Kyushu	314BW	29BG	22	13	0
105	April 26	Target of Opportunity	Kokubu A/F	29BG	in above	5	0	
105	April 26	Target of Opportunity	Chiran A/F	29BG	in above	1	0	
105	April 26	Target of Opportunity	Kanoya A/F	29BG	in above	1	0	
105	April 26	Target of Opportunity	Tanaga Shima	29BG	in above	2	0	
106	April 26	Kokubu Airfield	Kyushu	314BW	39BG	22	17	0
106	April 26	Targets of Opportunity	Miyakonojo A/F	39BG	in above	2	0	
106	April 26	Targets of Opportunity	Nagasaki Docks	39BG	in above	1	0	
106	April 26	Targets of Opportunity	Shibuski A/F	39BG	in above	1	0	
107	April 26	Miyakonojo Airfield	Kyushu	313BW	WING	21	17	0
107	April 26	Miyakonojo Airfield	Kyushu		504BG	9	8	0
107	April 26	Miyakonojo Air Field	Kyushu		330BG	20	2	0
107	April 26	Secondary Target	Miyazaki AF	330BG	in above	15	0	
107	April 26	Target of Opportunity	Yanikawa	330BG	in above	1	0	
107	April 26	Target of Opportunity	Makurazaki	330BG	in above	1	0	
108	April 27	Izumi Airfield	Kyushu	73BW	WING	22	21	1
108	April 27	Izumi Airfield	Kyushu	499BG	11	10	0	
108	April 27	Izumi Airfield	Kyushu	500BG	11	11	0	
109	April 27	Miyazaki Airfield	Kyushu	73BW	WING	23	21	0
109	April 27	Miyazaki Airfield	Kyushu	497BG	11	10	0	
109	April 27	Miyazaki Airfield	Kyushu	498BG	11	11	0	
110	April 27	Kokubu Airfield	Kyushu	313BW	WING	22	17	0
110	April 27	Target of Opportunity	Kanoya A/F	WING	in above	1	0	
110	April 27	Kokubu Airfield	Kyushu	9BG	10	9	0	
110	April 27	Kokubu Airfield	Kyushu	505BG	12	8	0	
111	April 27	Miyakonojo Airfield	Kyushu	313BW	WING	18	14	0
111	April 27	Miyakonojo Airfield	Kyushu	6BG	not avail	6	0	
111	April 27	Miyakonojo Airfield	Kyushu	504BG	not avail	8	0	
112	April 27	Kanoya Airfield	Kyushu	314BW	WING	21	21	0
112	April 27	Kanoya Airfield	Kyushu	19BG	11	10	0	
112	April 27	Kanoya Airfield	Kyushu	330BG	10	10	0	
113	April 27	Kushira Airfield	Kyushu	314BW	WING	19	15	1
113	April 27	Target of Opportunity	Kanoya A/F	WING	in above	1	0	
113	April 27	Target of Opportunity	Makurazaki A/F	WING	in above	1	0	
113	April 27	Kushira Airfield	Kyushu	29BG	7	4	0	
113	April 27	Kushira Airfield	Kyushu	39BG	12	11	1	
114	April 28	Izumi Airfield	Kyushu	73BW	500BG	25	23	0
115	April 28	Miyazaki Airfield	Kyushu	73BW	499BG	21	20	1
116	April 28	Kokubu Airfield	Kyushu	313BW	9BG	20	17	1
117	April 28	Miyakonojo Airfield	Kyushu	313BW	6BG	19	18	0
117	April 28	Target of Opportunity	Tanega Shima A/F	6BG	in above	1	0	
118	April 28	Kanoya Airfield	Kyushu	314BW	WING	23	22	1
118	April 28	Kanoya Airfield	Kyushu	19BG	11	11		
118	April 28	Kanoya Airfield	Kyushu	330BG	12	10	0	
119	April 28	Kushira Airfield	Kyushu	314BW	WING	23	22	2
119	April 28	Target of Opportunity	Kanoya A/F	WING	in above	1	0	
119	April 28	Target of Opportunity	Naka Fukara	WING	in above	1	0	
119	April 28	Kushira Airfield	Kyushu	29BG	11	10		
119	April 28	Kushira Airfield	Kyushu	39BG	12	11		
120	April 29	Miyazaki Airfield	Kyushu	73BW	497BG	22	19	2
121	April 29	Miyakonojo Airfield	Kyushu	73BW	498BG	23	22	0

MSN	Date	Target	Location	Wings	Grps	Number Up	Bombed	Lost
122	April 29	Kokubu Airfield	Kyushu	313BW	505BG	22	22	0
123	April 29	Kanoya East Airfield	Kyushu	313BW	504BG	15	14	0
124	April 29	Kanoya Airfield	Kyushu	314BW	WING	20	18	0
124	April 28	Kanoya Airfield	Kyushu	19BG	11	9	0	
124	April 28	Kanoya	Kyushu	330BG	9	9	0	
125	April 29	Kushira Airfield	Kyushu	314BW	WING	20	16	0
125	April 29	Kushira Airfield	Kyushu	29BG	9	9	0	
125	April29	Kushira Airfield	Kyushu	39BG	11	7	0	
127	April 30	Kanoya Airfield	Kyushu	314BW	19BG	11	10	0
128	April 30	Kanoya East Airfield	Kyushu	19BG	11	10	0	
129	April 30	Kokubu Airfield	Kyushu	39BG	10	7	0	
130	April 30	Oita Airfield	Kyushu	29BG	11	10	0	
131	April 30	Tomitaka Airfield	Kyushu	330BG	10	10	0	
132	April 30	Saeki Airfield	Kyushu	314BW	29BG	11	11	0
133	April 30	Tachiarai Airfield	Kyushu	314BW	29BG	11	9	0
134	April 30	Miyazaki Airfield	Kyushu	314BW	29BG	11	11	0
135	April 30	Miyakonojo Airfield	Kyushu	314BW	29BG	11	11	0
136	April 30	Kanoya Airfield	Kyushu	314BW	19BG	11	8	0
136	April 30	Target of Opportunity	Kokubu Airfield		19BG	in above	2	0
136	April 30	Target of Opportunity	Tanega Shima A/F		19BG	in above	1	0
137	May 3	Kanoya East Airfield	Kyushu	314BW	19BG	11	11	1
138	May 3	Kokubu Airfield	Kyushu	314BW	19BG	11	9	1
138	May 3	Target of Opportunity	Tanega Shima A/F		19BG	in above	1	0
140	May 3/4	Oita Airfield	Kyushu	314BW	39BG	22	11	1
141	May 4	Omura Airfield	Kyushu	314BW	330BG	10	10	0
141	May 4	Target of Opportunity	Kushikino A/F		330BG	in above	1	0
142	May 4	Saeki Airfield	Kyushu	314BW	39BG	9	9	0
143	May 4	Matsuyama Naval Air Station	Kyushu	314BW	330BG	19	17	0
143	May 4	Target of Opportunity	Lomuta		330BG	in above	1	0
143	May 4	Target of Opportunity	Wakagama		330BG	in above	1	0
144	May 4	Oita Airfield	Kyushu	314BW	29BG	17	17	0
145	May 4	Tachiari A/F	Kyushu	314BW	29BG	11	10	2
147	May 5	Kanoya Airfield	Kyushu	314BW	19BG	11	10	1
148	May 5	Chiran Airfield	Kyushu	314BW	19BG	11	8	0
149	May 5	Ibusuki Airfield	Kyushu	314BW	19BG	10	10	0
151	May 5/6	Kanoya Airfield	Kyushu	313BW	6BG	10	10	0
152	May 7	Ibusuki Airfield	Kyushu	313BW	6BG	11	10	0
153	May 7	Oita Airfield	Kyushu	313BW	505BG	10	10	2
154	May 7	Usa Airfield	Kyushu	313BW	505BG	11	11	1
155	May 7	Kanoya Airfield	Kyushu	314BW	504BG	21	17	0
155	May 7	Target of Opportunity	Omura City		504BG	in above	1	0
156	May 8	Miyakonojo Airfield	Kyushu	313BW	504BG	12	12	0
157	May 8	Oita Airfield	Kyushu	313BW	9BG	12	11	0
158	May 8	Matsuyama West Airfield	Shikoku	313BW	9BG	22	16	0
159	May 8	Matsuyama West Airfield	Shikoku	313BW	9BG	23	15	0
159	May 8	Target of Opportunity	Kumamota A/F		9BG	in above	2	0
160	May 10	Usa Airfield	Kyushu	313BW	6BG	22	15	0
160	May 10	Target of Opportunity	Matsuyama A/F		6BG	in above	2	0
160	May 10	Target of Opportunity	Nakamura		6BG	in above	1	0
160	May 10	Target of Opportunity	Nobeoka		6BG	in above	1	0
160	May 10	Target of Opportunity	Susaki Docks		6BG	in above	1	0
160	May 10	Kyushu Airfields	Unknown	313BW	9BG	23	16	0
160	May 10	Target of Opportunity	Unknown		9BG	in above	2	0
161	May 10	Miyazaki Airfield	Kyushu	313BW	505BG	12	7	0
161	May 10	Target of Opportunity	Tanoga Naval		505BG	in above	1	0
162	May 10	Kanoya Airfield	Kyushu	313BW	505BG	12	4	0
162	May 10	Target of Opportunity	Kyushu		505BG	in above	6	0
167	May 10	Oita Airfield	Kyushu	313BW	504BG	20	17	0
167	May 10	Target of Opportunity	Miyachi		504BG	in above	1	0
168	May 11	Saeki Airfield	Kyushu	313BW	505BG	11	7	0
168	May 11	Target of Opportunity	shimo Kawaguchi		505BG	in above	1	0
169	May 11	Nittagahara Airfield	Kyushu	313BW	6BG	11	5	0
169	May 11	Targets of Opportunity	Hoso Shima NB		6BG	in above	6	0
169	May 11	Targets of Opportunity	Kochi A/F		6BG	in above	1	0
169	May 11	Targets of Opportunity	Kubokawa		6BG	in above	1	0

MSN	Date	Target	Location	Wings	Grps	Number Up	Bombed	Lost
169	May 11	Targets of Opportunity	Shibushi		6BG	in above 1	0	
169	May 11	Targets of Opportunity	Tanega Shima		6BG	in above 1	0	
170	May 11	Miyazaki Airfield	Kyushu		9BG	12	11	0
171	May 11	Miyakonojo Airfield	Kyushu		9BG	11	10	0

OPERATION STARVATION: THE AERIAL MINING OF THE JAPANESE HOME WATERS

The Japanese Home Islands, Honshu, Kyushu, Shikoku and Hokkaido, contain very few natural resources, and the Japanese war economy depended on the importation of raw materials from conquered territory.

This fact had not escaped the U.S. Navy planners at Navy Pacific Headquarters at Pearl Harbor. Submarines were sent to the Far Eastern reaches of the Pacific, from the Home Islands on the north to the East Indies and Southeast Asia to the south, with orders to perform unrestricted warfare on the Japanese Merchant Marine.

The one area the submarines could not penetrate without extreme risk was the Sea of Japan and the Inland Sea, bounded by the south shore of Honshu, the northern side of Shikoku and the eastern side of Kyushu. Entry to this area from the Sea of Japan was through a chokepoint called the Shimonoseki Strait, a narrow body of water between Kyushu and Honshu. It was a premier target environment for submarines and they had no way to get to it.

The solution was to use the B-29s of the XXI Bomber Command to sow and replenish minefields not only in the approaches to the Shimonoseki Strait and the harbors of the Inland Sea, but also in ports on the Korean Peninsula.

Airborne delivery of marine mines had been successfully accomplished by the XX Bomber Command in mine-laying missions to Singapore, Saigon, Cam Ranh Bay, the Yangtze River and the Moesi River at Palembang in the East Indies.

By war's end the planes and men of the 313th had established 28 separate minefields. Between March 27, 1945, and war's end, they flew 46 missions, visiting up to nine minefields per mission.

Did it work? Yes it did. By July of 1945 ship movements across the Sea of Japan, through the Shimonoseki Strait and to the ports of the Inland Sea had practically stopped. And without those ships, Japan's war industry — what was left of it after the fire raids — was impotent.

For an explanation of the columns, see page 141.

MSN	Date	Target	Location	Grps	No. Up	Bombed	Lost
47	March 27/28	Mining Mission 1	Shimonoseki Strait	WING	unknown	105	3
47	March 27/28	Mine Field Mike	Shimonoseki Strait	505BG	24	24	2
47	March 27/28	Mine Field Mike	Shimonoseki	6BG	not avail	30	0
47	March 27/28	Mine Field Mike	Shimonoseki	504BG	49	20	1
47	March 27/28	Mine Field Love	Shimonoseki	9BG	31	31	0
49	March 30/31	Mining Mission 2	Shimonoseki	WING	95	87	2
49	March 30/31	Mine Field Item	Kure	6BG	23	23	0
49	March 30/31	Mine Field Item	Kure	9BG	8	in above	0
49	March 30/31	Mine Field Jig	Hiroshima Entry	9BG	11	10	1
49	March 30/31	Mine Field Love	East Shimonoseki	504BG	14	14	0
49	March 30/31	Mine Field Love	East Shimonoseki	505BG	29	in above	0
49	March 30/31	Mine Field Charlie	Hiroshima	9BG	1	1	0
49	March 30/31	Mine Field Roger	Sasebo Approach	504BG	3	3	0
49	March 30/31	Mine Field Roger	Sasebo Approach	9BG	4	4	0
49	March 30/31	Mine Field Roger	Sasebo Approach	6BG	2	2	0
50	March 31	Mine Field Item	Kyushu	WINGS	149	137	1
52	April 1/2	Mining Mission 3	Kure	9BG	6	6	0
53	April 2/3	Mining Mission 4 Mine Field Jig	Hiroshima	504BG	10	9	0
54	April 3/4	Mining Mission 5 Mine Field Item	Hiroshima	505BG	9	9	0
66	April 12/13	Mining Mission 7	Shimonoseki	9BG	6	5	0
139	May 3/4	Mining Mission 8	Shimonoseki	WING	100	91	0
139	May 3/4	Mine Field Able	Kobe/Osaka	504BG	16	16	0
139	May 3/4	Mine Field Able	Kobe/Osaka	505BG	22	19	0
139	May 3/4	Mine Field Mike	Shimonoseki	505BG	4	4	0

MSN	Date	Target	Location	Grps	Number Up	Bombed	Lost
139	May 3/4	Mine Field Mike	Shimonoseki	9BG	24	19	0
139	May 3/4	Secondary Target	Shimonoseki	9BG	in above	1	0
139	May 3/4	Mine Field Love	Moji Area	6BG	14	14	0
139	May 3/4	Mine Field Love	Suo Nada	6BG	20	19	0
150	May 5	Mining Mission 9	See Below	WING	99	90	0
150	May 5	Mine Field Able	Kobe-Osaka	505BG	12	in above	0
150	May 5/6	Mine Field Baker	Aki/Nada	6BG	5	in above	0
150	May 5/6	Mine Field Dog	Shoda Shima	9BG	28	in above	0
150	May 5/6	Mine Field Easy	Tokyo & Ise Bays	504BG	10	in above	0
150	May 5/6	Mine Field Fox	Bingo Nada	6BG	20	in above	0
150	May 5/6	Mine Field Jig/Item	Hiroshima/Kure	504BG	7	in above	0
150	May 5/6	Mine Field King	Tokuyama	505BG	8	in above	0
150	May 5/6	Mine Field Oboe	Tokyo	504BG	4	in above	0
150	May 5/6	Mine Field Tare	Nagoya	505BG	5	in above	0
173	May 13/14	Mining Mission 10	See Below	9BG	12	12	0
173	May 13/14	Mine Field Mike	Shimonoseki	9BG	3	3	0
173	May 13/14	Mine Field Love	Shimonoseki	9BG	5	5	0
173	May 13/14	Mine Field Uncle	Niigata	9BG	4	4	0
175	May 16/17	Mining Mission 11	Shimonoseki	WING	30	25	0
175	May 16/17	Secondary Target	Shimonoseki	WING	in above	2	0
175	May 16/17	Mine Field Mike	Shimonoseki	9BG	24	20	0
175	May 16/17	Mine Field Zebra	Maizuru	9BG	6	5	0
177	May 18/19	Mining Mission 12	See Below	9BG	34	30	0
177	May 18/19	Mine Field Love	Shimonoseki	9BG	22	18	0
177	May 18/19	Mine Field Zebra	Tsuruga	9BG	12	12	0
179	May 20/21	Mining-Mission 13	Shimonoseki	9BG	32	30	3
179	May 20/21	Mine Field Mike	Shimonoseki	9BG	23	22	0
179	May 20/21	Mine Field Love	Shimonoseki	9BG	5	4	0
179	May 20/21	Mine Field Zebra	Maizura Bay	9BG	4	4	0
180	May 22/23	Mining Mission 14	Shimonoseki	9BG	32	30	1
180	May 22/23	Mine Field Mike	Shimonoseki	9BG	9	8	0
180	May 22/23	Mine Field Mike	Shimonoseki	9BG	9	8	0
180	May 22/23	Mine Field Love	Shimonoseki	9BG	14	14	0
182	May 24/25	Mining Mission 15	Fushiki/Niigata	9BG	30	27	0
182	May 24/25	Secondary Targets	Fushiki/Niigata	9BG	in above	2	0
182	May 24/25	Mine Field Uncle	Niigata	9BG	11	9	0
182	May 24/25	Mine field Nan	Fushiki	9BG	10	9	0
182	May 24/25	Mine Field Mike	Shimonoseki	9BG	9	7	0
184	May 25/26	Mining Mission 16	See Below	WING	30	29	0
184	May 25/26	Mine Field Mike	Shimonoseki	9BG	2	2	0
184	May 25/26	Mine Field Nan	Fushiki	9BG	6	6	0
184	May 25/26	Mine Field Love	Shimonoseki	9BG	7	7	0
184	May 25/26	Mine Field Charlie	Fukuoka	9BG	15	14	0
185	May 27	Mining Mission 17	See Below	9BG	11	9	1
185	May 27	Mine Field Charlie	Fukuoka Bay	9BG	1	1	0
185	May 27	Mine Field Love	Shimonoseki	9BG	10	8	0
190	June 7/8	Mining Mission 18	See Below	WING	31	26	0
190	June 7/8	Mine Field Mike	Shimonoseki	505BG	10	9	0
190	June 7/8	Mine Field Love	Shimonoseki	505BG	10	7	0
190	June 7/8	Mine Field Charlie	Fukuoka Bay	505BG	5	4	0
190	June 7/8	Mine Field Mike	Shimonoseki	505BG	10	6	0
194	June 9/10	Mining Mission 19	Shimonoseki	WING	28	26	0
194	June 9/10	Mine Field Mike	Shimonoseki	505BG	20	19	0
194	June 9/10	Mine Field Love	Shimonoseki	505BG	8	7	0
201	June 11/12	Mining Mission 20	See Below	WING	26	26	0
201	June 11/12	Mine Field Mike	Shimonoseki	505BG	15	15	0
201	June 11/12	Mine Field Zebra	Tsuruga	505BG	11	11	0
202	June 13/14	Mining Mission 21	See Below	WING	30	29	0
202	June 13/14	Mine Field Love	Shimonoseki	505BG	17	17	0
202	June 13/14	Mine Field Uncle	Niigata	505BG	12	12	0
213	June 19/20	Mining Mission 24	See Below	WING	28	28	0
213	June 19/20	Mine Field Mike	Shimonoseki	505BG	10	10	0
213	June 19/20	Mine Field Uncle	Niigata	505BG	8	8	0
213	June 19/20	Mine Field Zebra	Miazuru/Miyazu	505BG	10	10	0
214	June 21/22	Mining Mission 25	See Below	WING	30	27	0

VII. The XXI Bomber Command

MSN	Date	Target	Location	Grps	Number Up	Bombed	Lost
214	June 21/22	Mine Field X-Ray	Senzaki	505BG	9	9	0
214	June 21/22	Mine Field Nan	Fushiki	505BG	10	10	0
214	June 21/22	Mime Field Able	Kobe/Osaka	505BG	6	6	0
221	June 23/24	Mining Mission 26	Fukuoka	WING	27	26	1
221	June 23/24	Mine Field Charlie	Fukuoka	505BG	9	9	
221	June 23/24	Mine Field Yoke	Sakai	505BG	8	8	
221	June 23/24	Mine Field Uncle	Niigata	505BG	9	9	
222	June 25/26	Mining Mission 27	See Below	WING	27	26	0
222	June 25/26	Mine Field Mike	Shimonoseki	505BG	11	11	0
222	June 25/26	Mine Field Love	Shimonoseki	505BG	6	6	0
222	June 25/26	Mine Field Zebra	Tsuruga Bay	505BG	9	9	0
233	June 27/28	Mining Mission 28	See Below	WING	30	29	0
233	June 27/28	Mine Field X-Ray	Hagi	505BG	14	14	0
233	June 27/28	Mine Field Able	Kobe/Osaka	505BG	8	8	0
233	June 27/28	Mine Field Uncle	Niigata	505BG	7	7	0
239	June 29/30	Mining - Mission 29	See Below	WING	29	25	0
239	June 29/30	Mine Field Mike	Yawata	505BG	12	12	0
239	June 29/30	Mine Field Zebra	Maizyuru	505BG	4	4	0
239	June 29/30	Mine Field Uncle	Miyahoura	505BG	9	9	0
244	July 1/2	Mining Mission 30	Shimonoseki City	WING	28	24	0
244	July 1/2	Mine Field Love	Shimonoseki City	505BG	9	2	0
244	July 1/2	Mine Field Nan	Fushiki, Nanao	505BG	15	22	0
246	July 3/4	Mining Mission 31	See Below	WING	31	28	0
246	July 3/4	Mine Field William	Funakoshi	505BG	6	10	0
246	July 3/4	Mine Field Zebra	Maizuru	505BG	4	3	0
246	July 3/4	Mine Field Mike	Shimonoseki	505BG	16	13	0
256	July 9/10	Mining Mission 32	See Below	WING	30	29	1
256	July 9/10	Mine Field Mike	Shimonoseki	6BG	14	not avail	not avail
256	July 9/10	Mine Field Uncle	Niigata/Nanao	6BG	9	not avail	not avail
256	July 9/10	Mine Field Nan	Fushiki	6BG	7	not avail	not avail
262	July 11	Mining Mission 33	See below	WING	30	27	0
262	July 11	Mine Field Mike	Shimonoseki	6BG	7	not avail	0
262	July 11	Mine Field Rashin	Korea	6BG	6	not avail	0
262	July 11	Mine Field Fusan	Korea	6BG	9	not avail	0
262	July 11	Mine Field Zebra	Maizuru	6BG	8	not avail	0
268	July 13	Mining Mission 34	See Below	WING	32	27	0
268	July 13	Mine Field Mike	Shimonoseki	6BG	9	not avail	0
268	July 13	Mine Field Seishin	Korea	6BG	6	not avail	0
268	July 13	Mine Field Masan-Reisu	Korea	6BG	13	not avail	0
268	July 15/16	Mine Field Charlie	Fukuoka	6BG	3	not avail	0
269	July 15/16	Mining Mission 35	See Below	6BG	28	26	0
269	July 15/16	Mine Field Uncle/Nan	Niigata/Naoetsu	6BG	9	not avail	0
269	July 15/16	Mine Field Rashin	Rashin, Korea	6BG	8	not avail	0
269	July 15/16	Mine Field Genzen-Konan	Genzen, Korea	6BG	4	not avail	0
269	July 15/16	Mine Field Fusan	Fusan, Korea	6BG	7	not avail	0
275	July 17/18	Mining Mission 36	See Below	WING	30	28	0
275	July 17/18	Mine Field Mike	Moji	6BG	10	not avail	0
275	July 17/18	Mine Field Seishin	Seishin, Korea	6BG	6	not avail	0
275	July 19	Mine Field Nan	Nanao Bay	6BG	14	not avail	0
276	July 19	Mining Mission 37	See Below	6BG	31	29	1
276	July 19	Mine Field Able	Shimonoseki	6BG	10	not avail	
276	July 19	Mine Field Uncle/Nan	Niigata	6BG	7	not avail	
276	July 19	Mine Field Zebra	Maizura/Miyama	6BG	9	not avail	
276	July 19	Mine Field Genzan - Konan	Korea	6BG	5	not avail	
282	July 22	Mining Mission 38	See Below	6BG	30	26	0
282	July 22	Mine Field Mike/Love	Shimonoseki	6BG	15	not avail	0
282	July 22	Mine Field Rashin	Korea	6BG	8	not avail	0
282	July 23/24	Mine Field Fusan/Musan	Korea	6BG	3/4	not avail	0
292	July 25/26	Mining Mission 39	See Below	WING	30	29	0
292	July 25/26	Mine Field Seishin	Seishin, Korea	504BG	6	5	0
292	July 25/26	Mine Field Fusan	Fusan, Korea	504BG	7	7	0
292	July 25/26	Mine Field Nan	Nanao Bay	504BG	11	11	0
292	July 25/26	Mine Field Zebra	Tsuruga Bay	504BG	6	6	0
296	July 27/28	Mining Mission 40	See Below	WING	31	24	2
296	July 26/27	Mine Field Mike	Shimonoseki	504BG	10	27	3

MSN	Date	Target	Location	Grps	Number Up	Bombed	Lost
296	July 26/27	Mine Field Uncle	Niigata	504BG	8	in above	
296	July 26/27	Mine Field Zebra	Maizura	504BG	10	in above	
296	July 26/27	Mine Field X-Ray	Senzaki	504BG	3	in above	
304	July 29/30	Mining Mission 41	See Below	WING	30	27	0
304	July 29/30	Mine Field Rashin	Raishin, Korea	504BG	14	in above	0
304	July 29/30	Mine Field Mike	Shimonoseki	504BG	10	in above	0
304	July 29/30	Mine Field Charlie	Fukuoka Bay	504BG	6	in above	0
305	August 1/2	Mining Mission 42	See Below	WING	45	38	0
305	August 1/2	Mine Field Rashin	Raishin, Korea	504BG	9	in above	0
305	August 1/2	Mine Field Seishin	Seishin, Korea	504BG	7	in above	0
305	August 1/2	Mine Field Mike	Moji Area	504BG	17	in above	0
305	August 1/2	Mine Field George	Hamada,	504BG	7	in above	0
305	August 1/2	Mine Field X-Ray	Hagi, Oura	504BG	5	in above	0
311	August 5/6	Mining Mission 43	See Below	504BG	30	27	0
311	August 5/6	Mine Field X-Ray	Hagi Harbor	504BG	6	in above	0
311	August 5/6	Mine Field Zebra	Tsuruga Harbor	504BG	10	in above	0
311	August 5/6	Mine Field Geijitsu	Geijitsu, Korea	504BG	5	in above	0
311	August 5/6	Mine Field Rashin	Rashin, Korea	504BG	9	in above	0
318	August 7/8	Mining Mission 44	See Below	504BG	27	27	0
318	August 7/8	Mine Field Rashin	Raishin, Korea	504BG	8	8	0
318	August 7/8	Mine Field Mike/Love	Shimonoseki	504BG	5	5	0
318	August 7/8	Mine Field Zebra	Maizuru	504BG	8	8	0
318	August 7/8	Mine Field Yoke	Sakai	504BG	6	6	0
324	August 9/10	Mining Mission 45	See Below	WING	32	32	0
324	August 9/10	Mine Field Rashin	Raishin, Korea	504BG	7	7	0
324	August 9/10	Mine Field Seishin	Seishin, Korea	504BG	7	7	0
324	August 9/10	Mine Field Mike	Shimonoseki	504BG	11	11	0
324	August 9/10	Mine Field Genzan	Genzan, Korea	504BG	7	7	0
331	August 14	Mining Mission 46	See Below	WING	41	38	0
331	August 14	Mine Field Nan	Nanao Bay	504BG	8	in above	0
331	August 14	Mine Field Mike	Shimonoseki	504BG	10	in above	0
331	August 14	Mine Field Zebra	Tsuruga Bay	504BG	17	in above	0
331	August 14	Mine Field George	Hamada Harbor	504BG	6	in above	0

"Operation Starvation" Minefield Locations

Minefield	Locations
ABLE	Kobe/Osaka approaches and harbors
BAKER	Aki Nada; main Inland Sea shipping lane between Hazuma Hana and Kajitoro Hana
CHARLIE	Fukuoka Bay and Harbor
DOG	North of Shoda Shima
EASY	Bisan Seto; south of Shodo Shima
FOX	Bingo Nada between Imabari and MiSaki
GEORGE	Hamada
HOW	Main harbor facilities at Kure Naval Base and the anchorage between Kureko and Nishinomi Shima
ITEM	Main shipping channel to Kure through Hashirashima Suido; southern entrance to Kure between Nishinomi and Kurahashi Shima
JIG	Entrance to Hiroshima Harbor between Itsuku Shima and Nishinomi Strait; northern approach to Hiroshima and Kure
KING	Channel leading to the Port of Tokuyama and anchorages of Port
LOVE	Outer eastern approaches to Shimonoseki Strait and Moji area
MIKE	Inner Shimonoseki approaches and strait; Moji Harbor; western approaches to Shimonoseki Kaikyo
NAN	Fushiki; Nanao Bay
OBOE	Tokyo Bay
ROGER	Western approach to Sasebo Naval Base and Mura Wan
TARE	Nagoya; across the mouth of Ise Wan
UNCLE	Niigata; Sakata
WILLIAM	Fungkoshi; Funakawa
X-RAY	Senzaki; Hagi; Oura
YOKE	Sakai
ZEBRA	Maizura Bay; Tsuruga Bay and harbor; Miyazu
RASHIN	Korea
SEISHIN	Korea
MASAN	Korea
FUSAN	Korea
PUSAN	Korea
GEIJITSU	Korea

VII. The XXI Bomber Command

"Pumpkin" Missions of the 509th Composite Group

The 509th Composite Group, on paper assigned to the 313th Bomb Wing, was, in reality, a completely independent unit reporting directly to 20th Air Force. Their reason for existence was, when the decision was made, to drop the atomic weapons.

To this end they ran practice missions against possible atomic targets, using bombs of the same general shape and weight as the "Fat Man." For lack of a better term, these were called Pumpkin missions.

Eighteen of these missions were flown, exclusive of the drops on Hiroshima and Nagasaki.

Date	Target	Location	Wings	Grps	Number Up	Bombed	Lost
July 20	Pumpkin Mission #1	Koriyama	313BW	509CG	3	3	0
July 20	Pumpkin Mission #2	Fukushima	313BW	509CG	2	2	0
July 20	Pumpkin Mission #3	Nagaoka	313BW	509CG	2	2	0
July 20	Pumpkin Mission $4	Toyama	313BW	509CG	3	3	0
July 24	Pumpkin Mission #5	Niihama	313BW	509CG	3	3	0
July 24	Pumpkin Mission #6	Kobe	313BW	509CG	4	4	0
July 24	Pumpkin Mission #7	Osaka	313BW	509CG	3	3	0
July 26	Pumpkin Mission #8	Nagaoka	313BW	509CG	4	4	0
Jily 26	Pumpkin Mission #9	Toyama	313BW	509CG	6	6	0
July 29	Pumpkin Mission #10	Ube	313BW	509CG	3	3	0
July 29	Pumpkin Mission #11	Koriyama	313BW	509CG	3	3	0
July 29	Pumpkin Mission #12	Yokkaichi	313BW	509CG	2	2	0
August 8	Pumpkin Mission #14	Osaka	313BW	509CG	4	4	0
August 8	Pumpkin Mission #15	Yokkaichi	313BW	509CG	2	2	0
August 14	Pumpkin Mission #17	Nagoya	313BW	509CG	4	4	0
August 14	Pumpkin Mission #18	Koroma	313BW	509GC	3	3	0

VIII

Strategic Air Command

The Strategic Air Command was created on 21 March 1946 to function as a long range nuclear delivery deterrent, as well as to maintain the capability to deliver conventional munitions as needed.

During the nine years the Strategic Air Command flew B-29s, 1946–1954, they maintained as many as 585 planes in thirteen bombardment wings (1951) and 83 RB-29s in four groups the same year.

B-29 Strength, by Year

Unit	1946	1947	1948	1949	1950	1951	1952	1953	1954	1955
2nd Bomb Group		30	45							
6th Bomb Wing						45				
7th Bomb Group	30	30	30							
9th Bomb Wing					45	45	45	45	45	
19th Bomb Group									30	
22nd Bomb Group			30	30	45					
22nd Bomb Wing						45	30	30	30	
28th Bomb Group	30	30	30							
40th Bomb Group	30									
40th Bomb Wing								30		
43rd Bomb Group	30	30								
44th Bomb Wing							45	30		
68th Bomb Wing								30	30	
90th Bomb Wing							45			
92nd Bomb Group		30	30	30	45					
93rd Bomb Group		30	45	45						

Unit	1946	1947	1948	1949	1950	1951	1952	1953	1954	1955
95th Bomb Wing							30			
97th Bomb Group	30	30	45	45						
98th Bomb Group		30	30	30	45					
98th Bomb Wing						45	30	30	30	
106th Bomb Wing						45	30			
301st Bomb Group		30	45	45	45					
301st Bomb Wing						45	45			
303rd Bomb Wing						45	30			
305th Bomb Wing						45				
306th Bomb Group			30	30	45					
307th Bomb Group	30	30	30	30	45					
307th Bomb Wing						45	30	30	30	
308th Bomb Wing						45	30	30		
310th Bomb Wing							30	30		
320th Bomb Wing							30			
340th Bomb Wing							30	30		
376th Bomb Wing						45	30	30		
444th Bomb Group	30									
448th Bomb Group	30									
449th Bomb Group	30									
467th Bomb Group	30									
485th Bomb Group	30									
498th Bomb Group	30									
498th Bomb group	30									
509th Bomb Group	30	30	30	45	45					
Total	360	360	430	330	360	585	480	285	165	

RB-29 Strength, by Year

Unit	1946	1947	1948	1949	1950	1951	1952	1953	1954	1955
1st Strategic Recon Sq			12							
5th Recon Group				36	36					
9th Recon Group				12						
16th Photo Recon Sq	12	12	12							
55th Recon Group				12						
68th Strategic Recon Wg							36			
90th Strategic Recon Wg								30	30	

Unit	1946	1947	1948	1949	1950	1951	1952	1953	1954	1955
91st Recon Group				24	12	12				
91st Strategic Recon Sq							10	10	10	
111th Strategic Recon Wg						35	30			
Total	12	12	24	84	48	83	70	40	10	

Bombardment and Reconnaissance Units Assigned B-29s

Organization	Tail Codes	Based At	Years*
2nd Bombardment Group, Medium	Empty Square	Chatham AFB GA	1947–1948
5th Strategic Reconnaissance Wing	Circle X	Mountain Home AFB ID	1949–1950
		Travis AFB CA	1949–1950
6th Bombardment Wing, Medium	Empty Triangle	Walker AFB NM	1951
7th Bombardment Group	Empty Triangle	Carswell AFB TX	1946–1948
9th Bombardment Wing	Circle R	Mountain Home AFB ID	1950–1954
16th Photo Recon Squadron	Unknown	Unknown	1946–1948
19th Bombardment Wing	Black Stripe	Kadena AB Okinawa	1950–1954
22nd Bombardment Group/Wing	Circle E	March AFB CA	1948–1954
28th Bombardment Group	Triangle S	Ellsworth AFB SD	1946–1948
32nd Composite Wing	Unknown	Kadena AB Okinawa	Unknown
40th Bombardment Group	Triangle S	Unknown	1946
40th Bombardment Wing	Triangle S	Schilling AFB KS	1953
43rd Bombardment Group	Circle K	Davis-Monthan AFB AZ	1946–1947
44th Bombardmant Wing	Circle T	Lake Charles AFB LA; B-29 Trainer	1951–1952
55th Reconnaissance Group	Unknown	Unknown	1949
55th Strategic Recon Wing	Square V	Forbes AFB KS	1951
68th Bombardment Wing	Unknown	Lake Charles AFB LA	1952–1953
90th Bombardment Wing	Unknown	Fairchild AFB WA	1951
90th Reconnaissance Group	Unkown	Unknown	1949–1951
91st Strategic Recon Wing	Square I	McGuire AFB NJ	1949–1954
		Barksdale AFB LA	
		Lockbourne AFB OH	
92nd Bombardment Group	Circle W	Fairchild AFB WA	1947–1951
93rd Bombardment Group	Circle M	Castle AFB CA	1947–1949
95th Bombardment Wing	Unknown	Unknown	1952
97th Bombardment Group	Triangle O	Eielsen AFB AK	1946–1949
		Smoky Hill AFB KS	
		Biggs AFB TX	
98th Bombardment Group	Square H	Fairchild AFB WA	1946-1950
	Square H	Yokota AB Japan	1950–1954
106th Bomb Wing	Circle A	Unknown	1951–1952
301st Bombardment Group	Square A	Smoky Hill AFB KS	1947–1950
301st Bombardment Wing	Square A	Barksdale AFB LA	1951–1952
303rd Bombardment Wing	Unknown	Davis-Monthan AFB AZ	1951–1952
305th Bombardment Wing	Unknown	MacDill AFB FL	1951
306th Bombardment Group	Square P	MacDill AFB FL	1948–1950
307th Bombardment Group	Square Y	MacDill AFB FL	1946–1950
	Square Y	Yokota AB Japan	1950–1954
308th Reconnaissance Group	Square O	Forbes AFB KS	Unknown
		Hunter AFB GA	Unknown
310th Bombardment Wing	Unknown	Forbes AFB KS	1952–1953
		Schilling AFB KS	

*Years in which the organization was equipped with B-29s.

VIII. Strategic Air Command

Organization	Tail Codes	Based At	Years
320th Bombardment Wing	Circle A	March AFB CA	1952
340th Bombardment Wing	Unknown	Unknown	1952–1953
376th Bombardment Wing	Unknown	Forbes AFB KS	1951–1953
	Unknown	Barksdale AFB LA	
444th Bomb Group	Triangle N	Unknown	1946
448th Bomb Group	Unknown	Unknown	1946
449th Bomb Group	Unknown	Unknown	1946
467th Bomb Group	Unknown	Unknown	1946
485th Bomb Group	Unknown	Unknown	1946
498th Bomb Group	Unknown	Unknown	1946
509th Bomb Group	Circle Arrow	Walker AFB NM	1946–1950

IX

THE B-29 IN KOREA

A detailed discussion of the use, and misuse, of the B-29 in the Korean Conflict is not within the parameters of this reference book. From the advent of the Chinese intervention in November 1950, the B-29s, which had essentially destroyed all strategic targets in North Korea, were used against tactical targets.

This resulted in small numbers of planes dropping partial bomb loads on numerous ground targets. The "Korean Tactical Drop Sample" list included in this section shows the 19th Bomb Group bombardment activity for the period 2 May 1951 through 14 May 1951. This activity was typical of all of the three B-29 bomb groups still active in Korea in 1951. In the twelve days and nights covered in this list, seventy-two planes made eighty-eight drops of between five and thirty-two bombs on a variety of tactical targets. In many cases the air crews did not know where they were to drop until over North Korean or Chinese territory.

This section also includes a brief discussion of each of the five B-29 Bomb Groups in Korea; plane rosters for those five groups; a list of known B-29 losses in the Korean Conflict; and a Korean Airfield Code List.

Bomb Groups in Korea

19th Bomb Group (Medium)

On 26 June, 1950, one day after the Inmum Gun (the North Korean Army) crossed the 38th Parallel, the 19th Bomb Group left its home base at Andersen AFB on Guam and staged to Kadena Air Base on Okinawa. On 28 June, four planes of the group attacked rail and marshalling yards in the vicinity of Seoul. This was the first of just under 650 combat missions during the course of the war.

The 19th, the only Bombardment Group not in the Strategic Air Command chain of command and, in 1950, the only Bombardment Group permanently stationed outside the continental limits of the United States, was under the operational control of the 20th Air Force until 8 July 1950. On that date it was attached to the FEAF Bomber Command (Provisional), where it would remain for the duration of the conflict. There were at least 33 reported aircraft losses in the 19th BG. Combat components of the 19th BG were the 28th, 30th and 93rd Bombardment Squadrons.

22nd Bomb Group (Medium)

The 22nd Bomb Group of SAC's 22nd Bomb Wing, March Air Force Base, was deployed on an 89-day TDY to Kadena Air Base on Okinawa, starting in early July, 1950. Organizationally, the 22nd was placed under the command of FEAF Bomber Command (Provisional) for its deployment, as was the 92nd BG.

The group flew its first mission, to Wonson, North Korea, on 13 July. When the 22nd returned to March AFB in late October or early November 1950, 335 sorties had been flown. One plane, 44-62279, was reported lost during operations from Okinawa. Combat components of the group were the 2nd, 19th and 33rd Bombardment Squadrons.

The 89 day TDY was a bureaucratic ploy to reduce the amount of money due to the airmen and officers sent on detached duty away from their home base. Regulations stated that detached service of 90 days or more was to be considered a permanent change of station (PCS), and the officer or airman thus ordered was entitled to moving costs and travel allowances for himself and his dependents. By restricting the span of temporary duty orders to 89 days, the bureaucracy insured that personnel were not eligible for moving costs and travel allowances.

92nd Bomb Group (Medium)

In early July 1950, the first planes of the 92nd Bomb Group began to arrive at Yokota Air Base, outside of Tachikawa, Japan, with deployment completed on 13 July. Targets of the 92nd over the next three months included factories, refineries, hydroelectric plants, air fields, bridges, tunnels, troop concentrations and other interdiction targets. The group returned to Spokane, Washington, in late October or early November 1950. Five planes — 44-61617, 44-61923, 44-62084, 44-62211 and one serial number unknown — were lost during the deployment.

98th Bomb Group (Medium)

The first planes of the 98th Bomb Group arrived at Yokota on 5 August 1950. The first mission — to Pyongyang — was flown two days later. The group's Korean Deployment concentrated mainly on targets of interdiction for the remainder of the war.

The last mission was flown on 25 July 1953. There were 34 known losses. Combat components of the group were the 343rd, 344th and 345th Bombardment Squadrons.

307th Bomb Group (Medium)

The third group that was to be in the Korean Conflict for its entirety was the 307th. It deployed from MacDill AFB on 1 August 1950 to Kadena Air Base on Okinawa, and by the end of hostilities in 1953 had flown over 5800 sorties. Twenty-two planes were reported lost. Combat components of the group were the 370th, 371st, and 372nd Bombardment Squadrons.

BOMB GROUP PLANE ROSTERS

19th Bomb Group

42-63557	42-93971	44-61657	44-61897	44-62063	44-62302	44-69866	44-86316	44-86414	45-21716
42-65272	42-94009	44-61669	44-61902	44-62071	44-62303	44-69959	44-86323	44-86422	45-21725
42-65306	42-94043	44-61693	44-61932	44-62099	44-69672	44-69999	44-86328	44-86433	45-21743
42-65333	42-94099	44-61705	44-61948	44-62110	44-69682	44-70007	44-86330	44-86446	45-21745
42-65352	44-27262	44-61718	44-61957	44-62152	44-69763	44-70012	44-86331	44-87591	45-21746
42-65357	44-27288	44-61749	44-61967	44-62170	44-69771	44-70041	44-86335	44-87596	44-21749
42-65361	44-61535	44-61751	44-62002	44-62183	44-69786	44-70042	44-86349	44-87597	
42-65369	44-61562	44-61790	44-62008	44-62201	44-69802	44-70077	44-86359	44-87618	
42-65370	44-61638	44-61815	44-62011	44-62218	44-69817	44-70125	44-86370	44-87657	
42-93874	44-61642	44-61830	44-62025	44-62224	44-69818	44-70134	44-86376	44-87661	
42-93903	44-61656	44-61835	44-62053	44-62253	44-69856	44-86254	44-86387	44-87734	

22nd Bomb Group

42-65275	44-27263	44-27278	44-61669	44-61950	44-62196	44-69661	44-69898	44-86366	44-21735
42-94038	44-27276	44-27292	44-61690	44-61954	44-62199	44-69661	44-70042	44-86414	45-21821
44-27260	44-27277	44-61661	44-61694	44-62060	44-62279	44-69746	44-86252	44-86441	

92nd Bomb Group

44-27330	44-61618	44-61923	44-62025	44-62102	44-62208	44-69805	44-84066	44-86438	
44-27326	44-61790	44-61925	44-62082	44-62111	44-62213	44-69980	44-86284	44-87620	
44-27332	44-61802	44-61951	44-62084	44-62114	44-62218	44-70073	44-86387	44-87760	
44-61617	44-61830	44-62010	44-62100	44-62188	44-62224	44-84032	44-86433		

98th Bomb Group

42-24834	42-93896	44-27341	44-61676	44-61815	44-62010	44-61874	44-61923	44-61938	44-62042
42-65352	42-93974	44-27288	44-61694	44-61830	44-61923	44-61878	44-61925	44-61953	44-62076
42-65353	44-27288	44-61562	44-61721	44-61923	44-61830	44-61894	44-61927	44-62009	44-62084
42-65357	44-27326	44-61617	44-61776	44-61925	44-61834	44-61896	44-61932	44-62010	44-62098
42-93880	44-27332	44-61657	44-61809	44-61951	44-61872	44-61897	44-61936	44-62041	44-62102

44-62103	44-62186	44-62253	44-69668	44-69812	44-83992	44-87760	44-86339	44-86392	44-87621
44-62016	44-62207	44-62261	44-69727	44-69894	44-84080	44-86290	44-86340	44-86400	44-87649
44-62108	44-62211	44-62270	44-69746	44-69909	44-86247	44-86295	44-86346	44-86415	45-21721
44-62141	44-62213	44-62279	44-69763	44-69944	44-86271	44-86316	44-86360	44-86433	45-21822
44-62166	44-62218	44-62281	44-69771	44-69977	44-86272	44-86327	44-86361	44-86436	45-21834
44-62167	44-62224	44-69656	44-69800	44-69998	44-86273	44-86330	44-86371	44-86446	
44-62173	44-62237	44-69667	44-69803	44-83934	44-87620	44-86335	44-86390	44-87341	

307th Bomb Group

42-65390	42-94062	44-61623	44-61816	44-62083	44-62287	44-83953	44-86357	44-86452
42-65392	42-94072	44-61676	44-61824	44-62102	44-69656	44-86268	44-86360	44-87760
42-93970	44-27287	44-61683	44-61872	44-62166	44-69909	44-86295	44-86387	45-21710
42-94031	44-27326	44-61694	44-61908	44-62209	44-69977	44-86318	44-86395	45-21814
42-94032	44-27347	44-61757	44-61940	44-62252	44-70021	44-86339	44-86422	
42-94045	44-61535	44-61802	44-62073	44-62270	44-70151	44-86343	44-86424	

KOREAN TACTICAL DROP SAMPLE: 19TH BOMB GROUP, MAY 2–14, 1951

Date	No. Up	Target	Bombs
5/2/51	5	RR bypass bridge at YD 2388	60 × 1000 LB GP
	1	Anak airfield at YC 1865	12 × 1000 LB GP
	2	Night Missions	
		CS 2884	10 × 500 LB GP by radar
		CT 1005	10 × 500 LB GP by radar
		CT 1106	10 × 500 LB GP by radar
		CT 2310	10 × 500 LB GP by radar
		CS 5674	10 × 500 LB GP by radar
		CS 5586	10 × 500 LB GP by radar
		CS 9993	10 × 500 LB GP by radar
		CS 9994	6 × 500 LB GP by radar
5/3/51	6	Sariwon supply center at YC 4065	72 × 1000 LB GP
	2	Troop concentrations at	
		CS 6077	10 × 500 LB GP by radar
		CS 6379	10 × 500 LB GP by radar
		DS6393	10 × 500 LB GP by radar
		CS 6481	10 × 500 LB GP by radar
		CS 4078	10 × 500 LB GP by radar
		CT 4608	10 × 500 LB GP by radar
		CT 3809	13 × 500 LB GP by radar
		CT 3610	7 × 500 LB GP by radar
5/5/51	8	Comm center at Sariwon YC 3966	320 × 500 LB GP
5/6/51	5	Supply center at Chinnampo YC 1090	190 ×500 LB incen cluster
	2	Night missions at	
		DT 3119	16 × 500 LB GP by radar
		DT 3219	16 × 500 LB GP by radar
		DT 2813	8 × 500 LB GP by radar
		DT 2510	16 × 500 LB GP by radar
		DT 3016	16 × 500 LB GP by radar
		DT 2813	8 × 500 LB GP by radar
		1 engine failure — salvoed EP 4075	40 × 500 LB IB
5/8/51	6	Sariwon YC3966	180 × 500 LB GP
	2	Troop concentrations at	
		DT 1615 - 1815	30 × 500 LB GP
		DT2222	10 × 500 LB GP
		CS1682	10 × 500 LB GP
		CS1680	13 × 500 LB GP
		CS 3883	17 × 500 LB GP

Date	No. Up	Target	Bombs
5/9/51	5	Ongjin Airfield YC 1200	200 × 500 LB GP
	1	Ongjin Airfield YC 100	40 × 500 LB GP
	2	Night mission	
		CS 2587	10 × 500 LB GP by radar
		CS1389	10 × 500 LB GP by radar
		CS 0794	10 × 500 LB GP by radar
		CS 1290	10 × 500 LB GP by radar
		CT 8321	10 × 500 LB GP by radar
		CT 8220	11 × 500 LB GP by radar
		CT 8412	8 × 500 LB GP by radar
		CT 8516	6 × 500 LB GP by radar
		CT 8519	5 × 500 LB GP by radar
5/10/51	4	Troop concentrations at	
		BS 9395	30 × 500 LB GP by radar
		BS 9398	30 × 500 LB GP by radar
		CS 9697	12 × 500 LB GP by radar
		BT 3119	10 × 500 LB GP by radar
		DT 2222	10 × 500 LB GP by radar
		CT 9603	10 × 500 LB GP by radar
		DT 0205	10 × 500 LB GP by radar
		CT 9903	10 × 500 LB GP by radar
		DT 0606	10 × 500 LB GP by radar
		DT 0603	10 × 500 LB GP by radar
		CS 7289	10 × 500 LB GP by radar
		DT 1414	8 × 500 LB GP by radar
5/12/51	6	Onjong-Ni Airfield XD 9408	237 × 500 LB GP
	2	Night mission	
		CT 7703	40 × 500 LB GP
		CT 9811 to 9412	40 × 500 LB GP
5/14/51	3	Sondok Airfield CU 6998	237 × 500 LB GP
	5	Night mission	
		CT 6203	9 × 500 LB GP
		CS 7599	20 × 500 LB GP
		CT 6602	11 × 500 LB GP
		CT 9801 to DT 0001	7 × 500 LB GP
		CS 9797 to CS 9997	6 × 500 LB GP
		CS 8899 to CS 9199	6 × 500 LB GP
		CT 8600 to CT 8901	6 × 500 LB GP
		CT 8903 to CT 9104	15 × 500 LB GP
		CT 5304	10 × 500 LB GP
		CT 5303	10 × 500 LB GP
		CT 5504	10 × 500 LB GP
		CT 5506	10 × 500 LB GP
		CT 1006	10 × 500 LB GP
		CS 2498	10 × 500 LB GP
		CS 0496	10 × 500 LB GP
		CS 1689	10 × 500 LB GP
		CT 9104	40 × 500 LB GP

Known B-29 Losses in Korea, 1950–1953

42-65306	44-27326	44-61810	44-61932	44-62108	44-62279	44-86247	44-86330	44-86436
42-65353	44-27332	44-61813	44-61940	44-62111	44-69682	44-86256	44-86343	44-87618
42-65357	44-61617	44-61815	44-61967	44-62152	44-69802	44-86268	44-86357	44-87734
42-65369	44-61656	44-61835	44-62011	44-62166	44-69803	44-86273	44-86370	45-21721
42-93974	44-61693	44-61867	44-62071	44-62167	44-69817	44-86280	44-86371	45-21725
42-94045	44-61749	44-61872	44-62073	44-62183	44-69818	44-86284	44-86392	45-21745
42-94072	44-61751	44-61894	44-62083	44-62211	44-69866	44-86295	44-86400	45-21749
44-27262	44-61776	44-61908	44-62084	44-62217	44-70042	44-86327	44-86414	45-21814
44-27288	44-61802	44-61923	44-62102	44-62252	44-70151	44-86328	44-86415	45-21822

South and North Korean Air Base Codes, 1950–1953

Code	South Korean Fields	Code	North Korean Fields
K-1	Pusan West Air base	K-17	Ongjin Airdrome
K-2	Taegu Air base	K-19	Haeju Airdrome
K-3	P'ohang Airdrome	K-20	Sinmak Airdrome
K-4	Sachon Airdrome	K-21	P'yonggang Airdrome
K-5	Taejon Airdrome	K-22	Onjong-Ni Airdrome
K-6	P'yong Taek Airdrome	K-23	P'yongyang Airdrome
K-7	Kwangju Airdrome	K-24	P'yongyang East Airdrome
K-8	Kunsan Air base	K-25	Wonsan Air base
K-9	Pusan East Air Base	K-26	Sondak Airdrome
K-10	Chinhae Air base	K-27	Yonpo Airdrome
K-11	Ulsan Air Base	K-28	Hamhung West Airdrome
K-12	Mangun Airdrome	K-28	Sinanju Airdrome
K-13	Suwon Air base	K-30	Sinuiju Airdrome
K-14	Kimpo Air base	K-31	Kiichu Airdrome
K-15	Mokp'o Airdrome	K-32	Oesichon-Dong Airdrome
K-16	Seoul Air Base	K-33	Hoemun Airdrome
K-18	Kangnung Airdrome	K-34	Ch'ongjin Airdrome
K-37	Taegu West Air Base	K-35	Hoeryong Airdrome
K-38	Wonju Airdrome	K-36	Kanggye #2 Airdrome
K-39	Cheju-Do Airdrome	K-50	Sokcho-Ri Airdrome
K-40	Cheju-Do Airdrome	K-52	Yanggu Airdrome
K-41	Ch'ungju Airdrome		
K-42	Andong Airdrome		
K-43	Kyongju Airdrome		
K-44	Changhowon-Ni Airdrome		
K-45	Yoju Airdrome		
K-46	Hoengsong Airdrome		
K-47	Ch'unch'on Airdrome		
K-48	Iri Airdrome		
K-49	Seoul East Airdrome		
K-51	Inje Airdrome		
K-55	Osan-ni Air base		
K-57	Kwang-ju Air base		

X

B-29s in the Royal Air Force

In 1950 the Royal Air Force was between planes to fill the heavy bombardment role. The World War II Avro Lancaster, with its top speed of 275 miles per hour, was sinking into obsolescence while the new jet bombers were far from being operational.

To assist the Royal Air Force in filling this equipment gap, the U. S. sent 92 B-29s to Great Britain on a loan basis. The B-29s were dropped from USAF inventory and, on arriving in the United Kingdom, were assigned RAF designators.

Two to four years later, when the majority of the planes were returned to USAF control, they were reentered in USAF inventory, with their original serial numbers, pending scrapping.

Redesignating the B-29 as "the Washington B Mark I," the RAF accepted delivery of the B-29s in 1950 and 1951, and returned approximately 67 in 1953 — 54. Two made their way halfway around the world to Australia. The remainder were operational losses while in British service.

The following list of B-29s loaned to the RAF contains a column of data not found in the Master List: the RAF serial number ("S/N").

USAF S/N	RAF S/N	USAF S/N	RAF S/N	USAF S/N	RAF S/N	USAF S/N	RAF S/N
42-65274	WF442	44-61894	WF502	44-62014	WF559	44-62129	WF554
42-93976	WF440	44-61895	WF492	44-61016	WF512	44-62135	WF566
42-94052	WF444	44-61897	WF436*	44-62019	WF561	44-62153	WF545
44-27342	WF438	44-61898	WF560**	44-62030	WF573	44-62154	WF548
44-61559	WF511	44-61937	WW346	44-62031	WF553	44-62155	WF494
44-61584	WF446	44-61938	WF504	44-62032	WF550	44-62159	WF447
44-61585	WF506	44-61952	WW345	44-62037	WF513	44-62198	WF491
44-61599	WF434	44-61963	WW354†	44-62043	WF500	44-62226	WF572
44-61634	WF439	44-61968	WW349	44-62046	WW344	44-62227	WW350
44-61642	WF493	44-61969	WF560**	44-62049	WW353‡	44-62231	WF503
44-61688	WF498	44-61978	WF558	44-62050	WF551	44-62234	WF507
44-61695	WF496	44-61982	WF501	44-62058	WW348	44-62235	WW343
44-61714	WF441	44-61983	WW347	44-62062	WF445	44-62236	WF562
44-61728	WF508	44-62001	WF505	44-62074	WF490	44-62238	WF547
44-61743	WF448	44-62003	WF509	44-62101	WF546	44-62239	WW355
44-61787	WF435	44-62005	WF510	44-62105	WF569	44-62241	WF570
44-61792	WF436*	44-62006	WF563	44-62109	WF521	44-62242	WW342
44-61883	WF443	44-62012	WF497	44-62117	WF557	44-62243	WF565
44-61889	WF499	44-62013	WF549	44-62128	WF495	44-62244	WF574

*RAF serial number WF436 issued to both 44-61792 and 44-61897.

**RAF serial number WF560 issued to both 44-61898 and 44-61969.

†WW354 (44-61963) was sent to Australia where it was issued Australian serial A76-2

‡WW344 (44-62049) was sent to Australia where it was issued Australian serial A76-1

USAF S/N	RAF S/N	USAF S/N	RAF S/N	USAF S/N	RAF S/N	USAF S/N	RAF S/N
44-62250	WF556	44-62257	WF571	44-62266	WF572	44-62296	WZ968
44-62254	WF555	44-62258	WW351	44-62280	WF514	44-62326	WF552
44-62255	WW352	44-62259	WF564	44-62282	WZ967	44-62328	WF547
44-62256	WF567§	44-62265	WF568	44-62283	WZ966	44-69680	WF437

§44-62256 appears to have been assigned two RAF serials — WF562 and WF567

Appendix: Serial Key Matrix

The researcher often finds himself delving into documents written during operational phases of a unit's history. The tendency of people keeping records in a war environment was to identify aircraft by the last three digits of the plane's serial number. This form of shorthand worked very well in the close-knit setting of a Bomb Group flight line.

However, decades later, the historian reading microfilmed copies of those same records has no idea if aircraft "267" was actually 42-6267, 44-86267 or several planes in between.

To help resolve this identity problem, this appendix provides a *serial key matrix* to identify the full serial numbers from the three digits so often found in historical documents. The best explanation is a demonstration.

Imagine that you find a reference to aircraft 267, reporting that it was lost in combat. Find the portion of the serial matrix listing the numbers 250–299. Run down the left column to the number 267. The digits to the right of 267 tell us that 267 must be one of the following:

42-6267
42-65267
44-27267
44-62267
44-86267

Checking these five serial numbers against the Master List, you will find that 42-6267 was reclaimed at Pyote AAF 12/21/49; 42-65267 ditched returning from a photo mission 1/22/45; 44-27267 was salvaged at Davis-Montham AFB 7/14/54; 44-62267 was reclaimed at McClellan AFB 5/7/54; and 44-86267 was reclaimed at Robins AFB 7/5/54. Of the five planes whose serial numbers end in 267, only one — number 42-65267 — could possibly be the plane reported lost in combat.

In the event that there were two or more planes lost in combat, try checking the "off inventory" date and the group to which the planes were assigned for further resolution.

000–049

	SN42-940XX	SN44-620XX	SN44-700XX	SN-840XX
000	94000	62000	70000	84000
001	94001	62001	70001	84001
002	94002	62002	70002	84002
003	94003	62003	70003	84003
004	94004	62004	70004	84004
005	94005	62005	70005	84005
006	94006	62006	70006	84006
007	94007	62007	70007	84007
008	94008	62008	70008	84008
009	94009	62009	70009	84009
010	94010	62010	70010	84010
011	94011	62011	70011	84011
012	94012	62012	70012	84012

	SN42-940XX	SN44-620XX	SN44-700XX	SN-840XX
013	94013	62013	70013	84013
014	94014	62014	70014	84014
015	94015	62015	70015	84015
016	94016	62016	70016	84016
017	94017	62017	70017	84017
018	94018	62018	70018	84018
019	94019	62019	70019	84019
020	94020	62020	70020	84020
021	94021	62021	70021	84021
022	94022	62022	70022	84022
023	94023	62023	70023	84023
024	94024	62024	70024	84024
025	94025	62025	70025	84025
026	94026	62026	70026	84026
027	94027	62027	70027	84027
028	94028	62028	70028	84028
029	94029	62029	70029	84029
030	94030	62030	70030	84030
031	94031	62031	70031	84031
032	94032	62032	70032	84032
033	94033	62033	70033	84033
034	94034	62034	70034	84034
035	94035	62035	70035	84035
036	94036	62036	70036	84036
037	94037	62037	70037	84037
038	94038	62038	70038	84038
039	94039	62039	70039	84039
040	94040	62040	70040	84040
041	94041	62041	70041	84041
042	94042	62042	70042	84042
043	94043	62043	70043	84043
044	94044	62044	70044	84044
045	94045	62045	70045	84045
046	94046	62046	70046	84046
047	94047	62047	70047	84047
048	94048	62048	70048	84048
049	94049	62049	70049	84049

050–099

	SN42-940XX	SN44-620XX	SN44-700XX	SN-840XX
050	94050	62050	70050	84050
051	94051	62051	70051	84051
052	94052	62052	70052	84052
053	94053	62053	70053	84053
054	94054	62054	70054	84054
055	94055	62055	70055	84055
056	94056	62056	70056	84056
057	94057	62057	70057	84057
058	94058	62058	70058	84058
059	94059	62059	70059	84059
060	94060	62060	70060	84060
061	94061	62061	70061	84061
062	94062	62062	70062	84062
063	94063	62063	70063	84063
064	94064	62064	70064	84064
065	94065	62065	70065	84065
066	94066	62066	70066	84066
067	94067	62067	70067	84067
068	94068	62068	70068	84068
069	94069	62069	70069	84069
070	94070	62070	70070	84070
071	94071	62071	70071	84071
072	94072	62072	70072	84072
073	94073	62073	70073	84073

Serial Key Matrix

074	94074	62074	70074	84074
075	94075	62075	70075	84075
076	94076	62076	70076	84076
077	94077	62077	70077	84077
078	94078	62078	70078	84078
079	94079	62079	70079	84079
080	94080	62080	70080	84080
081	94081	62081	70081	84081
082	94082	62082	70082	84082
083	94083	62083	70083	84083
084	94084	62084	70084	84084
085	94085	62085	70085	84085
086	94086	62086	70086	84086
087	94087	62087	70087	84087
088	94088	62088	70088	84088
089	94089	62089	70089	84089
090	94090	62090	70090	84090
091	94091	62091	70091	84091
092	94092	62092	70092	84092
093	94093	62093	70093	84093
094	94094	62094	70094	84094
095	94095	62095	70095	84095
096	94096	62096	70096	84096
097	94097	62097	70097	84097
098	94098	62098	70098	84098
099	94099	62099	70099	84099

100–149

	SN42-941XX	SN44-621XX	SN44-701XX	SN-841XX
100	94100	62100	70100	84100
101	94101	62101	70101	84101
102	94102	62102	70102	84102
103	94103	62103	70103	84103
104	94104	62104	70104	84104
105	94105	62105	70105	84105
106	94106	62106	70106	84106
107	94107	62107	70107	84107
108	94108	62108	70108	84108
109	94109	62109	70109	84109
110	94110	62110	70110	84110
111	94111	62111	70111	84111
112	94112	62112	70112	84112
113	94113	62113	70113	84113
114	94114	62114	70114	84114
115	94115	62115	70115	84115
116	94116	62116	70116	84116
117	94117	62117	70117	84117
118	94118	62118	70118	84118
119	94119	62119	70119	84119
120	94120	62120	70120	84120
121	94121	62121	70121	84121
122	94122	62122	70122	84122
123	94123	62123	70123	84123
124		62124	70124	84124
125		62125	70125	84125
126		62126	70126	84126
127		62127	70127	84127
128		62128	70128	84128
129		62129	70129	84129
130		62130	70130	84130
131		62131	70131	84131
132		62132	70132	84132
133		62133	70133	84133
134		62134	70134	84134

		62135	70135	84135
135		62135	70135	84135
136		62136	70136	84136
137		62137	70137	84137
138		62138	70138	84138
139		62139	70139	84139
140		62140	70140	Not B-29 #
141		62141	70141	84141
142		62142	70142	84142
143		62143	70143	84143
144		62144	70144	84144
145		62145	70145	84145
146		62146	70146	84146
147		62147	70147	84147
148		62148	70148	84148
149		62149	70149	84149

150–199

	SN44-621XX	*SN44-701XX*	*SN-841XX*
150	62150	70150	Not B-29 #
151	62151	70151	84151
152	62152	70152	84152
153	62153	70153	Not B-29 #
154	62154	70154	Not B-29 #
155	62155		84155
156	62156		
157	62157		
158	62158		
159	62159		
160	62160		
161	62162		
162	62162		
163	62163		
164	62164		
165	62165		
166	62166		
167	62167		
168	62168		
169	62169		
170	62170		
171	62171		
172	62172		
173	62173		
174	62174		
175	62175		
176	62176		
177	62177		
178	62178		
179	62179		
180	62180		
181	62181		
182	62182		
183	62183		
184	62184		
185	62185		
186	62186		
187	62187		
188	62188		
189	62189		
190	62190		
191	62191		
192	62192		
193	62193		
194	62194		
195	62195		
196	62196		

Serial Key Matrix

197	62197				
198	62198				
199	62199				

200–249

	SN42-62XX	*SN42-652XX*	*SN44-272XX*	*SN44-622XX*	*SN-862XX*
200				62200	
201				62201	
202		65202		62202	
203		65203		62203	
204		65204		62204	
205	6205	65205		62205	
206	6206	65206		62206	
207	6207	65207		62207	
208	6208	65208		62208	
209	6209	65209		62209	
210	6210	65210		62210	
211	6211	65211		62211	
212	6212	65212		62212	
213	6213	65213		62213	
214	6214	65214		62214	
215	6215	65215		62215	
216	6216	65216		62216	
217	6217	65217		62217	
218	6218	65218		62218	
219	6219	65219		62219	
220	6220	65220		62220	
221	6221	65221		62221	
222	6222	65222		62222	
223	6223	65223		62223	
224	6224	65224		62224	
225	6225	65225		62225	
226	6226	65226		62226	
227	6227	65227		62227	
228	6228	65228		62228	
229	6229	65229		62229	
230	6230	65230		62230	
231	6231	65231		62231	
232	6232	65232		62232	
233	6233	65233		62233	
234	6234	65234		62234	
235	6235	65235		62235	
236	6236	65236		62236	
237	6237	65237		62237	
238	6238	65238		62238	
239	6239	65239		62239	
240	6240	65240		62240	
241	6241	65241		62241	
242	6242	65242		62242	86242
243	6243	65243		62243	86243
244	6244	65244		62244	86244
245	6245	65245		62245	86245
246	6246	65246		62246	86246
247	6247	65247		62247	86247
248	6248	65248		62248	86248
249	6249	65249		62249	86249

250–299

	SN42-62XX	*SN42-652XX*	*SN44-272XX*	*SN44-622XX*	*SN-862XX*
250	6250	65250		62250	86250
251	6251	65251		62251	86251
252	6252	65252		62252	86252

253	6253	65253		62253	86253	
254	6254	65254		62254	86254	
255	6255	65255		62255	86255	
256	6256	65256		62256	86256	
257	6257	65257		62257	86257	
258	6258	65258		62258	86258	
259	6259	65259	27259	62259	86259	
260	6260	65260	27260	62260	86260	
261	6261	65261	27261	62261	86261	
262	6262	65262	27262	62262	86262	
263	6263	65263	27263	62263	86263	
264	6264	65264	27264	62264	86264	
265	6265	65265	27265	62265	86265	
266	6266	65266	27266	62266	86266	
267	6267	65267	27267	62267	86267	
268	6268	65268	27268	62268	86268	
269	6269	65269	27269	62269	86269	
270	6270	65270	27270	62270	86270	
271	6271	65271	27271	62271	86271	
272	6272	65272	27272	62272	86272	
273	6273	65273	27273	62273	86273	
274	6274	65274	27274	62274	86274	
275	6275	65275	27275	62275	86275	
276	6276	65276	27276	62276	86276	
277	6277	65277	27277	62277	86277	
278	6278	65278	27278	62278	86278	
279	6279	65279	27279	62279	86279	
280	6280	65280	27280	62280	86280	
281	6281	65281	27281	62281	86281	
282	6282	65282	27282	62282	86282	
283	6283	65283	27283	62283	86283	
284	6284	65284	27284	62284	86284	
285	6285	65285	27285	62285	86285	
286	6286	65286	27286	62286	86286	
287	6287	65287	27287	62287	86287	
288	6288	65288	27288	62288	86288	
289	6289	65289	27289	62289	86289	
290	6290	65290	27290	62290	86290	
291	6291	65291	27291	62291	86291	
292	6292	65292	27292	62292	86292	
293	6293	65293	27293	62293	86293	
294	6294	65294	27294	62294	86294	
295	6295	65295	27295	62295	86295	
296	6296	65296	27296	62296	86296	
297	6297	65297	27297	62297	86297	
298	6298	65298	27298	62298	86298	
299	6299	65299	27299	62299	86299	

300–349

	SN42-63XX	*SN42-633XX*	*SN42-653XX*	*SN44-273XX*	*SN44-623XX*	*SN44-863XX*
300	6300		65300	27300	62300	86300
301	6301		65301	27301	62301	86301
302	6302		65302	27302	62302	86302
303	6303		65303	27303	62303	86303
304	6304		65304	27304	62304	86304
305	6305		65305	27305	62305	86305
306	6306		65306	27306	62306	86306
307	6307		65307	27307	62307	86307
308	6308		65308	27308	62308	86308
309	6309		65309	27309	62309	86309
310	6310		65310	27310	62310	86310
311	6311		65311	27311	62311	86311
312	6312		65312	27312	62312	86312
313	6313		65313	27313	62313	86313
314	6314		Not B-29 #	27314	62314	86314

Serial Key Matrix

315	6315		65315	27315	62315	86315
316	6316		65316	27316	62316	86316
317	6317		65317	27317	62317	86317
318	6318		65318	27318	62318	86318
319	6319		65319	27319	62319	86319
320	6320		65320	27320	62320	86320
321	6321		65321	27321	62321	86321
322	6322		65322	27322	62322	86322
323	6323		65323	27323	62323	86323
324	6324		65324	27324	62324	86324
325	6325		65325	27325	62325	86325
326	6326		65326	27326	62326	86326
327	6327		65327	27327	62327	86327
328	6328		65328	27328	62328	86328
329	6329		65329	27329		86329
330	6330		65330	27330		86330
331	6331		65331	27331		86331
332	6332		65332	27332		86332
333	6333		65333	27333		86333
334	6334		65334	27334		86334
335	6335		65335	27335		86335
336	6336		65336	27336		86336
337	6337		65337	27337		86337
338	6338		65338	27338		86338
339	6339		65339	27339		86339
340	6340		65340	27340		86340
341	6341		65341	27341		86341
342	6342		65342	27342		86342
343	6343		65343	27343		86343
344	6344		65344	27344		86344
345	6345		65345	27345		86345
346	6346		65346	27346		86346
347	6347		65347	27347		86347
348	6348		65348	27348		86348
349	6349		65349	27349		86349

350–399

	SN42-63XX	SN42-633XX	SN42-653XX	SN44-273XX	SN44-863XX
350	6350		65350	27350	86350
351	6351		65351	27351	86351
352	6352	63352	65352	27352	86352
353	6353	63353	65353	27353	86353
354	6354	63354	65354	27354	86354
355	6355	63355	65355	27355	86355
356	6356	63356	65356	27356	86356
357	6357	63357	65357	27357	86357
358	6358	63358	65358	27358	86358
359	6359	63359	65359		86359
360	6360	63360	65360		86360
361	6361	63361	65361		86361
362	6362	63362	65362		86362
363	6363	63363	65363		86363
364	6364	63364	65364		86364
365	6365	63365	65365		86365
366	6366	63366	65366		86366
367	6367	63367	65367		86367
368	6368	63368	65368		86368
369	6369	63369	65369		86369
370	6370	63370	65370		86370
371	6371	63371	65371		86371
372	6372	63372	65372		86372
373	6373	63373	65373		86373
374	6374	63374	65374		86374
375	6375	63375	65375		86375
376	6376	63376	65376		86376

377	6377	63377	65377		86377
378	6378	63378	65378		86378
379	6379	63379	65379		86379
380	6380	63380	65380		86380
381	6381	63381	65381		86381
382	6382	63382	65382		86382
383	6383	63383	65383		86383
384	6384	63384	65384		86384
385	6385	63385	65385		86385
386	6386	63386	65386		86386
387	6387	63387	65387		86387
388	6388	63388	65388		86388
389	6389	63389	65389		86389
390	6390	63390	65390		86390
391	6391	63391	65391		86391
392	6392	63392	65392		86392
393	6393	63393	65393		86393
394	6394	63394	65394		86394
395	6395	63395	65395		86395
396	6396	63396	65396		86396
397	6397	63397	65397		86397
398	6398	63398	65398		86398
399	6399	63399	65399		86399

400–449

	SN42-64XX	SN42-244XX	SN42-634XX	SN42-654XX	SN44-864XX
400	6400		63400	65400	86400
401	6401		63401	65401	86401
402	6402		63402		86402
403	6403		63403		86403
404	6404		63404		86404
405	6405		63405		86405
406	6406		63406		86406
407	6407		63407		86407
408	6408		63408		86408
409	6409		63409		86409
410	6410		63410		86410
411	6411		63411		86411
412	6412		63412		86412
413	6413		63413		86413
414	6414		63414		86414
415	6415		63415		86415
416	6416		63416		86416
417	6417		63417		86417
418	6418		63418		86418
419	6419		63419		86419
420	6420	24420	63420		86420
421	6421	24421	63421		86421
422	6422	24422	63422		86422
423	6423	24423	63423		86423
424	6424	24424	63424		86424
425	6425	24425	63425		86425
426	6426	24426	63426		86426
427	6427	24427	63427		86427
428	6428	24428	63428		86428
429	6429	24429	63429		86429
430	6430	24430	63430		86430
431	6431	24431	63431		86431
432	6432	24432	63432		86432
433	6433	24433	63433		86433
434	6434	24434	63434		86434
435	6435	24435	63435		86435
436	6436	24436	63436		86436
437	6437	24437	63437		86437
438	6438	24438	63438		86438

	SN42-64XX	SN42-244XX	SN42-634XX	SN44-864XX
439	6439	24439	63439	86439
440	6440	24440	63440	86440
441	6441	24441	63441	86441
442	6442	24442	63442	86442
443	6443	24443	63443	86443
444	6444	24444	63444	86444
445	6445	24445	63445	86445
446	6446	24446	63446	86446
447	6447	24447	63447	86447
448	6448	24448	63448	86448
449	6449	24449	63449	86449

450–499

	SN42-64XX	SN42-244XX	SN42-634XX	SN44-864XX
450	6450	24450	63450	86450
451	6451	24451	63451	86451
452	6452	24452	63452	86452
453	6453	24453	63453	86453
454	6454	24454	63454	86454
455		24455	63455	86455
456		24456	63456	86456
457		24457	63457	86457
458		24458	63458	86458
459		24459	63459	86459
460		24460	63460	86460
461		24461	63461	86461
462		24462	63462	86462
463		24463	63463	86463
464		24464	63464	86464
465		24465	63465	86465
466		24466	63466	86466
467		24467	63467	86467
468		24468	63468	86468
469		24469	63469	86469
470		24470	63470	86470
471		24471	63471	86471
472		24472	63472	86472
473		24473	63473	86473
474		24474	63474	
475		24475	63475	
476		24476	63476	
477		24477	63477	
478		24478	63478	
479		24479	63479	
480		24480	63480	
481		24481	63481	
482		24482	63482	
483		24483	63483	
484		24484	63484	
485		24485	63485	
486		24486	63486	
487		24487	63487	
488		24488	63488	
489		24489	63489	
490		24490	63490	
491		24491	63491	
492		24492	63492	
493		24493	63493	
494		24494	63494	
495		24495	63495	
496		24496	63496	
497		24497	63497	
498		24498	63498	
499		24499	63499	

500–549

	SN42-245XX	SN42-635XX	SN44-615XX	SN44-875XX
500	24500	63500		
501	24501	63501		
502	24502	63502		
503	24503	63503		
504	24504	63504		
505	24505	63505		
506	24506	63506		
507	24507	63507		
508	24508	63508		
509	24509	63509		
510	24510	63510	61510	
511	24511	63511	61511	
512	24512	63512	61512	
513	24513	63513	61513	
514	24514	63514	61514	
515	24515	63515	61515	
516	24516	63516	61516	
517	24517	63517	61517	
518	24518	63518	61518	
519	24519	63519	61519	
520	24520	63520	61520	
521	24521	63521	61521	
522	24522	63522	61522	
523	24523	63523	61523	
524	24524	63524	61524	
525	24525	63525	61525	
526	24526	63526	61526	
527	24527	63527	61527	
528	24528	63528	61528	
529	24529	63529	61529	
530	24530	63530	61530	
531	24531	63531	61531	
532	24532	63532	61532	
533	24533	63533	61533	
534	24534	63534	61534	
535	24535	63535	61535	
536	24536	63536	61536	
537	24537	63537	61537	
538	24538	63538	61538	
539	24539	63539	61539	
540	24540	63540	61540	
541	24541	63541	61541	
542	24542	63542	61542	
543	24543	63543	61543	
544	24544	63544	61544	
545	24545	63545	61545	
546	24546	63546	61546	
547	24547	63547	61547	
548	24548	63548	61548	
549	24549	63549	61549	

550–599

	SN42-245XX	SN42-635XX	SN44-615XX	SN44-875XX
550	24550	63550	61550	
551	24551	63551	61551	
552	24552	63552	61552	
553	24553	63553	61553	
554	24554	63554	61554	
555	24555	63555	61555	
556	24556	63556	61556	
557	24557	63557	61557	

558	24558	63558	61558	
559	24559	63559	61559	
560	24560	63560	61560	
561	24561	63561	61561	
562	24562	63562	61562	
563	24563	63563	61563	
564	24564	63564	61564	
565	24565	63565	61565	
566	24566	63566	61566	
567	24567	63567	61567	
568	24568	63568	61568	
569	24569	63569	61569	
570	24570	63570	61570	
571	24571	63571	61571	
572	24572	63572	61572	
573	24573	63573	61573	
574	24574	63574	61574	
575	24575	63575	61575	
576	24576	63576	61576	
577	24577	63577	61577	
578	24578	63578	61578	
579	24579	63579	61579	
580	24580	63580	61580	
581	24581	63581	61581	
582	24582	63582	61582	
583	24583	63583	61583	
584	24584	63584	61584	87584
585	24585	63585	61585	87585
586	24586	63586	61586	87586
587	24587	63587	61587	87587
588	24588	63588	61588	87588
589	24589	63589	61589	87589
590	24590	63590	61590	87590
591	24591	63591	61591	87591
592	24592	63592	61592	87592
593	24593	63593	61593	87593
594	24594	63594	61594	87594
595	24595	63595	61595	87595
596	24596	63596	61596	87596
597	24597	63597	61597	87597
598	24598	63598	61598	87598
599	24599	63599	61599	87599

600–649

	SN42-244XX	SN42-636XX	SN44-616XX	SN44-876XX
600	24600	63600	61600	87600
601	24601	63601	61601	87601
602	24602	63602	61602	87602
603	24603	63603	61603	87603
604	24604	63604	61604	87604
605	24605	63605	61605	87605
606	24606	63606	61606	87606
607	24607	63607	61607	87607
608	24608	63608	61608	87608
609	24609	63609	61609	87609
610	24610	63610	61610	87610
611	24611	63611	61611	87611
612	24612	63612	61612	87612
613	24613	63613	61613	87613
614	24614	63614	61614	87614
615	24615	63615	61615	87615
616	24616	63616	61616	87616
617	24617	63617	61617	87617
618	24618	63618	61618	87618
619	24619	63619	61619	87619

620	24620	63620	61620	87620	
621	24621	63621	61621	87621	
622	24622	63622	61622	87622	
623	24623	63623	61623	87623	
624	24624	63624	61624	87624	
625	24625	63625	61625	87625	
626	24626	63626	61626	87626	
627	24627	63627	61627	87627	
628	24628	63628	61628	87628	
629	24629	63629	61629	87629	
630	24630	63630	61630	87630	
631	24631	63631	61631	87631	
632	24632	63632	61632	87632	
633	24633	63633	61633	87633	
634	24634	63634	61634	87634	
635	24635	63635	61635	87635	
636	24636	63636	61636	87636	
637	24637	63637	61637	87637	
638	24638	63638	61638	87638	
639	24639	63639	61639	87639	
640	24640	63640	61640	87640	
641	24641	63641	61641	87641	
642	24642	63642	61642	87642	
643	24643	63643	61643	87643	
644	24644	63644	61644	87644	
645	24645	63645	61645	87645	
646	24646	63646	61646	87646	
647	24647	63647	61647	87647	
648	24648	63648	61648	87648	
649	24649	63649	61649	87649	

650–699

	SN42-246XX	*SN42-636XX*	*SN44-616XX*	*SN44-696XX*	*SN44-876XX*	*SN45-216XX*
650	24650	63650	61650		87650	
651	24651	63651	61651		87651	
652	24652	63652	61652		87652	
653	24653	63653	61653		87653	
654	24654	63654	61654		87654	
655	24655	63655	61655	69655	87655	
656	24656	63656	61656	69656	87656	
657	24657	63657	61657	69657	87657	
658	24658	63658	61658	69658	87658	
659	24659	63659	61659	69659	87659	
660	24660	63660	61660	69660	87660	
661	24661	63661	61661	69661	87661	
662	24662	63662	61662	69662	87662	
663	24663	63663	61663	69663	87663	
664	24664	63664	61664	69664	87664	
665	24665	63665	61665	69665	87665	
666	24666	63666	61666	69666	87666	
667	24667	63667	61667	69667	87667	
668	24668	63668	61668	69668	87668	
669	24669	63669	61669	69669	87669	
670	24670	63670	61670	69670	87670	
671	24671	63671	61671	69671	87671	
672	24672	63672	61672	69672	87672	
673	24673	63673	61673	69673	87673	
674	24674	63674	61674	69674	87674	
675	24675	63675	61675	69675	87675	
676	24676	63676	61676	69676	87676	
677	24677	63677	61677	69677	87677	
678	24678	63678	61678	69678	87678	
679	24679	63679	61679	69679	87679	
680	24680	63680	61680	69680	87680	
681	24681	63681	61681	69681	87681	

Serial Key Matrix

682	24682	63682	61682	69682	87682	
683	24683	63683	61683	69683	87683	
684	24684	63684	61684	69684	87684	
685	24685	63685	61685	69685	87685	
686	24686	63686	61686	69686	87686	
687	24687	63687	61687	69687	87687	
688	24688	63688	61688	69688	87688	
689	24689	63689	61689	69689	87689	
690	24690	63690	61690	69690	87690	
691	24691	63691	61691	69691	87691	
692	24692	63692	61692	69692	87692	
693	24693	63693	61693	69693	87693	21693
694	24694	63694	61694	69694	87694	21694
695	24695	63695	61695	69695	87695	21695
696	24696	63696	61696	69696	87696	21696
697	24697	63697	61697	69697	87697	21697
698	24698	63698	61698	69698	87698	21698
699	24699	63699	61699	69699	87699	21699

700–749

	SN42-247XX	SN42-637XX	SN44-617XX	SN44-697XX	SN44-877XX	SN45-217XX
700	24700	63700	61700	69700	87700	21700
701	24701	63701	61701	69701	87701	21701
702	24702	63702	61702	69702	87702	21702
703	24703	63703	61703	69703	87703	21703
704	24704	63704	61704	69704	87704	21704
705	24705	63705	61705	69705	87705	21705
706	24706	63706	61706	69706	87706	21706
707	24707	63707	61707	69707	87707	21707
708	24708	63708	61708	69708	87708	21708
709	24709	63709	61709	69709	87709	21709
710	24710	63710	61710	69710	87710	21710
711	24711	63711	61711	69711	87711	21711
712	24712	63712	61712	69712	87712	21712
713	24713	63713	61713	69713	87713	21713
714	24714	63714	61714	69714	87714	21714
715	24715	63715	61715	69715	87715	21715
716	24716	63716	61716	69716	87716	21716
717	24717	63717	61717	69717	87717	21717
718	24718	63718	61718	69718	87718	21718
719	24719	63719	61719	69719	87719	21719
720	24720	63720	61720	69720	87720	21720
721	24721	63721	61721	69721	87721	21721
722	24722	63722	61722	69722	87722	21722
723	24723	63723	61723	69723	87723	21723
724	24724	63724	61724	69724	87724	21724
725	24725	63725	61725	69725	87725	21725
726	24726	63726	61726	69726	87726	21726
727	24727	63727	61727	69727	87727	21727
728	24728	63728	61728	69728	87728	21728
729	24729	63729	61729	69729	87729	21729
730	24730	63730	61730	69730	87730	21730
731	24731	63731	61731	69731	87731	21731
732	24732	63732	61732	69732	87732	21732
733	24733	63733	61733	69733	87733	21733
734	24734	63734	61734	69734	87734	21734
735	24735	63735	61735	69735	87735	21735
736	24736	63736	61736	69736	87736	21736
737	24737	63737	61737	69737	87737	21737
738	24738	63738	61738	69738	87738	21738
739	24739	63739	61739	69739	87739	21739
740	24740	63740	61740	69740	87740	21740
741	24741	63741	61741	69741	87741	21741
742	24742	63742	61742	69742	87742	21742
743	24743	63743	61743	69743	87743	21743

744	24744	63744	61744	69744	87744	21744
745	24745	63745	61745	69745	87745	21745
746	24746	63746	61746	69746	87746	21746
747	24747	63747	61747	69747	87747	21747
748	24748	63748	61748	69748	87748	21748
749	24749	63749	61749	69749	87749	21749

750–799

	SN42-247XX	SN42-637XX	SN44-617XX	SN44-697XX	SN44-877XX	SN45-217XX
750	24750	63750	61750	69750	87750	21750
751	24751	63751	61751	69751	87751	21751
752	24752		61752	69752	87752	21752
753	24753		61753	69753	87753	21753
754	24754		61754	69754	87754	21754
755	24755		61755	69755	87755	21755
756	24756		61756	69756	87756	21756
757	24757		61757	69757	87757	21757
758	24758		61758	69758	87758	21758
759	24759		61759	69759	87759	21759
760	24760		61760	69760	87760	21760
761	24761		61761	69761	87761	21761
762	24762		61762	69762	87762	21762
763	24763		61763	69763	87763	21763
764	24764		61764	69764	87764	21764
765	24765		61765	69765	87765	21765
766	24766		61766	69766	87766	21766
767	24767		61767	69767	87767	21767
768	24768		61768	69768	87768	21768
769	24769		61769	69769	87769	21769
770	24770		61770	69770	87770	21770
771	24771		61771	69771	87771	21771
772	24772		61772	69772	87772	21772
773	24773		61773	69773	87773	21773
774	24774		61774	69774	87774	21774
775	24775		61775	69775	87775	21775
776	24776		61776	69776	87776	21776
777	24777		61777	69777	87777	21777
778	24778		61778	69778	87778	21778
779	24779		61779	69779	87779	21779
780	24780		61780	69780	87780	21780
781	24781		61781	69781	87781	21781
782	24782		61782	69782	87782	21782
783	24783		61783	69783	87783	21783
784	24784		61784	69784		21784
785	24785		61785	69785		21785
786	24786		61786	69786		21786
787	24787		61787	69787		21787
788	24788		61788	69788		21788
789	24789		61789	69789		21789
790	24790		61790	69790		21790
791	24791		61791	69791		21791
792	24792		61792	69792		21792
793	24793		61793	69793		21793
794	24794		61794	69794		21794
795	24795		61795	69795		21795
796	24796		61796	69796		21796
797	24797		61797	69797		21797
798	24798		61798	69798		21798
799	24799		61799	69799		21799

800–849

	SN42-248XX	SN42-938XX	SN44-618XX	SN44-698XX	SN44-838XX	SN45-218XX
800	24800		61800	69800		21800
801	24801		61801	69801		21801

Serial Key Matrix

	SN42-248XX	SN42-938XX	SN44-618XX	SN44-698XX	SN44-838XX	SN45-218XX
802	24802		61802	69802		21802
803	24803		61803	69803		21803
804	24804		61804	69804		21804
805	24805		61805	69805		21805
806	24806		61806	69806		21806
807	24807		61807	69807		21807
808	24808		61808	69808		21808
809	24809		61809	69809		21809
810	24810		61810	69810		21810
811	24811		61811	69811		21811
812	24812		61812	69812		21812
813	24813		61813	69813		21813
814	24814		61814	69814		21814
815	24815		61815	69815		21815
816	24816		61816	69816		21816
817	24817		61817	69817		21817
818	24818		61818	69818		21818
819	24819		61819	69819		21819
820	24820		61820	69820		21820
821	24821		61821	69821		21821
822	24822		61822	69822		21822
823	24823		61823	69823		21823
824	24824	93824	61824	69824		21824
825	24825	93825	61825	69825		21825
826	24826	93826	61826	69826		21826
827	24827	93827	61827	69827		21827
828	24828	93828	61828	69828		21828
829	24829	93829	61829	69829		21829
830	24830	93830	61830	69830		21830
831	24831	93831	61831	69831		21831
832	24832	93832	61832	69832		21832
833	24833	93833	61833	69833		21833
834	24834	93834	61834	69834		21834
835	24835	93835	61835	69835		21835
836	24836	93836	61836	69836		21836
837	24837	93837	61837	69837		21837
838	24838	93838	61838	69838		21838
839	24839	93839	61839	69839		21839
840	24840	93840	61840	69840		21840
841	24841	93841	61841	69841		21841
842	24842	93842	61842	69842		21842
843	24843	93843	61843	69843		21843
844	24844	93844	61844	69844		21844
845	24845	93845	61845	69845		21845
846	24846	93846	61846	69846		21846
847	24847	93847	61847	69847		21847
848	24848	93848	61848	69848		21848
849	24849	93849	61849	69849		21849

850–899

	SN42-248XX	SN42-938XX	SN44-618XX	SN44-698XX	SN44-838XX	SN45-218XX
850	24850	93850	61850	69850		21850
851	24851	93851	61851	69851		21851
852	24852	93852	61852	69852		21852
853	24853	93853	61853	69853		21853
854	24854	93854	61854	69854		21854
855	24855	93855	61855	69855		21855
856	24856	93856	61856	69856		21856
857	24857	93857	61857	69857		21857
858	24858	93858	61858	69858		21858
859	24859	93859	61859	69859		21859
860	24860	93860	61860	69860		21860
861	24861	93861	61861	69861		21861
862	24862	93862	61862	69862		21862
863	24863	93863	61863	69863		21863

864	24864	93864	61864	69864		21864
865	24865	93865	61865	69865		21865
866	24866	93866	61866	69866		21866
867	24867	93867	61867	69867		21867
868	24868	93868	61868	69868		21868
869	24869	93869	61869	69869		21869
870	24870	93870	61870	69870		21870
871	24871	93871	61871	69871		21871
872	24872	93872	61872	69872		21872
873	24873	93873	61873	69873		
874	24874	93874	61874	69874		
875	24875	93875	61875	69875		
876	24876	93876	61876	69876		
877	24877	93877	61877	69877		
878	24878	93878	61878	69878		
879	24879	93879	61879	69879		
880	24880	93880	61880	69880		
881	24881	93881	61881	69881		
882	24882	93882	61882	69882		
883	24883	93883	61883	69883		
884	24884	93884	61884	69884		
885	24885	93885	61885	69885		
886	24886	93886	61886	69886		
887	24887	93887	61887	69887		
888	24888	93888	61888	69888		
889	24889	93889	61889	69889		
890	24890	93890	61890	69890	83890	
891	24891	93891	61891	69891	83891	
892	24892	93892	61892	69892	83892	
893	24893	93893	61893	69893	83893	
894	24894	93894	61894	69894	83894	
895	24895	93895	61895	69895	83895	
896	24896	93896	61896	69896	83896	
897	24897	93897	61897	69897	83897	
898	24898	93898	61898	69898	83898	
899	24899	93899	61899	69899	83899	

900–949

	SN42-249XX	SN42-938XX	SN44-619XX	SN44-699XX	SN44-839XX
900	24900	93900	61900	69900	83900
901	24901	93901	61901	69901	83901
902	24902	93902	61902	69902	83902
903	24903	93903	61903	69903	83903
904	24904	93904	61904	69904	83904
905	24905	93905	61905	69905	83905
906	24906	93906	61906	69906	83906
907	24907	93907	61907	69907	83907
908	24908	93908	61908	69908	83908
909	24909	93909	61909	69909	83909
910	24910	93910	61910	69910	83910
911	24911	93911	61911	69911	83911
912	24912	93912	61912	69912	83912
913	24913	93913	61913	69913	83913
914	24914	93914	61914	69914	83914
915	24915	93915	61915	69915	83915
916	24916	93916	61916	69916	83916
917	24917	93917	61917	69917	83917
918	24918	93918	61918	69918	83918
919	24919	93919	61919	69919	83919
920		93920	61920	69920	83920
921		93921	61921	69921	83921
922		93922	61922	69922	83922
923		93923	61923	69923	83923
924		93924	61924	69924	83924
925		93925	61925	69925	83925

Serial Key Matrix

926		93926	61926	69926	83926
927		93927	61927	69927	83927
928		93928	61928	69928	83928
929		93929	61929	69929	83929
930		93930	61930	69930	83930
931		93931	61931	69931	83931
932		93932	61932	69932	83932
933		93933	61933	69933	83933
934		93934	61934	69934	83934
935		93935	61935	69935	83935
936		93936	61936	69936	83936
937		93937	61937	69937	83937
938		93938	61938	69938	83938
939		93939	61939	69939	83939
940		93940	61940	69940	83940
941		93941	61941	69941	83941
942		93942	61942	69942	83942
943		93943	61943	69943	83943
944		93944	61944	69944	83944
945		93945	61945	69945	83945
946		93946	61946	69946	83946
947		93947	61947	69947	83947
948		93948	61948	69948	83948
949		93949	61949	69949	83949

950–999

	SN42-939XX	*SN44-619XX*	*SN44-699XX*	*SN44-839XX*
950	93950	61950	69950	83950
951	93951	61951	69951	83951
952	93952	61952	69952	83952
953	93953	61953	69953	83953
954	93954	61954	69954	83954
955	93955	61955	69955	83955
956	93956	61956	69956	83956
957	93957	61957	69957	83957
958	93958	61958	69958	83958
959	93959	61959	69959	83959
960	93960	61960	69960	83960
961	93961	61961	69961	83961
962	93962	61962	69962	83962
963	93963	61963	69963	83963
964	93964	61964	69964	83964
965	93965	61965	69965	83965
966	93966	61966	69966	83966
967	93967	61967	69967	83967
968	93968	61968	69968	83968
969	93969	61969	69969	83969
970	93970	61970	69970	83970
971	93971	61971	69971	83971
972	93972	61972	69972	83972
973	93973	61973	69973	83973
974	93974	61974	69974	83974
975	93975	61975	69975	83975
976	93976	61976	69976	83976
977	93977	61977	69977	83977
978	93978	61978	69978	83978
979	93979	61979	69979	83979
980	93980	61980	69980	83980
981	93981	61981	69981	83981
982	93982	61982	69982	83982
983	93983	61983	69983	83983
984	93984	61984	69984	83984
985	93985	61985	69985	83985
986	93986	61986	69986	83986
987	93987	61987	69987	83987

988	93988	61988	69988	83988
989	93989	61989	69989	83989
990	93990	61990	69990	83990
991	93991	61991	69991	83991
992	93992	61992	69992	83992
993	93993	61993	69993	83993
994	93994	61994	69994	83994
995	93995	61995	69995	83995
996	93996	61996	69996	83996
997	93997	61997	69997	83997
998	93998	61998	69998	83998
999	93999	61999	69999	83999

BIBLIOGRAPHY

Anderton, D.A. *B-29 Superfortress at War.* New York: Scribner's, 1978.

Andrade, J.M. *U.S. Military Designations and Serials.* Leicester: Midland Counties, 1979.

Baugher, J. *USAAF Serials.* http://home.att.net/%7ejbaugher/1942_1.html.

Bell, Dana. *Air Force Colors.* Vol III. Carollton TX: Squadron/Signal, 1997.

Birdsall, S. *B-29 Superfortress in Action.* Carrollton TX: Squadron/Signal, 1977.

_____. *Saga of the Superfortress.* New York: Doubleday, 1980.

Bradley, F.J. *No Strategic Targets Left.* Paducah KY: Turner Publishing Company, 1999

Burkett, Capt. P. *Unofficial History of the 499th Bombardment Group (VH).* Temple City, CA: Historical Aviation Album, 1981.

Campbell, John. *Boeing B-29 Superfortress.* Atglen, PA: Schiffer, 1977.

Carter, K.C., and Mueller, Robert. *Army Air Forces in World War II: Combat Chronology 1941–1945.* Washington DC: Office of Air Force History, USAF Headquarters, 1973.

"CBI Mission List, 58th Bomb Wing." *AAHS Journal,* 1962.

Chant, C., ed. *B-29 Superfort Profile.* Somerset, England: Haynes, 1983.

Fahey, J.C. *U.S. Army Aircraft 1909–1946.* New York: Ships and Aircraft, 1946.

_____. *USAF Aircraft 1947–1956.* Falls Church VA: Ships and Aircraft, 1956.

498th Bomb Group. *A Thumbnail Sketch of the 498th Bombardment Group, Saipan.*

Johnsen, F.A. *The B-29 Book.* Tacoma, WA: Bomber Books, 1978.

Kohn, G.C., and Rust, K.C. "The 313th Bomb Wing Journal." *AAHS Journal,* Vol. 9, No. 3 (Fall 1964).

Lloyd, Alwyn T. *B-29 Superfortress in Detail and Scale. Part 2: Derivatives.* Blue Ridge Summit PA: Tab, 1986.

Mann, Robert. *The Fireballs: An Unofficial History of the 54th Strategic Reconnaissance* Squadron (Medium) Weather. Privately published: 2000.

Marshall, C. *B-29 Superfortress.* Osceola WI: Motorbooks, 1993.

_____. *The Global Twentieth: An Anthology.* Vols. 1, 2 and 3. Memphis: Global, 1998.

_____. *Sky Giants Over Japan.* Winona MN: Apollo, 1984.

_____, and Thompson, W. *B-29 Photo Combat Diary.* North Branch MN: Specialty Press, 1996.

Maurer, M. *Air Force Combat Units of World War II.* New York: Franklin Watts, 1982.

_____. *Combat Squadrons of the Air Force, World War II.* Washington DC: Government Printing Office, 1969.

McGregor, C., Jr. *The Kagu Tsuchi Bomb Group.* Wichita Falls TX: Nortex Press, 1981.

Morrison, W. H. *The Birds From Hell (462nd Bomb Group).* Central Point OR: Hellgate Press, 2001.

_____. *Hellbirds: The Story of the B-29 in Combat.* Washington DC: Zenger, 1960.

_____. *Point of No Return: The Story of the 20th Air Force.* New York: Time Books, 1979.

Perry, Col. E.A. "An Account of a B-29 Mining Mission from Trincomalee, Ceylon, Against Palembang, Sumatra." *AAHS Journal,* 1977.

Pimlott, J. *B-29 Superfortress.* Englewood Cliffs NJ: Prentice Hall, 1980.

Rust, K.C. "Bomber Markings of the 20th Air Force." *AAHS Journal,* Fall 1962 and Winter 1962.

_____. *Twentieth Air Force Story.* Temple City CA: Historical Aviation Album, 1979.

Sakaida, H., and Takai, K. *B-29 Hunters of the JAAF (History of Air-to-Air Rammings).* Osceola WI: Motorbooks/Osprey, 2001.

6th Bomb Group. *Pirate's Log: Historical Record of the Sixth Bomb Group.* 6th Bombardment Group, 1945.

Smisek, S. *Narrative of 330BG Operations, April–August, 1945.* Online at http://b-29.org.

Smith, D., and Laird, R. *The 39th Bomb Group (VH).* Memory Press, 1996.

Smith, L., and Ashland, M. *The 9th Bomb Group (VH).* 9th Bomb Group, 1995. Now available online at www.9thbombgrouphistory.org/index.html.

Spencer, Otha. *Flying the Weather.* Campbell TX: The Country Studio, 1996.

20th Air Force and Keenan, R. *Twentieth Air Force Album.* Reprint by 20th Air Force Association, 2001.

United States Air Force. *Camouflage and Markings, B-29 Superfortress.* Washington DC: USAF Headquarters.

_____. Microfilm Roll AC-1, HG-839 Through 42-24879 Aircraft Record Cards, Historical Research Center, Maxwell AFB AL.

_____. Microfilm Roll AC-2, 42-24880 Through 42-72566 Aircraft Record Cards, Historical Research Center, Maxwell AFB AL.

_____. Microfilm Roll AC-4, 42-85938 Through 42-100820 Aircraft Record Cards, Historical Research Center, Maxwell AFB AL.

_____. Microfilm Roll AC-9, 44-16770 Through 44-29750 Aircraft Record Cards, Historical Research Center, Maxwell AFB AL.

_____. Microfilm Roll AC-13, 44-35963 Through 44-62004 Aircraft Record Cards, Historical Research Center, Maxwell AFB AL.

_____. Microfilm Roll AC-14, 44-62005 Through 44-64440 Aircraft Record Cards, Historical Research Center, Maxwell AFB AL.

_____. Microfilm Roll AC-15, 44-64442 Through 44-72015 Aircraft Record Cards, Historical Research Center, Maxwell AFB AL.

_____. Microfilm Roll AC-19, 44-81864 Through 44-84847 Aircraft Record Cards, Historical Research Center, Maxwell AFB AL.

_____. Microfilm Roll AC-20, 44-84850 Through 44-86393 Aircraft Record Cards, Maxwell AFB AL Historical Research Center, Maxwell AFB AL.

_____. Microfilm Roll AC-21, 44-86394 Through 44-89404 Aircraft Record Cards, Historical Research Center, Maxwell AFB AL.

_____. Microfilm Roll AC-24, 45-11598 Through 46-160 Aircraft Record Cards, Historical Research Center, Maxwell AFB AL.

_____. Microfilm Roll ACA-8, 44-35700 Through 44-73579 Aircraft Record Cards, Historical Research Center, Maxwell AFB AL.

_____. Microfilm Roll ACR-42, 42-4642 Through 42-6958, Aircraft Record Cards, Historical Research Center, Maxwell AFB AL.

_____. Microfilm Roll ACR 49, 42-23033 Through 42-25446, Aircraft Record Cards, Historical Research Center, Maxwell AFB AL.

_____. Microfilm Roll ACR-62, 42-58764 Through 42-64188, Aircraft Record Cards, Historical Research Center, Maxwell AFB AL.

_____. Microfilm Roll ACR-63, 42-64189 Through 42-67129, Aircraft Record Cards, Historical Research Center, Maxwell AFB AL.

_____. Microfilm Roll ACR-72, 42-92670 Through 42-95390, Aircraft Record Cards, Historical Research Center, Maxwell AFB AL.

_____. Microfilm Roll ACR-102, 44-26200 Through 44-30562, Aircraft Record Cards, Historical Research Center, Maxwell AFB AL.

_____. Microfilm Roll ACR-107, 44-50637 Through 44-63538, Aircraft Record Cards, Historical Research Center, Maxwell AFB AL.

_____. Microfilm Roll ACR-108, 44-63539 Through 44-70483, Aircraft Record Cards, Historical Research Center, Maxwell AFB AL.

_____. Microfilm Roll ACR-112, 44-80975 Through 44-84873, Aircraft Record Cards, Historical Research Center, Maxwell AFB AL.

_____. Microfilm Roll ACR-113, 44-84878 Through 44-88548, Aircraft Record Cards, Historical Research Center, Maxwell AFB AL.

_____. Microfilm Roll A7801, Tactical Mission Reports, XXI Bomber Command, 1945. Historical Research Center, Maxwell AFB AL.

_____. Microfilm Roll A7802, Tactical Mission Reports, XXI Bomber Command, 1945. Historical Research Center, Maxwell AFB AL.

_____. Microfilm Roll A7803, Tactical Mission Reports, XXI Bomber Command, 1945. Historical Research Center, Maxwell AFB AL.

_____. Microfilm Roll A7804, Tactical Mission Reports, XXI Bomber Command, 1945. Historical Research Center, Maxwell AFB AL.

_____. Microfilm Roll C0005, 58 BW History Sep 43–Jun 45, Historical Research Center, Maxwell AFB AL.

_____. Microfilm Roll C0006, 58 BW Support Docs Sep 43–Jun 45, Historical Research Center, Maxwell AFB AL.

_____. Microfilm Roll C0148, 313BW, 314BW 1944–1946 History and Supp Docs, Historical Research Center, Maxwell AFB AL.

_____. Microfilm Roll C0150, 314 BW History Dec 44–Aug 45, Historical Research Center, Maxwell AFB AL.

_____. Microfilm Roll C0151, 314 BW History May 45–May 46, Historical Research Center, Maxwell AFB AL.

_____. Microfilm Roll C0154, 315 BW History Jul 44–Sep 45, Historical Research Center, Maxwell AFB AL.

United States Army Air Forces. *Airplane Commander's Training Manual for the Superfortress*. AAF Manual 50-9. Washington DC: Office of Flying Safety, AAF Headquarters.

_____. *Pilot's Flight Operating Instructions for Army Model B-29 Airplanes*. Technical Manual AN 01-20EJ-1.

_____. *Erection and Maintenance Instructions for Army Model B-29 Airplanes*. Technical Manual AN 01-20EJ-2.

_____. *Operation CROSSROADS: The Official Pictorial Record*. Washington DC: Office of the United States Strategic Bombing Survey, Summary Report (Pacific War) 1946.

Personal Communications

The following persons were generous in communicating information valuable to my research. I note their names as well as the subjects on which we chiefly communicated.

Allen, J. 497th Bomb Group planes and missions.
Britton, T. Plane lists, nose art, and references.
Corradina, S. 40th Bomb Group aircraft histories.
Ellison, R., 497th and 498th Bomb Group mission lists.
Farrell, F. 19th Bomb Group planes and losses, Korea.
Hewlett, C. Washington Bombers of the RAF.
Weiler, P. 39th Bomb Group planes and names.